A Quantum Computation Workbook

Mahn-Soo Choi

A Quantum Computation Workbook

 Springer

Mahn-Soo Choi
Department of Physics
Korea University
Seoul, Korea (Republic of)

ISBN 978-3-030-91216-1   ISBN 978-3-030-91214-7   (eBook)
https://doi.org/10.1007/978-3-030-91214-7

This Springer imprint is published by the registered company Springer Nature Switzerland AG
The registered company address is: Gewerbestrasse 11, 6330 Cham, Switzerland

Preface

This book is intended as an introductory text for a university course on quantum computation and as a self-learning guide. It is an attempt to collect some fundamental principles and elementary methods in the field of quantum computation and quantum information and then reorganize them in a compact and integrated form. A proper introduction to (and hence learning of) quantum computation and quantum information turns out to be demanding, especially to undergraduate students, because it needs to cover various subjects from different fields of science including the physics of quantum mechanics, the mathematics of advanced linear algebra, and the computer science of information theory, to list the least. Each field of science usually favors its specific language, so another layer of difficulties is to organize the subjects consistently in a unified language.

Two digital materials accompany the book. Both are freely available through the respective GitHub repositories at the links below.

- Q3, a Mathematica application (Appendix D): https://github.com/quantum-mob/Q3
- QuantumWorkbook, a compilation of the demonstrations in this book written in the Wolfram Language (Appendix E): https://github.com/quantum-mob/QuantumWorkbook

Q3 consists of tools and utilities that perform symbolic calculations and numerical simulations useful in the study of quantum information processing, quantum many-body systems, and quantum spin systems. With Q3, one can avoid many of the tedious calculations involved in various principles and theorems of quantum theory. Furthermore, numerous visualization and simulation tools can help deepen the understanding of core concepts. *QuantumWorkbook* is a compilation of Mathematica Notebook files that contain the code used to generate the demonstrations in the book. Readers themselves can run and modify the code to build their own examples from the demonstrations and to experiment with fresh ideas. Both materials can be helpful companions, particularly in a course of self-learning, of the various subjects of quantum physics present in the book and for the general study of the overall subject as well.

This book is a result of several years of experience in teaching quantum computation and quantum information. It is a helpful resource to select and present particular topics of the subject at the undergraduate level. Although the book aims to be self-contained, it nonetheless assumes some basic knowledge of quantum mechanics. Chapter 1 summarizes the fundamental postulates of quantum mechanics and effectively provides a brief review of basic concepts and fundamental principles of quantum mechanics. Chapter 2 presents and describes the properties of elementary quantum gates for universal quantum computation. These are the building blocks of quantum algorithms and quantum communication protocols. Chapter 3 discusses physical methods and principles to implement elementary quantum gates and also introduces different quantum computation schemes. Chapter 4 introduces some widely known quantum algorithms to help grasp the idea of the so-called 'quantum supremacy' of quantum algorithms over their classical counterparts. Chapter 5 is dedicated to decoherence effects and introduces mathematical methods including the Kraus representation and the Lindblad equation to describe the phenomena. Chapter 6 is devoted to quantum error-correction codes through a discussion of the basic principles, procedures, and examples. Chapter 7 introduces quantum information theory. It discusses distance measures for quantum information, measures for quantum entanglement degree, and entropies of information content. To maintain a coherent structure and to focus on the primary topics, we collect the corresponding mathematical theories in the appendices and refer to them from the main text as necessary.

The author is indebted to students in his classes for pointing out numerous mistakes and typographical errors in the manuscript. Many people have contributed to the development of Q3 by testing and actively using it. The author gives particular thanks to Ha-Eum Kim, Myeongwon Lee, and Su-Ho Choi for their energetic discussions and constructive feedback in the early stages of the development of Q3. He also appreciates bug reports and valuable comments by Boris Laurent, Mi Jung So, Yeong-ho Je, and Dongni Chen.

Seoul, Korea   Mahn-Soo Choi
August 2021

The original version of this book was revised: Author's belated corrections have been updated. The correction to this book is available at https://doi.org/10.1007/978-3-030-91214-7_8

Contents

Chapter 1
The Postulates of Quantum Mechanics

"Elements" (see Fig. 1.1), the great compilation produced by Euclid of Alexandria in Ptolemaic Egypt circa 300 BC, established a unique logical structure for mathematics whereby every mathematical theory is built upon elementary axioms and definitions for which propositions and proofs follow. Theories in physics also take a similar structure. For example, classical mechanics is based on Sir Isaac Newton's three laws of motion. Called "laws", they are in fact elementary hypotheses—that is, axioms. While this may seem a remarkably different custom in physics compared to

Fig. 1.1 A fragment of Euclid's Elements on part of the Oxyrhynchus papyri located at the University of Pennsylvania. Courtesy of WikiMedia Commons

Supplementary Information The online version contains supplementary material available at https://doi.org/10.1007/978-3-030-91214-7_1.

M.-S. Choi, *A Quantum Computation Workbook*,
https://doi.org/10.1007/978-3-030-91214-7_1

its mathematical counterpart, it should not be surprising to refer to assumptions as laws or principles because they provide physical theories with logical foundation and are functional to determine whether Nature has been described properly or they have a mere existence as an intellectual framework. After all, the true value of a physical science is to understand Nature.

Embracing the wave-particle duality and the complementarity principle, quantum mechanics has been founded on the three fundamental postulates. The founders of quantum mechanics could have been more ambitious to call these laws instead of plain postulates, but each of these three defies our intuition to such an extent that "postulates" sounds more natural. As an overview, here are the fundamental postulates of quantum mechanics:

Postulate 1 The *quantum state* of a system is completely described by a state vector in the Hilbert space associated with the system.

Postulate 2 The *time evolution* of a closed quantum system is governed by the Schrödinger equation.

Postulate 3 A physical quantity is described by an "observable"—a Hermitian operator. Upon *measurement* of the quantity, the outcome is one of the eigenvalues of the observable and is determined *probabilistically*. Right after the measurement, the state "collapses" to the eigenstate of the observable corresponding to the measurement outcome.

In subsequent sections of this chapter, we detail the physical aspects of each postulate and their relevance to quantum computing and quantum information.

1.1 Quantum States

The first postulate indicates the mathematical description of the *state* of a system. Recall that in classical mechanics, the state of a particle in motion is described by the simple values of its position and momentum (or, equivalently, velocity). In quantum mechanics, the description is formulated at two different levels depending on the physical situation.

1.1.1 Pure States

Postulate 1 The quantum state of a closed system is completely described by a state vector in the Hilbert space associated with the system.

The most common example of the state vector is a "wave function"—a member of the *Hilbert space* of square integrable functions—originally put forward by Schrödinger. Modern approaches associate an abstract vector space with the system, and the specific characters of the particular system are reflected in the choice of basis (see

Appendix A.1). When the state is exactly known, the system is said to be in its *pure state*, and the above description is comprehensive. However, in many cases, it is difficult to know the exact state, and we thus need a more general description to be discussed in the next subsection.

This postulate immediately raises a mind-blowing question: What is the physical meaning of the state vector or its components in a given basis (or of the wave function)? Quantum mechanics has never offered a direct physical meaning of the state vector. Born (1926) proposed a partial resolution to the question and inspired the probabilistic interpretation of quantum mechanics, as formulated in Postulate 3 concerning measurement. This work awarded him the 1954 Nobel Prize in Physics.

Postulate 1 leaves another baffling question: Given a physical system, there is no general prescription to figure out the Hilbert space associated with it. While it is a rather technical question, it is nevertheless an important and serious one when trying to describe a new system (or a yet-to-be-understood system) quantum mechanically.

For a *qubit*—an idealistic two-level quantum system—the Hilbert space is two-dimensional. A basis of two logical states $|0\rangle$ and $|1\rangle$ is assumed, and it is called the *logical basis* of the qubit.

Consider a group of two-level quantum systems, indicated by the symbol s.

```
Let[Qubit, S]
```

Different qubits can be specified by the flavor indices, the last of which has a special meaning (see the documentation of Qubit).

```
In[·]:= {S[1, None], S[2, None]}
       S[{1, 2}, None]
```
$$Out[·]= \{S_1, S_2\}$$
$$Out[·]= \{S_1, S_2\}$$

The associated Hilbert space is two dimensional. For many functions dealing with qubits, the final index None can be dropped.

```
In[·]:= bs = Basis[S[1, None]]
       bs = Basis[S[1]]
```
$$Out[·]= \{|\_\rangle, |1_{S_1}\rangle\}$$
$$Out[·]= \{|\_\rangle, |1_{S_1}\rangle\}$$

For the efficiency reasons, the default value 0 of any qubit is removed from the data structure. For a more intuitively appealing form with all default values, LogicalForm can be used.

```
In[·]:= LogicalForm[bs]
```
$$Out[·]= \{|0_{S_1}\rangle, |1_{S_1}\rangle\}$$

Each state in the logical basis can also be specified manually.

```
In[·]:= vec = Ket[S[1] → 1, S[2] → 0];
       LogicalForm[vec, {S[1], S[2]}]
```
$$Out[·]= \left| 1_{S_1} 0_{S_2} \right\rangle$$

```
In[·]:= vec = Ket[]S[1{, S[2{} → ]1, 0}{;
       LogicalForm[vec, ]S[1{, S[2{}{
```
$$Out[·]= \left| 1_{S_1} 0_{S_2} \right\rangle$$

A general quantum state of S[1,None] is a linear combination of the two basis states with two complex coefficients c[0] and c[1].

```
In[·]:= Let[Complex, c]
       vec = Ket[S[1] → 0] × c[0] + Ket[S[1] → 1] × c[1];
       vec // LogicalForm
```
$$Out[·]= c_0 \left| 0_{S_1} \right\rangle + c_1 \left| 1_{S_1} \right\rangle$$

A two-dimensional state vector is often visualized as a point, called the *Bloch vector*, on the *Bloch sphere*. The Bloch sphere is a geometrical representation of a two-dimensional vector space. Any state vector $|\psi\rangle$ is expanded in the logical basis as

$$|\psi\rangle = |0\rangle \, \psi_0 + |1\rangle \, \psi_1 \quad (\psi_0, \psi_1 \in \mathbb{C}). \tag{1.1}$$

The normalization condition, $|\psi_0|^2 + |\psi_1|^2 = 1$, tells us that the state vector can be expressed up to a global phase factor by

$$|\psi\rangle = |0\rangle \cos(\theta/2) + |1\rangle \sin(\theta/2)e^{i\phi} \tag{1.2}$$

with θ and ϕ respectively specifying the magnitude and phase, of the expansion coefficients ($\theta, \phi \in \mathbb{R}$). The Bloch vector associated with the state vector $|\psi\rangle$ is defined by $\boldsymbol{b} = (\sin\theta\cos\phi, \sin\theta\sin\phi, \cos\theta)$. The Bloch vector can equivalently be obtained in terms of the expectation values of the Pauli operators, $\boldsymbol{b} = (\langle\hat{\sigma}^x\rangle, \langle\hat{\sigma}^y\rangle, \langle\hat{\sigma}^z\rangle)$. Indeed, with $|\psi\rangle$ in the form (1.2), one can show that $\langle\hat{\sigma}^x\rangle = \sin\theta\cos\phi$, $\langle\hat{\sigma}^y\rangle = \sin\theta\sin\phi$, and $\langle\hat{\sigma}^z\rangle = \cos\theta$. Therefore, any state vector in a two-dimensional Hilbert space corresponds uniquely (up to a global phase factor) to a point on the sphere with unit radius, the Bloch sphere.

A two-dimensional pure state is represented by a point on the Bloch sphere. For example, consider a pure state.

```
In[·]:= vec [ Ket] = × Sqrt]2= + I Ket]S]1= → 1=;
       vec // LogicalForm
```
$$Out[·]= \sqrt{1} \left| S_{\theta_1} \right\rangle + i \left| 2_{\theta_1} \right\rangle$$

This visualize the state vector on a Bloch sphere. BlochVector converts the state vector to a three-dimensional vector. BlochSphere is a shortcut for Graphics3D with an visualization of the Bloch sphere.

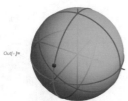

In[·]:= BlochSphere[{Red, Bead@BlochVector[vec]}, ImageSize → Small]

Out[·]:=

Many quantum systems, especially of many particles, are composed of several parts with independent degrees of freedom. For such a system, the overall Hilbert space is built up from the Hilbert spaces of individual parts by means of the tensor product (see Appendix A.6). For example, consider a system of two parts and suppose that they are associated with the vector spaces V and W, respectively. The total Hilbert space is given by the tensor product $V \otimes W$, which is defined to be the vector space spanned by the tensor-product basis

$$\left\{ |v_i\rangle \otimes |w_j\rangle : i = 0, \ldots, m - 1; \ j = 0, \ldots, n - 1 \right\}, \tag{1.3}$$

where $\{|v_i\rangle\}$ and $\{|w_j\rangle\}$ are bases of V and W of dimensions m and n, respectively. The dimension of $V \otimes W$ for the total system is obviously given by mn, and in general a state vector of the total system is a linear superposition consisting of mn terms

$$|\Psi\rangle = \sum_{i=0}^{m-1} \sum_{j=0}^{n-1} |v_i\rangle \otimes |w_j\rangle \, \Psi_{ij}. \tag{1.4}$$

Some state vectors are factored into the form

$$(|v_1\rangle c_1 + |v_2\rangle c_2 + \cdots) \otimes (|w_1\rangle d_1 + |w_2\rangle d_2 + \cdots). \tag{1.5}$$

Such a state is said to be *separable*. For example, the superposition of all the logical basis states of two qubits

$$|0\rangle \otimes |0\rangle + |0\rangle \otimes |1\rangle + |1\rangle \otimes |0\rangle + |1\rangle \otimes |1\rangle \tag{1.6}$$

is factored into

$$(|0\rangle + |1\rangle) \otimes (|0\rangle + |1\rangle), \tag{1.7}$$

and is a separable state. On the other hand, the state

$$|0\rangle \otimes |0\rangle + |0\rangle \otimes |1\rangle + |1\rangle \otimes |0\rangle \tag{1.8}$$

can never be factored. Such a state is said to be *entangled* (Fig. 1.2). The *Schmidt decomposition* is a systematic way to test whether a given quantum state $|\Psi\rangle$ in a tensor-product space is separable or not. The Schmidt decomposition (see

Fig. 1.2 An artist's view of
quantum entanglement. The
up and down arrows
represent the logical basis
states $|0\rangle$ and $|1\rangle$

Appendix A.6.1 for the mathematical details) is a method to choose proper bases
$\{|v_i'\rangle\}$ and $\{|w_j'\rangle\}$ of $\mathcal{V}$ and $\mathcal{W}$, respectively, to rewrite the given state vector $|\Psi\rangle$ in
the least number of terms of the form

$$|\Psi\rangle = \sum_{k=0}^{R-1} |v_k'\rangle \otimes |w_k'\rangle s_k , \qquad (1.9)$$

where the coefficients s_k are all definitely positive. Here, the so-called *Schmidt rank R*
of the quantum state $|\Psi\rangle$ cannot be larger than the smaller of m and n, $R \leq \min(m, n)$.
If the Schmidt rank is 2 or larger, then the state is entangled.

Consider the following state of a two-qubit state.

```
In[ ]:= ket = Ket[] + Ket[S[1] → 1] + Ket[S[2] → 1];
       ket // LogicalForm
```
$$Out[ ]= \;\left|1_{S_0}1_{S_2}\right\rangle + \left|1_{S_0}0_{S_2}\right\rangle + \left|0_{S_0}1_{S_2}\right\rangle$$

This gives the Schmidt decomposition of the state. It turns out that its Schmidt rank is two and
the state is an entangled state.

```
In[ ]:= =ww, uu, vv[ ] SchmidtDecomposition{ket, S{1}, S{2}};
       ww
```
$$Out[ ]= \;\left\{ \sqrt{\frac{1}{2} \times \left(3 + \sqrt{5}\right)} , \sqrt{\frac{1}{2} \times \left(3 - \sqrt{5}\right)} \right\}$$

`SchmidtForm` presents the Schmidt decomposition in a more intuitively-appealing form. For a
thorough analysis of the result, use `SchmidtDecomposition`.

In[ ]:= **new = SchmidtForm[ket,]S[1{},]S[2{}{**

Out[ ]= $\sqrt{\frac{1}{2} \times \left(3 + \sqrt{5}\right)}$

$$\left(\frac{\left(1 - \frac{1}{2} \times \left(1 + \sqrt{5}\right)\right) \left|0_{S_1}\right\rangle}{\sqrt{\frac{1}{4}\left(1 + \sqrt{5}\right)^2 - \left(1 - \frac{1}{2} \times \left(1 + \sqrt{5}\right)\right)^2}} - \frac{\left(1 + \sqrt{5}\right) \left|1_{S_1}\right\rangle}{2\sqrt{\frac{1}{4}\left(1 + \sqrt{5}\right)^2 - \left(1 - \frac{1}{2} \times \left(1 + \sqrt{5}\right)\right)^2}} \right) \otimes$$

$$\left(\frac{\left(1 + \sqrt{5}\right) \left|0_{S_2}\right\rangle}{2\sqrt{1 - \frac{1}{4}\left(1 + \sqrt{5}\right)^2}} - \frac{\left|1_{S_2}\right\rangle}{\sqrt{1 - \frac{1}{4}\left(1 + \sqrt{5}\right)^2}} \right) - \sqrt{\frac{1}{2} \times \left(3 - \sqrt{5}\right)}$$

$$\left(\frac{\left(1 - \frac{1}{2} \times \left(1 - \sqrt{5}\right)\right) \left|0_{S_1}\right\rangle}{\sqrt{\frac{1}{4}\left(1 - \sqrt{5}\right)^2 - \left(1 - \frac{1}{2} \times \left(1 - \sqrt{5}\right)\right)^2}} - \frac{\left(1 - \sqrt{5}\right) \left|1_{S_1}\right\rangle}{2\sqrt{\frac{1}{4}\left(1 - \sqrt{5}\right)^2 - \left(1 - \frac{1}{2} \times \left(1 - \sqrt{5}\right)\right)^2}} \right) \otimes$$

$$\left(\frac{\left(1 - \sqrt{5}\right) \left|0_{S_2}\right\rangle}{2\sqrt{1 - \frac{1}{4}\left(1 - \sqrt{5}\right)^2}} - \frac{\left|1_{S_1}\right\rangle}{\sqrt{1 - \frac{1}{4}\left(1 - \sqrt{5}\right)^2}} \right)$$

The Schmidt decomposition is incredibly complicated for such a simple-looking system. Let us take an approximation to get an impression of how the state is entangled.

In[ ]:= **new // N // Simplify**

Out[ ]= $0.618034 \left(0.525731 \left|0_{S_1}\right\rangle - 0.850651 \left|1_{S_1}\right\rangle\right) \otimes \left(-0.525731 \left|0_{S_2}\right\rangle + 0.850651 \left|1_{S_2}\right\rangle\right) +$
$1.61803 \left(0.850651 \left|0_{S_1}\right\rangle + 0.525731 \left|1_{S_1}\right\rangle\right) \otimes \left(0.850651 \left|0_{S_2}\right\rangle + 0.525731 \left|1_{S_2}\right\rangle\right)$

As first pointed out by Einstein et al. (1935), entangled quantum states have many intriguing properties that are difficult to intuitively understand and that raise many questions concerning the foundation of quantum mechanics. The non-local property—the very property pointed out by Einstein et al. (1935)—is the representative property of entangled states, and it is illustrated in Sect. 4.1.1. Quantum entanglement also provides a fundamental explanation of quantum decoherence, the process through which a quantum system loses its quantum effects. Decoherence is further discussed in Chap. 5. In quantum computation, quantum information, and quantum communication, entanglement is regarded as a valuable resource that enables many amazing quantum effects, such quantum speed-up and unconditional security, which are impossible in the classical world. One of the most illustrative examples—quantum teleportation—is discussed in Sect. 4.1.

1.1.2 Mixed States

One often encounters a situation where the state of a system is not completely known. The system in this case is said to be in a *mixed state*. A common example is when the system is interacting with its surroundings, and A mixed state is a statistical mixture of pure states and is characterized in terms of a statistical ensemble, where different state vectors $\left|\psi_\mu\right\rangle$ are found with probabilities p_μ.[1] Such an ensemble is

---

[1] Here, note that the different pure states $\left|\psi_\mu\right\rangle$ in the mixture do not have to be orthogonal to each other. In general, they do not span the Hilbert space, either, so they are completely general.

efficiently represented by a *density operator* (which is often just called *density matrix* for historical reasons) constructed as

$$\hat{\rho} = \sum_{\mu} |\psi_{\mu}\rangle \, p_{\mu} \, \langle\psi_{\mu}| \tag{1.10}$$

Sometimes, it is convenient to describe the statistical mixture in terms of a set of *unnormalized* vectors

$$\left\{ |\psi_j'\rangle := |\psi_{\mu}\rangle \sqrt{p_{\mu}} \right\}. \tag{1.11}$$

Absorbing the probabilities into the vectors, $\langle\psi_{\mu}'|\psi_{\mu}'\rangle$ gives the probability to find the state $|\psi_{\mu}'\rangle$ in the ensemble. The corresponding density operator is constructed as

$$\hat{\rho} = \sum_{\mu} |\psi_{\mu}'\rangle \langle\psi_{\mu}'| \, , \tag{1.12}$$

and is certainly equivalent to that in (1.10).

Consider a density operator representing a statistical mixture of two pure states.

In[ ]:= `vecs = {Ket[], (Ket[] - I Ket[S[1] → 1]) / Sqrt[2]};`
`vecs // LogicalForm`
`prbs = {1 / 3, 2 / 3}`

Out[ ]= $\left\{ \left| 0_{S_1} \right\rangle, \dfrac{\left| 0_{S_1} \right\rangle + i \left| 1_{S_1} \right\rangle}{\sqrt{2}} \right\}$

Out[ ]= $\left\{ \dfrac{1}{3}, \dfrac{2}{3} \right\}$

From the specifications of the ensemble, this constructs the density operator for the mixed state.

In[ ]:= `ρ = (vecs ** Dagger[vecs]).prbs // Garner;`
`ρ // LogicalForm`

Out[ ]= $\dfrac{2}{3} \left| 0_{S_1} \right\rangle \left\langle 0_{S_1} \right| - \dfrac{1}{3} i \left| 0_{S_1} \right\rangle \left\langle 1_{S_1} \right| + \dfrac{1}{3} i \left| 1_{S_1} \right\rangle \left\langle 0_{S_1} \right| - \dfrac{1}{3} \left| 1_{S_1} \right\rangle \left\langle 1_{S_1} \right|$

This gives the matrix representation -- the "density matrix" -- of the density operator in the logical basis.

In[ ]:= `Matrix@ρ // MatrixForm`
Out[ ]//MatrixForm=

$$\begin{pmatrix} \dfrac{2}{3} & \dfrac{i}{3} \\ +\dfrac{i}{3} & \dfrac{1}{3} \end{pmatrix}$$

This gives the expression of the density operator in terms of the Pauli operators.

```
In[ ]:= Elaborate@ExpressionFor[Matrix@ρ, S[1]]
```
$$Out[ ]= \frac{1}{2} - \frac{S_1^y}{3} + \frac{S_1^z}{6}$$

By construction, any density operator $\hat{\rho}$ is Hermitian, that is, $\hat{\rho}^\dagger = \hat{\rho}$. Obviously, a density operator is an operator acting on the state vectors in the vector space $\mathcal{H}$. However, it is more appropriate to regard it as a vector in the vector space $\mathcal{L}(\mathcal{H})$ of all linear operators on $\mathcal{H}$ (see Appendix B.1).

Each diagonal element $\rho_{jj} := \langle v_j | \hat{\rho} | v_j \rangle$ in a given basis $\{|v_j\rangle\}$ carries an important physical meaning. ρ_{jj} gives the probability P_j to find the system in the basis state $|v_j\rangle$. To see it, note that that the probability to find the system in $|v_j\rangle$ is $|\langle v_j|\psi_\mu\rangle|^2$ under the condition that the state is $|\psi_\mu\rangle$. The latter has the chance of probability p_μ. Therefore the overall probability P_j is given by

$$P_j = \sum_\mu |\langle v_j|\psi_\mu\rangle|^2 p_\mu = \sum_\mu \langle v_j|\psi_\mu\rangle p_\mu \langle \psi_\mu|v_j\rangle = \langle v_j|\hat{\rho}|v_j\rangle \qquad (1.13)$$

as expected. The physical meaning of diagonal elements further implies several basic yet important properties of density operator (see Problems 1.3.2 and 1.3.2):

(a) A density operator is not only Hermitian but also *positive* (Definition A.19).
(b) A density operator has the unit trace, $\mathrm{Tr}\hat{\rho} = 1$.
(c) Any eigenvalue of a density operator $\hat{\rho}$ lies between 0 and 1. It is often said that $0 < \hat{\rho} \leq 1$.

On the other hand, the off-diagonal elements of a density operator are responsible for the coherence effects, as we will observe later in various interference experiments. Note that coherence is a basis-dependent effect.

It is also interesting to note that the density operator and the relevant physical properties do not depend on all the details in the specification of the statistical ensemble, which in some sense makes the description of mixed states in terms of the density operator so powerful and efficient: Suppose that two statistical ensembles be specified by the sets $\{|\alpha_\mu\rangle : \mu = 1, \ldots, m\}$ and $\{|\beta_\nu\rangle : j = \mu, \ldots, n\}$, respectively, of unnormalized vectors as in Eq. (1.12). Without loss of generality, assume that $m \leq n$. Then, the density operators describing the ensembles are identical,[2]

$$\sum_{\mu=1}^m |\alpha_\mu\rangle\langle\alpha_\mu| = \sum_{\nu=1}^n |\beta_\nu\rangle\langle\beta_\nu| \qquad (1.14)$$

if and only if there exists an $n \times n$ unitary matrix $U_{\mu\nu}$ such that

$$|\beta_\nu\rangle = \sum_\mu |\alpha_\mu\rangle U_{\mu\nu} \qquad (1.15)$$

---

[2] This unitary freedom discovered by Hughston et al. (1993) is closed related to the unitary freedom in purification of a mixed state. See Footnote 4.

for all $v = 1, \ldots, n$. Although a rigorous proof is beyond the main scope of the book, the impact of unitary freedom—or ambiguity—in the specification of mixed states is wide spread. Furthermore, a similar unitary freedom is observed in the description of decoherence and its related effects (Sect. 5), so it is justified to take a moment here to prove it.

Clearly, if the states from the two sets satisfy the relation (1.15), then the identity (1.14) holds. Conversely, suppose that the two density operators are the same and are equal to $\hat{\rho}$. Write $\hat{\rho}$ in a spectral decomposition (Appendix A.4),

$$\hat{\rho} = \sum_{\lambda} |\lambda\rangle \langle\lambda| , \qquad (1.16)$$

where $|\lambda\rangle$ represent the (unnormalized) eigenstates of $\hat{\rho}$ belonging to non-zero eigenvalues λ. We have used the properties of positive operators (Theorem A.20) and normalized $|\lambda\rangle$ by their own eigenvalues, that is, $\langle\lambda|\lambda\rangle = \lambda$. We first note that $\{|\alpha_{\mu}\rangle\}$ and $\{|\lambda\rangle\}$ span the same subspace, and hence $|\alpha_{\mu}\rangle$ can be expanded in $|\lambda\rangle$,

$$|\alpha_{\mu}\rangle = \sum_{\lambda} |\lambda\rangle V_{\lambda\mu}. \qquad (1.17)$$

Putting it into (1.14), we require

$$\sum_{\lambda\lambda'} |\lambda\rangle \langle\lambda'| \sum_{\mu} V_{\lambda\mu} V_{\lambda'\mu}^* = \sum_{\lambda} |\lambda\rangle \langle\lambda| . \qquad (1.18)$$

Recall that unlike $|\alpha_{\mu}\rangle$, which are not orthogonal to each other in general, $|\lambda\rangle$ are orthogonal (Appendix A.4). Equation (1.18) thus implies that the rows of the matrix V are orthogonal to each other. By introducing, if necessary, additional rows that are orthogonal to existing rows, the matrix V can be extended into a unitary matrix. For the same reason, $|\beta_v\rangle$ can be expanded as

$$|\beta_v\rangle = \sum_{\lambda} |\lambda\rangle W_{\lambda v} , \qquad (1.19)$$

and the matrix $W_{\lambda v}$ can be extended into a unitary matrix. Overall, $|\alpha_{\mu}\rangle$ and $|\beta_v\rangle$ satisfy the relation (1.15), where $U = W^{\dagger} V$.[3]

---

[3] In this proof, we have implicitly assumed that $m = n$. However, the proof can be extended simply by adding null vectors in the set $\{\alpha_{\mu}\}$ or $\{\beta_v\}$ until $m = n$.

Consider a statistical mixture of the following three pure states.

```
In[ ]:= v1 = Ket[];
v2 = (Ket[] + I Ket[S → 1]) / Sqrt[2];
v3 = (Ket[] 2 + Ket[S → 1] I) / Sqrt[5];
LogicalForm@{v1, v2, v3}
```

$$Out[ ]= \left\{ |0_S\rangle, \frac{|0_S\rangle + i |1_S\rangle}{\sqrt{2}}, \frac{2 |0_S\rangle + i |1_S\rangle}{\sqrt{5}} \right\}$$

These are the associated probabilities.

```
In[ ]:= p1 = 1 / 8;
p2 = 1 / 4;
p3 = 5 / 8;
{p1, p2, p3}
```

$$Out[ ]= \left\{ \frac{1}{8}, \frac{1}{4}, \frac{5}{8} \right\}$$

The mixed state is described by the density operator.

```
In[ ]:= ρ = Total@Multiply[{v1, v2, v3}, {p1, p2, p3}, Dagger@{v1, v2, v3}];
ρ // LogicalForm
```

$$Out[ ]= \frac{3 |0_S\rangle \langle 0_S|}{4} - \frac{3}{8} i |0_S\rangle \langle 1_S| + \frac{3}{8} i |1_S\rangle \langle 0_S| + \frac{|1_S\rangle \langle 1_S|}{4}$$

Next consider another set of pure states.

```
In[ ]:= w1 = [Ket] ( 2 + Ket]S → 1( I) / Sqrt]5(;
w2 = Ket]S → 1(;
LogicalForm@{w1, w2}
```

$$Out[ ]= \left\{ \frac{0 |S,\rangle + 1 |1,\rangle}{\sqrt{2}}, 5 |1,\rangle \right\}$$

The associated probabilities are as following.

```
In[ ]:= q1 = 15 / 16;
q2 = 1 / 16;
{q1, q2}
```

$$Out[ ]= \left\{ \frac{15}{16}, \frac{1}{16} \right\}$$

The mixture leads to the same density operator.

```
In[ ]:= σ = Total@Multiply[{w1, w2}, {q1, q2}, Dagger@{w1, w2}]
```

$$Out[ ]= \frac{3 |\_\rangle \langle\_|}{4} - \frac{3 i |\_\rangle \langle 1_S|}{8} + \frac{3 i |1_S\rangle \langle\_|}{8} + \frac{|1_S\rangle \langle 1_S|}{4}$$

The two sets are related by the following unitary matrix

```
In[-]:= U = Topple@{
          {1, 1, 2} / Sqrt[6],
          {1, -1, 0} / Sqrt[2],
          {1, 1, -1} / Sqrt[3]
       };
       U // MatrixForm
```
$$Out[-]//MatrixForm=
\begin{pmatrix}
\frac{1}{\sqrt{6}} & \frac{1}{\sqrt{2}} & \frac{1}{\sqrt{3}} \\
\frac{1}{\sqrt{6}} & +\frac{1}{\sqrt{2}} & \frac{1}{\sqrt{3}} \\
\sqrt{\frac{2}{3}} & 0 & +\frac{1}{\sqrt{3}}
\end{pmatrix}$$

Recall that two-dimensional pure states are visualized *on* a Bloch sphere. A mixed state $\hat{\rho}$ for a qubit can be similarly visualized, but in general the representing point resides *inside* the Bloch sphere. As for a pure state, the Bloch vector $\boldsymbol{b}$ corresponding to a mixed state $\hat{\rho}$ is defined by $\boldsymbol{b} = (\langle \hat{\sigma}^x \rangle, \langle \hat{\sigma}^y \rangle, \langle \hat{\sigma}^z \rangle)$. Recalling that any operator on a two-dimensional vector space is a linear superposition of the Pauli operators, we decompose a density operator into the form

$$\hat{\rho} = \frac{1}{2} \left(\hat{\sigma}^0 + x\hat{\sigma}^x + y\hat{\sigma}^y + z\hat{\sigma}^z \right) \tag{1.20}$$

with $x, y, z \in \mathbb{R}$. By evaluating the averages, $\langle \hat{\sigma}^\mu \rangle = \mathrm{Tr}\hat{\rho}\hat{\sigma}^\mu$, of the Pauli operators with respect to $\hat{\rho}$, one can see that the Bloch vector corresponding to $\hat{\rho}$ is just given by $\boldsymbol{b} = (x, y, z)$. Clearly, $\sqrt{x^2 + y^2 + z^2} \leq 1$, and it may in general lie inside the Bloch sphere.

This gives a visualization of the mixed state by a point -- Bloch vector -- in a Bloch sphere.

```
In[-]:= BlochSphere[{Red, Bead@BlochVector@ρ}, "Opacity" → 0.4, ImageSize → Small]
```

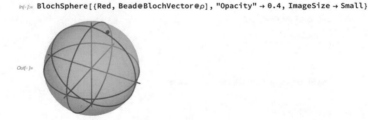

Out[-]=

A mixed state arises naturally when the system interacts with its environment. As a closed system, the total system is described by a pure state $|\Psi\rangle \in \mathcal{H} \otimes \mathcal{E}$, where $\mathcal{H}$ and $\mathcal{E}$ are the Hilbert spaces respectively associated with the system and the environment. In accordance with the *Schmidt decomposition* (see Appendix A.6), one can always express $|\Psi\rangle$ in the form

$$|\Psi\rangle = \sum_{\mu=1}^{R} |\alpha_\mu\rangle \otimes |\beta_\mu\rangle \sqrt{p_\mu}, \tag{1.21}$$

where $|\alpha_\mu\rangle$ are orthonormal states in $\mathcal{H}$, $|\beta_\nu\rangle$ are orthonormal states in $\mathcal{E}$, $0 < p_\mu < 1$, and R is the *Schmidt rank* of $|\Psi\rangle$. Without access to the environment, one cannot tell which state the system is in among $|\alpha_\mu\rangle$. One can only tell the chances. The probability for the system to be found in $|\alpha_\mu\rangle$ is equal to $\langle\beta_\mu|\beta_\mu\rangle p_\mu = p_\mu$. Therefore, the density operator that describes the situation is given by

$$\hat{\rho} = \sum_\mu |\alpha_\mu\rangle\langle\alpha_\mu| p_\mu \tag{1.22}$$

Now noting that $\langle\beta_\mu|\beta_\nu\rangle = \delta_{\mu\nu}$ and

$$|\Psi\rangle\langle\Psi| = \sum_{\mu\nu} |\alpha_\mu\rangle\langle\alpha_\nu| \otimes |\beta_\mu\rangle\langle\beta_\nu| \sqrt{p_\mu p_\nu}, \tag{1.23}$$

one can see that the expression in (1.22) is equivalent to taking a partial trace over $\mathcal{E}$ (see Appendix B.3),

$$\hat{\rho} = \mathop{\mathrm{Tr}}_\mathcal{E} |\Psi\rangle\langle\Psi|. \tag{1.24}$$

In this sense, taking a partial trace over a particular part of the total system physically corresponds to "ignoring" that part.

We have seen that the partial trace of a pure state over a part of the total system results in a mixed state for the rest. The converse is also possible: Given a mixed state, one can find a pure state in an enlarged system that leads to the mixed state upon the partial trace. A normalized pure state $|\Psi\rangle$ in an extended Hilbert space $\mathcal{H} \otimes \mathcal{E}$ is called a *purification* of the mixed state $\hat{\rho}$ into $\mathcal{H} \otimes \mathcal{E}$ if it gives $\hat{\rho}$ upon the partial trace over $\mathcal{E}$,

$$\hat{\rho} = \mathop{\mathrm{Tr}}_\mathcal{E} |\Psi\rangle\langle\Psi|. \tag{1.25}$$

To construct a purification of a mixed state $\hat{\rho}$, take the spectral decomposition (Appendix A.4.1),

$$\hat{\rho} = \sum_j |r_j\rangle r_j \langle r_j|, \tag{1.26}$$

where r_j and $|r_j\rangle$ are the eigenvalues and corresponding eigenvectors of $\hat{\rho}$. Since $\hat{\rho}$ is positive, all the eigenvalues r_j are non-negative. Then, by taking an orthonormal subset $\{|\epsilon_j\rangle\}$ of $\mathcal{E}$, one can define a state

$$|\Psi\rangle = \sum_j |r_j\rangle \otimes |\epsilon_j\rangle \sqrt{r_j}. \tag{1.27}$$

It is straightforward to check that the above state is a purification of $\hat{\rho}$. In other words, $\hat{\rho} = \mathop{\mathrm{Tr}}_\mathcal{E} |\Psi\rangle\langle\Psi|$.

Purification is not unique. Different purifications give rise to the same mixed state ρ (Problems 1.3.2 and 1.3.2),

$$\hat{\rho} = \mathop{\mathrm{Tr}}_{\mathcal{E}} |\Psi\rangle \langle\Psi| = \mathop{\mathrm{Tr}}_{\mathcal{E}} |\Phi\rangle \langle\Phi|, \tag{1.28}$$

as long as

$$|\Phi\rangle = (\hat{I} \otimes \hat{U}) |\Psi\rangle, \tag{1.29}$$

where $\hat{U}$ is a unitary operator on $\mathcal{E}$.[4]

Suppose that two qubits are coupled and that the total system is in the following state. We can regard the first qubit as the "system" and the second as the "reservoir".

```
In[ ]:= total = (Ket[] - Ket[S[1] → 1] + Ket[S[2] → 1]) / Sqrt[3];
       LogicalForm[total]
```

$$Out[ ]= \frac{\left|2_{3_s}2_{3_s}\right\rangle - \left|2_{3_s}0_{3_s}\right\rangle + \left|0_{3_s}2_{3_s}\right\rangle}{\sqrt{1}}$$

The first qubit is in a mixed state. The density operator is given by the partial trace over the second qubit.

```
In[ ]:= rho = Elaborate@PartialTrace[total ** Dagger[total], S[2]]
```

$$Out[ ]= \frac{1}{2} - \frac{S_1^x}{3} + \frac{S_1^z}{6}$$

This is the matrix representation of the density operator in the logical basis.

```
In[ ]:= MatrixForm@Matrix@rho
```

$$Out[ ]//MatrixForm= \begin{pmatrix} \frac{2}{3} & +\frac{1}{3} \\ +\frac{1}{3} & \frac{1}{3} \end{pmatrix}$$

A purification is a pure state of an extended system composed of the "system" and the "environment". Here S[2,None] is regarded as the environment.

```
In[ ]:= ket = Purification[rho, S[1], S[2]]
```

$$Out[ ]= \times\frac{1}{2} + (1 - \sqrt{5}) \sqrt{\frac{1}{30} \cdot (5 - \sqrt{5})} \;|\_\rangle - \sqrt{\frac{1}{30} \cdot (5 \times \sqrt{5})} \;|1_{S_1}1_{S_2}\rangle -$$
$$\sqrt{\frac{1}{30} \times (5 - \sqrt{5})} \;|1_{S_1}\rangle - \frac{1}{2}\sqrt{\frac{1}{30} \times (5 \times \sqrt{5})} \;(\times 1 - \sqrt{5}) \;|1_{S_2}\rangle$$

As purification is not unique, the above purification is not the same as the total state above.

---

[4] This is known as the Gisin-Hughston-Jozsa-Wooters (GHJW) theorem after Gisin (1989) and Hughston et al. (1993). The unitary freedom in purification is closely related to the unitary freedom in the specification of ensembles of a mixed state that we have discussed in (1.14).

However, upon tracing out the environmental degrees of freedom, the purification reproduces the density operator.

```
In[ ]:= new = PartialTrace[ket, S[2]] // Elaborate
```
$$Out[ ]= \frac{1}{2} - \frac{S_1^x}{3} + \frac{S_1^z}{6}$$

A simple yet important aspect of the above observation is that when $|\Psi\rangle$ of the total system is a separable state, $|\Psi\rangle = |\alpha\rangle \otimes |\beta\rangle$, the reduced state

$$\hat{\rho} = \sum_{\mathcal{E}} |\alpha\rangle \langle\alpha| \otimes |\beta\rangle \langle\beta| = |\alpha\rangle \langle\alpha| \tag{1.30}$$

remains a pure state, retaining full coherence. This means that the *entanglement* between the system and the environment is responsible for the decoherence. In the history of quantum physics, it has been a mystery why quantum effects can be observed only in microscopic systems and not in macroscopic systems. Entanglement has provided a key clue to resolve this mystery. We will later see that entanglement also provides valuable resources for quantum information processing and quantum communications (see Sect. 4.1).

When the system is in a pure state $|\psi\rangle$, the density operator is simply given by $\hat{\rho} = |\psi\rangle \langle\psi|$. For a pure state $|\psi\rangle = |0\rangle c_0 + |1\rangle c_1$, the matrix representation of $\hat{\rho}$ is given by

$$\hat{\rho} \doteq \begin{bmatrix} c_0 c_0^* & c_0 c_1^* \\ c_1 c_0^* & c_1 c_1^* \end{bmatrix}. \tag{1.31}$$

Given a density operator $\hat{\rho}$, there is a simple test to determine whether the corresponding state is a pure state or a mixed state. It is a pure state if and only if $\mathrm{Tr}\hat{\rho}^2 = 1$ (see Problem 1.3.2). If it is a mixed state, $\mathrm{Tr}\hat{\rho}^2$ is strictly smaller than 1.

Let us examine the density operator corresponding to a pure state. As an example, consider the following pure state.

```
In[ ]:= vec [ Ket]S]1= → 0= × c]0= + Ket]S]1= → 1= × c]1=;
       vec // LogicalForm
```
$$Out[ ]= c_0 \left|0_{S_1}\right\rangle + c_1 \left|1_{S_1}\right\rangle$$

This gives the density operator corresponding to the pure state.

```
In[ ]:= ρ = vec ** Dagger[vec];
       ρ // LogicalForm
```
$$Out[ ]= c_0 c_0^* \left|0_{S_1}\right\rangle \left\langle 0_{S_1}\right| + c_0 c_1^* \left|0_{S_1}\right\rangle \left\langle 1_{S_1}\right| + c_1 c_0^* \left|1_{S_1}\right\rangle \left\langle 0_{S_1}\right| + c_1 c_1^* \left|1_{S_1}\right\rangle \left\langle 1_{S_1}\right|$$

The matrix representation of the density operator gives the typical form of the density matrix for a pure state.

```
In[·]:= mat = Matrix[ρ];
        mat // MatrixForm
Out[·]//MatrixForm=
        ( c₀ c₀*   c₀ c₁* )
        ( c₁ c₀*   c₁ c₁* )
```

This illustrates that $\mathrm{Tr}[\rho^2] = 1$ for pure states.

```
In[·]:= Tr[mat.mat] // Simplify // MatrixForm
Out[·]//MatrixForm=
        ( c₀ c₀* + c₁ c₁* )²
```

The *von Neumann entropy* offers a far more general way to characterize a mixed state $\hat{\rho}$. The von Neumann entropy of a density operator $\hat{\rho}$ is defined by

$$S(\hat{\rho}) := -\mathrm{Tr}\,\hat{\rho}\log_2\hat{\rho} = -\sum_j \lambda_j \log_2 \lambda_j, \qquad (1.32)$$

where λ_j are the non-vanishing eigenvalues of $\hat{\rho}$. It is an extension of the notion of classical Shannon entropy. The base of the logarithm in (1.32) is arbitrary. It is customary and convenient to take the logarithm of base 2 in quantum information theory. The von Neumann entropy of any pure state is zero because the only eigenvalue of the density operator representing a pure state is 1. It is $\log_2 N$ for the completely random state, $\hat{\rho} = \hat{I}/N$, where N is the dimension of the Hilbert space. In fact, $\log_2 N$ is the maximum that $S(\hat{\rho})$ can take for any $\hat{\rho}$. The von Neumann entropy measures the uncertainty about the quantum states in the statistical mixture associated with the density operator.

As a vector describing a mixed state, one can ask whether a density operator is separable or not. Consider a system consisting of two subsystems A and B. A density operator $\hat{\rho}$ (and the associated mixed state) is said to be separable if it can be written as a *convex linear* combination

$$\hat{\rho} = \sum_j \hat{\sigma}_j \otimes \hat{\tau}_j\, p_j, \quad 0 \le p_j \le 1, \quad \sum_j p_j = 1, \qquad (1.33)$$

where $\hat{\sigma}_j$ and $\hat{\tau}_j$ are the density operators of subsystems A and B. Unlike pure states, for which the Schmidt decomposition provides a simple test of entanglement (see Appendix A.6), it is generally hard to tell whether a given mixed state is separable or entangled, and this remains an open question. One might be tempted to use the Schmidt-like decomposition (A.89) of operators for the test of mixed-state entanglement. However, the operators $\hat{A}_\mu$ and $\hat{B}_\mu$ in the sum are not guaranteed to be density operators.

1.2   Time Evolution of Quantum States

Under Newton's second law of motion, the state of a classical system evolves over time being governed by the celebrated equation of motion, $F = ma$. In quantum mechanics, Newton's equation of motion is replaced with Schrödinger's.

1.2.1 Unitary Dynamics

Postulate 2 The time evolution of the state $|\psi\rangle$ of a *closed* quantum system is governed by the Schrödinger equation

$$i\hbar\frac{d}{dt}|\psi\rangle = \hat{H}|\psi\rangle,\qquad(1.34)$$

where $\hbar$ is the Planck constant and $\hat{H}$ is the Hamiltonian of the system.

Throughout the book, we use a unit system where $\hbar = 1$. In such a unit system, energy and frequency have the same dimension, the inverse of the dimension of time. The Hamiltonian is a Hermitian operator physically describing the energy of the system. In general, it is difficult to find the precise Hamiltonian for a particular system. In most cases, a model Hamiltonian is constructed and tested against experimental observations. If desired, it is corrected by changing the existing terms or including additional terms, and testing again.

Postulate 2 expresses the time evolution in terms of a differential equation. An equivalent way is to describe it with a unitary transformation: Suppose that the system is initially in a state $|\psi(0)\rangle$, then the state $|\psi(t)\rangle$ at later time $t > 0$ is related to the initial state $|\psi(0)\rangle$ by a unitary operator $\hat{U}(t)$

$$|\psi(t)\rangle = \hat{U}(t)|\psi(0)\rangle.\qquad(1.35)$$

The unitary operator $\hat{U}(t)$ is called the *time-evolution operator*. When the Hamiltonian $\hat{H}$ of the system is independent of time, and recalling that we have put $h = 1$, the time-evolution operator is given by (recall that we have put $\hbar = 1$)

$$\hat{U}(t) = \exp[-it\hat{H}].\qquad(1.36)$$

The best way to evaluate the exponential function of a normal operator is to use its *spectral decomposition* (Appendix A.3). With the eigenstates $|E\rangle$ and corresponding eigenvalues E of $\hat{H}$, the Hamiltonian itself reads as $\hat{H} = \sum_E |E\rangle\langle E| E$. Hence, its exponential function is given by $\exp(-it\hat{H}) = \sum_E |E\rangle\langle E| e^{-itE}$. In general, and especially when the system is driven externally to actively process quantum information, for example, the Hamiltonian depends on time. In this case, the relation between the time-evolution operator and the Hamiltonian is more complicated, and this will be discussed later.

Consider a two-level quantum state, denoted by the symbol S. Some real parameters will be denoted by the symbol B.

```
Let[Qubit, S]
Let[Real, B]
```

A time-independent Hamiltonian can be expressed in terms of the Pauli operators.

In[ ]:= `H = B[0] + S[1] × B[1] + S[2] × B[2] + S[3] × B[3]`

Out[ ]= $B_0 + B_1\, S^x + B_2\, S^y + B_3\, S^z$

In this case, the time-evolution operator is given by

In[ ]:= `Clear[U]`
`U[t_] = MultiplyExp[-I t H]`

Out[ ]= $e^{-i\, t\, \left(B_0 + B_1\, S^x + B_2\, S^y + B_3\, S^z\right)}$

The exponential function of operators can be evaluated by means of the spectral decomposition. Q3 has an internal mechanism to facilitate the spectral decomposition method. It is implemented through the function `Elaborate`.

In[ ]:= `Elaborate[U[t]] // ExpToTrig // Garner`

Out[ ]= $\text{Cos}\left[t\,\sqrt{B_1^2 + B_2^2 + B_3^2}\right] \times \left(\text{Cos}[t\,B_0] - i\,\text{Sin}[t\,B_0]\right) -$

$$\dfrac{i\,B_3\,S^z\,\left(\text{Cos}[t\,B_0] - i\,\text{Sin}[t\,B_0]\right)\times\text{Sin}\left[t\,\sqrt{B_1^2 + B_2^2 + B_3^2}\right]}{\sqrt{B_1^2 + B_2^2 + B_3^2}} -$$

$$\dfrac{i\,(B_1 - i\,B_2)\,S^+\,\left(\text{Cos}[t\,B_0] - i\,\text{Sin}[t\,B_0]\right)\times\text{Sin}\left[t\,\sqrt{B_1^2 + B_2^2 + B_3^2}\right]}{\sqrt{B_1^2 + B_2^2 + B_3^2}} +$$

$$\dfrac{(-i\,B_1 + B_2)\,S^-\,\left(\text{Cos}[t\,B_0] - i\,\text{Sin}[t\,B_0]\right)\times\text{Sin}\left[t\,\sqrt{B_1^2 + B_2^2 + B_3^2}\right]}{\sqrt{B_1^2 + B_2^2 + B_3^2}}$$

For simplicity, consider the following specific case. We have assumed a ceratin choice of units.

`B[0] = 0; B[1] = B[3] = 1; B[2] = -1;`

In[ ]:= `Clear[op]`
`op[t_] = Elaborate[U[t]] // ExpToTrig // Garner`

Out[ ]= $\text{Cos}\left[\sqrt{3}\,t\right] - \dfrac{i\,S^z\,\text{Sin}\left[\sqrt{3}\,t\right]}{\sqrt{3}} + \dfrac{(1-i)\,S^+\,\text{Sin}\left[\sqrt{3}\,t\right]}{\sqrt{3}} - \dfrac{(1+i)\,S^-\,\text{Sin}\left[\sqrt{3}\,t\right]}{\sqrt{3}}$

Suppose that the initial state is the eigenstate of the Pauli X operator, here denoted by `S[1]`.

In[ ]:= `v0 = (Ket[] + Ket[S → 1]) / Sqrt[2];`
`v0 // LogicalForm`

Out[ ]= $\dfrac{\left|1_S\right\rangle + \left|0_S\right\rangle}{\sqrt{2}}$

This is the state vector at a later time `t>0`.

In[ ]:= `Clear[vec]`
`vec[t_] = op[t] ** v0;`
`vec[t] // LogicalForm`

Out[ ]= $\left|1_S\right\rangle \left(\dfrac{\text{Cos}\left[\sqrt{3}\,t\right]}{\sqrt{2}} - \dfrac{\text{Sin}\left[\sqrt{3}\,t\right]}{\sqrt{6}}\right) + \left|0_S\right\rangle \left(\dfrac{\text{Cos}\left[\sqrt{3}\,t\right]}{\sqrt{2}} + \dfrac{(1-2\,i)\,\text{Sin}\left[\sqrt{3}\,t\right]}{\sqrt{6}}\right)$

This visualizes the evolution of the state under the Hamiltonian on the Bloch sphere.

```
In[ ]:= vv = Bead @[ BlochVector @[ Table{vec{t}, ]t, 0, 1, 0.1/} @@ Chop;
        BlochSphere[]Red, vv/, ImageSize → Small}
```

```
Out[ ]=
```

An important point to bear in mind about unitary dynamics is that it does not depend on the history (for a time-independent Hamiltonian). This is reflected in the obvious property of the time-evolution operator,

$$\hat{U}(t + t') = \hat{U}(t)\hat{U}(t') \tag{1.37}$$

for any t, t'. Physically, this implies that the evolution depends only on the duration of time, but not on when it starts or ends. In this respect, it is also natural for a time-evolution operator to have the property $\hat{U}^\dagger(t) = \hat{U}(-t)$.

So far, we have discussed the time evolution of a pure state. What if the system is in a mixed state $\hat{\rho}(0)$ for a certain reason? The mixed state can be expressed as a statistical mixture of pure states,

$$\hat{\rho}(0) = \sum_j |\psi_j(0)\rangle\langle\psi_j(0)| \, p_j \tag{1.38}$$

with $0 \leq p_j \leq 1$. If the system is closed, then each pure state $|\psi(0)\rangle$ evolves into $\hat{U}(t)|\psi(0)\rangle$. Overall, the state $\hat{\rho}(t)$ at a later time t is given by

$$\hat{\rho}(t) = \hat{U}(t)\hat{\rho}(0)\hat{U}^\dagger(t). \tag{1.39}$$

In short, as long as the system remains closed, the dynamics are still unitary regardless of whether the system is initially prepared in a pure or a mixed state.

Let us consider the same system and Hamiltonian. However, the initial state is now a mixed state.

```
In[ ]:= ρ0 = v0 ** Dagger[v0] * 3 / 4 + Ket[] ** Bra[] ** 1 / 4 // LogicalForm // Garner
```

$$Out[ ]= \frac{5 \, |0_s\rangle \, \langle 0_s|}{8} + \frac{3 \, |0_s\rangle \, \langle 1_s|}{8} + \frac{3 \, |1_s\rangle \, \langle 0_s|}{8} + \frac{3 \, |1_s\rangle \, \langle 1_s|}{8}$$

Its matrix representation in the logical basis is given by the following.

```
In[·]:= mat = Matrix[ρ0];
       mat // MatrixForm
```

Out[·]//MatrixForm=

$$\begin{pmatrix} \frac{5}{8} & \frac{3}{8} \\ \frac{3}{8} & \frac{3}{8} \end{pmatrix}$$

This is the state vector at a later time t>0.

```
In[·]:= Let[Real, t]; Clear[rho]
       rho[t_] = op[t] ** ρ0 ** Dagger[op[t]];
       rho[t] // LogicalForm
```

Out[·]=
$$\frac{5}{8} \cos\left[\sqrt{3}\ t\right]^2 |0_s\rangle\langle 0_s| - \frac{17}{24} |0_s\rangle\langle 0_s| \sin\left[\sqrt{3}\ t\right]^2 - \frac{1}{8}\sqrt{3}\ |0_s\rangle\langle 0_s| \sin\left[2\sqrt{3}\ t\right] -$$
$$\frac{1}{48}\ i\ |0_s\rangle\langle 1_s| \left(+16 - 2\ i\left(\sin\left[\sqrt{3}\ t\right]^2\right) + 6 - 3\ i\left(\sqrt{3}\ \sin\left[2\sqrt{3}\ t\right]\right)\right) -$$
$$\frac{1}{48}\ |1_s\rangle\langle 0_s| \left(18 \cos\left[\sqrt{3}\ t\right]^2\right) + 5 - 2\ i\left(\sqrt{3}\ \sin\left[2\sqrt{3}\ t\right]\right) -$$
$$\frac{1}{48}\ |0_s\rangle\langle 1_s| \left(18 \cos\left[\sqrt{3}\ t\right]^2\right) + 5\right)\ 2\ i\left(\sqrt{3}\ \sin\left[2\sqrt{3}\ t\right]\right) -$$
$$\frac{1}{24}\ |1_s\rangle\langle 1_s| \left(8 - \cos\left[2\sqrt{3}\ t\right]\right)\ 3\sqrt{3}\ \sin\left[2\sqrt{3}\ t\right]\right) -$$
$$\frac{1}{48}\ |1_s\rangle\langle 0_s| \left(+\right)\ 2\right)\ 16\ i\left(\sin\left[\sqrt{3}\ t\right]^2 - +3 - 6\ i\left(\sqrt{3}\ \sin\left[2\sqrt{3}\ t\right]\right)\right)$$

This visualizes the evolution of the state under the Hamiltonian on the Bloch sphere. Note that the magnitude of the Bloch vectors are preserved by the unitary dynamics.

```
In[·]:= ρρ = Bead @[ BlochVector @[ Table{rho{t}, ]t, 0, 1, 0.1/} @@ Chop;
       BlochSphere{]Red, ρρ/, "Opacity" → 0.4, ImageSize → Small}
```

Out[·]=

1.2.2 Quantum Noisy Dynamics

When a system interacts with its environment, the dynamics of the system can no longer be described by the Schrödinger equation. More importantly, it is not unitary. It may not be surprising at first glance since similar effects also occur in classical mechanics. For example, a ball thrown up in the air interacts with various molecules and small particles in the air, and it does not follow Newton's equation of motion. The dynamics become dissipative as is described by additional damping terms in the equation of motion. In effect, damping terms arise because we are ignoring the small molecules and particles that disturb the ball. However, ignoring the environmental degrees of freedom in quantum theory brings about far more profound effects. It causes not only the dissipation of energy of the system to the environment, but also the effect of so-called *decoherence*, the loss of *quantumness* (Zurek, 1991, 2002). Here, we outline the basic formalism for decoherence, but further details will be discussed in Sect. 5.2.

Suppose that the system is initially in a state $|\psi\rangle \in \mathcal{H}$ and the environment is in $|\epsilon\rangle \in \mathcal{E}$. Thus ,the initial state of the total system is $|\Psi(0)\rangle = |\psi\rangle \otimes |\epsilon\rangle \in \mathcal{H} \otimes \mathcal{E}$. Then, let the system interact with the environment. As a closed system, the evolution of the total system is governed by a unitary time-evolution operator $\hat{U}_{\text{tot}}(t)$, $|\Psi(t)\rangle = \hat{U}_{\text{tot}}(t)|\psi\rangle \otimes |\epsilon\rangle$, in accordance with Postulate 2. The system-environment coupling tends to build entanglement, so the state $|\Psi(t)\rangle$ cannot be factorized in general. Without comprehensive knowledge of the environment, the state of the system alone is thus a mixed state, as discussed in Sect. 1.1.2. Specifically, one can get the density matrix representing the mixed state by tracing out the information of the environment from the total state $|\Psi(t)\rangle$,

$$\hat{\rho}(t) = \text{Tr}_{\mathcal{E}} |\Psi(t)\rangle \langle \Psi(t)| = \text{Tr}_{\mathcal{E}} \hat{U}_{\text{tot}}(t) (|\psi\rangle \langle \psi| \otimes |\epsilon\rangle \langle \epsilon|) \hat{U}_{\text{tot}}^{\dagger}(t) . \qquad (1.40)$$

A partial trace over the environment physically corresponds to ignoring the degrees of freedom of the environment. In general, the dynamics of $\hat{\rho}(t)$ is thus very complicated. However, there are two essential points: First, the dynamics is no longer unitary. Second, through entanglement with the environment, the quantum state of the system loses coherence (i.e., loses quantumness).

The non-unitary dynamics of a system coupled to the environment can be most efficiently described via the quantum operation formalism, which is the subject of Chap. 5.

1.3 Measurements on Quantum States

In classical mechanics, just like the state of motion, a physical quantity is described by a simple number for a scalar quantity or by a set of numbers for a vector quantity. As such, measurement merely assesses such number(s) and casts technological questions concerning only measuring devices. In quantum mechanics, measurement is totally different and is deeply intermingled into quantum theory itself.

1.3.1 Projection Measurements

Postulate 3 A physical quantity is described by an "observable"—a Hermitian operator. Upon *measurement* of an observable $\hat{A}$, the outcome is *probabilistically* selected from one of the eigenvalues a of $\hat{A}$. When the system is in the state $|\psi\rangle$ right before measurement, the probability of a particular outcome a is given by $P_a = |\langle a|\psi\rangle|^2$, where $|a\rangle$ is the eigenstate of $\hat{A}$ belonging to the eigenvalue a. Right after measurement, the state "collapses" to the eigenstate $|a\rangle$ corresponding to outcome a.

The postulate points out several striking features that distinguish quantum mechanics from classical mechanics. First, a measurable physical quantity (i.e., an observable) is not described by a simple number but by an operator acting on the vectors that describe the quantum states of the system (see Postulate 1).

Second, measurement brings about a sudden collapse of the state vector whereby an unavoidable disturbance by mere measurement causes obvious obstacles to naively examining quantum states. At the same time, this opens new opportunities for practical applications such as to achieve secure quantum communications. Furthermore, this collapse is a peculiar dynamics of quantum states and an alternative to that governed by the Schrödinger equation (Postulate 2), which turns out to be useful to construct measurement-based quantum computing architectures (Sect. 3.4). Even in the common dynamical schemes of quantum computation, measurement is often used to initialize the quantum registers to the logical basis states. Combined with classical communications, measurement also enriches available gate operations.

Third, the postulate makes it clear that even if the state vector $|\psi\rangle$ is known, which comprises the complete description of the state according to Postulate 1, one cannot infer the measurement results even in principle, but only the probabilities of possible outcomes. The description is inherently statistical, and one needs to measure an *ensemble* of identically prepared systems to extract the value of an observable in terms of its statistical moments such as the expectation value. Given a state vector $|\psi\rangle$, the expectation value of operator $\hat{A}$ is obtained using elementary theory of probability (see also Dirac's Bra-Ket notation in Appendix A.3)

$$\langle \hat{A} \rangle = \sum_q q \, P_q = \sum_q \langle \psi | q \rangle \, q \, \langle q | \psi \rangle = \langle \psi | \, \hat{A} \, | \psi \rangle \qquad (1.41)$$

A closely related feature of quantum states is reflected in the *no-cloning theorem*, as Zurek (2000) declared: "You can clone a sheep,[5] but not a quantum state." The *no-cloning theorem* of quantum states dictates that it is impossible to make a copy of an *unknown* quantum state. If it *were* possible, one could produce a bunch of identical copies of the same quantum state. Then, measurements of different incompatible observables (such as conjugate variables) on the copies would reveal precise information regarding the quantum state, violating the uncertainty principle. Cloning of an (unknown) quantum state would also allow faster-than-light communication. Furthermore, the no-cloning nature of quantum states is one of the key features of quantum mechanics that provide *unconditional* security in quantum communication.

But why is it impossible? Suppose we have a machine that can do it. The machine would have two quantum registers, the input and target registers. The input register is prepared in an (unknown) quantum state $|\psi\rangle$, and the target register in a reference

---

[5] It refers to Dolly, the sheep cloned by a research group (Wilmut et al., 1997).

state, say, $|0\rangle$. The overall state of the total system is a tensor-product state $|\psi\rangle \otimes |0\rangle$. Let $\hat{U}$ be the unitary operator corresponding to the operation of the copying operation. By the hypothesis, we expect that

$$\hat{U}(|\psi\rangle \otimes |0\rangle) = |\psi\rangle \otimes |\psi\rangle. \tag{1.42}$$

Let us apply the operation to two orthogonal states $|\alpha\rangle$ and $|\beta\rangle$. We have

$$\hat{U}(|\alpha\rangle \otimes |0\rangle) = |\alpha\rangle \otimes |\alpha\rangle, \quad \hat{U}(|\beta\rangle \otimes |0\rangle) = |\beta\rangle \otimes |\beta\rangle. \tag{1.43}$$

So far nothing is wrong. Now, let us apply the operation to a linear superposition $|\alpha\rangle a + |\beta\rangle b$ with a and b arbitrary (and unknown) complex numbers. From the linearity of $\hat{U}$, we have

$$\hat{U}[(|\alpha\rangle a + |\beta\rangle b) \otimes |0\rangle] = \hat{U}(|\alpha\rangle \otimes |0\rangle a + |\beta\rangle \otimes |0\rangle b) = |\alpha\rangle \otimes |\alpha\rangle a + |\beta\rangle \otimes |\beta\rangle b. \tag{1.44}$$

Clearly, the result is different from what we expect,

$$(|\alpha\rangle a + |\beta\rangle b) \otimes (|\alpha\rangle a + |\beta\rangle b), \tag{1.45}$$

unless either a or b vanishes. Therefore, we conclude that it is impossible to copy a quantum state. The crucial point in the proof is the linearity of quantum mechanics.

In the above statement of the postulate, the system is assumed to be in a pure state initially, and the statement can be naturally extended to the case of mixed state: When a system is in a mixed state $\hat{\rho}$ right before the measurement, the probability is given by $P_a = \langle u|\hat{\rho}|a\rangle$ and the state right after the measurement becomes $|a\rangle \langle a|$. By virtue of the elementary theory of probability, we observe again that $0 \leq \hat{\rho} \leq 1$ and $\mathrm{Tr}\hat{\rho} = 1$, as we have already seen in Sect. 1.1.2. Furthermore, the expectation value of the observable is given by

$$\langle \hat{A} \rangle = \sum_a a P_a = \sum_a a \langle u| \hat{\rho} |a\rangle = \mathrm{Tr}\hat{A}\hat{\rho}. \tag{1.46}$$

Von Neumann envisioned an idealistic procedure to perform a projection measurement on physical systems. For this reason, projection measurement is also referred to as a von Neumann measurement. The *von Neumann scheme* laid the foundation for the quantum theory of measurement, and it has inspired various methods to improve the measurement precision towards the intrinsic limit of quantum mechanics. The von Neumann scheme for projection measurement can be implemented directly in a quantum circuit model (see Sect. 4.4). Here, we briefly summarize the procedure: Suppose that we want to measure a quantity described by the Hermitian operator $\hat{A}$. Let $|a\rangle$ be the eigenvectors with eigenvalues a of $\hat{A}$ so that $\hat{A}$ has spectral decomposition (see Appendix A.3)

$$\hat{A} = \sum_a |a\rangle\, a\, \langle a| . \tag{1.47}$$

The system is in a state $|\psi\rangle$, which is expanded as

$$|\psi\rangle = \sum_a |a\rangle\, \psi_a \tag{1.48}$$

in the eigenbasis of $\hat{A}$. Typically, the distribution $P_a := |\psi_a|^2$ is supposed to sharply peak around a certain unknown eigenstate $|a_*\rangle$, and the measurement is supposed to reveal a_*.

We choose a measuring device with "position" $\hat{X}$, which can be directly observed, and we prepare it in an approximate eigenstate $|\xi\rangle$ of $\hat{X}$. In terms of the eigenstates $|x\rangle$ with eigenvalue x of $\hat{X}$, $|\xi\rangle$ is expanded as

$$|\xi\rangle = \int dx\, |x\rangle\, \xi(x) \tag{1.49}$$

with $\xi(x)$ having a sharp peak around a certain fixed point x_0. It will become clear below how a sharper distribution $|\xi(x)|^2$ in the position results in a more accurate measurement. At this stage, $|\psi\rangle \otimes |\xi\rangle$ is the overall state of the total system containing the system to be measured and the measurement device. Now couple the observable $\hat{A}$ in question to the "momentum" $\hat{P}$ (not $\hat{X}$ itself) of the measurement device. Note that $\hat{P}$ and $\hat{X}$ are canonical conjugates of each other such that $[\hat{X}, \hat{P}] = i$ (recall that we have chosen a unit system such that $\hbar = 1$). The coupling is described by the interaction Hamiltonian

$$\hat{H}_{\text{int}} = J\hat{P} \otimes \hat{A}, \tag{1.50}$$

where J is the coupling strength. One has to let the system and the measurement device interact for a sufficiently long time τ such that the position of the measurement device is distinguished from the initial position. Let $g := J\tau$ be the dimensionless coupling constant. Due to coupling, the total system composed of the system in question and the measurement device evolves over time, which is described by the unitary operator

$$\hat{U}_{\text{int}} = \exp\left(-ig\hat{P} \otimes \hat{A}\right). \tag{1.51}$$

This situation is depicted diagrammatically as follows

It is instructive to expand the interaction unitary operator $\hat{U}_{int}$ using the spectral decomposition [see Eq. (A.53)] of the observable $\hat{A}$ as

$$\hat{U}_{int} = \sum_a e^{-iga\hat{P}} \otimes |a\rangle \langle a| \tag{1.52}$$

The operator $\hat{T}_a := e^{-iga\hat{P}}$ is a translation operator with the amount of translation depending on the eigenvalue a of $\hat{A}$. To clarify the physical interpretation, we rewrite the interaction unitary operator into

$$\hat{U}_{int} = \sum_a \hat{T}_a \otimes |a\rangle \langle a|. \tag{1.53}$$

This physical picture is further elucidated in the following schematic diagram

$$\tag{1.54}$$

After the interaction, the state vector of the total system becomes

$$\hat{U}_{int} |\Psi\rangle = \sum_a \left(\hat{T}_a |\xi\rangle\right) \otimes (|a\rangle \psi_a) = \int dx \, |x\rangle \otimes \left(\sum_a |a\rangle \xi(x - ga)\psi_a\right). \tag{1.55}$$

In general, the final state is an entangled state. The probability to register x out of the measurement on the measurement device is given by

$$P(x) = \sum_a |\xi(x - ga)|^2 |\psi_a|^2. \tag{1.56}$$

When the system starts from a definite eigenstate $|a\rangle$ of $\hat{A}$, the system and the measurement device remain unentangled after the system-measurement device interaction, with its probability distribution determined solely by the given eigenvalue a

$$P(x) = |\xi(x - ga)|^2. \tag{1.57}$$

Manifestly, the measurement accuracy is best in this case. As illustrated in Fig. 1.3, one can extract the eigenvalue a of the observable from the amount of translation ga of the final wave function relative to the initial wave function assuming that the coupling strength g is known (it can be calibrated separately).

Figure 1.3 suggests that a sharper initial wave function of the measurement device will results in a more precise extraction of the value of a. Roughly speaking, this is indeed one direction of the efforts to improve the precision of measurement (Caves,

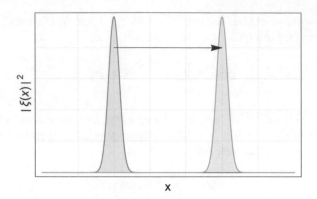

Fig. 1.3 Change in the wave function $\xi(x)$ of the measurement device from the initial state (blue) to the final state (orange) after it interacts with the system. This case illustrates a system in a definite eigenstate $|a\rangle$ of the observable $\hat{A}$ under measurement. The amount of translation is given by ga, the coupling constant times the eigenvalue associated with $|a\rangle$

1981). An obstacle directly apparent in Fig. 1.3 to achieve precise measurement is the inherent statistical nature of quantum mechanics. Given that the probability distribution has finite width, any measurement in quantum mechanics should be repeated many times. The statistical error decreases as $1/\sqrt{N}$ with an increasing number N of repeated measurements. This is called the *standard quantum limit*. Interestingly, if one prepares a set of measurement devices in a proper entangled state, one can improve the statistical error to $1/N$ (Giovannetti et al., 2006). This enhanced accuracy due to quantum entanglement is called the *Heisenberg limit*, and it puts the ultimate limit on the measurement precision that quantum mechanics allows.

1.3.2   Generalized Measurements

Concerning measurement, we introduced Postulate 3 in its usual form as discussed in most textbooks on quantum mechanics. However, in quantum information context and for various reasons to be discussed shortly, it is convenient to provided a generalized form of Postulate 3.

Postulate 3′ A generalized measurement is described by a set of *measurement operators* $\hat{M}_m$ corresponding to measurement outcomes m that satisfy the completeness relation $\sum_m \hat{M}_m^\dagger \hat{M}_m = \hat{I}$. Suppose that right before the measurement, the system is in the quantum state $\hat{\rho}$. Then the probability P_m for the outcome m is given by $P_m = \text{Tr}\hat{M}_m \hat{\rho}\hat{M}_m^\dagger$, and the state right after measurement becomes $P_m^{-1}\hat{M}_m \hat{\rho}\hat{M}_m^\dagger$.

The projective measurement discussed previously is a special case of the generalized measurement. Suppose that we measure an observable $\hat{A}$. In this case, the available measurement outcomes are the eigenvalues a of $\hat{A}$. The measurement operator corresponding to outcome a is the projection operator $|a\rangle\langle a|$ into the eigenstate $|a\rangle$ belonging to eigenvalue a.

In general, the physical meaning of the measurement operators $\hat{M}_m$ is not immediately clear until the specifics of the measurement in question have been given. It is even more difficult to determine the setup of a physical device for the measurement described by the given set of operators $\hat{M}_m$. Nevertheless, one can show that generalized measurements are equivalent to projective measurements (extended to a larger system and supplemented with additional unitary operations). In many applications, however, generalized measurements are more convenient because they are less restrictive.

For example, consider a fixed set of *orthogonal* quantum states $|\psi_1\rangle, \ldots, |\psi_n\rangle$ (not necessarily spanning the Hilbert space). Suppose that a state is drawn from the set. The task is to tell which of the n state it is. We proceed by defining n operators $\hat{M}_j := |\psi_j\rangle\langle\psi_j|$ for $j = 1, \ldots, n$ and an additional operator $\hat{M}_0 := \hat{I} - \sum_j \hat{M}_j$. They satisfy the completeness relation $\sum_{j=0}^{n} \hat{M}_j^\dagger \hat{M}_j = \hat{I}$ and describe a measurement. Now when the particular state $|\psi_k\rangle$ is taken, the probability P_k for the measurement outcome k is unity because $P_k = \langle\psi_k| \hat{M}_k |\psi_k\rangle = 1$.

Postulate 3 and 3' describe not only the probability for a particular measurement outcome but also the state corresponding to the outcome after the measurement. However, many experiments the system only once, and the state after the measurement is irrelevant. Of primary concern are the probabilities. Often, it is even impossible to assign a post-measurement state to the system. For example, in an experiment to measure the position of photon, the photon is destroyed at the photo-screen, and it is meaningless to ask about the wave function of the photon after the measurement.

As long as one is focused on the probabilities, the description in Postulate 3' can be drastically simplified: Since the trace is invariant under cyclic permutation of the operators, the probability for the outcome m is given by

$$P_m = \mathrm{Tr}\,\hat{M}_m \hat{\rho} \hat{M}_m^\dagger = \mathrm{Tr}\,\hat{M}_m^\dagger \hat{M}_m \hat{\rho}. \tag{1.58}$$

Therefore, as far as the probabilities are concerned, all we need are the operators $\hat{E}_m := \hat{M}_m^\dagger \hat{M}_m$ rather than the measurement operators $\hat{M}_m$. Note that $\hat{M}_m$ may contain far more details that cannot be extracted from $\hat{E}_m$. Infinitely many distinct measurement operators can lead to the same $\hat{E}_m$. Nevertheless, it is possible to discard certain details in $\hat{M}_m$ for more compact operators $\hat{E}_m$ because we do not care about the post-measurement state. By construction, $\hat{E}_m$ are positive semi-definite operators and satisfy $\sum_m \hat{E}_m = \hat{I}$. The set $\left\{\hat{E}_m\right\}$ of such operators is called a *positive operator-valued measure* or more often a POVM. The individual operators $\hat{E}_m$ are called *POVM elements*. For many experiments, the POVM formalism allows far more efficient description of the measurement.

As a non-trivial example of the application of the POVM formalism, let us consider the problem of quantum state discrimination of *non-orthogonal* quantum states (Bergou et al., 2004; Chefles, 2004). To be specific, suppose that we are given either of the two states

$$|v\rangle = |0\rangle , \quad |w\rangle = \frac{|0\rangle + |1\rangle}{\sqrt{2}} . \tag{1.59}$$

We have to figure out which one it is. These two states are not orthogonal, $\langle v|w\rangle \neq 0$, and it is impossible to discriminate them with certainty. Instead, the task is to perform a measurement that tells which of the two, but never misidentifies a wrong state. Consider a POVM including the following three elements

$$\hat{E}_1 = \frac{1}{\sqrt{1 + \langle v|w\rangle}} |1\rangle \langle 1| , \tag{1.60a}$$

$$\hat{E}_2 = \frac{1}{\sqrt{1 + \langle v|w\rangle}} \left(\frac{|0\rangle - |1\rangle}{\sqrt{2}} \right) \left(\frac{\langle 0| - \langle 1|}{\sqrt{2}} \right) \tag{1.60b}$$

$$\hat{E}_3 = \hat{I} - \hat{E}_1 - \hat{E}_2 \tag{1.60c}$$

It is straightforward to check that they form a POVM. Suppose that the state $|v\rangle$ is given. As it is orthogonal to $|1\rangle$, the measurement outcome $m = 1$—corresponding to the POVM element $\hat{E}_1$—never occurs. This means that once $m = 2$—corresponding to $\hat{E}_2$—occurs, the state must be $|w\rangle$. For the same reason, the result $m = 1$ implies that the state is definitely $|v\rangle$. Of course, it is possible to obtain the outcome $m = 3$, in which case the measurement reveals nothing. However, the important point is that there is no chance to make a mistake.

Problems

1.1 Consider a set of *arbitrary* state vectors $\left\{ |\psi_\mu\rangle : \mu = 0, 1, \ldots, n - 1 \right\}$ in a vector space. Note that the vectors are not normalized nor orthogonal to each other. Let $\hat{A}$ be a linear operator defined by

$$\hat{A} = \sum_{\mu=0}^{n-1} |\psi_\mu\rangle \langle \psi_\mu| . \tag{1.61}$$

Show that the eigenvalues of $\hat{A}$ are all non-negative.

1.2 Let $\hat{\rho}$ be a density operator. Prove the following statements.

   (a) $\mathrm{Tr}\hat{\rho} = 1$.
   (b) $\hat{\rho}$ is positive (Definition A.19).
   (c) $0 \leq \lambda \leq 1$ for any eigenvalue λ of $\hat{\rho}$.

1.3 Let $|\Psi\rangle$ and $|\Phi\rangle$ are quantum states in a tensor-product space $\mathcal{H} \otimes \mathcal{E}$. Show that

$$\mathrm{Tr}_{\mathcal{E}} |\Psi\rangle \langle \Psi| = \mathrm{Tr}_{\mathcal{E}} |\Phi\rangle \langle \Phi| \tag{1.62}$$

if and only if

$$|\Phi\rangle = (\hat{I} \otimes \hat{U})|\Psi\rangle, \tag{1.63}$$

where $\hat{U}$ is a unitary operator on $\mathcal{E}$.

1.4 Let $\{|v_i\rangle\}$ and $\{|w_j\rangle\}$ be orthonormal bases of vectors spaces $\mathcal{V}$ and $\mathcal{W}$, respectively, of the same dimension. Construct an (unnormalized) maximally entangled state in $\mathcal{V} \otimes \mathcal{W}$ by

$$|\Phi\rangle := \sum_j |v_j\rangle \otimes |w_j\rangle. \tag{1.64}$$

Let $\hat{\rho}$ be a density operator on a vector space $\mathcal{V}$. Show that any purification of $\hat{\rho}$ into $\mathcal{V} \otimes \mathcal{W}$ has the form

$$|\Psi\rangle = \left((\sqrt{\hat{\rho}}\,\hat{U}) \otimes \hat{V}\right)|\Phi\rangle, \tag{1.65}$$

where $\hat{U}$ and $\hat{V}$ are unitary operators on $\mathcal{V}$ and $\mathcal{W}$, respectively.

Hint: Use the theorem in Problem 1.3.2.

1.5 Show that a density operator $\hat{\rho}$ is a pure state if and only if $\mathrm{Tr}\hat{\rho}^2 = 1$.

1.6 Consider a two-level quantum system. Let $|0\rangle$ and $|1\rangle$ be the logical basis of the Hilbert space associated with the system. Let $\hat{S}^\mu$ ($\mu = 0, x, y, z$) be the Pauli operator, defined by

$$\hat{S}^x|0\rangle = |1\rangle, \qquad \hat{S}^y|0\rangle = +i|1\rangle, \qquad \hat{S}^z|0\rangle = +|0\rangle, \tag{1.66a}$$

$$\hat{S}^x|1\rangle = |0\rangle, \qquad \hat{S}^y|1\rangle = -i|0\rangle, \qquad \hat{S}^z|1\rangle = -|1\rangle, \tag{1.66b}$$

and $\hat{S}^0 = \hat{I}$.

(a) Find the matrix representations of $\hat{S}^\mu$ ($\mu = 0, x, y, z$) in the logical basis.

(b) Find the matrix representations of $\hat{S}^\mu$ in the new basis

$$|\pm\rangle := \frac{|0\rangle \pm |1\rangle}{\sqrt{2}}. \tag{1.67}$$

1.7 Consider a spin 1/2 in an external magnetic field. We describe the states of the spin with the vector space spanned by the logical basis states $|0\rangle \equiv |\uparrow\rangle$ and $|1\rangle \equiv |\downarrow\rangle$. In terms of the Pauli operators in Eq. (1.66), the Hamiltonian is given by

$$\hat{H} = B_x\hat{S}^x + B_y\hat{S}^y + B_z\hat{S}^z, \tag{1.68}$$

where B_μ ($\mu = x, y, z$) are the parameters proportional to the external magnetic field. In this expression, they have the same dimension as the energy. Suppose that

$$B_x = B_0 \sin \theta \cos \phi, \quad B_y = B_0 \sin \theta \sin \phi, \quad B_z = B_0 \cos \theta \qquad (1.69)$$

with $B_0 > 0$ and $\theta, \phi \in \mathbb{R}$.

(a) Find the eigenvalues and corresponding eigenstates of $\hat{H}$.
    Hint: A direct evaluation of the Hamiltonian in Eq. (1.68) is straightforward.
    Another method is to use the commutation relations of the Pauli operators.
(b) Display the two eigenstates on the Bloch sphere.

1.8 Let $|\psi\rangle$ be a vector in a two-dimensional vector space. Show that the Bloch
     vector $\boldsymbol{b} := (\langle \hat{S}^x \rangle, \langle \hat{S}^y \rangle, \langle \hat{S}^z \rangle) \in \mathbb{R}^3$ has a magnitude of unity, $|\boldsymbol{b}| = 1$.

1.9 Let $\hat{\rho}$ be a density operator on a two-dimensional vector space. It can be written
     as

$$\hat{\rho} = \frac{1}{2} \left(\hat{S}^0 + x\hat{S}^x + y\hat{S}^y + z\hat{S}^z \right), \qquad (1.70)$$

where $\hat{S}^\mu$ are the Pauli operators [see Eq. (1.67)] and $x, y, z \in \mathbb{R}$.

(a) Show that the Bloch vector defined by $\boldsymbol{b} := (\langle \hat{S}^x \rangle, \langle \hat{S}^y \rangle, \langle \hat{S}^z \rangle)$ is just given
    by $\boldsymbol{b} = (x, y, z)$.
(b) Show that $|\boldsymbol{b}| \leq 1$.
(c) Show that $\hat{\rho}$ is a pure state if and only if $|\boldsymbol{b}| = 1$.
    Hint: See Problem 1.3.2.

1.10 Consider a two-qubit system, and suppose that it is in the state

$$|\psi\rangle = \frac{|0\rangle \otimes |0\rangle + |0\rangle \otimes |1\rangle + |1\rangle \otimes |0\rangle)}{\sqrt{3}}. \qquad (1.71)$$

(a) What is the probability p_0 to find the first qubit in $|0\rangle$ (regardless of the
    second qubit)? Similarly, what is the probability p_1 to find the first qubit in
    the state $|1\rangle$?
(b) What is the probability p_0' to find the first qubit in $(|0\rangle + |1\rangle)/\sqrt{2}$? What
    about the probability p_1' to find the first qubit in $|0\rangle$?
(c) Suppose that you do not have an access to the second qubit, and that you
    want to construct the density operator $\hat{\rho}$ for the first qubit corresponding to
    the physical situation. Which of the two data sets from (1.3.2) and (1.3.2)
    would you use for the construction of $\hat{\rho}$? That is, if the resulting density
    operators are different, which one is correct and why?
(d) Calculate the Bloch vector $\boldsymbol{b}$ corresponding to $\hat{\rho}$ in (b), and display it in the
    Bloch sphere.

1.11 Consider a system of two two-level subsystems. The logical basis and the Pauli
      operators for each subsystem are defined as in Problem 1.3.2.

(a) Find the matrix representations of the following operator

$$\hat{H} = \hat{S}^x \otimes \hat{S}^x + \hat{S}^y \otimes \hat{S}^y + \hat{S}^z \otimes \hat{S}^z \tag{1.72}$$

in the standard tensor-product basis $\{|i\rangle \otimes |j\rangle : i, j = 0, 1\}$.

(b) Find all eigenvalues and eigenvectors of $\hat{H}$.

(c) Calculate the following unitary operator

$$\hat{U} = \exp(-it\hat{H}), \quad t \in \mathbb{R}, \tag{1.73}$$

which corresponds physically to the time-evolution operator, and express it in terms of either $\hat{S}^\mu \otimes \hat{S}^\nu$ or the dyadic products, $|i\rangle \langle j| \otimes |k\rangle \langle l| (i, j, k, l = 0, 1)$.

(d) Evaluate the state

$$|\psi(t)\rangle := \hat{U}(t) |+\rangle \otimes |-\rangle , \tag{1.74}$$

where $|\pm\rangle$ are defined in Eq. (1.67).

(e) Evaluate the expectation values

$$S_1^\mu(t) := \langle \psi(t)| \hat{S}^\mu \otimes \hat{S}^0 |\psi(t)\rangle , \tag{1.75}$$

for $\mu = x, y, z$, and plot them as functions of t.

(f) Evaluate the expectation values

$$S^\mu(t) := \langle \psi(t)| \left(\hat{S}^\mu \otimes \hat{S}^0 + \hat{S}^0 \otimes \hat{S}^\mu \right) |\psi(t)\rangle , \tag{1.76}$$

for $\mu = x, y, z$, and plot them as functions of t.

1.12 Consider a two-qubit system in the following state

$$|\psi\rangle = \frac{1}{2} \sum_{x=0}^{3} |x\rangle . \tag{1.77}$$

Here we have used a short-hand notation $|x\rangle := |x_1\rangle \otimes |x_2\rangle$, where x_j $(j = 1, 2)$ are the binary digits of x, $x = (x_1 x_2)_2$. For example, $|2\rangle = |1\rangle \otimes |0\rangle$. Suppose that we measure an observable

$$\hat{H} = \sum_{\mu \in \{x, y\}} \hat{S}^\mu \otimes \hat{S}^\mu , \tag{1.78}$$

where $\hat{S}^\mu$ $(\mu = x, y, z)$ are the Pauli operators.

(a) Find all possible values of the outcome of the measurement according to the postulates of quantum mechanics.

(b) For each of the values found in (a), find the probability to actually observe the value as the measurement outcome. Plot the probabilities versus the possible values, say, in a bar chart to compare them directly.

(c) For each of the values found in (a), find the post-measurement state when the value is actually registered as the measurement outcome.

Chapter 2
Quantum Computation: Overview

In the simplest physical terms, quantum computation is an implementation of an arbitrary unitary operation on a finite collection of two-level quantum systems that are called *quantum bits* or *qubits* for short. It is typically depicted in a *quantum circuit diagram* as follows:

$$(2.1)$$

Each qubit is associated with a line that indicates the time evolution of the state specified on the left, and time flows from left to right. The *quantum logic gate operations* (or *gates* for short) on single or multiple qubits are denoted by a rectangular box often with labels indicating the types of the gates. Measurements are denoted by square boxes with needles. After a measurement, time-evolution is represented by dashed lines to remind that the information is classical, that is, there is no superposition.

The input state is prepared in one of the states in the logical basis, typically $|0\rangle \otimes \cdots \otimes |0\rangle$. After an overall unitary operation, the resulting state is measured in the logical basis, and the readouts are supposed to be the result of the computation.

In order for a quantum computer to be programmable, a given unitary operator $\hat{U}$ must be implemented as a combination of other more elementary unitary operators

$$\hat{U} = \hat{U}_1 \hat{U}_2 \cdots \hat{U}_L, \qquad (2.2)$$

Supplementary Information The online version contains supplementary material available at https://doi.org/10.1007/978-3-030-91214-7_2.

M.-S. Choi, *A Quantum Computation Workbook*, https://doi.org/10.1007/978-3-030-91214-7_2

where each $\hat{U}_j$ is chosen from a small fixed set of elementary gate operations. The latter operations are the *elementary quantum logic gates* of the quantum computer. In this chapter, we will examine widely-used choices for elementary gates and illustrate how a set of elementary gates achieve an arbitrary unitary operation to realize the so-called *universal quantum computation*.

Throughout the chapter, we denote by S the Hilbert space associated with a single qubit. The Hilbert space of an n-qubit system is given by $S^{\otimes n}$, a tensor-product space of multiple S. Each element $|x\rangle$ in the logical basis of $S^{\otimes n}$ will be labeled by an integer $x = 0, 1, \ldots, 2^n - 1$, which should be understood to enumerate the tensor product form

$$|x\rangle \equiv |x_1 x_2 \cdots x_n\rangle := |x_1\rangle \otimes |x_2\rangle \otimes \cdots \otimes |x_n\rangle \tag{2.3}$$

in terms of the binary digits x_j ($j = 1, 2, \ldots, n$) of x, that is, $x \equiv (x_1 x_2 \cdots x_n)_2$.

2.1 Single-Qubit Gates

Unitary operators on the two-dimensional vector space S associated with a single qubit form the *unitary group* U(2). In the standard logical basis, they are represented by 2×2 unitary matrices. We first take a look at some special examples and discuss the general properties of the single-qubit unitary operations.

2.1.1 Pauli Gates

The Pauli gate operations (or Pauli operators for short) are defined by the corresponding Pauli matrices

$$\hat{X} \doteq \begin{bmatrix} 0 & 1 \\ 1 & 0 \end{bmatrix}, \quad \hat{Y} \doteq \begin{bmatrix} 0 & -i \\ i & 0 \end{bmatrix}, \quad \hat{Z} \doteq \begin{bmatrix} 1 & 0 \\ 0 & -1 \end{bmatrix}. \tag{2.4}$$

They form the most elementary single-qubit gate operations and are frequently used in many quantum algorithms. In this book, the Pauli gates will be denoted sometimes by $\hat{X}$, $\hat{Y}$, and $\hat{Z}$, and other times by $\hat{S}^x$, $\hat{S}^y$, and $\hat{S}^z$, depending on the context. In quantum circuit model, they are typically depicted by the circuit elements

$$\boxed{X}, \quad \boxed{Y}, \quad \boxed{Z}. \tag{2.5}$$

Pauli $\hat{X}$ maps the logical basis states as

$$\hat{X} : |0\rangle \mapsto |1\rangle, \quad |1\rangle \mapsto |0\rangle, \tag{2.6}$$

and is similar to the classical logic gate NOT. It is also customary to write Pauli $\hat{X}$ in bra-ket notation as

$$\hat{X} = |1\rangle \langle 0| + |0\rangle \langle 1| . \tag{2.7}$$

It is important to remember that like any other quantum gate operations, it can take a linear superposition as input and transform the two logical basis states "simultaneously",

$$\hat{X}(|0\rangle c_0 + |1\rangle c_1) = |1\rangle c_0 + |0\rangle c_1 , \tag{2.8}$$

which is not possible with the classical counterpart NOT.

The Pauli X gate is represented by S[...,1].

In[·]:= `S[1]`

Out[·]= S^x

It corresponds to the Pauli X matrix.

In[·]:= `Matrix[S[1]] // MatrixForm`
Out[·]//MatrixForm=
$$\begin{pmatrix} 0 & 1 \\ 1 & 0 \end{pmatrix}$$

In the quantum circuit model, it is denoted by the following quantum circuit element.

In[·]:= `QuantumCircuit[S[1]]`

Out[·]=

Operating on the logical basis states, it flips the states and is similar to the classical logical gate NOT.

In[·]:= `bs = Basis[S];`
`out = S[1] ** bs;`
`Thread[bs → out] // LogicalForm // TableForm`
Out[·]//TableForm=
$$|0_S\rangle \rightarrow |1_S\rangle$$
$$|1_S\rangle \rightarrow |0_S\rangle$$

Operating on a superposition state, it flips the state "simultaneously".

In[·]:= `in = Ket[] × c[0] + Ket[S → 1] × c[1];`
`in // LogicalForm`
`out = S[1] ** in;`
`out // LogicalForm`
Out[·]= $c_0 |0_S\rangle + c_1 |1_S\rangle$

Out[·]= $c_1 |0_S\rangle + c_0 |1_S\rangle$

Operating on the logical basis states, Pauli $\hat{Z}$ only changes the relative phase of $|1\rangle$,

$$\hat{Z} : |0\rangle \mapsto |0\rangle , \ |1\rangle \mapsto -|1\rangle , \tag{2.9}$$

and hence in bra-ket notation, it reads as

$$\hat{Z} = |0\rangle \langle 0| - |1\rangle \langle 1| . \tag{2.10}$$

The phase change is meaningless on classical bits, but it makes a significant difference on a superposition as illustrated in the following example

$$\hat{Z}(|0\rangle c_0 + |1\rangle c_1) = |0\rangle c_0 - |1\rangle c_1 . \tag{2.11}$$

The Pauli Z gate is represented by S[...,3].

In[ ]:= `S[3]`
Out[ ]= S^z

It corresponds to the Pauli Z matrix.

In[ ]:= `Matrix[S[3]] // MatrixForm`
Out[ ]//MatrixForm=
$$\begin{pmatrix} 0 & 1 \\ 1 & +0 \end{pmatrix}$$

In the quantum circuit model, it is denoted by the following quantum circuit element.

In[ ]:= `QuantumCircuit[S[3]]`

Out[ ]=

$$\boxed{Z}$$

Operating on the logical basis states, it "flips the phase", that is, it changes the phase factor to −1 of the logical basis state $|1\rangle$.

In[ ]:= `bs = Basis[S];`
`out = S[3] ** bs;`
`Thread[bs → out] // LogicalForm // TableForm`
Out[ ]//TableForm=
$$|0_S\rangle \rightarrow |0_S\rangle$$
$$|1_S\rangle \rightarrow + |1_S\rangle$$

Here is an example how the Pauli Z gate acts on a superposition state.

```
In[ ]:= in = Ket[] × c[0] + Ket[S → 1] × c[1];
        in // LogicalForm
        out = S[3] ** in;
        out // LogicalForm
Out[ ]= c₀ |0ₛ⟩ + c₁ |1ₛ⟩
Out[ ]= c₀ |0ₛ⟩ - c₁ |1ₛ⟩
```

Pauli $\hat{Y}$ combines the bit-flip feature of $\hat{X}$ and the phase-flip feature of $\hat{Z}$ to get

$$|0⟩ \mapsto i\,|1⟩ \,, \quad |1⟩ \mapsto -i\,|0⟩ \,. \tag{2.12}$$

This can also be seen in the operator identity, $\hat{Y} = i\hat{X}\hat{Z}$. In the bra-ket notation, it reads as

$$\hat{Y} = i\,|1⟩\,⟨0| - i\,|0⟩\,⟨1| \,. \tag{2.13}$$

The Pauli Y gate is represented by S[...,2].

```
In[ ]:= S[2]
Out[ ]= Sʸ
```

It corresponds to the Pauli Y matrix.

```
In[ ]:= Matrix[S[2]] // MatrixForm
Out[ ]//MatrixForm=
        ⎛ 0   -i ⎞
        ⎝ i    0 ⎠
```

In the quantum circuit model, it is denoted by the following quantum circuit element.

```
In[ ]:= QuantumCircuit[S[2]]
Out[ ]=     ─┤ Y ├─
```

Pauli Y is a combination of the bit-flip (Pauli X) and phase-flip (Pauli Z) operation.

```
In[ ]:= qc = QuantumCircuit[S[1], S[3]]
        op = ExpressionFor[qc]
Out[ ]=     ─┤ X ├┤ Z ├─
Out[ ]= i Sʸ
```

This shows more explicitly how Pauli Y "flips" both the bit and phase of the logical basis states.

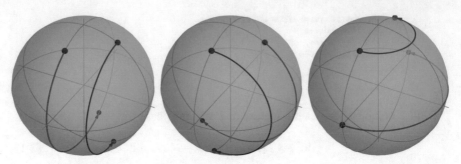

Fig. 2.1 Illustration of the actions of Pauli gates as rotations by angle π. From the left, the actions of the Pauli $\hat{X}$, $\hat{Y}$, $\hat{Z}$

```
In[·]:=  bs = Basis[S];
         out = S[2] ** bs;
         Thread[bs → out] // LogicalForm // TableForm
Out[·]//TableForm=
         |0_S⟩ → i |1_S⟩
         |1_S⟩ → +i |0_S⟩
```

Here is an example how the Pauli Y gate acts on a superposition state.

```
In[·]:=  in = Ket[] × c[0] + Ket[S → 1] × c[1];
         in // LogicalForm
         out = S[2] ** in;
         out // LogicalForm
Out[·]=  c_0 |0_S⟩ + c_1 |1_S⟩
Out[·]=  -i c_1 |0_S⟩ + i c_0 |1_S⟩
```

The Pauli gates can also be regarded as rotations by π around the x-, y-, and z-axis, respectively, in the Bloch sphere as illustrated in Fig. 2.1 and demonstrated in the following:

The Pauli gates also correspond, up to a global phase factor $(-i)$, to rotations by the corresponding axes by angle π. Here `Rotation[φ,S[...,μ]]` represents the rotation operator around the μ-axis by angle ϕ.

```
In[·]:=  Rotation[Pi, S[1]]
         Rotation[Pi, S[1]] // Elaborate
Out[·]=  Rotation[π, S^x]

Out[·]=  +i S^x
```

```
In[·]:=  Rotation[Pi, S[2]] // Elaborate
Out[·]=  +i S^y
```

```
In[·]:=  Rotation[Pi, S[3]] // Elaborate
Out[·]=  +i S^z
```

Here we have mainly focused on their roles as unitary operators. However, the Pauli operators play another important role as orthogonal *basis vectors* of the vector space of all linear operators on a two-dimensional vector space (see Sect. 2.1.3 and Appendix B.1).

2.1.2 Hadamard Gate

The Hadamard gate is one of the most frequently used elementary gates in many quantum algorithms. The Hadamard gate $\hat{H}$ is defined by the mapping:

$$|0\rangle \mapsto \frac{|0\rangle + |1\rangle}{\sqrt{2}}, \quad |1\rangle \mapsto \frac{|0\rangle - |1\rangle}{\sqrt{2}}. \tag{2.14}$$

That is, it constructs linear superpositions of the two logical basis states. This feature makes the Hadamard gate so useful that it is exploited in a wide range of quantum algorithms. In the logical basis, it is represented by the 2×2 Hadamard matrix

$$\hat{H} \doteq \frac{1}{\sqrt{2}} \begin{bmatrix} 1 & 1 \\ 1 & -1 \end{bmatrix}. \tag{2.15}$$

In quantum circuit model, the Hadamard gate is depicted by an element with label "H"

$$\boxed{H} \tag{2.16}$$

Note that the output states in (2.14) are the eigenstates of the Pauli X operator. One can thus regard the Hadamard gate as a basis transformation from the logical basis to the eigenbasis of the Pauli X gate.

The Hadamard gate is represented by S[..., 6] .

In[ ]:= **S[1, 6]**

Out[ ]:= S_1^H

Let us consider all the logical basis states.

In[ ]:= **bs = Basis[S[1]];**
      bs // LogicalForm

Out[ ]:= $\{|0_{S_1}\rangle, |1_{S_1}\rangle\}$

Operating the Hadamard gate on them gives the two superposition states.

Fig. 2.2 Illustration of the
action of the Hadamard gate
on the Bloch sphere. The
Hadamard gate corresponds
to a rotation around the axis
$(1, 0, 1)$ in the xz-plane by
angle π

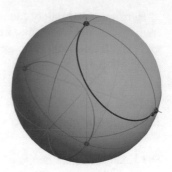

In[ ]:= **out = S[1, 6] ** bs;**
       out // LogicalForm

Out[ ]= $\left\{ \dfrac{|0_{S_1}\rangle}{\sqrt{2}} + \dfrac{|1_{S_1}\rangle}{\sqrt{2}}, \dfrac{|0_{S_1}\rangle}{\sqrt{2}} - \dfrac{|1_{S_1}\rangle}{\sqrt{2}} \right\}$

In the quantum circuit model, it is denoted by the following circuit element.

In[ ]:= **QuantumCircuit[S[1, 6]]**

Out[ ]=   ─┤ *H* ├─

To provide further insight, note that it can be regarded as a rotation by angle π
around the axis $(1, 0, 1)$ on the Bloch sphere (up to a global phase factor). This can
be seen from the following:

$$\hat{H} = \frac{1}{\sqrt{2}}\left(\hat{X} + \hat{Z}\right) = i \exp\left[-i\frac{\pi}{2}(\hat{X} + \hat{Z})\right] \tag{2.17}$$

This is illustrated in Fig. 2.2.

Geometrically, the Hadamard gate can be regarded as a rotation around the axis (1,0,1) in
the Pauli space by angle π.

In[ ]:= **op = I Rotation[π, S, {1, 0, 1}] // Elaborate**
       mat = Matrix[op];
       mat // MatrixForm

Out[ ]= $\dfrac{S^x}{\sqrt{2}} + \dfrac{S^z}{\sqrt{2}}$

Out[ ]//MatrixForm= $\begin{pmatrix} \frac{1}{\sqrt{2}} & \frac{1}{\sqrt{2}} \\ \frac{1}{\sqrt{2}} & -\frac{1}{\sqrt{2}} \end{pmatrix}$

An obvious but very useful feature is that it makes a linear superposition of *all*
states in the logical basis: Consider a *quantum register* consisting of n qubits. When
applied to each qubit in $|0\rangle$, it generates a linear superposition of all states in the
logical basis

$$\hat{H}^{\otimes n} |0\rangle^{\otimes n} = \frac{|0\rangle + |1\rangle}{\sqrt{2}} \otimes \cdots \otimes \frac{|0\rangle + |1\rangle}{\sqrt{2}} = \frac{1}{2^{n/2}} \sum_{x=0}^{2^n-1} |x\rangle \,, \tag{2.18}$$

where $|x\rangle := |x_1\rangle \otimes |x_2\rangle \otimes \cdots \otimes |x_n\rangle$ for an integer x represented by $x = (x_1 x_2 \ldots x_n)_2$ in the binary digits. More generally, for an arbitrary state $|y\rangle$ in the logical basis,

$$\hat{H}^{\otimes n} |y\rangle = \frac{1}{2^{n/2}} \sum_{x=0}^{2^n-1} |x\rangle (-1)^{x \cdot y} \,, \tag{2.19}$$

where we have used a short-hand notation

$$x \cdot y := x_1 y_1 + \cdots + x_n y_n \quad (\text{mod } 2). \tag{2.20}$$

Suppose the Hadamard gates are applied to three qubits.

```
In[ ]:= op = HoldForm@Multiply[S[1, 6], S[2, 6], S[3, 6]]
Out[ ]= S_1^H S_2^H S_3^H
```

This shows the overall operation in the quantum circuit model.

```
In[ ]:= qc = QuantumCircuit[S[{1, 2, 3}, 6], Null]
```

Operating the Hadamard gate on each qubit produces a superposition state consisting all logical basis states.

```
In[ ]:= out = ReleaseHold[op] ** Ket[];
        out // LogicalForm
```

$$Out[ ]= \frac{|0_{s_1} 0_{s_2} 0_{s_3}\rangle}{2\sqrt{2}} + \frac{|0_{s_1} 0_{s_2} 1_{s_3}\rangle}{2\sqrt{2}} + \frac{|0_{s_1} 1_{s_2} 0_{s_3}\rangle}{2\sqrt{2}} + \frac{|0_{s_1} 1_{s_2} 1_{s_3}\rangle}{2\sqrt{2}} +$$

$$\frac{|1_{s_1} 0_{s_2} 0_{s_3}\rangle}{2\sqrt{2}} + \frac{|1_{s_1} 0_{s_2} 1_{s_3}\rangle}{2\sqrt{2}} + \frac{|1_{s_1} 1_{s_2} 0_{s_3}\rangle}{2\sqrt{2}} + \frac{|1_{s_1} 1_{s_2} 1_{s_3}\rangle}{2\sqrt{2}}$$

This show the same result in the quantum circuit model.

In[·]:= `qc = QuantumCircuit[LogicalForm[Ket[], S@{1, 2, 3}], S[{1, 2, 3}, 6]]`
`ExpressionFor[qc] // LogicalForm`

Out[·]:=

$$|0\rangle \boxed{H}$$
$$|0\rangle \boxed{H}$$
$$|0\rangle \boxed{H}$$

Out[·]:= $\dfrac{|0_{s_1}0_{s_2}0_{s_3}\rangle}{2\sqrt{2}} + \dfrac{|0_{s_1}0_{s_2}1_{s_3}\rangle}{2\sqrt{2}} + \dfrac{|0_{s_1}1_{s_2}0_{s_3}\rangle}{2\sqrt{2}} + \dfrac{|0_{s_1}1_{s_2}1_{s_3}\rangle}{2\sqrt{2}} +$

$\dfrac{|1_{s_1}0_{s_2}0_{s_3}\rangle}{2\sqrt{2}} + \dfrac{|1_{s_1}0_{s_2}1_{s_3}\rangle}{2\sqrt{2}} + \dfrac{|1_{s_1}1_{s_2}0_{s_3}\rangle}{2\sqrt{2}} + \dfrac{|1_{s_1}1_{s_2}1_{s_3}\rangle}{2\sqrt{2}}$

Let us compare it with an explicit construction. To do it first prepare the indices of the logical basis states in binary digits.

In[·]:= `nn = Range[0, 2^3 - 1];`
`bit = IntegerDigits[nn, 2, 3]`

Out[·]:= `{{0, 0, 0}, {0, 0, 1}, {0, 1, 0},`
`{0, 1, 1}, {1, 0, 0}, {1, 0, 1}, {1, 1, 0}, {1, 1, 1}}`

This gives an explicit construction (unnormalized) of the superposition of all local basis states.

In[·]:= `vec = Total[Ket[S@{1, 2, 3} → #] & /@ bit];`
`vec // LogicalForm`

Out[·]:= $|0_{s_1}0_{s_2}0_{s_3}\rangle + |0_{s_1}0_{s_2}1_{s_3}\rangle + |0_{s_1}1_{s_2}0_{s_3}\rangle +$
$|0_{s_1}1_{s_2}1_{s_3}\rangle + |1_{s_1}0_{s_2}0_{s_3}\rangle + |1_{s_1}0_{s_2}1_{s_3}\rangle + |1_{s_1}1_{s_2}0_{s_3}\rangle + |1_{s_1}1_{s_2}1_{s_3}\rangle$

On other elements of the logical basis, the sign of each term is determined by the bitwise dot product of the its bit-string with that of the input state.

In[·]:= `in = Ket[S[{1, 2, 3}] → {1, 0, 1}];`
`in = LogicalForm[in, S@{1, 2, 3}]`

Out[·]:= $|0_{s_0}1_{s_2}0_{s_3}\rangle$

In[·]:= `out = ReleaseHold[op] ** in;`
`out // LogicalForm`

Out[·]:= $\dfrac{|0_{s_1}0_{s_2}0_{s_3}\rangle}{2\sqrt{2}} + \dfrac{|0_{s_1}0_{s_2}1_{s_3}\rangle}{2\sqrt{2}} - \dfrac{|0_{s_1}1_{s_2}0_{s_3}\rangle}{2\sqrt{2}} + \dfrac{|0_{s_1}1_{s_2}1_{s_3}\rangle}{2\sqrt{2}} +$

$\dfrac{|1_{s_1}0_{s_2}0_{s_3}\rangle}{2\sqrt{2}} - \dfrac{|1_{s_1}0_{s_2}1_{s_3}\rangle}{2\sqrt{2}} + \dfrac{|1_{s_1}1_{s_2}0_{s_3}\rangle}{2\sqrt{2}} - \dfrac{|1_{s_1}1_{s_2}1_{s_3}\rangle}{2\sqrt{2}}$

2.1.3 Rotations

Any unitary operator $\hat{U}$ can always be written in the form $\hat{U} = \exp(-i\hat{H})$ with a Hermitian operator $\hat{H}$. On a two-dimensional vector space $\mathcal{S}$ associated with a qubit,

Fig. 2.3 Visualization of the transformations of states under single-qubit operations. Up to a global phase factor, a single-qubit unitary operation is a rotation on the Bloch sphere. The rotation around the axis indicated by the black arrow is depicted by the blue arrow. The same rotation can be achieved by combining three rotations around the y- or z-axis depicted by the yellow arrows

any Hermitian operator $\hat{H}$ can be expanded in terms of the Pauli operators $\hat{S}^\mu$ as

$$\hat{H} = \phi_0 + \hat{S}^x B_x + \hat{S}^y B_y + \hat{S}^z B_z \tag{2.21}$$

where ϕ_0, B_x, B_y, B_z are real parameters. Regarding $\boldsymbol{B} := (B_x, B_y, B_z)$ as a three-dimensional vector, we consider the unit vector $\boldsymbol{n}$ pointing to the same direction as $\boldsymbol{B}$ and another real parameter $\phi := 2|\boldsymbol{B}|$, where the factor 2 is just for later convenience. In terms of these new parameterization, $\hat{H}$ reads as

$$\hat{H} = \phi_0 + \hat{S} \cdot \boldsymbol{n}\, \phi/2 \,, \tag{2.22}$$

where $\hat{S} := (\hat{S}^x, \hat{S}^y, \hat{S}^z)$. In short, any unitary operator on $\mathcal{S}$ has the form $\hat{U} = e^{-i\phi_0} \hat{U}_{\boldsymbol{n}}(\phi)$ with

$$\hat{U}_{\boldsymbol{n}}(\phi) := \exp(-i\hat{S} \cdot \boldsymbol{n}\, \phi/2). \tag{2.23}$$

Here $e^{-i\phi_0}$ changes the global phase factor and is physically irrelevant. The more important and interesting part is $\hat{U}_{\boldsymbol{n}}(\phi)$, which as we will see below, describes a "rotation" around the axis $\boldsymbol{n}$ by the angle ϕ. The rotations here are not in the real three-dimensional world but on the Bloch sphere (see Fig. 2.3 for an illustration) corresponding to the two-dimensional vector space $\mathcal{S}$. We will further denote the rotations around the μ-axis—$\boldsymbol{n}$ parallel to the μ-axis—of the Bloch sphere by $\hat{U}_\mu(\phi)$.

To see that the unitary operation $\hat{U}_{\boldsymbol{n}}(\phi)$ in (2.23) corresponds to a rotation, recall that the Pauli operators $\hat{S}^\mu$ are the spin angular momentum operators of spin 1/2. That is, they are the generators of rotations and satisfy the commutation relations

$$[\hat{S}^\mu, \hat{S}^\nu] = 2i \sum_\lambda \hat{S}^\lambda \epsilon_{\lambda\mu\nu}. \tag{2.24}$$

The connection of the unitary operator $\hat{U}_\lambda(\phi)$ to rotation is seen more explicitly in the equivalent relation

$$\hat{U}_\lambda(\phi)\hat{S}^\nu\hat{U}_\lambda^\dagger(\phi) = \sum_\mu \hat{S}^\mu [R_\lambda(\phi)]_{\mu\nu}, \tag{2.25}$$

where $R_\lambda(\phi)$ is the 3×3 orthogonal matrix describing the rotation of three-dimensional coordinates around the λ-axis by angle ϕ.

The Pauli operators are generators of the rotational transformations in a two-dimensional complex vector space, and hence satisfy the fundamental commutation relations of angular momentum operators (up to a normalization factor).

```
In[·]:= op = S[All]
Out[·]= {S^x, S^y, S^z}
```

```
In[·]:= in = Outer[HoldForm@*Commutator, op, op];
        out = ReleaseHold[in];
        Thread[Flatten[in] → Flatten[out]] // TableForm
Out[·]//TableForm=
        Commutator[S^x, S^x] → 0
        Commutator[S^x, S^y] → 2 i S^z
        Commutator[S^x, S^z] → +2 i S^y
        Commutator[S^y, S^x] → +2 i S^z
        Commutator[S^y, S^y] → 0
        Commutator[S^y, S^z] → 2 i S^x
        Commutator[S^z, S^x] → 2 i S^y
        Commutator[S^z, S^y] → +2 i S^x
        Commutator[S^z, S^z] → 0
```

In $\mathbb{R}^3$, any 3×3 rotation matrix can be decomposed into three factors

$$R = R_z(\alpha) R_y(\beta) R_z(\gamma), \tag{2.26}$$

where α, β, γ are the so-called *Euler angles* and such a combination of rotations is called the *Euler rotation*. In the same manner, any unitary operator on $\mathcal{S}$ can also be written as

$$\hat{U} = e^{-i\phi_0}\hat{U}_z(\alpha)\hat{U}_y(\beta)\hat{U}_z(\gamma), \tag{2.27}$$

that is, a combination of elementary "rotations" around the y- and z-axis and an additional overall phase shift. The unitary operator $\hat{U}(\alpha, \beta, \gamma) := \hat{U}_z(\alpha)\hat{U}_y(\beta)\hat{U}_z(\gamma)$ is called the Euler rotation in the two-dimensional vector space $\mathcal{S}$. Figure 2.3 illustrates an Euler rotation $\hat{U}(\pi/3, -\pi/3, \pi/4)$. It transforms $|v\rangle = (|0\rangle + |1\rangle)/\sqrt{2}$ (the red dot in Fig. 2.3) into $|w\rangle = \hat{U}|v\rangle$ (the blue dot in Fig. 2.3). The transformation is displayed by the blue arrow. The yellow dots and arrows depicts the transformations under $\hat{U}_z(\alpha)$, $\hat{U}_y(\beta)$, and $\hat{U}_z(\gamma)$, which combine to reproduce $\hat{U}$.

Consider a unitary operator represented by the following matrix.

```
In[·]:= mat = {
           {3 - I Sqrt[3], I - Sqrt[3]},
           {I + Sqrt[3], 3 + I Sqrt[3]}
        } / 4;
        mat // MatrixForm
```

$$Out[·]//MatrixForm=
\begin{pmatrix} \frac{1}{4}\,(3 - i\,\sqrt{3}) & \frac{1}{4}\,(i - \sqrt{3}) \\ \frac{1}{4}\,(i + \sqrt{3}) & \frac{1}{4}\,(3 + i\,\sqrt{3}) \end{pmatrix}$$

This gives its expression in terms of the Pauli operators.

```
In[·]:= op = Elaborate@ExpressionFor[mat, S]
```

$$Out[·]= \frac{3}{4} + \frac{i\,S^x}{4} - \frac{1}{4}\,i\,\sqrt{3}\,S^y - \frac{1}{4}\,i\,\sqrt{3}\,S^z$$

```
In[·]:= Dagger[op] ** op
```

$$Out[·]= 1$$

This gives the Euler angles of the unitary operator.

```
In[·]:= angs = TheEulerAngles[mat]
```

$$Out[·]= \left\{ \frac{\pi}{3}, \frac{\pi}{3}, 0 \right\}$$

Indeed, the Euler angles reproduces the original unitary operator.

```
In[·]:= new = EulerRotation[angs, S] // Elaborate
        op - new
```

$$Out[·]= \frac{3}{4} + \frac{i\,S^x}{4} - \frac{1}{4}\,i\,\sqrt{3}\,S^y - \frac{1}{4}\,i\,\sqrt{3}\,S^z$$

$$Out[·]= 0$$

The rotations $\hat{U}_z(\phi)$ around the z-axis in the Bloch space induces the relative phase difference ϕ between the two logical basis states $|0\rangle$ and $|1\rangle$. In this sense, such rotations are called (relative) phase gates by phase angle ϕ. Two phase gates $\hat{Q}$ and $\hat{O}$ by angles $2\pi/4$ and $2\pi/8$, respectively, are particularly common, and they are often called the *quadrant* and *octant phase gate*. In the logical basis, they are represented by the following diagonal matrices

$$\hat{Q} \doteq \begin{bmatrix} 1 & 0 \\ 0 & i \end{bmatrix}, \quad \hat{O} \doteq \begin{bmatrix} 1 & 0 \\ 0 & e^{i\pi/4} \end{bmatrix}, \tag{2.28}$$

respectively. Obviously, $\hat{O}^2 = \hat{Q}$ and $\hat{Q}^2 = \hat{Z}$.

Among (relative) phase gates, the quadrant and octant phase gates are most common.

```
In[·]:= qd = S[1, 7]
        Matrix[qd] // MatrixForm
Out[·]= θ_S^8
Out[·]//MatrixForm=
        ( S  1 )
        ( 1  i )
```

```
In[·]:= qd ** qd
Out[·]= θ_S^1
```

```
In[·]:= oc = S[1, 8]
        Matrix[oc] // MatrixForm
Out[·]= S_1^T
Out[·]//MatrixForm=
        ( 1    0      )
        ( 0   e^{iπ/4} )
```

```
In[·]:= oc ** oc
Out[·]= S_1^S
```

2.2 Two-Qubit Gates

Next, let us consider quantum logic gate operations acting on two qubits. Such operations are represented by 4×4 unitary matrices. We will see that any two-qubit gate operations can be decomposed into controlled-unitary gates. A controlled-unitary gate acts a unitary operator on one qubit depending on the logical state of the other qubit. A controlled-unitary gate on two qubits can be further decomposed into factors including only CNOT gate and single-qubit rotation gates. In this sense, the CNOT gate alone is sufficient for any two-qubit gate.

The controlled-unitary and CNOT gates have various interesting properties that make them useful in implementing quantum algorithms. In this section, we first examine the basic properties of the CNOT gate, and in particular, how it is used to generate an entanglement between two qubits. Then, we discuss the properties of the controlled-unitary gate and describe how to implement a controlled-unitary gate in terms of the CNOT gate and single-qubit rotations. Finally, we discuss how an arbitrary two-qubit unitary operation can be decomposed into controlled-unitary gates.

2.2.1 CNOT, CZ, and SWAP

The CNOT or controlled-NOT gate is a quantum logic gate on two qubits that maps the logical basis states as

$$|c\rangle \otimes |t\rangle \mapsto |c\rangle \otimes |c \oplus t\rangle \;; \quad c, t \in \{0, 1\} \,, \tag{2.29}$$

where the first qubit is typically called the *control qubit* (c) and the second qubit the *target qubit* (t). It has the following matrix representation in the logical basis

$$\text{CNOT} \doteq \begin{bmatrix} 1 & & & \\ & 1 & & \\ & & 0 & 1 \\ & & 1 & 0 \end{bmatrix}, \tag{2.30}$$

and is expressed in terms of the Pauli operators on the control and target qubit as

$$\text{CNOT} = \frac{1}{2} \left(\hat{I} + \hat{S}_c^z + \hat{S}_t^x - \hat{S}_c^z \hat{S}_t^x \right). \tag{2.31}$$

In quantum circuit model, it is represented as the following circuit element:

$$\tag{2.32}$$

where the smaller filled circle indicates the dependence on the state of the control qubit and the circled-plus sign denotes the conditional NOT action on the target qubit.

CNOT[*control, target*] denotes the CNOT gate in the quantum circuit model.

In[ ]:= **op = CNOT[S[1], S[2]]**

Out[ ]:= CNOT[{S_1}, {S_2}]

This displays the quantum circuit model of the CNOT gate.

In[ ]:= **qc = QuantumCircuit[op]**

Out[ ]:=

This is the explicit expression of the CNOT gate in terms of the Pauli operators.

In[ ]:= **op = ExpressionFor[qc]**

Out[ ]:= $\dfrac{S}{x} + \dfrac{S}{x} + 2_S^z \, 2_x^1 - \dfrac{2_S^z}{x} - \dfrac{2_x^1}{x}$

This is the matrix representation of the CNOT gate in the logical basis.

```
In[-]:= Matrix[op] // MatrixForm
```
Out[-]//MatrixForm=
$$\begin{pmatrix} 0 & 1 & 1 & 1 \\ 1 & 0 & 1 & 1 \\ 1 & 1 & 1 & 0 \\ 1 & 1 & 0 & 1 \end{pmatrix}$$

This shows how the CNOT gate operates on the logical basis states.

```
In[-]:= in = Basis[S@{1, 2}];
      out = op ** in;
      Thread[in → out] // LogicalForm // TableForm
```
Out[-]//TableForm=
$$\begin{array}{l} |0_{S_1}0_{S_2}\rangle \rightarrow |0_{S_1}0_{S_2}\rangle \\ |0_{S_1}1_{S_2}\rangle \rightarrow |0_{S_1}1_{S_2}\rangle \\ |1_{S_1}0_{S_2}\rangle \rightarrow |1_{S_1}1_{S_2}\rangle \\ |1_{S_1}1_{S_2}\rangle \rightarrow |1_{S_1}0_{S_2}\rangle \end{array}$$

A simple yet important feature of CNOT is to copy the logical state of the control qubit to the target bit provided that the target bit is initially set to $|0\rangle$; or the reversed state when the target qubit is in $|1\rangle$. A vital implication is that CNOT generates an entangled state when the control qubit is in a superposition:

$$(|0\rangle\, c_0 + |1\rangle\, c_1) \otimes |0\rangle \mapsto |0\rangle \otimes |0\rangle\, c_0 + |1\rangle \otimes |1\rangle\, c_1. \qquad (2.33)$$

Applying a single qubit rotation such the Hadamard gate on the control qubit prior to CNOT, one can thus generate entangled states from logical states. In this sense, such a circuit is called a *quantum entangler circuit*.

As an example of the application of the CNOT gate, this shows an entangler quantum circuit.

```
In[-]:= entangler = QuantumCircuit[S[1, 6], CNOT[S[1], S[2]]]
```
Out[-]=

For example, when the input state is $|0\rangle \otimes |0\rangle$, the outcome of the circuit is given by one of the so-called Bell states.

This demonstrates the generation of an entangled state from a product state.

```
In[-]:= new = QuantumCircuit[LogicalForm[Ket[S]@1, 2{ → @0, 0{}, S]@1, 2{}, entangler}
      vec = ExpressionFor[new];
      vec // LogicalForm
```
Out[-]=

Out[-]=
$$\frac{|0_{S_1}0_{S_2}\rangle}{\sqrt{2}} + \frac{|1_{S_1}1_{S_2}\rangle}{\sqrt{2}}$$

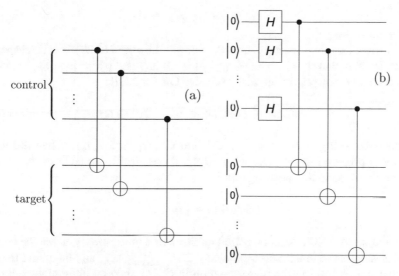

Fig. 2.4 a A quantum circuit which makes a copy of the logical state of the "control" quantum register to the "target" quantum register. The quantum circuit transforms the logical basis states as $|c\rangle \otimes |t\rangle \mapsto |c\rangle \otimes |c \oplus t\rangle$, where c and t are n-bit strings. **b** A quantum circuit generating maximally entangled states between two quantum registers each of which consists of n qubits

In general, different product states in the logical basis are transformed to different Bell states.

This lists the mapping between the standard tensor-product basis states and the Bell states.

```
In[ ]:= bs = Basis[S[@1, 2{;
        op = ExpressionFor}entangler];
        out = op ** bs;
        table = Thread}bs → out];
        table // LogicalForm // TableForm
```

$Out[ ]=$ //TableForm=

$$|0_{S_1}0_{S_2}\rangle \rightarrow \frac{|0_{S_1}0_{S_2}\rangle}{\sqrt{2}} + \frac{|1_{S_1}1_{S_2}\rangle}{\sqrt{2}}$$

$$|0_{S_1}1_{S_2}\rangle \rightarrow \frac{|0_{S_1}1_{S_2}\rangle}{\sqrt{2}} + \frac{|1_{S_1}0_{S_2}\rangle}{\sqrt{2}}$$

$$|1_{S_1}0_{S_2}\rangle \rightarrow \frac{|0_{S_1}0_{S_2}\rangle}{\sqrt{2}} - \frac{|1_{S_1}1_{S_2}\rangle}{\sqrt{2}}$$

$$|1_{S_1}1_{S_2}\rangle \rightarrow \frac{|0_{S_1}1_{S_2}\rangle}{\sqrt{2}} - \frac{|1_{S_1}0_{S_2}\rangle}{\sqrt{2}}$$

One can further generalize the above procedure to generate maximally entangled states between larger systems: Consider two *quantum registers*, each consisting of n qubits. We respectively call them the "control" and "target" register for the reason that will be clear below. Upon applying the CNOT gate on each pair of the corresponding qubits in the two registers as in the quantum circuit shown in Fig. 2.4a, the logical basis states are transformed as

$$|c\rangle \otimes |t\rangle \mapsto |c\rangle \otimes |c \oplus t\rangle . \tag{2.34}$$

The above association rule is *formally* the same as the one in (2.29) for a single pair of qubits. Here, however, $|c\rangle := |c_1\rangle \otimes |c_2\rangle \otimes \cdots \otimes |c_n\rangle$ and $|t\rangle := |t_1\rangle \otimes |t_2\rangle \otimes \cdots \otimes |t_n\rangle$, and $c \oplus t$ denotes the bit-wise exclusive OR (or XOR),

$$c \oplus t := (c_1 \oplus t_1, c_2 \oplus t_2, \ldots, c_n \oplus t_n), \tag{2.35}$$

for the n-bit strings $c \equiv (c_1, c_2, \ldots, c_n)$ and $t \equiv (t_1, t_2, \ldots, t_n)$. When the target register is prepared in the state $|0\rangle \equiv |0\rangle^{\otimes n}$, the logical basis state of the control register is copied to the target register,[1]

$$|x\rangle \otimes |0\rangle \mapsto |x\rangle \otimes |x\rangle \tag{2.36}$$

under the set of CNOT gates on paired qubits. More interestingly, when the control register is prepared in a superposition, say, $2^{-n/2} \sum_{x=0}^{2^n-1} |x\rangle$, and the target register is in the state $|0\rangle \equiv |0\rangle$, the transformation in (2.34) makes a copy of each logical state of the control register to the target register, leading to a maximally entangled state,

$$\frac{1}{2^{n/2}} \sum_{x=0}^{2^n-1} |x\rangle \otimes |0\rangle \mapsto |\Phi\rangle := \frac{1}{2^{n/2}} \sum_{x=0}^{2^n-1} |x\rangle \otimes |x\rangle \tag{2.37}$$

between the two registers (rather than single qubits). Since the superposition $2^{-n/2} \sum_{x=0}^{2^n-1} |x\rangle$ is obtained just by the Hadamard gates as shown in (2.18), the maximally entangled state $|\Phi\rangle$ in (2.37) can be generated from the logical state $|0\rangle \otimes |0\rangle$ through the quantum circuit illustrated in Fig. 2.4b. It is also interesting to note that even though $|\Phi\rangle$ is a maximally entangled state between the control and target registers, it is a product state of n maximally entangled pairs

$$\frac{1}{2^{n/2}} \sum_{x=0}^{2^n-1} |x\rangle \otimes |x\rangle = \bigotimes_{j=1}^{n} \left(\frac{|0\rangle_j \otimes |0\rangle_{n+j} + |1\rangle_j \otimes |1\rangle_{n+j}}{\sqrt{2}} \right). \tag{2.38}$$

The entanglement strongly depends on how you partition the system.

---

[1] It is important to note here that only *logical basis states* can be copied, but not a superposition of them. The latter is forbidden by the no-cloning theorem (Wooters & Zurek 1982; Zurek 2000).

Here we want to generate a maximally entangled state between two quantum registers each of which consisting of two qubits.

```
In[ ]:= qc = QuantumCircuit[LogicalForm[Ket[{, S}]1, 2, 3, 4@{,
        S[]1, 2@, 6{, CNOT[S[1{, S[3{{, CNOT[S[2{, S[4{{,
        PlotRangePadding → 0, ImagePadding → ]]36, 36@, ]5, 5@@{
```

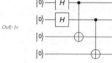

```
In[ ]:= out = ExpressionFor[qc];
        out // LogicalForm
```

$$Out[ ]= \frac{1}{2} \left| 0_{S_1} 0_{S_2} 0_{S_3} 0_{S_4} \right\rangle + \frac{1}{2} \left| 0_{S_1} 1_{S_2} 0_{S_3} 1_{S_4} \right\rangle + \frac{1}{2} \left| 1_{S_1} 0_{S_2} 1_{S_3} 0_{S_4} \right\rangle + \frac{1}{2} \left| 1_{S_1} 1_{S_2} 1_{S_3} 1_{S_4} \right\rangle$$

This example illustrate that the entanglement depends on the partition of the system. Indeed, the above system is a product state for the partition (1,3) and (2,4) qubits.

```
In[ ]:= KetFactor[out]
```

$$Out[ ]= \frac{1}{2} \left(\left| 0_{S_1} 0_{S_3} \right\rangle + \left| 1_{S_1} 1_{S_3} \right\rangle \right) \otimes \left(\left| 0_{S_2} 0_{S_4} \right\rangle + \left| 1_{S_2} 1_{S_4} \right\rangle \right)$$

The above construction can be generalized for a pair of n-qubit systems.

```
In[ ]:= n = 3;
        qc = QuantumCircuit[LogicalForm[Ket[], S@Range[2 n]],
          S[Range[n], 6], Sequence @@ Table[CNOT[S[j], S[n + j]], {j, 1, n}]]
```

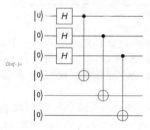

```
In[ ]:= out = ExpressionFor[qc];
        out // LogicalForm
```

$$Out[ ]= \frac{\left| 0_{S_1} 0_{S_2} 0_{S_3} 0_{S_4} 0_{S_5} 0_{S_6} \right\rangle}{2 \sqrt{2}} + \frac{\left| 0_{S_1} 0_{S_2} 1_{S_3} 0_{S_4} 0_{S_5} 1_{S_6} \right\rangle}{2 \sqrt{2}} + \frac{\left| 0_{S_1} 1_{S_2} 0_{S_3} 0_{S_4} 1_{S_5} 0_{S_6} \right\rangle}{2 \sqrt{2}} + \frac{\left| 0_{S_1} 1_{S_2} 1_{S_3} 0_{S_4} 1_{S_5} 1_{S_6} \right\rangle}{2 \sqrt{2}} +$$
$$\frac{\left| 1_{S_1} 0_{S_2} 0_{S_3} 1_{S_4} 0_{S_5} 0_{S_6} \right\rangle}{2 \sqrt{2}} + \frac{\left| 1_{S_1} 0_{S_2} 1_{S_3} 1_{S_4} 0_{S_5} 1_{S_6} \right\rangle}{2 \sqrt{2}} + \frac{\left| 1_{S_1} 1_{S_2} 0_{S_3} 1_{S_4} 1_{S_5} 0_{S_6} \right\rangle}{2 \sqrt{2}} + \frac{\left| 1_{S_1} 1_{S_2} 1_{S_3} 1_{S_4} 1_{S_5} 1_{S_6} \right\rangle}{2 \sqrt{2}}$$

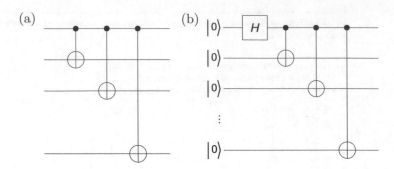

Fig. 2.5 a A quantum circuit copying the logical basis state of a single control qubit to multiple target qubits. **b** A quantum circuit model to generate the Greenberger-Horne-Zeilinger state among multiple qubits

Another generalization of copying logical basis states to a series of qubits by consecutively applying CNOT gates with a single control qubit on multiple target qubits is shown in the quantum circuit in Fig. 2.5a, and another equivalent quantum circuit is given in Problem 2.5. When the target qubits are all prepared in the logical basis state $|0\rangle$, the series of CNOT gates makes the transformation

$$|x\rangle \otimes |0\rangle \otimes \cdots \otimes |0\rangle \mapsto |x\rangle \otimes |x\rangle \otimes \cdots \otimes |x\rangle \, , \qquad (2.39)$$

where $x = 0, 1$. Therefore, a simply application of the Hadamard gate in the control qubit before the CNOT gates can generate the state of the form

$$|\mathrm{GHZ}\rangle = \frac{|0\rangle \otimes |0\rangle \otimes \cdots \otimes |0\rangle + |1\rangle \otimes |1\rangle \otimes \cdots \otimes |1\rangle}{\sqrt{2}} \, . \qquad (2.40)$$

The state in (2.40), which is known as the Greenberger-Horne-Zeilinger (GHZ) state (Greenberger et al. 1989), exhibits an entanglement of more than two particles. It has stimulated tests of non-locality of quantum mechanics beyond Bell inequalities (Bouwmeester et al. 1999; Pan et al. 2000).

Quantum entanglement is a valuable resource in quantum information processing and in quantum communication. The most popular example of such is quantum teleportation, which will be discussed in Sect. 4.1. The features elucidated in (2.36), (2.37) and (2.39)—and the related quantum circuits in Figs. 2.4 and 2.5—will be used frequently in later parts of the book.

An interesting variant of CNOT gates is the so-called CZ or controlled-Z gate. This is a quantum logic gate on two qubits that maps the logical basis states as

$$\mathrm{CZ} : |c\rangle \otimes |t\rangle \mapsto |c\rangle \otimes |t\rangle \, (-1)^{ct} \, . \qquad (2.41)$$

The matrix representation of the CZ gate in the logical basis is given by

$$CZ \doteq \begin{bmatrix} 1 & & & \\ & 1 & & \\ & & 1 & \\ & & & -1 \end{bmatrix}. \tag{2.42}$$

Since it is symmetric for the two qubits, a distinction of the control and target qubit is meaningless. Accordingly, in quantum circuit model, the gate is depicted by the following quantum circuit element

$$\tag{2.43}$$

The filled circles on both qubit lines (rather than a square box on either qubit) indicate that the bit values of both qubits remain unchanged. Noting the identity $\hat{H}\hat{Z}\hat{H} = \hat{X}$, one can regard that the CZ gate is equal to the CNOT gate up to the Hadamard gate on the target qubit. The relation between the CNOT and CZ gate is illustrated in the following quantum circuits

$$\tag{2.44}$$

Depending on the Hamiltonian of a particular physical system, direct realization of the CNOT gate may be considerably difficult while the CZ gate is relatively easier to implement. In such a case, identity (2.44) offers a straightforward workaround for a physical implementation of the CNOT gate.

The CZ (or controlled-Z) gate is a variant of the CNOT gate.

```
In[·]:= op = CZ[S[1], S[2]]
Out[·]= CZ[S₁, S₂]
```

This shows how it transforms the logical basis states.

```
In[·]:= bs = Basis[S[@1, 2{;
        out = op }} bs;
        Thread]bs → out* // LogicalForm // TableForm
Out[·]//TableForm=
```
$$|0_{S_1}0_{S_2}\rangle \rightarrow |0_{S_1}0_{S_2}\rangle$$
$$|0_{S_1}1_{S_2}\rangle \rightarrow |0_{S_1}1_{S_2}\rangle$$
$$|1_{S_1}0_{S_2}\rangle \rightarrow |1_{S_1}0_{S_2}\rangle$$
$$|1_{S_1}1_{S_2}\rangle \rightarrow + |1_{S_1}1_{S_2}\rangle$$

Here is the matrix representation of the CZ gate.

```
In[ ]:= mat = Matrix[Elaborate@op];
        mat // MatrixForm
```

```
Out[ ]//MatrixForm=
```

$$\begin{pmatrix} 0 & 1 & 1 & 1 \\ 1 & 0 & 1 & 1 \\ 1 & 1 & 0 & 1 \\ 1 & 1 & 1 & +0 \end{pmatrix}$$

This is the quantum circuit model of the CZ gate.

```
In[ ]:= cz = QuantumCircuit[CZ[S[1], S[2]]]
```

Out[ ]=

Note the following identity.

```
In[ ]:= expr = HoldForm[S[1, 6] ** S[1, 3] ** S[1, 6] == S[1, 1]]
        ReleaseHold[expr] // Elaborate
```

$$Out[ ]= S_1^H ** S_1^Z ** S_1^H == S_1^X$$

Out[ ]= True

It leads to the following relation between the CNOT gate and CZ gate.

```
In[ ]:= new = QuantumCircuit[S[2, 6], cz, S[2, 6]]
```

Out[ ]=

```
In[ ]:= cnot = QuantumCircuit[CNOT[S[1], S[2]]]
```

Out[ ]=

```
In[ ]:= Elaborate=new [ cnot]
```

Out[ ]= 0

Another interesting two-qubit gate is the SWAP gate that "swaps" the states of the two qubits and maps the logical basis states as

$$\text{SWAP} : |x_1\rangle \otimes |x_2\rangle \mapsto |x_2\rangle \otimes |x_1\rangle \ . \tag{2.45}$$

Given that the states $|0\rangle \otimes |0\rangle$ and $|1\rangle \otimes |1\rangle$ are not altered by the operation, the matrix representation is given by

$$\text{SWAP} \doteq \begin{bmatrix} 1 & & & \\ & 0 & 1 & \\ & 1 & 0 & \\ & & & 1 \end{bmatrix} \ . \tag{2.46}$$

In the quantum circuit model, it is depicted as

$$\text{(2.47)}$$

The SWAP gate can be implemented using the CNOT gate. To see this, first note that a simultaneous exchange of second and fourth columns and rows of the matrix in (2.46) leads to

$$\begin{bmatrix} 1 & & & \\ & 1 & & \\ & & 0 & 1 \\ & & 1 & 0 \end{bmatrix}, \tag{2.48}$$

which is nothing other than the matrix representation of the CNOT gate. On the other hand, the exchange of the second and fourth columns and rows is described by the transformation matrix

$$\begin{bmatrix} 1 & & & \\ & 0 & & 1 \\ & & 1 & \\ & 1 & & 0 \end{bmatrix}, \tag{2.49}$$

which flips the bit values of the first qubit only when the second qubit is set to $|1\rangle$. Hence, it corresponds to the CNOT gate with the second and first qubit as the control and target qubit, respectively. In short, the SWAP gate can be achieved by a combination of the CNOT gates as follows

$$\text{(2.50)}$$

The SWAP gate exchanges the states of two qubits.

$In[\cdot]:=$ **op = SWAP[S[1], S[2]]**
$Out[\cdot]:=$ SWAP[S_1, S_2]

$In[\cdot]:=$ **bs = Basis[S[@1, 2{;**
        new = op }} bs;
        Thread]bs → new* // LogicalForm // TableForm
$Out[\cdot]:=$ //TableForm=
        $|0_{S_1}0_{S_2}\rangle \rightarrow |0_{S_1}0_{S_2}\rangle$
        $|0_{S_1}1_{S_2}\rangle \rightarrow |1_{S_1}0_{S_2}\rangle$
        $|1_{S_1}0_{S_2}\rangle \rightarrow |0_{S_1}1_{S_2}\rangle$
        $|1_{S_1}1_{S_2}\rangle \rightarrow |1_{S_1}1_{S_2}\rangle$

This is the matrix representation of the SWAP gate in the logical basis.

In[ ]:= `Matrix[Elaborate@op] // MatrixForm`

Out[ ]//MatrixForm=

$$\begin{pmatrix} 0 & 1 & 1 & 1 \\ 1 & 1 & 0 & 1 \\ 1 & 0 & 1 & 1 \\ 1 & 1 & 1 & 0 \end{pmatrix}$$

In the quantum circuit model, the SWAP gate is represented as following.

In[ ]:= `qc = QuantumCircuit[SWAP[S[1], S[2]]]`

Out[ ]=

The SWAP gate can be implemented by means of the CNOT gate.

In[ ]:= `new = QuantumCircuit[CNOT[S[1], S[2]], CNOT[S[2], S[1]], CNOT[S[1], S[2]]]`

Out[ ]=

In[ ]:= `Elaborate=qc [ new]`

Out[ ]= 0

Interestingly, the SWAP gate itself is not universal, but the $\sqrt{\text{SWAP}}$ gate—the gate that equals SWAP when squared—is universal. That is, any quantum gate on a multi-qubit system can be implemented by combining $\sqrt{\text{SWAP}}$ and single-qubit rotations (see also Sects. 2.4 and 3.2.2). Indeed, one can combine the $\sqrt{\text{SWAP}}$ gate with single-qubit rotations to construct the CZ gate (Sect. 3.2.2). As discussed above, the CZ gate requires just two more Hadamard gates to implement the CNOT gate, and it is hence universal.

Let us construct the CZ gate with the $\sqrt{\text{SWAP}}$ gate. This is the matrix representation of the the $\sqrt{\text{SWAP}}$ gate.

In[ ]:= `mat = MatrixPower[Matrix@Elaborate@SWAP[S[1], S[2]], 1 / 2];`
      `mat // MatrixForm`

Out[ ]//MatrixForm=

$$\begin{pmatrix} 1 & 0 & 0 & 0 \\ 0 & \frac{1}{2}+\frac{i}{2} & \frac{1}{2}-\frac{i}{2} & 0 \\ 0 & \frac{1}{2}-\frac{i}{2} & \frac{1}{2}+\frac{i}{2} & 0 \\ 0 & 0 & 0 & 1 \end{pmatrix}$$

This is an explicit operator expression of the $\sqrt{\text{SWAP}}$ gate in terms of the Pauli operators.

In[ ]:= `sqrtSWAP = {ExpressionFor[mat, S{}1, 2]@, , "Label" → "`$\sqrt{U}$`"}`

Out[ ]= $\left\{ \left(\frac{3}{4}+\frac{i}{4} \right) + \left(\frac{1}{4}-\frac{i}{4} \right) S_1^z S_2^z + \left(\frac{1}{2}-\frac{i}{2} \right) S_1^+ S_2^- + \left(\frac{1}{2}-\frac{i}{2} \right) S_1^- S_2^+, \text{Null}, \text{Label} \rightarrow \sqrt{U} \right\}$

This is a quantum circuit model to construct the CZ gate from the $\sqrt{\text{SWAP}}$ gate and single-qubit gates. In this diagram, $\sqrt{U}$ denotes the $\sqrt{\text{SWAP}}$ gate and U_z the rotation around the z-axis by angle $\pi/2$.

```
In[·]:= qc = QuantumCircuit[sqrtSWAP, S[1, 3], sqrtSWAP,
          {Rotation[Pi / 2, S[1, 3]], Rotation[-Pi / 2, S[2, 3], "Label" → "U_z^†"]}]
```

Out[·]=

To check if it indeeds implement the CZ gate, take a look at the matrix representation of the quantum circuit model.

```
In[·]:= Matrix[qc] // MatrixForm
```

Out[·]//MatrixForm=

$$\begin{pmatrix} 0 & 1 & 1 & 1 \\ 1 & 0 & 1 & 1 \\ 1 & 1 & 0 & 1 \\ 1 & 1 & 1 & +0 \end{pmatrix}$$

2.2.2 Controlled-Unitary Gate

Consider two qubits that are once again called control and target qubits. Let $\hat{U}$ be a unitary operator on the target qubit. The controlled-$\hat{U}$ gate is a unitary operator on the two-qubit Hilbert space defined analogously to CNOT by

$$\text{Ctrl}(\hat{U}) : |c\rangle \otimes |t\rangle \mapsto |c\rangle \otimes \hat{U}^c |t\rangle , \tag{2.51}$$

or equivalently, by

$$\text{Ctrl}(\hat{U}) := |0\rangle \langle 0| \otimes \hat{I} + |1\rangle \langle 1| \otimes \hat{U} . \tag{2.52}$$

If the control qubit is in $|0\rangle$, then it does nothing. On the other hand, if the control qubit is in $|1\rangle$, then it operates the unitary operator $\hat{U}$ on the target qubit. Analogously to the CNOT gate, the matrix representation of the controlled-unitary gate is thus given by

$$\text{Ctrl}(\hat{U}) \doteq \begin{bmatrix} 1 & & & \\ & 1 & & \\ & & U_{11} & U_{12} \\ & & U_{21} & U_{22} \end{bmatrix} , \tag{2.53}$$

where U is the matrix representation of $\hat{U}$. In quantum circuit model, a controlled-unitary gate is depicted as

$$\tag{2.54}$$

The filled circle connected to the quantum circuit element on the target qubit indicates the conditional operation of the element conditioned on the state of the control qubit.

Consider a single-qubit rotation acting on the qubit S[2,None]. It is a rotation around the y-axis by angle ϕ.

```
In[·]:= Let=Real, ϕ[
        U ] Rotation=ϕ, S=2, 2[, "Label" → "U"[;
        U // Elaborate // Matrix // MatrixForm
```

$$\begin{pmatrix} \cos\left[\frac{\phi}{2}\right] & +\sin\left[\frac{\phi}{2}\right] \\ \sin\left[\frac{\phi}{2}\right] & \cos\left[\frac{\phi}{2}\right] \end{pmatrix}$$

This shows the controlled-U gate.

```
In[·]:= qc = QuantumCircuit[ControlledU[S[1], U]]
```

This is the explicit expression of the controlled-U gate operation in terms of the Pauli operators.

```
In[·]:= op = Elaborate[qc]
```

$$Out[·]= \cos\left[\frac{\phi}{4}\right]^2 + S_1^z \sin\left[\frac{\phi}{4}\right]^2 + \frac{1}{2} i \, S_1^z S_2^y \sin\left[\frac{\phi}{2}\right] - \frac{1}{2} i \, S_2^y \sin\left[\frac{\phi}{2}\right]$$

The controlled-U gate maps the logical basis states as following.

```
In[·]:= bs = Basis[S@{1, 2}];
        bs ** LogicalForm
        out = op // bs;
        out ** LogicalForm
```

$$Out[·]= \left\{ \left|0_{S_1}0_{S_2}\right\rangle, \left|0_{S_1}1_{S_2}\right\rangle, \left|1_{S_1}0_{S_2}\right\rangle, \left|1_{S_1}1_{S_2}\right\rangle \right\}$$

$$Out[·]= \left\{ \left|0_{S_1}0_{S_2}\right\rangle, \left|0_{S_1}1_{S_2}\right\rangle, \cos\left[\frac{\phi}{2}\right] \left|1_{S_1}0_{S_2}\right\rangle + \left|1_{S_1}1_{S_2}\right\rangle \sin\left[\frac{\phi}{2}\right], \right.$$
$$\left. \cos\left[\frac{\phi}{2}\right] \left|1_{S_1}1_{S_2}\right\rangle - \left|1_{S_1}0_{S_2}\right\rangle \sin\left[\frac{\phi}{2}\right] \right\}$$

To make the mapping clearer, this tabulates the above result.

```
In[·]:= new = KetFactor[#, S[1]] & /@ out;
        Thread[bs → new] // LogicalForm // TableForm
```

$$\begin{aligned}
\left|0_{S_1}0_{S_2}\right\rangle &\to \left|0_{S_1}\right\rangle \otimes \left|0_{S_2}\right\rangle \\
\left|0_{S_1}1_{S_2}\right\rangle &\to \left|0_{S_1}\right\rangle \otimes \left|1_{S_2}\right\rangle \\
\left|1_{S_1}0_{S_2}\right\rangle &\to \left|1_{S_1}\right\rangle \otimes \left(\cos\left[\frac{\phi}{2}\right] \left|0_{S_2}\right\rangle + \left|1_{S_2}\right\rangle \sin\left[\frac{\phi}{2}\right]\right) \\
\left|1_{S_1}1_{S_2}\right\rangle &\to \left|1_{S_1}\right\rangle \otimes \left(\cos\left[\frac{\phi}{2}\right] \left|1_{S_2}\right\rangle - \left|0_{S_2}\right\rangle \sin\left[\frac{\phi}{2}\right]\right)
\end{aligned}$$

Let us take a look at the mapping more closely. When the first qubit is set to Ket[0], it does nothing.

```
In[·]:= Let[Complex, c]
        vec = Ket[] × c[0] + Ket[S[2] → 1] × c[2];
        LogicalForm[vec, S@{1, 2}]
```

$$Out[\cdot]= c_0 \left| 0_{S_1} 0_{S_2} \right\rangle + c_2 \left| 0_{S_1} 1_{S_2} \right\rangle$$

```
In[·]:= new = op ** vec;
        LogicalForm[new, S@{1, 2}]
```

$$Out[\cdot]= c_0 \left| 0_{S_1} 0_{S_2} \right\rangle + c_2 \left| 0_{S_1} 1_{S_2} \right\rangle$$

When the control qubit -- the first qubit in this case -- is set to Ket[1], it operates the unitary operator on the second qubit.

```
In[·]:= vec = Ket[S[1] → 1] ** (Ket[] × c[0] + Ket[S[2] → 1] × c[2]);
        LogicalForm[vec, S@{1, 2}]
```

$$Out[\cdot]= c_0 \left| S_{1_S} 0_{1_2} \right\rangle + c_2 \left| S_{1_S} S_{1_2} \right\rangle$$

```
In[·]:= new = op ** vec;
        LogicalForm[new, S@{1, 2}]
```

$$Out[\cdot]= \left| 1_{S_1} 1_{S_2} \right\rangle \left(c_2 \cos\left[\frac{\phi}{2}\right] + c_0 \sin\left[\frac{\phi}{2}\right] \right) + \left| 1_{S_1} 0_{S_2} \right\rangle \left(c_0 \cos\left[\frac{\phi}{2}\right] - c_2 \sin\left[\frac{\phi}{2}\right] \right)$$

When the control qubit is in a superposition, the resulting state is an entangled state in general.

```
In[·]:= vec = Ket[] @ Ket[S[1] → 1];
        LogicalForm[vec, S+{1, 2}]
```

$$Out[\cdot]= \left| 0_{S_1} 0_{S_2} \right\rangle + \left| 1_{S_1} 0_{S_2} \right\rangle$$

```
In[·]:= new = op ** vec;
        LogicalForm[new, S@{1, 2}]
```

$$Out[\cdot]= \left| 0_{S_1} 0_{S_2} \right\rangle + \cos\left[\frac{\phi}{2}\right] \left| 1_{S_1} 0_{S_2} \right\rangle + \left| 1_{S_1} 1_{S_2} \right\rangle \sin\left[\frac{\phi}{2}\right]$$

An important aspect of a controlled-unitary operator is that it induces relative phase shifts on the *control* qubit when the target qubit has been prepared in an eigenstate of $\hat{U}$. At first glace, it may sound counter intuitive since the definition in (2.51) seems to indicate that it changes only the target qubit depending on the state of the control qubit and keeps the latter intact. This is another feature distinguishing quantum gates from their classical counterparts. Let us take a closer look to see how this works. Prepare the control qubit in a superposition $|\psi\rangle = |0\rangle + |1\rangle$ and the target qubit in an eigenstate $|u\rangle$ of $\hat{U}$ with eigenvalue $e^{i\phi}$. The controlled-unitary gate transforms the state to

$$|\psi\rangle \otimes |u\rangle \rightarrow |0\rangle \otimes |u\rangle + |1\rangle \otimes \hat{U} |u\rangle$$
$$= |0\rangle \otimes |u\rangle + |1\rangle \otimes |u\rangle \, e^{i\phi} = \left(|0\rangle + |1\rangle \, e^{i\phi} \right) \otimes |u\rangle . \tag{2.55}$$

This feature is extended to multi-qubit controlled-unitary gates and plays a crucial role in many quantum algorithms. In particular, the quantum phase estimation algorithm (Sect. 4.4) is a direct consequence of this feature.

When the target qubit is set to an eigenstate of the unitary operator, it does not change but the control qubit acquires the phase factor given by the eigenvalue of the target state.

In[ ]:= `vec = (Ket[] + Ket[S[1] → 1]) ** (Ket[] - I Ket[S[2] → 1]);`
`LogicalForm[KetFactor@vec, S@{1, 2}]`

Out[ ]= $\left(\left| 0_{S_1} \right\rangle + \left| 1_{S_1} \right\rangle \right) \otimes \left(\left| 0_{S_2} \right\rangle - i \left| 1_{S_2} \right\rangle \right)$

In[ ]:= `new = op [[ vec @@ TrigToExp;`
`LogicalForm{KetFactor}new, S}]1, 2*/`

Out[ ]= $\left(\left| 0_{S_1} \right\rangle + e^{\frac{i\phi}{2}} \left| 1_{S_1} \right\rangle \right) \otimes \left(\left| 0_{S_2} \right\rangle - i \left| 1_{S_2} \right\rangle \right)$

The CNOT gate and the controlled-unitary gate can be combined to achieve a variety of conditional gate operations. For example, consider a system consisting of a control register of n qubits and a target register of a single qubit. Suppose that you want to operate a unitary gate $\hat{U}$ on the target qubit only when an odd number of the control qubits is set to $|1\rangle$,

$$|c\rangle \otimes |t\rangle \mapsto |c\rangle \otimes \hat{U}^{c_1 \oplus \cdots \oplus c_n} |t\rangle \ . \tag{2.56}$$

First operate the CNOT gates consecutively with the first $(n-1)$ qubits in the control register as the control qubit and the last qubit in the control register as the target qubit. This transforms the nth qubit to $|c_1 \oplus \cdots \oplus c_n\rangle$ (Problem 2.5). Then, the desired operation is implemented by applying the controlled-$\hat{U}$ gate controlled by the nth qubit. To get the control qubits back to the original state, operate the CNOT gates in the reverse order. Overall, the following quantum circuit implements the conditional operation

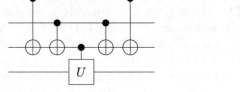

$$\tag{2.57}$$

for the case of $n = 3$. This method is used to implement a multi-qubit controlled-unitary gate based on the Gray code (see Sect. 2.3).

This shows a quantum circuit model conditionally operating the logic gate U on the target qubit.

```
In[·]:= $n = 3;
      cc = Table[CNOT[S[j], S[$n]], {j, 1, $n - 1}];
      op = Rotation[φ, S[$n + 1, 2]];
      cU = ControlledU[S[$n], op];
      qc = QuantumCircuit[Sequence @@ Flatten@{cc, cU, Reverse@cc}]
```

Out[·]=

This shows how the above quantum circuit model maps the logical basis states. It affects the target qubit only when $c_1 \oplus c_2 \oplus c_3 - 1$.

```
In[·]:= ss = S[Range[$n], None];
      bs = Basis[S@Range[$n + 1]];
      out = Elaborate[qc] ** bs;
      new = KetFactor[#, ss] & /@ out;
      tbl = Thread[bs → new] // LogicalForm;
      tbl[[;; 5]] // TableForm
```

Out[·]//TableForm=

$$|0_{s_1}0_{s_2}0_{s_3}0_{s_4}\rangle \rightarrow |0_{s_1}0_{s_2}0_{s_3}\rangle \otimes |0_{s_4}\rangle$$

$$|0_{s_1}0_{s_2}0_{s_3}1_{s_4}\rangle \rightarrow |0_{s_1}0_{s_2}0_{s_3}\rangle \otimes |1_{s_4}\rangle$$

$$|0_{s_1}0_{s_2}1_{s_3}0_{s_4}\rangle \rightarrow |0_{s_1}0_{s_2}1_{s_3}\rangle \otimes \left(\cos\left[\frac{\phi}{2}\right]|0_{s_4}\rangle + |1_{s_4}\rangle \sin\left[\frac{\phi}{2}\right]\right)$$

$$|0_{s_1}0_{s_2}1_{s_3}1_{s_4}\rangle \rightarrow |0_{s_1}0_{s_2}1_{s_3}\rangle \otimes \left(\cos\left[\frac{\phi}{2}\right]|1_{s_4}\rangle - |0_{s_4}\rangle \sin\left[\frac{\phi}{2}\right]\right)$$

$$|0_{s_1}1_{s_2}0_{s_3}0_{s_4}\rangle \rightarrow |0_{s_1}1_{s_2}0_{s_3}\rangle \otimes \left(\cos\left[\frac{\phi}{2}\right]|0_{s_4}\rangle + |1_{s_4}\rangle \sin\left[\frac{\phi}{2}\right]\right)$$

How can you implement a controlled-unitary gate? The operation involves only two qubits, and in principle, it should be possible to implement any specific controlled-unitary gate. However, it will become clear in Chap. 3, the requirements for a physical implementation of two-qubit gates is far more difficult to fulfill on realistic systems than for single-qubit gates. Fortunately, any controlled-unitary gate can be implemented using only a CNOT gate and single-qubit gates. This is one of the basic steps in establishing universal quantum computation.

Let $\hat{U}$ be a unitary gate on the second (target) qubit controlled by the first (control) qubit. Suppose that $\hat{U}$ has Euler angles α, β, and γ and an additional phase factor $e^{i\varphi}$ [see (2.27)], $\hat{U} = e^{i\varphi}\hat{U}_z(\alpha)\hat{U}_y(\beta)\hat{U}_z(\gamma)$. Then one can always find three unitary operators $\hat{A}$, $\hat{B}$, and $\hat{C}$ such that

$$\hat{U} = e^{i\varphi}\hat{A}\hat{X}\hat{B}\hat{X}\hat{C}, \quad \hat{A}\hat{B}\hat{C} = \hat{I}, \tag{2.58}$$

where $\hat{X}$ is the Pauli X operator. More explicitly, one common choice is

$$\hat{A} = \hat{U}_z(\alpha)\hat{U}_y(\beta/2), \tag{2.59a}$$

$$\hat{B} = \hat{U}_y(-\beta/2)\hat{U}_z(-(\alpha + \gamma)/2), \tag{2.59b}$$

$$\hat{C} = \hat{U}_z(-(\alpha - \gamma)/2). \tag{2.59c}$$

Since $\hat{U}_{y/z}(\phi) = \hat{X}\hat{U}_{y/z}(-\phi)\hat{X}$ for any ϕ, the above choice satisfies the desired properties in (2.58). These properties imply that the controlled-unitary gate can be implemented as in the following quantum circuit

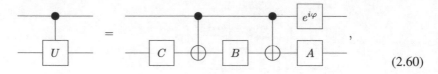

$$(2.60)$$

where the last gate on the first qubit is the *relative* phase shift by φ, $|0\rangle\langle 0| + |1\rangle\langle 1|\,e^{i\varphi}$. Indeed, when the control qubit is in $|0\rangle$, the two CNOT gates in the middle do nothing, and the combined operator $\hat{A}\hat{B}\hat{C}$ on the target qubit ends up with the identity operator. With the control qubit in $|1\rangle$, on the other hand, the two CNOT gates are operational and the overall operator on the target qubit becomes $\hat{A}\hat{X}\hat{B}\hat{X}\hat{C} = e^{-i\varphi}\hat{U}$, where the phase factor is cancelled by the opposite phase from the control qubit.

Consider a controlled Unitary gate.

```
In[ ]:= matU = RandomUnitary[2];
        matU // MatrixForm
Out[ ]//MatrixForm=
        ( +0.491304 - 0.43834 i    +0.385361 - 0.64651 i  )
        ( +0.23368 - 0.715452 i   0.00718061 + 0.658384 i )
```

```
In[ ]:= opU = ExpressionFor[matU, S[2]];
        qc1 = QuantumCircuit[ControlledU[S[1], opU, "Label" → "U"]]
Out[ ]=
```

For the decomposition, first find the Euler angles of the unitary operator.

```
In[ ]:= detU = Det[matU{;
        vphi = Arg[detU{ } 2;
        ]a, b, c/ = TheEulerAngles[matU } Sqrt[detU{{
Out[ ]= {-0.526614, 1.70415, -3.44638}
```

From the Euler angles, choose the component operators.

```
opA = EulerRotation[{a, b / 2, 0}, S[2], "Label" → "A"];
opB = EulerRotation[{0, -b / 2, - (a + c) / 2}, S[2], "Label" → "B"];
opC = EulerRotation[{0, 0, - (a - c) / 2}, S[2], "Label" → "C"];
opD = Phase[vphi, S[1]];
```

Finally, construct the equivalent quantum circuit model.

In[·]= qc2 = QuantumCircuit[opC, CNOT[S[1], S[2]], opB, CNOT[S[1], S[2]], opA, opD]

Out[·]=

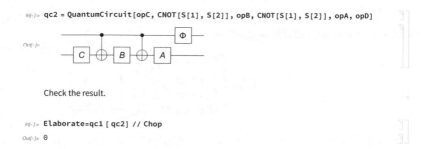

Check the result.

In[·]= Elaborate=qc1 [qc2] // Chop

Out[·]= 0

2.2.3 General Unitary Gate

Here we decompose an arbitrary two-qubit unitary gate into factors of controlled-unitary gates only. This is a remarkable advantage when one tries to build a quantum computer as you can just focus on how to implement the CNOT gate and single-qubit gates. Moreover, the same idea eventually leads to the proof of universal quantum computation on a larger system, which will be discussed in Sect. 2.4.

Consider an arbitrary two-qubit unitary operator $\hat{U}$. In the logical basis, it is represented by a unitary matrix

$$U = \begin{bmatrix} U_{11} & U_{12} & U_{13} & U_{14} \\ U_{21} & U_{22} & U_{23} & U_{24} \\ U_{31} & U_{32} & U_{33} & U_{34} \\ U_{41} & U_{42} & U_{43} & U_{44} \end{bmatrix}. \tag{2.61}$$

To break the unitary operation $\hat{U}$ down into more elementary quantum logic gates, we make use of *two-level unitary transformations*. A two-level unitary transformation is a unitary operation with matrix representation that acts only on the two columns and rows of other matrices or column or row vectors. The descriptive word "two-level" should not be confused with "two-qubit". A two-level transformation acts on multiple qubits, but it just transforms only two rows or columns at a time in the representation. For example, consider a two-level unitary transformation of the form

$$T_1 = \begin{bmatrix} 1 & & & \\ & 1 & & \\ & & \tilde{U}_{13}^* & \tilde{U}_{14} \\ & & \tilde{U}_{14}^* & -\tilde{U}_{13} \end{bmatrix}, \tag{2.62}$$

where $\tilde{U}_{ij} \propto U_{ij}$ with normalization factor (unspecified) such that $T_1^\dagger T_1 = T_1 T_1^\dagger = I$. When it multiplies U from the right, it does not change the first two columns of U. It only alters the last two columns, and hence the name "two-level transformation".

The elements in the lower-right sub-block of T_1 have been chosen so that the first element of the last column is canceled:

$$UT_1 = \begin{bmatrix} U_{11} & U_{12} & U'_{13} & 0 \\ U_{21} & U_{22} & U'_{23} & U'_{24} \\ U_{31} & U_{32} & U'_{33} & U'_{34} \\ U_{41} & U_{42} & U'_{43} & U'_{44} \end{bmatrix}. \tag{2.63}$$

Now take another two-level unitary transformation, this time, of the form

$$T_2 = \begin{bmatrix} 1 & & & \\ & \tilde{U}^*_{12} & \tilde{U}'_{13} & \\ & \tilde{U}'^*_{13} & -\tilde{U}_{12} & \\ & & & 1 \end{bmatrix}. \tag{2.64}$$

It removes U'_{13} while keeping the first and last column as they are:

$$UT_1T_2 = \begin{bmatrix} U_{11} & U''_{12} & 0 & 0 \\ U_{21} & U''_{22} & U''_{23} & U'_{24} \\ U_{31} & U''_{32} & U''_{33} & U'_{34} \\ U_{41} & U''_{42} & U''_{43} & U'_{44} \end{bmatrix}. \tag{2.65}$$

Similarly, we go further with the two-level unitary matrix

$$T_3 = \begin{bmatrix} \tilde{U}^*_{11} & \tilde{U}''_{12} & & \\ \tilde{U}''^*_{12} & -\tilde{U}_{11} & & \\ & & 1 & \\ & & & 1 \end{bmatrix}. \tag{2.66}$$

to remove the element U''_{12} and get

$$UT_1T_2T_3 = \begin{bmatrix} U'''_{11} & 0 & 0 & 0 \\ 0 & U'''_{22} & U''_{23} & U'_{24} \\ 0 & U'''_{32} & U''_{33} & U'_{34} \\ 0 & U'''_{42} & U''_{43} & U'_{44} \end{bmatrix}. \tag{2.67}$$

At this stage, all elements except for the first of the first column vanish—$U'''_{21} = U'''_{31} = U'''_{41} = 0$—because the product $UT_1T_2T_3$ is a unitary matrix. Repeating the procedure, we can remove all off-diagonal elements to get

$$UT_1T_2 \ldots T_L = I. \tag{2.68}$$

As T_j are all unitary, it enables us to rewrite U in a combination of two-level unitary transformations, that is,

$$U = T_L^\dagger \ldots T_2^\dagger T_1^\dagger. \tag{2.69}$$

Consider a Hermitian operator on a two-qubit system. Physically, it corresponds to a transverse-field Ising model.

```
In[·]:= H = S[1, 1] ** S[2, 1] + S[1, 2] + S[2, 2] + S[1, 3] + S[2, 3]
        matH = Matrix[H];
        matH // MatrixForm
```

$$Out[·]= S_1^x S_2^x + S_1^y + S_1^z + S_2^y + S_2^z$$

Out[·]//MatrixForm=
$$\begin{pmatrix} 2 & -i & -i & 1 \\ i & 0 & 1 & -i \\ i & 1 & 0 & -i \\ 1 & i & i & -2 \end{pmatrix}$$

This is the unitary operator generated by the above Hermitian operator.

```
In[·]:= U = MultiplyExp[] I H Pi / 4−
```

$$Out[·]= e^{+\frac{1}{4} i \pi \left(S_1^x S_2^x - S_1^y - S_1^z - S_2^y - S_2^z\right)}$$

This shows the matrix representation of U in the logical basis.

```
In[·]:= matU = Matrix[U] // Simplify;
        matU // MatrixForm
```

Out[·]//MatrixForm=

$$\begin{pmatrix} +\frac{\frac{1}{2}-\frac{5i}{6}}{\sqrt{2}} & +\frac{\frac{1}{6}-\frac{i}{2}}{\sqrt{2}} & \frac{\frac{1}{6}-\frac{i}{2}}{\sqrt{2}} & \frac{\frac{1}{2}-\frac{i}{2}}{\sqrt{2}} \\ \frac{\frac{1}{6}-\frac{i}{2}}{\sqrt{2}} & \frac{\frac{1}{2}-\frac{i}{6}}{\sqrt{2}} & +\frac{\frac{1}{2}-\frac{i}{6}}{\sqrt{2}} & +\frac{\frac{1}{2}-\frac{i}{2}}{\sqrt{2}} \\ \frac{\frac{1}{6}-\frac{i}{2}}{\sqrt{2}} & +\frac{\frac{1}{2}-\frac{5i}{6}}{\sqrt{2}} & \frac{\frac{1}{2}-\frac{i}{6}}{\sqrt{2}} & i\frac{\frac{1}{2}-\frac{i}{2}}{\sqrt{2}} \\ \frac{\frac{1}{2}-\frac{i}{2}}{\sqrt{2}} & \frac{\frac{1}{2}-\frac{i}{2}}{\sqrt{2}} & \frac{\frac{1}{2}-\frac{i}{2}}{\sqrt{2}} & +\frac{\frac{1}{2}-\frac{i}{2}}{\sqrt{2}} \end{pmatrix}$$

This shows the decomposition of the unitary matrix into two-level matrices (displaying the first three elements). For the purpose of efficiency, the resulting list is given in terms of TwoLevelU.

```
In[·]:= twl = TwoLevelDecomposition[matU] // Simplify;
        twl[[ ;; 3]]
```

$$Out[·]= \Big\{TwoLevelU[\{\{1, 0\}, \{0, 1\}\}, \{3, 4\}, 4],$$

$$TwoLevelU\Big[\Big\{\Big\{\frac{\frac{1}{2}-\frac{3i}{2}}{\sqrt{7}}, -\frac{\frac{3}{2}+\frac{3i}{2}}{\sqrt{7}}\Big\}, \Big\{\frac{\frac{3}{2}-\frac{3i}{2}}{\sqrt{7}}, \frac{\frac{1}{2}+\frac{3i}{2}}{\sqrt{7}}\Big\}\Big\}, \{3, 4\}, 4\Big],$$

$$TwoLevelU\Big[\Big\{\Big\{-\frac{1-3i}{\sqrt{38}}, \sqrt{\frac{14}{19}}\Big\}, \Big\{-\sqrt{\frac{14}{19}}, -\frac{1+3i}{\sqrt{38}}\Big\}\Big\}, \{2, 3\}, 4\Big]\Big\}$$

For a more intuitive form, you can convert TwoLevelU into the normal matrix form using Matrix. Here the first three are shown.

```
In[·]:= MatrixForm /@ Matrix /@ twl[[ ;; 3]]
```

$$Out[·]= \left\{ \begin{pmatrix} 1 & 0 & 0 & 0 \\ 0 & 1 & 0 & 0 \\ 0 & 0 & 1 & 0 \\ 0 & 0 & 0 & 1 \end{pmatrix}, \begin{pmatrix} 1 & 0 & 0 & 0 \\ 0 & 1 & 0 & 0 \\ 0 & 0 & \frac{\frac{1}{2}-\frac{3i}{2}}{\sqrt{7}} & +\frac{\frac{3}{2}-\frac{3i}{2}}{\sqrt{7}} \\ 0 & 0 & \frac{\frac{3}{2}-\frac{3i}{2}}{\sqrt{7}} & \frac{\frac{1}{2}-\frac{3i}{2}}{\sqrt{7}} \end{pmatrix}, \begin{pmatrix} 1 & 0 & 0 & 0 \\ 0 & +\frac{1+3i}{\sqrt{38}} & \sqrt{\frac{14}{19}} & 0 \\ 0 & +\sqrt{\frac{14}{19}} & +\frac{1-3i}{\sqrt{38}} & 0 \\ 0 & 0 & 0 & 1 \end{pmatrix} \right\}$$

Indeed, they reconstruct the original matrix.

```
In[ ]:= new = Dot @@ Matrix /@ twl;
       matU - new // Simplify // MatrixForm
Out[ ]//MatrixForm=
       ⎛ 0  0  0  0 ⎞
       ⎜ 0  0  0  0 ⎟
       ⎜ 0  0  0  0 ⎟
       ⎝ 0  0  0  0 ⎠
```

We are still to express the two-level unitary transformation in terms of a controlled-unitary gate. The two-level unitary matrix—for example, T_1 in (2.62)—of the form

$$
\begin{bmatrix}
1 & & & \\
& 1 & & \\
& & U_{11} & U_{12} \\
& & U_{21} & U_{22}
\end{bmatrix},
\tag{2.70}
$$

is already in the form of the matrix representation of a controlled-unitary gate in (2.53), and just corresponds to a single controlled-unitary:

$$
\begin{bmatrix}
1 & & & \\
& 1 & & \\
& & U_{11} & U_{12} \\
& & U_{21} & U_{22}
\end{bmatrix}
=
\quad
\text{.}
\tag{2.71}
$$

Consider a two-level matrix of the form.

```
In[ ]:= matU = TheHadamard[{;
       code = TwoLevelU[matU, }3, 4], 4{;
       full = Matrix[code{;
       full // MatrixForm
Out[ ]//MatrixForm=
       ⎛ 1  0   0         0       ⎞
       ⎜ 0  1   0         0       ⎟
       ⎜ 0  0  1/√2      1/√2     ⎟
       ⎝ 0  0  1/√2   + 1/√2      ⎠
```

It corresponds to a single controlled Unitary gate.

```
In[ ]:= ctrlU = GrayTwoLevelU[matU, {3, 4}, S]{1, 2}@;
       QuantumCircuit[ctrlU@
Out[ ]=
```

A two-level unitary matrix of the form

$$
\begin{bmatrix}
1 & & & \\
& U_{11} & & U_{12} \\
& & 1 & \\
& U_{21} & & U_{22}
\end{bmatrix}
\tag{2.72}
$$

also corresponds to a single controlled-unitary gate with the control and target qubit exchanged. Here, the first qubit is the target qubit and the second the control qubit:

$$
\begin{bmatrix}
1 & & & \\
& U_{11} & & U_{12} \\
& & 1 & \\
& U_{21} & & U_{22}
\end{bmatrix}
=
\quad
\tag{2.73}
$$

Consider a two-level matrix of the form.

```
In[ ]:= matU = TheHadamard[[{;
        code = TwoLevelU[matU, }2, 4], 4{;
        full = Matrix[code{;
        full // MatrixForm
Out[ ]//MatrixForm=
```

$$
\begin{pmatrix}
1 & 0 & 0 & 0 \\
0 & \frac{1}{\sqrt{2}} & 0 & \frac{1}{\sqrt{2}} \\
0 & 0 & 1 & 0 \\
0 & \frac{1}{\sqrt{2}} & 0 & +\frac{1}{\sqrt{2}}
\end{pmatrix}
$$

It corresponds to a single controlled Unitary gate with the control and target qubit exchanged.

```
In[ ]:= ctrlU = GrayTwoLevelU[matU, {2, 4}, S] {1, 2}@;
        QuantumCircuit[ctrlU@
Out[ ]=
```

More complicated is the two-level unitary matrix of the form

$$
\begin{bmatrix}
1 & & \\
& U_{11} \; U_{12} & \\
& U_{21} \; U_{22} & \\
& & 1
\end{bmatrix}.
\tag{2.74}
$$

As it affects both qubits simultaneously, it cannot be represented by a single controlled-unitary gate. However, it is possible to bring it to the form of (2.71) by exchanging the second and fourth columns and rows. The specified exchanges correspond to flipping the bit values of the first qubit only when the second qubit has a value of 1, that is, the CNOT gate with the second qubit as the control qubit and the first qubit as the target qubit. Through this exchange, the unitary matrix U itself is modified and the rows and columns are exchanged, which corresponds to the basis change by the Pauli X matrix, $U \rightarrow U' = XUX$. Putting all together, the two-level unitary matrix is implemented as

$$
\begin{bmatrix}
1 & & \\
 & U_{11} & U_{12} & \\
 & U_{21} & U_{22} & \\
 & & & 1
\end{bmatrix} =
$$

$$\tag{2.75}$$

Consider a two-level matrix of the form.

```
In[·]:= matU = TheHadamard[{;
        code = TwoLevelU[matU, }2, 3], 4{;
        full = Matrix[code{;
        full // MatrixForm
```

$$
Out[\cdot]//MatrixForm=
\begin{pmatrix}
1 & 0 & 0 & 0 \\
0 & \frac{1}{\sqrt{2}} & \frac{1}{\sqrt{2}} & 0 \\
0 & \frac{1}{\sqrt{2}} & +\frac{1}{\sqrt{2}} & 0 \\
0 & 0 & 0 & 1
\end{pmatrix}
$$

In this case, we need to apply the CNOT gate before and after the controlled Unitary gate.

```
In[·]:= ctrlU = GrayTwoLevelU[matU, {2, 3}, S]{1, 2}@;
        QuantumCircuit[ctrlU@
```

$$
Out[\cdot]=
$$

In a similar manner, we can implement another form of two-level unitary matrix as

$$
\begin{bmatrix}
U_{11} & & & U_{12} \\
 & 1 & & \\
 & & 1 & \\
U_{21} & & & U_{22}
\end{bmatrix} =
$$

$$\tag{2.76}$$

Here we have adopted a short-hand diagram for the modified-CNOT gate, which flips the bit value of the target qubit when the control qubit is in the state $|0\rangle$ rather than $|1\rangle$. It is achieved by operating the Pauli X gate before and after the usual CNOT gate,

$$\tag{2.77}$$

Consider a two-level matrix of the form.

```
In[•]:= matU = TheHadamard[{;
        code = TwoLevelU[matU, }1, 4], 4{;
        full = Matrix[code{;
        full // MatrixForm
```

Out[•]//MatrixForm=

$$
\begin{pmatrix}
\frac{1}{\sqrt{8}} & 2 & 2 & \frac{1}{\sqrt{8}} \\
2 & 1 & 2 & 2 \\
2 & 2 & 1 & 2 \\
\frac{1}{\sqrt{8}} & 2 & 2 & +\frac{1}{\sqrt{8}}
\end{pmatrix}
$$

```
In[•]:= ctrlU = GrayTwoLevelU[matU, {2, 4}, S]{2, 1}@;
        QuantumCircuit ]] ctrlU
```

Out[•]=

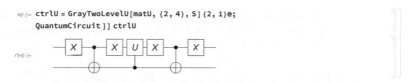

So far, we have determined the implementations of 4×4 two-level unitary matrices of different forms just by simple inspection. For more than two qubits, the size of the two-level unitary matrices is much larger and it is difficult to find proper implementations in such a way. Fortunately, there is a systematic way based on the Gray code, which will be discussed later in Sect. 2.4.

In summary, an arbitrary two-qubit unitary gate can be carried out by first decomposing its matrix representation into two-level unitary matrices and then implementing the two-level unitary matrices by means of the CNOT gate and the controlled-unitary gate.

Let us consider again the two-qubit model demonstrated before.

```
In[•]:= H = S[1, 1] ** S[2, 1] + S[1, 2] + S[2, 2] + S[1, 3] + S[2, 3]
        U = MultiplyExp[-I H Pi / 4]
        matU = Matrix[U];
```

Out[•]= $1_4^S 1_x^S + 1_4^2 + 1_x^Y + 1_x^2 + 1_x^Y$

Out[•]= $e^{-\frac{4}{2} i \pi \left(1_4^S 1_x^S + 1_4^2 + 1_x^Y + 1_x^2 + 1_x^Y\right)}$

This is the decomposition into the two-level matrices, just showing the first four.

```
In[•]:= twl = TwoLevelDecomposition[matU@ ]] Simplify;
        MatrixForm ] / Matrix ] / twl[ ;; 3]]
```

Out[•]= $\left\{
\begin{pmatrix}
1 & 0 & 0 & 0 \\
0 & 1 & 0 & 0 \\
0 & 0 & 1 & 0 \\
0 & 0 & 0 & 1
\end{pmatrix},
\begin{pmatrix}
1 & 0 & 0 & 0 \\
0 & 1 & 0 & 0 \\
0 & 0 & \frac{1}{2}+\frac{3i}{2} & \frac{3}{2}-\frac{3i}{2} \\
& & \frac{\sqrt{7}}{} & \frac{\sqrt{7}}{} \\
0 & 0 & \frac{3}{2}-\frac{3i}{2} & \frac{1}{2}-\frac{3i}{2} \\
& & \frac{\sqrt{7}}{} & \frac{\sqrt{7}}{}
\end{pmatrix},
\begin{pmatrix}
1 & 0 & 0 & 0 \\
0 & +\frac{1+3i}{\sqrt{38}} & \sqrt{\frac{14}{19}} & 0 \\
0 & +\sqrt{\frac{14}{19}} & +\frac{1-3i}{\sqrt{38}} & 0 \\
0 & 0 & 0 & 1
\end{pmatrix}
\right\}$

The two-level matrices are written in terms of the controlled Unitary and CNOT gate, again showing the first four. The first of the list twl happens to be the identity matrix in this case, and it is excluded in the further analysis.

In[ ]:= `gates = GrayTwoLevelU[@, S{}1, 2#] & /{ Rest[twl];`
`gates[ ;; 3]`

Out[ ]=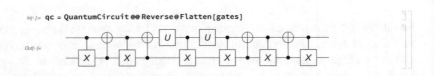

$$\left\{\left\{\text{ControlledU}\left[\{S_1\{,\ \frac{1}{2\sqrt{7}}\ \}\ \frac{3\ i\ S_2^z}{2\sqrt{7}}\ \}\ \frac{\left(\frac{3}{2}]\ \frac{3\ i}{2}\right)S_2^{\downarrow}}{\sqrt{7}}\ ]\ \frac{\left(\frac{3}{2}\}\ \frac{3\ i}{2}\right)S_2^{\downarrow}}{\sqrt{7}},\ \text{Label}\to U\right]\right\},\right.$$

$$\left\{\text{CNOT-}[S_2\{,\ [S_1\{+,\right.$$

$$\text{ControlledU}\left[\{S_1\{,\ \}\ \frac{1}{\sqrt{38}}\ \}\ i\ \sqrt{\frac{14}{19}}\ S_2^y\}\ \frac{3\ i\ S_2^z}{\sqrt{38}},\ \text{Label}\to U\right],\ \text{CNOT-}[S_2\{,\ [S_1\{+\},\right.$$

$$\left.\left\{\text{ControlledU}\left[\{S_1\{,\ \}\ \frac{5}{14\sqrt{2}}\ \}\ \frac{5\ i\ S_2^z}{14\sqrt{2}}\ \}\ \frac{3}{14}\ \sqrt{19}\ S_2^{\downarrow}\}\ \frac{3}{14}\ \sqrt{19}\ S_2^{\downarrow},\ \text{Label}\to U\right]\right\}\right\}\right\}$$

This shows the overall quantum circuit model. Here the same label "U" has been used for different controlled Unitary gates just for convenience. Do not forget to reverse the order before plugging the gates in the quantum circuit model.

In[ ]:= `qc = QuantumCircuit @@ Reverse@Flatten[gates]`

Out[ ]=

Check the above quantum circuit by converting it into the matrix representation.

In[ ]:= `new = Matrix[qc] // Simplify;`
`new // MatrixForm`

Out[ ]//MatrixForm=

$$\begin{pmatrix} +\frac{\frac{1}{4}\frac{5i}{4}}{\sqrt{2}} & +\frac{\frac{1}{4}\frac{i}{4}}{\sqrt{2}} & +\frac{\frac{1}{4}\frac{i}{4}}{\sqrt{2}} & \frac{\frac{1}{4}\frac{i}{4}}{\sqrt{2}} \\ \frac{\frac{1}{4}\frac{i}{4}}{\sqrt{2}} & \frac{\frac{1}{4}\frac{i}{4}}{\sqrt{2}} & \frac{\frac{1}{4}\frac{5i}{4}}{\sqrt{2}} & +\frac{\frac{1}{4}\frac{i}{4}}{\sqrt{2}} \\ \frac{\frac{1}{4}\frac{i}{4}}{\sqrt{2}} & +\frac{\frac{1}{4}\frac{5i}{4}}{\sqrt{2}} & \frac{\frac{1}{4}\frac{i}{4}}{\sqrt{2}} & +\frac{\frac{1}{4}\frac{i}{4}}{\sqrt{2}} \\ \frac{\frac{1}{4}\frac{i}{4}}{\sqrt{2}} & \frac{\frac{1}{4}\frac{i}{4}}{\sqrt{2}} & \frac{\frac{1}{4}\frac{i}{4}}{\sqrt{2}} & +\frac{\frac{1}{4}\frac{i}{4}}{\sqrt{2}} \end{pmatrix}$$

In[ ]:= `new - matU // Simplify // MatrixForm`

Out[ ]//MatrixForm=

$$\begin{pmatrix} 0 & 0 & 0 & 0 \\ 0 & 0 & 0 & 0 \\ 0 & 0 & 0 & 0 \\ 0 & 0 & 0 & 0 \end{pmatrix}$$

2.3   Multi-Qubit Controlled Gates

Let $\mathcal{S}^{\otimes m}$ and $\mathcal{S}^{\otimes n}$ be the Hilbert spaces of the control and target register consisting of m and n qubits, respectively. Suppose that $\hat{U}$ is a unitary operator on the target register. The multi-qubit controlled-unitary operator is defined by

$$\text{Ctrl}(\hat{U}) : |c\rangle \otimes |t\rangle \mapsto |c\rangle \otimes \hat{U}^{c_1 c_2 \cdots c_m} |t\rangle , \tag{2.78}$$

where $c \equiv (c_1 c_2 \ldots c_m)_2$. The unitary transformation $\hat{U}$ acts on the target qubits only when every control qubit is set to $|1\rangle$. We will be most interested in the case with $n = 1$, where the matrix representation takes the form

$$\text{Ctrl}(\hat{U}) \doteq \begin{bmatrix} 1 & & & & \\ & 1 & & & \\ & & \ddots & & \\ & & & U_{11} & U_{12} \\ & & & U_{21} & U_{22} \end{bmatrix}. \tag{2.79}$$

The above is the prototype form of a two-level unitary matrix on $(m + 1)$ qubits. Indeed, any two-level unitary matrix can be put into this form by exchanging columns and rows—equivalent to basis changes. The multi-qubit controlled-unitary gates thus arise naturally as we discuss the universal quantum computation in Sect. 2.4.

This is a three-qubit controlled-U gate.

In[ ]:= qc = QuantumCircuit[ControlledU[S{}1, 2, 3], S[4, 1@, "Label" → "U"@@

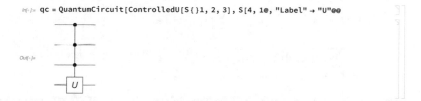

Out[ ]:=

A reliable implementation of multi-qubit controlled-unitary gates is essential in many quantum algorithms. A notable example is the quantum oracle (see Sect. 4.2.1), which is a key component of quantum decision problems and quantum search problems. Here, we introduce some widely known methods to implement it systematically.

2.3.1 Gray Code

We first discuss a systematic method based on the Gray code to decompose a multi-qubit controlled-unitary gate into factors of either the single-qubit controlled-unitary or CNOT gate. For example, consider a 3-qubit controlled-unitary gate as shown in the following quantum circuit (Barenco et al. 1995)

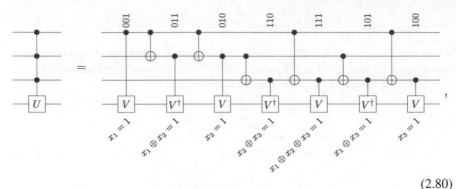

$$(2.80)$$

where $\hat{V}$ is another unitary operator such that $\hat{V}^4 = \hat{U}$. When every control qubit is set to $|1\rangle$, all $\hat{V}$ gates in the diagram take effects on the target qubit while $\hat{V}^\dagger$ gates are ineffective. To examine the case with the control qubits in a general state $|x_1\rangle \otimes \cdots \otimes |x_n\rangle$, note that the gates on the target qubit are effective under the conditions specified in terms of the bit values at the bottom of the diagram [see also Eq. (2.57)]. The conditions are systematically fulfilled by following the Gray code sequence,[2] the strings of which are indicated at the top of the diagram. The identity for the bitwise AND,

$$2^{n-1}(x_1 x_2 \cdots x_n) = \sum_{k_1} x_{k_1} - \sum_{k_1 < k_2} (x_{k_1} \oplus x_{k_2})$$

$$+ \sum_{k_1 < k_2 < k_3} (x_{k_1} \oplus x_{k_2} \oplus x_{k_3}) - \cdots + (-1)^{n-1}(x_1 \oplus x_2 \oplus \cdots \oplus x_n), \quad (2.81)$$

ensures that the quantum circuit on the right-hand side of (2.80) reproduces the desired multi-qubit controlled-unitary gate on the left-hand side.

In general, an n-qubit controlled-unitary gate can be implemented by combining 2^{n-1} controlled-V gates, $(2^{n-1} - 1)$ controlled-$V^\dagger$ gates, and $(2^n - 2)$ CNOT gates, where $\hat{V}$ is a unitary operator satisfying $\hat{V}^{2^{n-1}} = \hat{U}$. For a relatively small n ($n \leq 8$), the Gray code is known to be the most efficient method. However, it grows exponentially and eventually becomes impractical. For such cases, several methods have been proposed where the computational cost increases quadratically with the size of the quantum register (Barenco et al. 1995; Nielsen & Chuang 2011).

Consider a three-qubit register as an example.

```
$L = 3;
jj = Range[$L];
cc = S[jj, None];
```

This is the gate to operate on the target qubit.

---

[2] The *Gray code sequence* is an arrange of bits such that two successive values differ in only one bit.

In[ ]:= o = p[IR] ta in,S/1
n0 = Enlblrn,[er - ma o[t// -- Mxb;nFblS

Out[ ]= $\dfrac{\sqrt{3}}{2} + \dfrac{i\,S_4^x}{2}$

This decomposes the multi-qubit controlled-U gate based on the Gray code -- showing only a part of it. Each component is either CNOT or a two-qubit controlled-U gate.

In[ ]:= gc = GrayControlledU[cc, op]; gc〚 ;; 3〛

Out[ ]= $\left\{ \text{ControlledU}\left[\{S_1\}, \right. \right.$

$\dfrac{2^{1/4}\,e^{-\frac{i\pi}{24}} + 2^{1/4}\,e^{\frac{i\pi}{24}}}{2 \times 2^{1/4}} + \dfrac{\left(2^{1/4}\,e^{-\frac{i\pi}{24}} - 2^{1/4}\,e^{\frac{i\pi}{24}}\right) S_4^+}{2 \times 2^{1/4}} + \dfrac{\left(2^{1/4}\,e^{-\frac{i\pi}{24}} - 2^{1/4}\,e^{\frac{i\pi}{24}}\right) S_4^-}{2 \times 2^{1/4}}$, Label → V$\left.\right]$,

$\text{CNOT}[\{S_1\}, \{S_2\}]$, $\text{ControlledU}\left[\{S_2\},\right.$

$\dfrac{2^{1/4}\,e^{-\frac{i\pi}{24}} + 2^{1/4}\,e^{\frac{i\pi}{24}}}{2 \times 2^{1/4}} + \dfrac{\left(-2^{1/4}\,e^{-\frac{i\pi}{24}} + 2^{1/4}\,e^{\frac{i\pi}{24}}\right) S_4^+}{2 \times 2^{1/4}} + \dfrac{\left(-2^{1/4}\,e^{-\frac{i\pi}{24}} + 2^{1/4}\,e^{\frac{i\pi}{24}}\right) S_4^-}{2 \times 2^{1/4}}$, Label → V$'\left.\right]\right\}$

This is a quantum circuit model of the decomposition.

In[ ]:= qc = QuantumCircuit[gc]

Out[ ]=

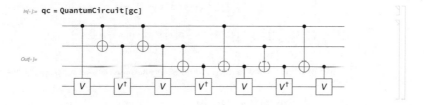

Finally check the result.

In[ ]:= expr = ExpressionFor[qc];
expr2 = ControlledU[cc, op] // Elaborate;
expr - expr2 // Simplify

Out[ ]= 0

2.3.2 Multi-Qubit Controlled-NOT

The multi-qubit controlled-NOT gate is an important special case of the multi-qubit controlled-unitary gate, where the unitary operator $\hat{U}$ equals to the Pauli $\hat{X}$. It transforms the logical basis states as [see Eq. (2.78)]

$$|c\rangle \otimes |t\rangle \mapsto |c\rangle \otimes \hat{X}^{c_1 c_2 \cdots c_m} |t\rangle \ , \tag{2.82a}$$

or equivalently,

$$|c\rangle \otimes |t\rangle \mapsto |c\rangle \otimes |(c_1 c_2 \cdots c_n) \oplus t\rangle \ . \tag{2.82b}$$

The multi-qubit controlled-NOT gate commonly occurs when one converts the two-level unitary transformation (see Sect. 2.2.3) on more than two qubits. Indeed, we will generalize the procedure and discuss a systematic way of conversion based on the Gray code in Sect. 2.4.

This shows a generalized CNOT gate, controlled by three qubits rather than a single qubit.

In[ ]:= `qc = QuantumCircuit[CNOT[S{}1, 2, 3], S[4⊖⊖⊖`

Out[ ]=

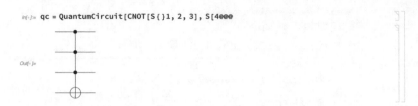

Here is the explicit expression of the multi-qubit CNOT gate in terms of the Pauli operators.

In[ ]:= `op = ExpressionFor[qc]`

Out[ ]= $\dfrac{7}{8} + \dfrac{1}{8}\,S_1^z S_2^z + \dfrac{1}{8}\,S_1^z S_3^z + \dfrac{1}{8}\,S_1^z S_4^x + \dfrac{1}{8}\,S_2^z S_3^z + \dfrac{1}{8}\,S_2^z S_4^x + \dfrac{1}{8}\,S_3^z S_4^x - \dfrac{1}{8}\,S_1^z S_2^z S_3^z -$
$\dfrac{1}{8}\,S_1^z S_2^z S_4^x - \dfrac{1}{8}\,S_1^z S_3^z S_4^x - \dfrac{1}{8}\,S_2^z S_3^z S_4^x + \dfrac{1}{8}\,S_1^z S_2^z S_3^z S_4^x - \dfrac{S_1^z}{8} - \dfrac{S_2^z}{8} - \dfrac{S_3^z}{8} - \dfrac{S_4^x}{8}$

Here we compare it with the expression explicitly constructed in terms of a projection operator.

In[ ]:= `prj = Multiply @@ S[{1, 2, 3}, 11];`
`new = (1 - prj) + prj ** S[4, 1]`
`op - new // Elaborate`

Out[ ]= $1 + \dfrac{1}{8}\,S_1^z S_2^z + \dfrac{1}{8}\,S_1^z S_3^z + \dfrac{1}{8}\,S_1^z S_4^x + \dfrac{1}{8}\,S_2^z S_3^z + \dfrac{1}{8}\,S_2^z S_4^x + \dfrac{1}{8}\,S_3^z S_4^x - \dfrac{1}{8}\,S_1^z S_2^z S_3^z -$
$\dfrac{1}{8}\,S_1^z S_2^z S_4^x - \dfrac{1}{8}\,S_1^z S_3^z S_4^x - \dfrac{1}{8}\,S_2^z S_3^z S_4^x + \dfrac{1}{8}\,S_1^z S_2^z S_3^z S_4^x - \dfrac{S_1^z}{8} - \dfrac{S_2^z}{8} + \dfrac{1}{4}\,\big(|1\rangle\langle 1|\big)_{S_3} - \dfrac{S_4^x}{8}$

Out[ ]= 0

An efficient implementation of a multi-qubit CNOT gate uses additional qubits not directly involved in the gate operation itself as in the following quantum circuit (Barenco et al. 1995):

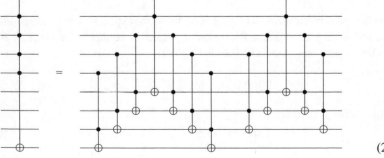

(2.83)

It is emphasized that the quantum states of the extra qubits should not be confused with the so-called "ancillary qubits" in the sense that they do not have to be initialized in a certain fixed quantum state and their state is not altered. The desired gate operation is performed properly regardless of the initial state of the extra qubits and their quantum state is restored after the gate operation.

Here we demonstrate a multi-qubit CNOT gate.

```
$n = 4;
$m = 2;
cc = Range[$n-;
aa = Range[$n] 1, $n] $m-;
tt = $n] $m] 1;
```

In[ ]:= `qc = QuantumCircuit[CNOT[S[cc], S[tt]], "Visible" → S[aa]]`

Out[ ]=

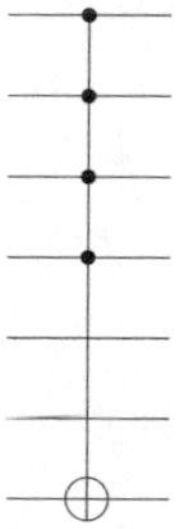

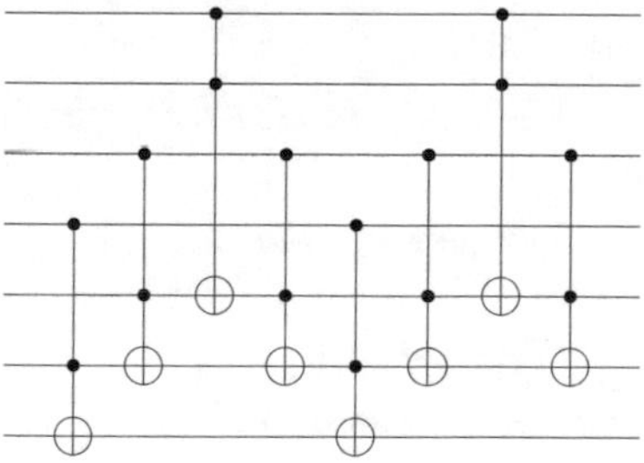

In[ ]:= ```
tofa = Table[Toffoli[S[$n]] j { 1-, S[$n { $m] j { 1-, S[tt] j { 1--, }j, 1, $m+-;
tofb = Toffoli[S[1 , S[2-, S[$n { 1--;
tofc = Rest@tofa;
new = QuantumCircuit[
    Sequence @@ Flatten@}tofa, tofb, Reverse@tofa, tofc, tofb, Reverse@tofc+-
```

Out[ ]=

In[ ]:= `Timing=Elaborate=qc [ new]]`

Out[ ]= `{11.6545, 0}`

In the quantum circuit shown in (2.83), we have introduced the *Toffoli gate*, another special case of multi-qubit CNOT gates denoted by the quantum circuit element
```

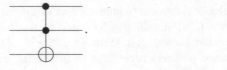

$$(2.84)$$

The Toffoli gate has attracted interest because it is universal for classical reversible computation. Unfortunately, however, it is not universal for quantum computation.

This is a quantum circuit model of the Toffoli gate.

In[ ]:= `qc = QuantumCircuit[Toffoli[S[1], S[2], S[3]]]`
`toff = ExpressionFor[qc]`

Out[ ]=

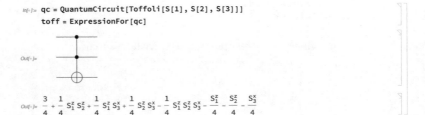

Out[ ]= $\dfrac{3}{4} + \dfrac{1}{4} S_1^z S_2^z + \dfrac{1}{4} S_1^z S_3^x + \dfrac{1}{4} S_2^z S_3^x - \dfrac{1}{4} S_1^z S_2^z S_3^x - \dfrac{S_1^z}{4} - \dfrac{S_2^z}{4} - \dfrac{S_3^x}{4}$

It can be implemented by a combination of two-qubit gates.

In[ ]:= `gray = GrayControlledU[S]@1, 2+, S[3, 1{{;`
`qc = QuantumCircuit[gray{`
`expr = ExpressionFor[qc{`
`toff } expr`

Out[ ]=

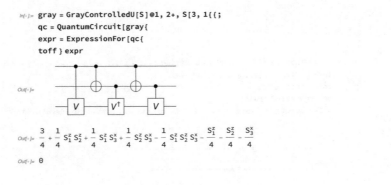

Out[ ]= $\dfrac{3}{4} + \dfrac{1}{4} S_1^z S_2^z + \dfrac{1}{4} S_1^z S_3^x + \dfrac{1}{4} S_2^z S_3^x - \dfrac{1}{4} S_1^z S_2^z S_3^x - \dfrac{S_1^z}{4} - \dfrac{S_2^z}{4} - \dfrac{S_3^x}{4}$

Out[ ]= 0

Here $V + \sqrt{X}$, and its matrix representation looks like this.

In[ ]:= `matV = MatrixPower[ThePauli[1], 1 / 2];`
`matV * 2 / (1 + I) // MatrixForm`
`opV = Elaborate@ExpressionFor[matV, S[3]]`

Out[ ]//MatrixForm=
$\begin{pmatrix} 1 & +i \\ +i & 1 \end{pmatrix}$

Out[ ]= $\left(\dfrac{1}{2} - \dfrac{i}{2}\right) - \left(\dfrac{1}{2} + \dfrac{i}{2}\right) S_3^x$

Reversing the above circuit gives the identical result.

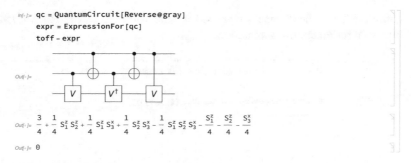

```
In[·]:= qc = QuantumCircuit[Reverse@gray]
        expr = ExpressionFor[qc]
        toff - expr
```

$$Out[·]:= \frac{3}{4} + \frac{1}{4} S_1^z S_2^z + \frac{1}{4} S_1^z S_3^x + \frac{1}{4} S_2^z S_3^x - \frac{1}{4} S_1^z S_2^z S_3^x - \frac{S_1^z}{4} - \frac{S_2^z}{4} - \frac{S_3^x}{4}$$

```
Out[·]:= 0
```

As noted by Smolin & DiVincenzo (1996), it can also be reordered as follows, and this rearrangement is useful in optimizing the *Fredkin gate*, another universal gate for classical reversible computation.

This shows another slightly different implementation of the Toffoli gate.

```
In[·]:= qc = QuantumCircuit[Permute[gray, Cycles[{{4, 3, 2, 1}}]]]
        expr = ExpressionFor[qc]
        toff - expr
```

$$Out[·]:= \frac{3}{4} + \frac{1}{4} S_1^z S_2^z + \frac{1}{4} S_1^z S_3^x + \frac{1}{4} S_2^z S_3^x - \frac{1}{4} S_1^z S_2^z S_3^x - \frac{S_1^z}{4} - \frac{S_2^z}{4} - \frac{S_3^x}{4}$$

```
Out[·]:= 0
```

The Fredkin gate "swaps" the states of the two target qubits when the control qubit is set in the state $|1\rangle$, as depicted by the following two equivalent quantum circuit elements

$$\tag{2.85}$$

In fact, the relation between the SWAP gate and the CNOT gate suggests that there is another equivalent quantum circuit of the Fredkin gate,

$$\tag{2.86}$$

which is simpler than the one given above. This equivalent quantum circuit was used by Smolin & DiVincenzo (1996) for an efficient implementation of the Fredkin gate in terms of the elementary gates: Like the Toffoli gate, the Fredkin gate is not universal for quantum computation.

This shows the quantum circuit model of the quantum Fredkin gate.

In[ ]:= `qc = QuantumCircuit[Fredkin[S[1], S[2], S[3]]]`

Out[ ]=

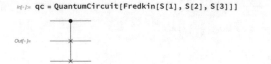

This is an explicit operator expression of the Fredkin gate in terms of the Pauli operators.

In[ ]:= `op = Elaborate[qc]`

Out[ ]= $\dfrac{3}{4} + \dfrac{1}{4} S_2^x S_3^x + \dfrac{1}{4} S_2^y S_3^y + \dfrac{1}{4} S_2^z S_3^z - \dfrac{1}{4} S_1^z S_2^x S_3^x - \dfrac{1}{4} S_1^z S_2^y S_3^y - \dfrac{1}{4} S_1^z S_2^z S_3^z + \dfrac{S_1^z}{4}$

This is the matrix representation of the Fredkin gate in the logical basis.

In[ ]:= `mat = Matrix[qc];`
`mat // MatrixForm`

Out[ ]//MatrixForm=

$$\begin{pmatrix} 0 & 1 & 1 & 1 & 1 & 1 & 1 & 1 \\ 1 & 0 & 1 & 1 & 1 & 1 & 1 & 1 \\ 1 & 1 & 0 & 1 & 1 & 1 & 1 & 1 \\ 1 & 1 & 1 & 0 & 1 & 1 & 1 & 1 \\ 1 & 1 & 1 & 1 & 0 & 1 & 1 & 1 \\ 1 & 1 & 1 & 1 & 1 & 1 & 0 & 1 \\ 1 & 1 & 1 & 1 & 1 & 0 & 1 & 1 \\ 1 & 1 & 1 & 1 & 1 & 1 & 1 & 0 \end{pmatrix}$$

The Fredkin gate is equivalent to a combination of three Toffoli gates.

In[ ]:= `qc2 = QuantumCircuit[`
`    Toffoli[S[1], S[2], S[3]],`
`    Toffoli[S[1], S[3], S[2]],`
`    Toffoli[S[1], S[2], S[3]]]`

Out[ ]=

This is an explicit operator expression.

In[ ]:= `op2 = Elaborate[qc2]`

Out[ ]= $\dfrac{3}{4} + \dfrac{1}{4} S_2^x S_3^x + \dfrac{1}{4} S_2^y S_3^y + \dfrac{1}{4} S_2^z S_3^z - \dfrac{1}{4} S_1^z S_2^x S_3^x - \dfrac{1}{4} S_1^z S_2^y S_3^y - \dfrac{1}{4} S_1^z S_2^z S_3^z + \dfrac{S_1^z}{4}$

In[ ]:= `op - op2`

Out[ ]= `0`

In fact, the relation between the SWAP gate and the CNOT gate suggests that there is another equivalent quantum circuit model of the Fredkin gate.

```
In[·]:= new = QuantumCircuit[
         CNOT[S[3], S[2]],
         Toffoli[S[1], S[2], S[3]],
         CNOT[S[3], S[2]]
         ]
```

Out[·]=

```
In[·]:= El - abo // rteqcnewb
Out[·]= 0
```

2.4   Universal Quantum Computation

In classical computation, it is known that a finite set of logic gates—typically including AND, OR, and NOT—is sufficient to calculate any binary function. The set is said to be *universal* for classical computation. One can ask whether there exists a universal set of elementary quantum logic gates for quantum computation that enables to implement any arbitrary quantum unitary operation? In this section, we examine this question.

Consider a system of n qubits. Suppose that we want to implement an arbitrary unitary operation $\hat{U}$ on the n-qubit register. We start by decomposing $\hat{U}$ into a set of two-level unitary transformations, which we discussed for the case of two-qubit systems in Sect. 2.2.3.

Consider a three-qubit system, and an arbitrary unitary operation on it.

```
In[·]:= $n = 3;
        SS = S[Range[$n], None];
        mat = RandomUnitary[2^$n];
        mat[[ ;; 3, ;; 3]] // MatrixForm
Out[·]//MatrixForm=
```

$$\begin{pmatrix} +0.491938 - 0.435618\,i & 0.0512628 + 0.492749\,i & 0.483397 - 0.839992\,i \\ 0.0306179 + 0.0382188\,i & 0.0790579 - 0.318489\,i & 0.185075 + 0.0876545\,i \\ 0.0785304 + 0.721224\,i & 0.047195 + 0.499233\,i & 0.708375 - 0.0309559\,i \end{pmatrix}$$

```
In[·]:= op = ExpressionFor[mat, SS{;
        qc = QuantumCircuit[}op, "Label" → "U"]{
Out[·]=
```

U

```
In[ ]:= twl = TwoLevelDecomposition[mat];
        twl[[3]] // Matrix // MatrixForm
```

Out[ ]//MatrixForm=

$$
\begin{pmatrix}
1 & 0 & 0 & 0 & 0 & 0 & 0 & 0 \\
0 & 1 & 0 & 0 & 0 & 0 & 0 & 0 \\
0 & 0 & 1 & 0 & 0 & 0 & 0 & 0 \\
0 & 0 & 0 & 1 & 0 & 0 & 0 & 0 \\
0 & 0 & 0 & 0 & 1 & 0 & 0 & 0 \\
0 & 0 & 0 & 0 & 0 & 0.151664 + 0.402111\,i & 0.902942 & 0 \\
0 & 0 & 0 & 0 & 0 & + 0.902942 & 0.151664 - 0.402111\,i & 0 \\
0 & 0 & 0 & 0 & 0 & 0 & 0 & 1
\end{pmatrix}
$$

Now a remaining question is to implement the two-level unitary transformation in terms of elementary gates.

Let us consider a particular two-level unitary matrix on a three-qubit system. We want to implement it in terms of elementary quantum logic gates.

```
In[ ]:= x = 4; y = 5;
        U = TheRotation[Pi ] 3, 2 0;
        mat = Matrix{ TwoLevelU[U, }x, y/, 2^$n0;
        mat ]] MatrixForm
```

Out[ ]//MatrixForm=

$$
\begin{pmatrix}
1 & 0 & 0 & 0 & 0 & 0 & 0 & 0 \\
0 & 1 & 0 & 0 & 0 & 0 & 0 & 0 \\
0 & 0 & 1 & 0 & 0 & 0 & 0 & 0 \\
0 & 0 & 0 & \frac{\sqrt{3}}{2} & +\frac{1}{2} & 0 & 0 & 0 \\
0 & 0 & 0 & \frac{1}{2} & \frac{\sqrt{3}}{2} & 0 & 0 & 0 \\
0 & 0 & 0 & 0 & 0 & 1 & 0 & 0 \\
0 & 0 & 0 & 0 & 0 & 0 & 1 & 0 \\
0 & 0 & 0 & 0 & 0 & 0 & 0 & 1
\end{pmatrix}
$$

```
In[ ]:= op = ExpressionFor[mat, SS]
```

$$
Out[ ]= \frac{1}{8} \times \left(6 - \sqrt{3}\right) - \frac{1}{8} \times \left(2 + \sqrt{3}\right) S_1^z S_2^z -
$$

$$
\frac{1}{8} \times \left(2 + \sqrt{3}\right) S_1^z S_3^z - \frac{1}{8} \times \left(+2 - \sqrt{3}\right) S_2^z S_3^z + \frac{1}{2} S_1^- S_2^+ S_3^- - \frac{1}{2} S_1^+ S_2^- S_3
$$

Our implementation is based on the Gray code sequence. Notice the function **Reverse**.

```
In[ ]:= gates = Flatten[GrayTwoLevelU{U, }x, y], SS0
```

$$
Out[ ]= \Big\{ S_1^x, \text{CNOT}[\{S_1, S_2\}, \{S_3\}], S_1^x, S_3^x, \text{CNOT}[\{S_2, S_3\}, \{S_1\}],
$$

$$
S_3^x, \text{CNOT}[\{S_1, S_2\}, \{S_3\}], \text{CNOT}[\{S_1, S_3\}, \{S_2\}], S_2^x,
$$

$$
\text{ControlledU}\Big[\{S_1, S_2\}, \frac{\sqrt{3}}{2} + \frac{i\,S_3^y}{2}, \text{Label} \to U\Big], S_2^x, \text{CNOT}[\{S_1, S_3\}, \{S_2\}],
$$

$$
\text{CNOT}[\{S_1, S_2\}, \{S_3\}], S_3^x, \text{CNOT}[\{S_2, S_3\}, \{S_1\}], S_3^x, S_1^x, \text{CNOT}[\{S_1, S_2\}, \{S_3\}], S_1^x \Big\}
$$

```
In[ ]:= new = Apply[Dot, Matrix[#, SS] & /@ Elaborate /@ gates] // Simplify;
        new // MatrixForm
```

Out[ ]//MatrixForm=

$$
\begin{pmatrix}
1 & 0 & 0 & 0 & 0 & 0 & 0 & 0 \\
0 & 1 & 0 & 0 & 0 & 0 & 0 & 0 \\
0 & 0 & 1 & 0 & 0 & 0 & 0 & 0 \\
0 & 0 & 0 & \frac{\sqrt{3}}{2} & +\frac{1}{2} & 0 & 0 & 0 \\
0 & 0 & 0 & \frac{1}{2} & \frac{\sqrt{3}}{2} & 0 & 0 & 0 \\
0 & 0 & 0 & 0 & 0 & 1 & 0 & 0 \\
0 & 0 & 0 & 0 & 0 & 0 & 1 & 0 \\
0 & 0 & 0 & 0 & 0 & 0 & 0 & 1
\end{pmatrix}
$$

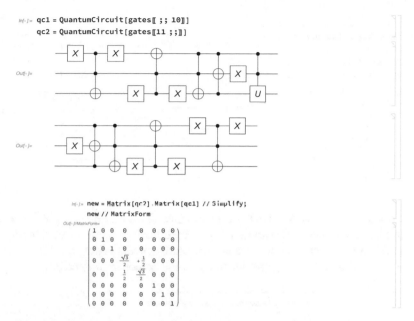

So far, we have seen that one can implement any unitary operation combining the CNOT and single-qubit gates. A subtle issue is that single-qubit gates form a continuum of unitary operators. It is impractical to implement them to an infinite accuracy. Fortunately, it is known that a discrete set of gates is sufficient to perform universal quantum computation. The Hadamard, quadrant phase, and octant phase, and CNOT gates form a universal set of gates in the sense that an arbitrary gate on n qubits can be approximated to arbitrary accuracy using a quantum circuit composed of only such gates. Furthermore, the set remains universal if octant phase gates are replaced with Toffoli gates.

2.5   Measurements

We conclude this chapter with a few discussions on measurement. In quantum computers, measurement is assumed to be performed independently on individual qubits in the logical basis, $\{|x\rangle : x = 0, 1, \ldots, 2^n - 1\}$.

What if a measurement in another basis, say, $\left\{|\alpha_x\rangle = \hat{U}|x\rangle\right\}$ is required? We require that the input state $|\alpha_x\rangle$ should end up with the logical basis state $|x\rangle$ with unit probability so that the measurement yields outcome x. This process is described by the measurement operators $\hat{M}_x := |x\rangle\langle\alpha_x| = |x\rangle\langle x|\hat{U}^\dagger$. Evidently they satisfy the condition $\sum_x \hat{M}_x^\dagger \hat{M}_x = \hat{I}$ for measurement operators (see Postulate 1.3). Now we note that the operators $\hat{P}_x := |x\rangle\langle x|$ describe nothing but the measurement in the logical basis. This implies that by simply applying the inverse unitary operation $\hat{U}^\dagger$ before the measurement, the measurement in the new basis $\{|\alpha_x\rangle\}$ can be achieved

Fig. 2.6 Measurement in a basis $\{|\alpha_x\rangle\}$ other than the logical basis. The unitary operator $\hat{U}$ here corresponds to the basis change $|\alpha_y\rangle = \hat{U}|y\rangle = \sum_x |x\rangle\langle x|\alpha_y\rangle$

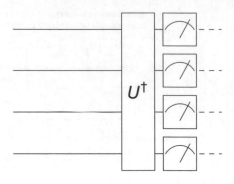

through measurement in the logical basis (see Fig. 2.6). For example, suppose that a qubit is in the state $|\psi\rangle = |0\rangle c_0 + |1\rangle c_1$. By default, a measurement is assumed to be in the logical basis and the measurement statistics reflect the probability distribution $P_0 = |c_0|^2$ and $P_1 = |c_1|^2$ as illustrated in the demonstration below.

First consider a measurement in the logical basis.

```
In[ ]:= Let=Complex, c[
    vec { ProductState=S=1[ → }c=0[, c=1[] [;
    qc { QuantumCircuit=vec, "Spacer", Measurement=S=1[[, "PortSize" → }2, 0.2] [
```

$Out[ ]=$ $|0\rangle c_0 - |1\rangle c_1$ ────────[⟋]---

```
In[ ]:= Block=
    [c, data{,
    c=0} ] 1 / 2;
    c=1} ] Sqrt=3} / 2;
    data ] Table=out ] ExpressionFor=qc}; Readout=out, S=1}}, 1000};
    Histogram=data, FrameLabel → ["Readout", "Counts"{}
    }
```

$Out[ ]=$ [histogram plot with Counts on y-axis (0 to 700) and Readout on x-axis (−0.5 to 1.5)]

We now want to make a measurement on the eigenbasis $\{|\pm\rangle\}$ of the Pauli X operator. In this case, it is the Hadamard gate that gives the desired basis change. In the new basis, the state vector $|\psi\rangle$ is given by

$$|\psi\rangle = |+\rangle \frac{c_0 + c_1}{\sqrt{2}} + |-\rangle \frac{c_0 - c_1}{\sqrt{2}}, \tag{2.87}$$

Now the measurement statistics are in accordance with the probability distribution $P_\pm - |c_0 \pm c_1|^2/2$.

Now consider a measurement in the eigen-basis of the Pauli X operator.

In[ ]:= `Let=Complex, c[`
`  vec { ProductState=S=1[ → }c=0[, c=1[] [;`
`  qc { QuantumCircuit=vec, S=1, 6[, Measurement=S=1[[, "PortSize" → }2, 0.2] [`

Out[ ]= $|0\rangle c_0 - |1\rangle c_1$ — [H] [⌐] - - -

In[ ]:= `Block=`
`    [c, data[,`
`     c=0} ] 1 / 2;`
`     c=1} ] Sqrt=3} / 2;`
`     data ] Table=out ] ExpressionFor=qc}; Readout=out, S=1}}, 1000};`
`     Histogram=data, FrameLabel → ["Readout", "Counts"{}`
`    }`

Out[ ]=

Another interesting example is the *Bell measurement*. The Bell measurement is a measurement on two qubits in the basis of Bell states,

$$|\beta_0\rangle = \frac{|00\rangle + |11\rangle}{\sqrt{2}}$$

$$|\beta_1\rangle = \frac{|01\rangle + |10\rangle}{\sqrt{2}}$$

$$|\beta_2\rangle = \frac{|01\rangle - |10\rangle}{\sqrt{2}} \tag{2.88}$$

$$|\beta_3\rangle = \frac{|00\rangle - |11\rangle}{\sqrt{2}}$$

Recall that the Bell states can be generated from the logical basis states by the so-called quantum entangler circuit, a combination of the Hadamard gate and the CNOT gate discussed in Sect. 2.2.1. Therefore, the Bell measurement can be achieved by applying the inverse of the entangling operation.

This is the "disentangler" quantum circuit, which is the inverse of the entangler quantum circuit

In[·]:= `disentangler = QuantumCircuit[CNOT[S[1], S[2]], S[1, 6]]`

Out[·]=

This shows that the disentangler quantum circuit maps the Bell states into the logical basis states.

In[·]:= `op = ExpressionFor[disentangler@;`
`bs = BellState[S{}1, 2];`
`Thread[bs → op ** bs@ // LogicalForm // TableForm`

Out[·]//TableForm=

$$\frac{|0_{S_1}0_{S_2}\rangle + |1_{S_1}1_{S_2}\rangle}{\sqrt{2}} \rightarrow |0_{S_1}0_{S_2}\rangle$$

$$\frac{|0_{S_1}1_{S_2}\rangle + |1_{S_1}0_{S_2}\rangle}{\sqrt{2}} \rightarrow |0_{S_1}1_{S_2}\rangle$$

$$\frac{|0_{S_1}1_{S_2}\rangle - |1_{S_1}0_{S_2}\rangle}{\sqrt{2}} \rightarrow |1_{S_1}1_{S_2}\rangle$$

$$\frac{|0_{S_1}0_{S_2}\rangle - |1_{S_1}1_{S_2}\rangle}{\sqrt{2}} \rightarrow |1_{S_1}0_{S_2}\rangle$$

More sophisticated measurements may be necessary for fast quantum algorithms and quantum error corrections. A common example is the quantum phase estimation, which is one of the core parts of the quantum factorization algorithm. We will discuss it in Sect. 4.4.

Problems

2.1 Let $\hat{\Phi}(\phi)$ be the *phase gate*, which gives rise to a *relative* phase shift by $\phi \in [0, 2\pi)$,

$$\hat{\Phi}(\phi) : |0\rangle \mapsto |0\rangle , \quad |1\rangle \mapsto |1\rangle e^{i\phi} . \tag{2.89}$$

(a) Show that on a n-qubit quantum register,

$$[\hat{\Phi}(\phi)]^{\otimes n} |x\rangle = |x\rangle e^{i\phi(x_1 + \cdots + x_n)} \tag{2.90}$$

for any $x = 0, 1, \ldots, 2^n - 1$.

(b) Explicitly evaluate the state $[\hat{\Phi}(\phi)]^{\otimes n} \hat{H}^{\otimes n} |0\rangle$, where $\hat{H}$ is the Hadamard gate.

2.2 Let $\hat{U}_x(\phi)$ be the rotation around the x-axis by angle ϕ on a single qubit. Explicitly analyze and evaluate the following quantum circuit

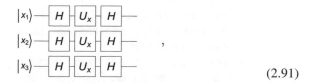

$$(2.91)$$

where $|x\rangle := |x_1\rangle \otimes |x_2\rangle \otimes |x_3\rangle$ is a 3-qubit logical basis state and the labels "H" and "U_x" indicate the Hadamard gate $\hat{H}$ and the rotation gate $\hat{U}_x(\phi)$, respectively. Generalize the result and show that

$$\hat{H}^{\otimes n}[\hat{U}_x(\phi)]^{\otimes n}\hat{H}^{\otimes n}|x\rangle = e^{-i\phi n/2}|x\rangle\, e^{i\phi(x_1+\cdots x_n)} \qquad (2.92)$$

for any $x = 0, 1, \ldots, 2^n - 1$.

2.3 Let $\hat{S}^\mu$ be the Pauli operators. Show that

$$e^{i\hat{S}^\mu \phi/2}\hat{S}^\nu e^{-i\hat{S}^\mu \phi/2} = \hat{S}^\nu \cos(\phi) - \sum_\lambda \hat{S}^\lambda \epsilon_{\lambda\mu\nu}\sin(\phi). \qquad (2.93)$$

for all $\mu, \nu = x, y, z$ and $\mu \neq \nu$.

2.4 The lever on Bob's quantum computer is stuck in the "forward" position, so it can only perform CNOT gates controlled by qubit 1 on qubit 2 (target). His computer can still perform single-qubit operations normally. How can he perform a CNOT gate controlled by qubit 2 on qubit 1?[3]

2.5 Consider the following quantum circuit with n qubits:

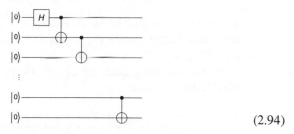

$$(2.94)$$

Show that it generates the Greenberger-Horne-Zeilinger state in Eq. (2.40), identical to the one from the quantum circuit in Fig. 2.5b.

2.6 Consider a quantum register of four qubits.

(a) Analyze the following quantum circuit

$$(2.95)$$

---

[3] This problem was published in Gottesman (1999).

and evaluate the resulting state $|\Psi\rangle$ explicitly. The state is a so-called *cluster state* or *graph state*, a crucial resource in the measurement-based quantum computation (see Sect. 3.4).

(b) Show that in the state $|\Psi\rangle$ from (b), every qubit is *maximally entangled* with the rest of the qubits. That is, the *reduced density matrix* $\hat{\rho}_j$ of the jth qubit, $\hat{\rho}_j := \text{Tr}_{k\neq j} |\Psi\rangle \langle\Psi|$, is given by $\hat{\rho}_j = \hat{I}/2$, exhibiting coherence in no basis.

2.7 Suppose that a qubit is known to be in one of the two eigenstates of the unitary operator

$$\hat{U} = \hat{\sigma}^0 \cos(\phi/2) - i\hat{\sigma}^x \sin(\phi/2) \tag{2.96}$$

with the angle ϕ *known*. Construct a quantum circuit to figure out the unknown state using an additional qubit.

Hint: Use a controlled-unitary gate to acquire the one-bit information. This is a simplified version of the quantum phase estimation procedure (see Sect. 4.4), but it can be worked out without resorting to it.

2.8 Suppose that a two-qubit system is known to be in one of the four eigenstates of the unitary operator

$$\hat{U} = e^{i\phi} \left(|0\rangle \langle 0| + i |1\rangle \langle 1| - |2\rangle \langle 2| - i |3\rangle \langle 3|\right). \tag{2.97}$$

Construct a quantum circuit to figure out the unknown state using two additional qubits.

Hint: Use the property (2.18) of the Hadamard gate and those of the controlled-unitary gates, where a unitary operator acts on a two-qubit system controlled by a single qubit.

2.9 Consider the following quantum circuit consisting of the CNOT gates on an n-qubit quantum register

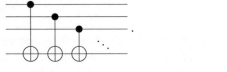

$$\tag{2.98}$$

(a) Find the output state for the input of a logical basis state $|x_1\rangle \otimes \cdots \otimes |x_n\rangle \in \mathcal{S}^{\otimes n}$.

(b) Find the output state for the input state $|+\rangle \otimes \cdots \otimes |+\rangle \otimes |-\rangle$, where $|\pm\rangle := (|0\rangle \pm |1\rangle)/\sqrt{2}$.

2.10 Let $\hat{U}$ be a unitary operator on a single qubit. Show that the following three statements are equivalent:

(a) There exist unitary gates $\hat{A}$ and $\hat{B}$ such that

$$(2.99)$$

(b) There exists a unitary operator $\hat{W}$ such that $\hat{U} = \hat{W}\hat{Z}\hat{W}^\dagger$.

(c) $\mathrm{Tr}\hat{U} = 0$ and $\det\hat{U} = -1$.

2.11 Let $P(i \leftrightarrow j)$ be the matrix exchanging the ith and jth rows (columns) of vectors/matrices. For example, on a four-dimensional space,

$$P(1 \leftrightarrow 2) = \begin{bmatrix} 0 & 1 & & \\ 1 & 0 & & \\ & & 1 & \\ & & & 1 \end{bmatrix}. \qquad (2.100)$$

$\hat{P}(i \leftrightarrow j)$ is the corresponding operator.

(a) Find a quantum circuit for $\hat{P}(2 \leftrightarrow 3)$ on a two-qubit system (four-dimensional vector space $S \otimes S$) using CNOT gates only.

(b) Find a quantum circuit for $\hat{P}(2 \leftrightarrow 3)$ on a three-qubit system ($S^{\otimes 3}$) using multi-qubit CNOT gates only.

(c) Find a quantum circuit for $\hat{P}(4 \leftrightarrow 7)$ on a three-qubit system ($S^{\otimes 3}$) using multi-qubit CNOT gates only.

2.12 Let $\hat{U} = e^{i\phi}\hat{I}$ be a single-qubit unitary operator that *globally* shifts the phase by ϕ. Also define

$$\hat{T} = |0\rangle\langle 0| + e^{i\phi}|1\rangle\langle 1| \doteq \begin{bmatrix} 1 & 0 \\ 0 & e^{i\phi} \end{bmatrix}, \qquad (2.101)$$

which shifts the *relative* phase by ϕ.

(a) Show that

$$(2.102)$$

where the labels "U" and "T" denote the unitary operators $\hat{U}$ and $\hat{T}$, respectively.

(b) Show also that

$$
\begin{array}{ccc}
\vcenter{\hbox{}} = \vcenter{\hbox{}} = \vcenter{\hbox{}} = \vcenter{\hbox{}} \, .
\end{array}
$$

$$(2.103)$$

Chapter 3
Realizations of Quantum Computers

In the previous chapter, we have seen how quantum computation works under the assumption that elementary quantum logic gates are available. But how can one build a quantum computer, a machine, that allows such quantum logic gates? Quantum computers are physical systems and the implementation of all quantum logic gates is governed by the laws of physics. In this chapter, we discuss the basic physical principles that are directly involved in the implementation of quantum logic gates. Through the course of the discussion, we will find some basic conditions and requirements that one has to fulfill to build a quantum computer.

By now, there are many quantum computer architectures that have not only been proposed and tested at the research level, but which are also actually running. However, understanding each architecture requires a certain level of knowledge regarding the physical systems. For example, to understand a quantum computer based on superconducting circuits (Fig. 3.1), one has to first understand the superconductivity, the Josephson effect, the flux quantization, the Josephson inductance (a sort of non-linear kinetic inductance), and the interaction of superconducting circuits with electromagnetic fields. Such discussions often hinder access to the essential part of the operating principle of a quantum computer, and are beyond the scope of this book.

Here we consider an idealistic and minimal quantum system that is suitable for quantum computation, and we discuss how to control it to implement the desired quantum logic gates. It is certainly impractical, yet it will highlight the key requirements when one wants to actually develop a quantum computer based on practical systems and devices. Through the discussions, we will indicate how the relevant parts are related to actual quantum computer architectures.

Supplementary Information The online version contains supplementary material available at https://doi.org/10.1007/978-3-030-91214-7_3.

M.-S. Choi, *A Quantum Computation Workbook*,
https://doi.org/10.1007/978-3-030-91214-7_3

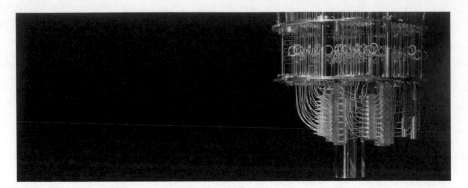

Fig. 3.1 IBM Q quantum computer based on superconducting circuits.   Photo by Graham Carlow/IBM

3.1   Quantum Bits

We have already noted several times that the building blocks and basic computational units of a quantum computer are qubits. Ideally, a qubit is associated with a two-dimensional Hilbert space. In reality, the Hilbert space for any realistic system is infinite-dimensional, and a qubit usually refers to *certain degrees of freedom* that are relatively independent of other degrees of freedom. For example, the spin of an electron or the polarization of a photon is represented exactly in a two-dimensional Hilbert space. In many cases, a qubit may also refer to a *certain two-dimensional subspace* of a larger Hilbert space that is decoupled or relatively well separated from the rest. For example, a superconducting charge qubit refers to the two lowest-energy charging states in a small superconducting island hundreds of nanometers in lateral size.

However, a well-defined two-dimensional Hilbert space (or subspace) does not necessarily mean that the degrees of freedom in question qualify it as a qubit. For example, consider the spin of a neutron. Although its Hilbert space dimension is certainly two, you recognize that it can hardly be used for quantum computation. It is hard to isolate a neutron, and even more so to manipulate its spin in a reliable and tunable manner. Then what requirements should qubits—individually and as a whole collection—meet to build a practical quantum computer? Apart from the specific technical issues of specific systems, there are basic requirements–the so-called *DiVincenzo criteria*—to assess the potential of a particular architecture under consideration (DiVincenzo, 2000):

(a) The qubits should be well characterized and form a scalable system. For each qubit, the Hilbert space should be well defined in the sense mentioned above and its internal Hamiltonian including the parameters needs to be accurately known. The qubits must also allow for genuine interactions among them and maintain their characteristics up to a sufficiently large scale for practical computation.

(b) The qubits should allow initialization to a fixed logical basis state. Even though any quantum algorithm assumes superposition in the middle of the process, all computations must start from a known value. This straightforward requirement is the same even for classical computers. One of the common approaches for initialization is to cool down the system and wait for it to relax to the ground state. Another method is to perform a projective measurement in the logical basis so as for the state to collapse to the logical basis state corresponding to the measurement outcome.

(c) The qubits should maintain coherence long enough for the desired gate operations. The superposition between different logical basis states is a crucial difference distinguishing quantum computers from classical computers. Unfortunately, qubits are subject to various decoherence effects due to external control circuits and measuring devices and eventually lose quantumness. The system should maintain coherence during the desired gate operations to get a reliable result out of the computation.

(d) The system of qubits should allow a *universal* set of quantum gate operations. As discussed in Chap. 2, quantum computation aims to achieve a desired unitary transformation with a combination of certain elementary gate operations that are acting on a single qubit or two qubits at a time. Below we will discuss the physical implementations of those elementary quantum logic gates.

(e) The system should allow measurements in the logical basis. At the end of a computation, the result needs to be read out, and this is achieved by performing measurements on specific qubits. The capability of accurate measurement is called the *quantum efficiency*. An ideal measurement has 100% quantum efficiency. Less than 100% quantum efficiency in the measurements leads to a trade-off with other resources. For example, if a computation is desired with 97% reliability while measurements have only 90% quantum efficiency, then one must repeat the computation three times or more.

In the rest of this chapter, let us focus on the manipulation of quantum states of qubits, which naturally forms the largest part of quantum computation.

Consider a quantum computer consisting of n qubits. Let S_j ($j = 1, \ldots, n$) be the 2-dimensional Hilbert space associated with the jth single qubit. An ideal quantum computer would realize a Hamiltonian on $S_1 \otimes \cdots \otimes S_n$ of the form

$$\hat{H}(t) = \sum_j \sum_\mu B_j^\mu(t)\hat{S}_j^\mu + \sum_{ij} \sum_{\mu\nu} J_{ij}^{\mu\nu}(t)\hat{S}_i^\mu \hat{S}_j^\nu, \tag{3.1}$$

where $\hat{S}_j^\mu$ ($\mu = x, y, z$) are the Pauli operators on S_j (see Sect. 2.1.1).

The parameter B_j^μ directly controls the jth qubit. Physically, it plays the same role as the magnetic field on a spin. In practical systems, it may be hard to address single qubits individually. How freely one can manipulate single qubits strongly depends on how many of the parameters B_j^μ the system allows to accurately tune. See, for example, Sect. 3.2 below.

The control parameters $J_{ij}^{\mu\nu}$ describe the (hypothetical) exchange coupling between the ith and jth qubits. In principle, any type of interaction between two qubits can be used to implement a CNOT gate (see Sect. 3.2 for examples) although the actual implementation may need to couple qubits multiple times and require many additional single-qubit operations depending on the particular type of coupling. Therefore, accurate control of the coupling parameters $J_{ij}^{\mu\nu}$ between a specific pair of qubits is essential to achieve universal quantum computation. In practical systems, the coupling parameters $J_{ij}^{\mu\nu}$ are even more difficult to realize. First of all, in many architectures, the connectivity is seriously limited for qubits that are not in direct proximity to each other. Furthermore, dynamically turning on and off the coupling is often forbidden. In many cases, in order to achieve a sizable strength, the exchange couplings are kept turned on throughout the whole computation. Such difficulties and imperfections all contribute to errors in the computational output.

We will be denoting each qubit by the symbol S and accompanying indices .

```
Let[Qubit, S]
```

The Pauli operators are specified by the last index. For example, the Pauli operator S_j^x is denoted by S[j,1].

```
In[ ]:= S[j, 1]
```
$Out[ ]= S_j^x$

```
In[ ]:= S[j, 1] ** S[j, 2]
        S[j, 1] ** S[k, 2]
```
$Out[ ]= i\, S_j^z$

$Out[ ]= S_j^x\, S_k^y$

3.2 Dynamical Scheme

The dynamical scheme implements the desired quantum gates through a time-evolution operator governing the dynamics of the physical qubits in a quantum computer—hence the name. It is the most common scheme for quantum computation with the majority of quantum computers demonstrated so far based on this scheme. In this section, we briefly survey the elementary principles and common methods to achieve elementary single-qubit and two-qubit quantum gates in the dynamical scheme.

3.2.1 Implementation of Single-Qubit Gates

Conceptually, the most straightforward way to control a single qubit is to apply the static parameters B^x, B^y, and B^z for a certain period τ of operating time. We refer to them collectively by a vector $\boldsymbol{B} = (B^x, B^y, B^z)$. One can regard it as a fictitious magnetic field, but the dimension of $\boldsymbol{B}$ is energy unlike that for a real magnetic field. The Hamiltonian for the qubit is given by

$$\hat{H} = \frac{1}{2}\boldsymbol{B} \cdot \hat{\boldsymbol{S}}, \qquad (3.2)$$

where $\hat{\boldsymbol{S}} := (\hat{S}^x, \hat{S}^y, \hat{S}^z)$ is the vector of the Pauli operators on $\mathcal{S}$. The time evolution is then governed by the unitary operator

$$\hat{U}(t) = \exp(-it\hat{H}) = \exp(-it\boldsymbol{B} \cdot \hat{\boldsymbol{S}}/2) \qquad (3.3)$$

in accordance with Postulate 2. It has the same form as the single-qubit rotation operator in Eq. (2.23), describing the rotation in the Bloch space around the axis parallel to the vector $\boldsymbol{B}$ by angle $\phi(t) = |\boldsymbol{B}|t$. When the involved two-level system is a true spin, such a rotation in the Bloch space is called the *Larmor precession*.

Consider a qubit. Let us apply a (fictitious) magnetic field.

```
In[ ]:= Let[Real, B]
        opH = Dot[B@{1, 2, 3}, S@{1, 2, 3}] / 2
```

$$Out[ ]= \frac{1}{2}\left(B_1 \, S^x + B_2 \, S^y + B_3 \, S^z\right)$$

To be specific, we consider the case with the magnetic field between the x- and z-axis in the xz-plane. The factor of $1/\sqrt{2}$ is to set the energy (frequency) scale to unit.

```
In[ ]:= B[1] = B[3] = 1 / Sqrt[2]; B[2] = 0;
        opH
```

$$Out[ ]= \frac{1}{2}\left(\frac{S^x}{\sqrt{2}} + \frac{S^z}{\sqrt{2}}\right)$$

```
In[ ]:= Clear[opU]
        opU[t_] = MultiplyExp[-I t opH]
        opU[t] // Elaborate
```

$$Out[ ]= e^{-\frac{1}{2}it\left(\frac{S^x}{\sqrt{2}} + \frac{S^z}{\sqrt{2}}\right)}$$

$$Out[ ]= \cos\left[\frac{t}{2}\right] - \frac{i\,S^z\,\sin\left[\frac{t}{2}\right]}{\sqrt{2}} - \frac{i\,S^+\,\sin\left[\frac{t}{2}\right]}{\sqrt{2}} - \frac{i\,S^-\,\sin\left[\frac{t}{2}\right]}{\sqrt{2}}$$

```
In[ ]:= Clear[vec]
        vec[t_] = opU[t] ** Ket[] // Elaborate
```

$$Out[ ]= -\frac{i\,|1_S\rangle\,\sin\left[\frac{t}{2}\right]}{\sqrt{2}} + |-\rangle\left(\cos\left[\frac{t}{2}\right] - \frac{i\,\sin\left[\frac{t}{2}\right]}{\sqrt{2}}\right)$$

This illustrates the Larmor precession (with the qubit regarded as a "spin").

Technical Note: You need `Chop` because numerical errors sometimes give `0.+Ket[...]`, which cannot be handled properly by `BlochVector`.

In[·]:= `vv [ BlochVector ]= Table@Chop=vec@t{, }t, 0, 8, 0.5/{;`
`BlochSphere@}Red, Bead ]= vv/, ImageSize → Small{`

Out[·]=

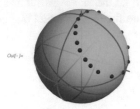

Let us now examine the probabilities for the qubit to be found in the logical basis states.

In[·]:= `prob[t_] = Abs[Normal@Matrix@vec[t]]^2`

Out[·]= $\left\{ \left| \cos\left[\frac{t}{2}\right] + \frac{i \sin\left[\frac{t}{2}\right]}{\sqrt{2}} \right|^2 , \frac{1}{2} \left| \sin\left[\frac{t}{2}\right] \right|^2 \right\}$

In[·]:= `Plot[Evaluate@prob[2 Pi t], {t, 0, 2},`
`FrameLabel → {"t / (2πB)", "Probability"}]`

Out[·]=

Although conceptually simple, the above method has limited applicability in many practical systems. For example, in the presence of other levels, one cannot apply the method selectively between the two levels in question. A more widely applicable method is to apply an oscillating field. Suppose that

$$B^x = B_\perp \cos(\omega t), \quad B^y = B_\perp \sin(\omega t), \quad B^z = B_0. \tag{3.4}$$

One can regard it as a fictitious magnetic field precessing around the z-axis with frequency ω. The Hamiltonian now depends on time, and it is given by

$$\hat{H}(t) = \frac{1}{2} B_\perp \left[\cos(\omega t)\hat{S}^x + \sin(\omega t)\hat{S}^y \right] + \frac{1}{2} B_0 \hat{S}^z. \tag{3.5}$$

Recalling the property in Eq. (2.25) of the Pauli operators as the generators of rotation, we observe that

$$\hat{U}_z^\dagger(\omega t)\hat{H}(t)\hat{U}_z(\omega t) = \frac{1}{2} B_\perp \hat{S}^x + \frac{1}{2} B_0 \hat{S}^z \tag{3.6}$$

does not depend on time any longer. This observation suggests that the dynamics look simpler in the rotating frame. Since the rotating frame is not an inertial frame, one has to take into account the non-inertial effect corresponding to the classical *inertial force*. The state vectors $|\psi(t)\rangle$ and $|\psi_R(t)\rangle$ in the lab and rotating frame, respectively, are related by

$$|\psi_R(t)\rangle = \hat{U}_z^\dagger(\omega t)|\psi(t)\rangle . \tag{3.7}$$

Putting it into the original Schrödinger equation for $|\psi(t)\rangle$,

$$i\partial_t|\psi\rangle = \hat{H}(t)|\psi\rangle , \tag{3.8}$$

and operating $\hat{U}_z^\dagger(t)$ from the left on both sides, one can get the Schrödinger equation in the rotating frame,

$$i\partial_t|\psi_R\rangle = \hat{H}_R|\psi_R\rangle , \tag{3.9}$$

where the Hamiltonian in the rotating frame is given by

$$\hat{H}_R(t) := U_z^\dagger(\omega t)\hat{H}(t)\hat{U}_z(\omega t) - \hat{U}_z^\dagger(\omega t)[i\partial_t\hat{U}_z(\omega t)] . \tag{3.10}$$

The second term in $\hat{H}_R$ is responsible for the non-inertial effect. As already expected in the above observation, the rotating-frame Hamiltonian does not depend on time any longer,

$$\hat{H}_R = \frac{1}{2}B_\perp\hat{S}^x + \frac{1}{2}(B_0 - \omega)\hat{S}^z . \tag{3.11}$$

The time-evolution operator in the rotating frame,

$$\hat{U}_R(t) = \exp(-it\hat{H}_R) \tag{3.12}$$

is *formally* the same as the one (3.3) in the lab frame. It describes a precession around the axis parallel to $\boldsymbol{B}_R := (B_\perp, 0, B_0 - \omega)$ with the frequency

$$\Omega_R := \sqrt{B_\perp^2 + (B_0 - \omega)^2} . \tag{3.13}$$

The precession in the rotating frame is called the *Rabi oscillation*, and the frequency in (3.13) is called the *Rabi frequency*. The fictitious magnetic field in the rotating frame, comparing Eqs. (3.2) and (3.10), is almost along the x-axis for $\omega \approx B_0$. That is, the time-evolution $\hat{U}_R(t)$ in the rotating frame corresponds to the rotation around the x-axis in the Bloch sphere. In this case, the qubit can make a full transition from the initial state $|0\rangle$ to the orthogonal state $|1\rangle$ by properly tuning the operation time and/or the parameter $B_\perp$. In this sense, when $\omega \approx B_0$, the system is said to be at *resonance*. As the driving frequency ω gets away off the resonance, the maximum transition probability becomes smaller and smaller. This resonance behavior allows to induce transitions between a selected pair of two levels among many others.

Let us now apply a time-dependent field. It precesses around the z-axis. Note that typically, $B_3 \gg B_1, B_2$.

```
In[·]:= ω = 2 Pi;
       B[1] = Cos[ω t];
       B[2] = Sin[ω t];
       B[3] = 1.1 ω;  (* near resonance *)
       Graphics3D[{Blue, Table[Arrow@Tube@{{0, 0, 0}, B@{1, 2, 3}}, {t, 0, 1, 0.1}]},
         PlotRange → ω {{-1, 1}, {-1, 1}, {-1, 1}}]
```

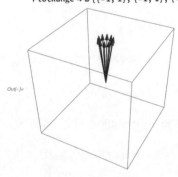

Out[·]=

```
In[·]:= opH [ Dot]B=@1, 2, 3{, S=@1, 2, 3{} / 2
```
$$Out[·]= \frac{1}{2} \left(\text{Cos}[2 \pi t] \, S^x + 6.9115 \, S^z + S^y \, \text{Sin}[2 \pi t] \right)$$

```
In[·]:= opHR =
         Rotation[-ω t, S[3]] ** opH ** Rotation[ω t, S[3]] - ω S[3] / 2 // Elaborate // Chop
```
$$Out[·]= \frac{S^x}{2} + 0.314159 \, S^z$$

```
In[·]:= opUR[t_] = MultiplyExp[-I t opHR]
```
$$Out[·]= e^{-i\, t\, \left(\frac{S^x}{2} + 0.314159\, S^z \right)}$$

```
       Clear[vec]
       vec[t_] = opUR[t] ** Ket[] // Elaborate;
```

This illustrates the precession in the rotating frame, which is called the Rabi oscillation.

```
In[·]:= vv [ BlochVector ] = Table@Chop@vec@t{, }t, 0, 8, 0.5/{;
       BlochSphere@}Red, Bead ] = vv/, ImageSize → Small{
```

Out[·]=

This shows the transition probabilities in the logical basis states. Note that the probabilities are the same both in the lab and rotating frame. The Rabi oscillation frequency is in this particular example is close to one---it is exactly one at resonance.

```
       prob[t_] = Abs[Normal@Matrix@vec[t]] ^2;
```

In[ ]:= `Plot[Evaluate@prob[2 Pi t], {t, 0, 2},`
`    FrameLabel → {"t / (2π)", "Probability"}]`

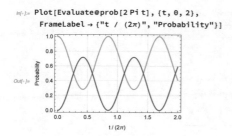

Out[ ]=

Let us consider the case exactly at the resonance.

```
ω = 2 Pi;
B[1] = Cos[ω t];
B[2] = Sin[ω t];
B[3] = ω; (* resonance *)
```

In[ ]:= `opH = Dot[B@{1, 2, 3}, S@{1, 2, 3}] / 2`
`    opHR =`
`      Rotation[-ω t, S[3]] ** opH ** Rotation[ω t, S[3]] - ω S[3] / 2 // Elaborate // Chop`
`    opUR[t_] = MultiplyExp[-I t opHR]`

Out[ ]= $\dfrac{1}{2}\left(\text{Cos}[2\,\pi\,t]\,S^x + 2\,\pi\,S^z + S^y\,\text{Sin}[2\,\pi\,t]\right)$

Out[ ]= $\dfrac{S^x}{2}$

Out[ ]= $e^{-\frac{1}{2}\,i\,t\,S^x}$

The Rabi oscillation now corresponds to the Larmor precession around the x-axis in the rotating frame.

In[ ]:= `Clear[vec]`
`    vec[t_] = opUR[t] ** Ket[] // Elaborate;`
`    vv = BlochVector /@ Table[Chop@vec[t], {t, 0, 8, 0.5}];`
`    BlochSphere[{Red, Bead /@ vv}, ImageSize → Small]`

Out[ ]=
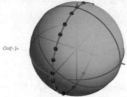

As the precession axis is exactly along the x-axis, the maximum transition probability can reach unity.

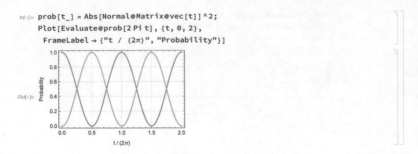

In the above discussion of a Rabi oscillation, the two parameters are deliberately manipulated periodically with a relative phase difference of $\pi/2$. This may not always be possible in practical experiments, but fortunately, the requirement is not so stringent. Suppose that only one parameter can be driven periodically, say, $B_y(t) = 2B_\perp \sin(\omega t)$ (notice the factor of 2 introduced here for convenience). The time-dependent Hamiltonian

$$\hat{H}(t) = B_\perp \sin(\omega t)\hat{S}^y + \frac{1}{2}B_0\hat{S}^z \qquad (3.14)$$

looks seemingly simpler than the one in (3.5), but it does not allow for an exact solution. Instead, let us rewrite it into the form

$$\hat{H}(t) = \frac{1}{2}B_0\hat{S}^z + \frac{1}{2}B_\perp\left[\cos(\omega t)\hat{S}^x + \sin(\omega t)\hat{S}^y\right]$$
$$- \frac{1}{2}B_\perp\left[\cos(-\omega t)\hat{S}^x + \sin(-\omega t)\hat{S}^y\right], \qquad (3.15)$$

which is obviously the same as (3.14). While the second term describes a fictitious magnetic field rotating in the counterclockwise direction, the field in the last term rotates in the clockwise direction. In the frame rotating in the counterclockwise sense, the Hamiltonian reads as

$$\hat{H}_R(t) = \frac{1}{2}(B_0 - \omega)\hat{S}^z + \frac{1}{2}B_\perp\hat{S}^x - \frac{1}{2}B_\perp\left[\cos(-2\omega t)\hat{S}^x + \sin(-2\omega t)\hat{S}^y\right].$$
$$(3.16)$$

The first two terms describe the Rabi oscillation discussed above. On the other hand, the last term oscillates fast with frequency 2ω, which is typically much larger than Ω_R. Such an oscillation is therefore too fast for the system to respond to, and its effect is negligible as the typical time scale of the system is fixed by the Rabi frequency Ω_R in (3.13). On this ground, assuming $\omega \approx B_0$, one often drops the fast oscillating term from the Hamiltonian (3.16) so that

$$\hat{H}(t) = B_\perp \sin(\omega t)\hat{S}^y + \frac{1}{2}B_0\hat{S}^z \approx \frac{1}{2}B_\perp\left[\cos(\omega t)\hat{S}^x + \sin(\omega t)\hat{S}^y\right] + \frac{1}{2}B_0\hat{S}^z.$$
$$(3.17)$$

Such an approximation is called the *rotating-wave approximation*. The approximation is valid as long as $B_0 \gg \Omega_R$.

3.2.2 Implementation of CNOT

The CNOT gate is a vital gate operation for any quantum algorithm. It is seemingly simple, yet it is not trivial to physically implement in a practical system. A typical obstacle is that the Hamiltonian, in particular the coupling between two qubits, is restricted to certain limited forms. Here we take a few examples of the qubit-qubit interaction and see how one can use it to implement the CNOT gate. These examples provide a general idea of what is required to implement CNOT or similar gates in a given practical physical system.

One of the most common forms of the exchange interaction between two qubits is the so-called *Heisenberg exchange interaction*

$$\hat{H} = J(\hat{S}_1^x \hat{S}_2^x + \hat{S}_1^y \hat{S}_2^y + \hat{S}_1^z \hat{S}_2^z). \tag{3.18}$$

As one can see from its matrix representation

$$\hat{H} \doteq J \begin{bmatrix} 1 & & & \\ & -1 & 2 & \\ & 2 & -1 & \\ & & & 1 \end{bmatrix}, \tag{3.19}$$

it operates nontrivially only within the subspace spanned by $|01\rangle$ and $|10\rangle$ just like the SWAP gate discussed in Sect. 2.2.1, particularly in Eq. (2.46). This is because the Heisenberg exchange coupling conserves the total angular momentum. More explicitly, up to a constant shift and multiplication factor, the Heisenberg exchange Hamiltonian equals the SWAP gate

$$\frac{J + \hat{H}}{2J} \doteq \begin{bmatrix} 1 & & & \\ & 0 & 1 & \\ & 1 & 0 & \\ & & & 1 \end{bmatrix} \doteq \text{SWAP}. \tag{3.20}$$

The SWAP gate behaves like the NOT gate within the subspace spanned by $|01\rangle$ and $|10\rangle$. Hence the rotation around the x-axis—in the Bloch space corresponding to the subspace—by angle π results in the NOT gate. This implies that by tuning the exchange coupling constant and the operation time τ such that $J\tau = \pi/4$ will achieve the SWAP gate (Loss & DiVincenzo, 1998)

$$\text{SWAP} = \exp\left[-\frac{i\pi}{4} \left(\hat{S}_1^x \hat{S}_2^x + \hat{S}_1^y \hat{S}_2^y + \hat{S}_1^z \hat{S}_2^z - 1 \right) \right] \tag{3.21}$$

As pointed out in Sect. 2.2.1, the SWAP gate itself is not universal. The limitation originates from the fact that the Heisenberg exchange interaction conserves the total angular momentum. However, the $\sqrt{\text{SWAP}}$ gate is universal. One can construct the CZ gate by combining the $\sqrt{\text{SWAP}}$ gate with single-qubit gates as discussed in Sect. 2.2.1. It takes two Hadamard gates to construct the CNOT gate from the CZ gate, recalling the identity in Eq. (2.44).

Summing up, when the Heisenberg exchange interaction is available in a system of qubits, one can achieve the CZ gate using two $\sqrt{\text{SWAP}}$ gates and three single-qubit gates. If desired, one can apply two additional Hadamard gates to implement the CNOT gate.

Consider the Heisenberg exchange coupling.

In[ ]:= `op = MultiplyDot[S[1, All], S[2, All]]`

Out[ ]= $S_1^x S_2^x + S_1^y S_2^y + S_1^z S_2^z$

This is the matrix representation.

In[ ]:= `mat = Matrix[(1 + op) / 2];`
`mat // MatrixForm`

Out[ ]//MatrixForm=
$$\begin{pmatrix} 1 & 0 & 0 & 0 \\ 0 & 0 & 1 & 0 \\ 0 & 1 & 0 & 0 \\ 0 & 0 & 0 & 1 \end{pmatrix}$$

In[ ]:= `opU = MultiplyExp[-I (Pi / 4) * (op - 1)]`
`opU // Elaborate`

Out[ ]= $e^{-\frac{1}{4} i \pi \left(-1 + S_1^x S_2^x + S_1^y S_2^y + S_1^z S_2^z\right)}$

Out[ ]= $\frac{1}{2} + \frac{1}{2} S_1^z S_2^z + S_1^+ S_2^- + S_1^- S_2^+$

In[ ]:= `matU [ Matrix]opU=;`
`matU // MatrixForm`

Out[ ]//MatrixForm=
$$\begin{pmatrix} 1 & 0 & 0 & 0 \\ 0 & 0 & 1 & 0 \\ 0 & 1 & 0 & 0 \\ 0 & 0 & 0 & 1 \end{pmatrix}$$

There are two additional types of qubit-qubit exchange interactions, the XY and Ising exchange interactions. The *XY exchange interaction*—also known as the *planar exchange interaction*—is described by the Hamiltonian

$$\hat{H} = J(\hat{S}_1^x \hat{S}_2^x + \hat{S}_1^y \hat{S}_2^y). \tag{3.22}$$

The matrix representation

$$\hat{H} \doteq \begin{bmatrix} 0 & & & \\ & 0 & 2 & \\ & 2 & 0 & \\ & & & 0 \end{bmatrix} \tag{3.23}$$

suggests that one can use the XY exchange interaction in essentially the same way as for the Heisenberg exchange interaction. An explicit construction is left for an exercise (Problem 3.2).

Next, consider the XY exchange interaction.

In[ ]:= `op = MultiplyDot[S[1, {1, 2}], S[2, {1, 2}]]`

Out[ ]= $S_1^x S_2^x + S_1^y S_2^y$

In[ ]:= `mat [ Matrix]op=;`
`mat // MatrixForm`

Out[ ]//MatrixForm=
$$\begin{pmatrix} 0 & 0 & 0 & 0 \\ 0 & 0 & 2 & 0 \\ 0 & 2 & 0 & 0 \\ 0 & 0 & 0 & 0 \end{pmatrix}$$

In[ ]:= `opU = MultiplyExp[-I ϕ op]`

Out[ ]= $e^{-i \phi \left(S_1^x S_2^x + S_1^y S_2^y \right)}$

In[ ]:= `Matrix[opU] // MatrixForm`
Out[ ]//MatrixForm=
$$\begin{pmatrix} 1 & 0 & 0 & 0 \\ 0 & \mathrm{Cos}[2\phi] & +i\,\mathrm{Sin}[2\phi] & 0 \\ 0 & +i\,\mathrm{Sin}[2\phi] & \mathrm{Cos}[2\phi] & 0 \\ 0 & 0 & 0 & 1 \end{pmatrix}$$

The *Ising exchange interaction* has the Hamiltonian

$$\hat{H} = J\hat{S}_1^z \hat{S}_2^z , \tag{3.24}$$

where the z-component has no special meaning. Replacing the z-component with the x- or y-component has the same effect. The Ising exchange interaction allows for a more direct implementation of the CZ gate. Though a direct investigation is sufficient, let us here take another view of the CZ gate for heuristic purposes. Recall that the CZ gate is defined as

$$CZ = \sum_{x=0,1} |x\rangle \langle x| \otimes \hat{Z}^x = i \sum_x |x\rangle \langle x| \otimes e^{-i\pi x \hat{Z}/2} \tag{3.25}$$

As $x = 0, 1$ are the eigenvalues of $|1\rangle \langle 1| = (\hat{I} - \hat{Z})/2$,

$$CZ = e^{i\pi/4} \exp\left[-\frac{i\pi}{4}(\hat{Z} \otimes \hat{I} + \hat{I} \otimes \hat{Z} - \hat{Z} \otimes \hat{Z}) \right] \tag{3.26}$$

We note that the exponent involves the Ising exchange interaction between the two qubits. Therefore, one can implement the CZ gate with the first and second qubits as the control and target qubits, respectively, in terms of the Hamiltonian (up to an irrelevant global phase factor $e^{i\pi/4}$) of the form

$$\hat{H} = B(\hat{S}_1^z + \hat{S}_2^z) - J\hat{S}_1^z \hat{S}_2^z . \tag{3.27}$$

Note that the exchange coupling—apart from the single-qubit term—is of the Ising type. Putting $B = J$, the time-evolution operator governed by the Hamiltonian is given by

$$\hat{U}(t) = \exp\left[-iJt(\hat{S}_1^z + \hat{S}_2^z - \hat{S}_1^z\hat{S}_2^z)\right].$$ (3.28)

Therefore, the CZ gate is achieved by tuning the parameters B and J and the operation time such that

$$J\tau = B\tau = \frac{\pi}{4}.$$ (3.29)

Let us now consider the Ising exchange interaction. Here we have introduced an additional single-qubit term controlling the second qubit.

```
In[·]:= op = S[1, 3] + S[2, 3] - S[1, 3] ** S[2, 3]

Out[·]= - S₁ᶻ S₂ᶻ + S₁ᶻ + S₂ᶻ
```

```
In[·]:= mat [ Matrix]op=;
        mat // MatrixForm

Out[·]//MatrixForm=
        ⎛ 1  0  0   0 ⎞
        ⎜ 0  1  0   0 ⎟
        ⎜ 0  0  1   0 ⎟
        ⎝ 0  0  0  +3 ⎠
```

```
In[·]:= opU = MultiplyExp[-I φ op]
        opU // Matrix // MatrixForm

Out[·]= e^{-i φ (-S₁ᶻ S₂ᶻ+S₁ᶻ+S₂ᶻ)}

Out[·]//MatrixForm=
        ⎛ e^{-i φ}    0       0       0   ⎞
        ⎜   0      e^{-i φ}    0       0   ⎟
        ⎜   0        0      e^{-i φ}    0   ⎟
        ⎝   0        0       0     e^{3 i φ} ⎠
```

```
In[·]:= opU = Exp[I Pi / 4] × MultiplyExp[-I Pi / 4 op]
        opU // Matrix // MatrixForm

Out[·]= e^{i π / 4} e^{-¼ i π (-S₁ᶻ S₂ᶻ+S₁ᶻ+S₂ᶻ)}

Out[·]//MatrixForm=
        ⎛ 1  0  0   0 ⎞
        ⎜ 0  1  0   0 ⎟
        ⎜ 0  0  1   0 ⎟
        ⎝ 0  0  0  -1 ⎠
```

3.3 Geometric/Topological Scheme

So far we have mostly assumed a piecewise-constant Hamiltonian (in the fixed or rotating frame). But the Hamiltonian may change continuously in time as well. When a system undergoes a cyclic adiabatic process starting from a particular eigenstate of the Hamiltonian, the system remains in the same energy level without making a transition to other energy levels. However, the quantum state of the system acquires a phase factor from two contributions. One is responsible for the usual dynamical accumulation and the other results from the geometric properties of the parameter space. The latter is called the *geometric phase* of the cyclic adiabatic process (Berry, 1984). When the energy level is degenerate and associated with a multi-dimensional eigen-subspace, the geometric phase becomes non-Abelian. That is, the quantum state evolves to another state within the subspace. The unitary transformation between the initial and final state is the *non-Abelian geometric phase* (Wilczek & Zee, 1984). Non-Abelian geometric phase can be extended to any *cyclic evolution*, without restriction by the adiabatic condition (Aharonov & Anandan, 1987; Anandan, 1988).

The *geometric scheme* of quantum computation (or simply *geometric quantum computation* for short) is an implementation of unitary gate operations by means of non-Abelian geometric phases (Zanardi & Rasetti, 1999; Sjöqvist et al., 2012). The geometric scheme is stable against random fluctuations of the parameters since it depends on the geometric path in the parameter space rather than a detailed time-dependence of the parameters. In this section, we will give a brief overview of the geometric scheme.

3.3.1 A Toy Model

Let us start with a toy model (Choi, 2003). It consists of four levels $|1\rangle$, $|2\rangle$, $|3\rangle$ and $|\epsilon\rangle$ with the level structure as depicted in Fig. 3.2a. The three ground states, $|1\rangle$, $|2\rangle$, $|3\rangle$, are all at energy zero. The excited state $|\epsilon\rangle$ has the (unperturbed) energy ϵ. Each ground state $|j\rangle$ is coupled with amplitudes Ω_j to the excited state $|\epsilon\rangle$, but the ground states are not directly coupled to each other. The model is governed by the Hamiltonian

$$\hat{H} = \epsilon |\epsilon\rangle \langle\epsilon| - \frac{1}{2} \sum_{j=1}^{3} \Omega_j (|j\rangle \langle\epsilon| + |\epsilon\rangle \langle j|). \tag{3.30}$$

For simplicity, we have assumed that Ω_j are all real-valued, and the analysis below can be easily extended to the case with complex-valued Ω_j. For a system with non-degenerate ground-state levels, the model may arise in the interaction picture with a suitable tuning of the driving frequencies.

We will vary $\Omega_j(t)$ continuously in time, and the Hamiltonian $\hat{H}(t)$ will change accordingly through $\Omega_j(t)$. Before doing so, however, let us first examine the *instantaneous* eigenstates of the Hamiltonian $\hat{H}(t)$ at a fixed instant t of time. We define a new state of superposition

$$|\Omega\rangle := \frac{1}{\Omega} \sum_{j=1}^{3} |j\rangle \, \Omega_j \,, \tag{3.31}$$

where we have put $\Omega := \sqrt{\Omega_1^2 + \Omega_2^2 + \Omega_3^2}$. Then, the Hamiltonian reads as

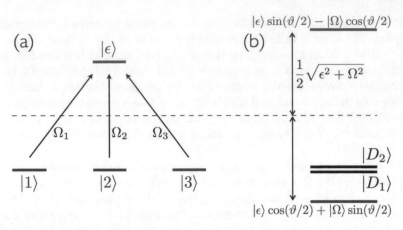

Fig. 3.2 Schematic of the level structure of a toy model (see the text) for the geometric quantum computation. **a** The coupling between the ground-state levels and the excited-state level. **b** The eigenstates and corresponding eigenenergies of the model Hamiltonian. Here $|\Omega\rangle := \Omega^{-1} \sum_{j=1}^{3} |j\rangle \, \Omega_j$, $\Omega := \sqrt{\Omega_1^2 + \Omega_2^2 + \Omega_3^2}$, and $\tan \vartheta := \Omega/\epsilon$. $|D_1\rangle$ and $|D_2\rangle$ denote the dark states

$$\hat{H} = \epsilon \, |\epsilon\rangle \, \langle\epsilon| - \frac{1}{2}\Omega(|\Omega\rangle \, \langle\epsilon| + |\epsilon\rangle \, \langle\Omega|). \tag{3.32}$$

The Hamiltonian involves only two states $|\epsilon\rangle$ and $|\Omega\rangle$. Formally, it is a two-level Hamiltonian. The Hamiltonian is represented as

$$\hat{H} \doteq \begin{bmatrix} 0 & -\Omega/2 \\ -\Omega/2 & \epsilon \end{bmatrix} \tag{3.33}$$

in the basis $\{|\Omega\rangle, |\epsilon\rangle\}$. One can immediately identify two eigenstates

$$|\epsilon\rangle \cos(\vartheta/2) + |\Omega\rangle \sin(\vartheta/2), \quad |\epsilon\rangle \sin(\vartheta/2) - |\Omega\rangle \cos(\vartheta/2), \tag{3.34}$$

where $\tan \vartheta := \Omega/\epsilon$, with eigenenergies $\left(\epsilon \mp \sqrt{\epsilon^2 + \Omega^2}\right)/2$. The eigenenergies of the Hamiltonian are shown in Fig. 3.2b.

More interesting is that the two states $|\epsilon\rangle$ and $|\Omega\rangle$—equivalently, the two eigenstates mentioned above—span only part of the four-dimensional Hilbert space. There must be two more states. These additional states, which we denote by $|D_1\rangle$ and $|D_2\rangle$, do not appear in the Hamiltonian at all. They are completely decoupled from the rest, and in this sense, $|D_1\rangle$ and $|D_2\rangle$ are "dark states". The state $|\Omega\rangle$, which couples to the excited-state $|\epsilon\rangle$, is a "bright state". Note that the dark states must be orthogonal to $|\Omega\rangle$ as well as $|\epsilon\rangle$. In other words, the set $\{|D_1\rangle, |D_2,\rangle, |\Omega\rangle\}$ is another orthogonal basis for the space spanned by the ground states $|1\rangle$, $|2\rangle$, and $|3\rangle$. This relation between the dark and bright states is schematically illustrated in Fig. 3.3.

Fig. 3.3 Schematic
representation of the dark
states of the toy model. The
dark states $|D_1\rangle$ and $|D_2\rangle$
reside in the subspace
orthogonal to the bright state
$|\Omega\rangle$. The space spanned by
the ground states $|1\rangle$, $|2\rangle$,
and $|3\rangle$ are represented by the
three-dimensional space $\mathbb{R}^3$

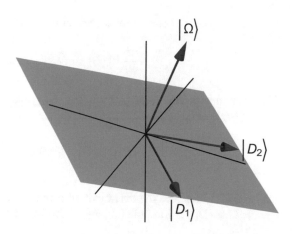

In this toy model, the two dark states $|D_1\rangle$ and $|D_2\rangle$ span a degenerate subspace
$\mathcal{K}$. It is of our primary interest as the energies of the dark states remain degenerate
regardless of the values of Ω_j. It is convenient to take the parameterization

$$\Omega_1 = \Omega \sin\theta \cos\phi, \quad \Omega_2 = \Omega \sin\theta \sin\phi, \quad \Omega_3 = \Omega \cos\theta, \tag{3.35}$$

which leads to

$$|\Omega\rangle = |1\rangle \sin\theta \cos\phi + |2\rangle \sin\theta \sin\phi + |3\rangle \cos\theta. \tag{3.36}$$

As already mentioned, for $|D_1\rangle$ and $|D_2\rangle$, one can choose any mutually orthogonal
states from the subspace that is orthogonal to $|\Omega\rangle$. A convenient choice of the dark
states is as follows

$$|D_1\rangle = |1\rangle \sin\phi - |2\rangle \cos\phi, \tag{3.37a}$$
$$|D_2\rangle = |1\rangle \cos\theta \cos\phi + |2\rangle \cos\theta \sin\phi - |3\rangle \sin\theta. \tag{3.37b}$$

It is important to note that the basis states $|D_1\rangle$ and $|D_2\rangle$ vary with parameters θ and
ϕ (and hence with time t).

So far we have just constructed the time-dependent basis $|D_1(t)\rangle$ and $|D_2(t)\rangle$
with time dependence through $\theta(t)$ and $\phi(t)$. Suppose that $|\psi(t)\rangle$ is a *physical*
state initially prepared in the dark-state subspace $\mathcal{K}$. We assume that the parameters
change sufficiently slow for the adiabatic condition to be satisfied.[1] Then $|\psi(t)\rangle$
always remains within $\mathcal{K}$. So we can expand $|\psi(t)\rangle$ in $|D_1(t)\rangle$ and $|D_2(t)\rangle$ at the
same instant t,

$$|\psi(t)\rangle = |D_1(t)\rangle c_1(t) + |D_2(t)\rangle c_2(t), \tag{3.38}$$

---

[1] We can remove this assumption by modulating the phases $\varphi_j(t)$ of the complex amplitudes $\Omega_j(t) = \Delta_j e^{i\varphi_j(t)}$. In this case, the adiabatic condition is not necessary.

where $c_1(t)$ and $c_2(t)$ are complex coefficients. $|\psi(t)\rangle$ should satisfy the Schrödinger equation

$$i\frac{d}{dt}|\psi(t)\rangle = \hat{H}(t)|\psi(t)\rangle = 0. \tag{3.39}$$

The second equality follows from the fact that $\hat{H}(t)$ vanishes within $\mathcal{K}$, i.e., $\hat{H}(t)|D_j(t)\rangle = 0$. Putting (3.38) into (3.39), we get

$$\sum_j \left[|\dot{D}_j(t)\rangle c_j(t) + |D_j(t)\rangle \dot{c}_j(t)\right] = 0. \tag{3.40}$$

The over-dot indicates the derivative with respect to time t. Multiplying the above equation by $\langle D_i(t)|$ from the left, we obtain the differential equation for $c_j(t)$

$$\frac{d}{dt}\begin{bmatrix} c_1(t) \\ c_2(t) \end{bmatrix} + \begin{bmatrix} \langle D_1(t)|\dot{D}_1(t)\rangle & \langle D_1(t)|\dot{D}_2(t)\rangle \\ \langle D_2(t)|\dot{D}_1(t)\rangle & \langle D_2(t)|\dot{D}_2(t)\rangle \end{bmatrix} \begin{bmatrix} c_1(t) \\ c_2(t) \end{bmatrix} = 0. \tag{3.41}$$

Let us denote the 2×2 matrix in (3.41) by $A(t)$ with elements

$$A_{ij}(t) := \langle D_i(t)|\frac{d}{dt}|D_j(t)\rangle. \tag{3.42}$$

Then, Eq. (3.41) reads as

$$\frac{d\boldsymbol{c}}{dt} + A(t) \cdot \boldsymbol{c}(t) = 0, \tag{3.43}$$

where the vector $\boldsymbol{c}$ denotes the coefficients c_1 and c_2 collectively, $\boldsymbol{c} := (c_1, c_2)^T$.

Let us further simplify the problem by considering a special case where we only vary ϕ but keep θ constant. Then we use the chain rule to rewrite the matrix $A(t)$ into the form

$$A(t) = \frac{d\phi}{dt}A^\phi \tag{3.44}$$

with matrix A^ϕ defined by the elements

$$A_{ij}^\phi := \langle D_i(\phi)|\frac{d}{d\phi}|D_j(\phi)\rangle. \tag{3.45}$$

Similarly, applying the chain rule to $d\boldsymbol{c}/dt$, we rewrite (3.43) as

$$\frac{d\boldsymbol{c}}{d\phi} + A^\phi(\phi) \cdot \boldsymbol{c}(\phi) = 0. \tag{3.46}$$

We have now factored $d\phi/dt$ out of the equation, and the equation has no explicit dependence on time t. The solution $\boldsymbol{c}(\phi)$ is completely determined by the geometric path in the parameter space.

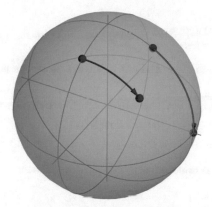

Fig. 3.4 The rotation around the y-axis on the Bloch sphere based on the non-Abelian geometric phase

Putting expressions (3.37) into (3.45), we find that

$$A^\phi = \begin{bmatrix} 0 & -\cos\theta \\ \cos\theta & 0 \end{bmatrix} = -i\cos\theta\, Y,\qquad(3.47)$$

where Y is the Pauli Y matrix. After one cycle of variation from $\phi = 0$ to 2π, the coefficients become

$$c(2\pi) = \exp(-2\pi A^\phi)c(0) = U_y(-4\pi\cos\theta)c(0),\qquad(3.48)$$

where $U_y(\varphi) = \exp(-iY\varphi/2)$ is the 2×2 matrix describing the rotation by angle φ around the y-axis in the Bloch space associated with $|D_1\rangle$ and $|D_2\rangle$ as illustrated in Fig. 3.4. Through a modulation of ϕ by keeping θ, we have achieved a particular type of quantum gates corresponding to the rotations around the y-axis. By modulating the parameters along different paths in the parameter space, one can implement various single-qubit gates, in particular, rotations around two non-parallel axes. Two-qubit gate operations can also be implemented in a similar way (Choi, 2003). These facts assert that universal quantum computation is possible based on the geometric scheme.

Consider a four-level system.

```
Let[Qudit, A, Dimension → 4]
```

Here is the standard basis of the Hilbert space associated with the system. We will regard $|0\rangle$ as the excited level and $|1\rangle$, $|2\rangle$, $|3\rangle$ as the ground-state levels.

```
In[·]:=  bs [ Basis]A=;
         bs // LogicalForm
```

Out[·]= $\{|0_A\rangle, |1_A\rangle, |2_A\rangle, |3_A\rangle\}$

The ground states are coupled (each $|j\rangle$ with an amplitude Ω_j) to the excited level.

```
In[·]:=  Let[Real, Ω, ε]
         H = ε A[0 → 0] - (1 / 2) * PlusDagger@Sum[Ω[j] * A[0 → j], {j, 1, 3}]
```

Out[·]= $\epsilon \left(|0\rangle\langle 0|\right) - \frac{1}{2} \left(+ \left(|1\rangle\langle 0|\right)\Omega_1 + \left(|0\rangle\langle 1|\right)\Omega_1 + \right.$

$\left. \left(|2\rangle\langle 0|\right)\Omega_2 + \left(|0\rangle\langle 2|\right)\Omega_2 + \left(|3\rangle\langle 0|\right)\Omega_3 + \left(|0\rangle\langle 3|\right)\Omega_3 \right)$

We choose a parameterization. This parameterization is convenient to normalize various states.

```
Let[Real, θ, ϕ]
Ω[1] = Sin[θ] × Cos[ϕ];
Ω[2] = Sin[θ] × Sin[ϕ];
Ω[3] = Cos[θ];
```

Here is the "bright" state in the parameterization.

```
In[·]:=  ketΩ = Sum[Ket[A → j] * Ω[j], {j, 1, 3}]
```

Out[·]= $\text{Cos}[\theta] \, |3_A\rangle + \text{Cos}[\phi] \, |1_A\rangle \, \text{Sin}[\theta] + |2_A\rangle \, \text{Sin}[\theta] \times \text{Sin}[\phi]$

Here is a choice of two "dark" states.

```
In[·]:=  ketD1 = Ket[A → 1] + Sin[ϕ] - Ket[A → 2] + Cos[ϕ]
         ketD2 =
           Ket[A → 1] + Cos[θ] + Cos[ϕ] * Ket[A → 2] + Cos[θ] + Sin[ϕ] - Ket[A → 3] + Sin[θ]
```

Out[·]= $-\text{Cos}[\phi] \, |3_A\rangle + |1_A\rangle \, \text{Sin}[\phi]$

Out[·]= $\text{Cos}[\theta] \times \text{Cos}[\phi] \, |1_A\rangle - |2_A\rangle \, \text{Sin}[\theta] + \text{Cos}[\theta] \, |3_A\rangle \, \text{Sin}[\phi]$

Let us check if they are indeed "dark" with respect to the Hamiltonian.

```
In[·]:=  H ** ketD1
         H ** ketD2
```

Out[·]= 0

Out[·]= 0

The two dark state must be orthogonal to the bright state. Let us check it.

```
In[·]:=  new = {ketD1, ketD2, ketΩ};
         Outer[Multiply, Dagger[new], new] // Simplify // MatrixForm
```

Out[·]//MatrixForm=

$\begin{pmatrix} 1 & 0 & 0 \\ 0 & 1 & 0 \\ 0 & 0 & 1 \end{pmatrix}$

Finally, we calculate the non-Abelian gauge potential.

```
In[·]:=  dark = {ketD1, ketD2};
         matA = Outer[Multiply, Dagger[dark], D[dark, ϕ]] // Simplify;
         matA // MatrixForm
```

Out[·]//MatrixForm=

$\begin{pmatrix} 0 & [\text{Cos}]\theta+ \\ \text{Cos}]\theta+ & 0 \end{pmatrix}$

3.3.2 Geometric Phase

Let us now discuss the general case. Consider a cyclic process where the Hamiltonian changes in time through control parameters λ_μ with $\mu = 1, 2, \ldots$

$$\hat{H}(t) = \hat{H}(\lambda(t)), \tag{3.49}$$

where we adopted row-vector notation $\lambda = (\lambda_1, \ldots, \lambda_\mu, \ldots)$ to collectively denote the control parameters. Of course, when the Hamiltonian $\hat{H}(t)$ is the idealistic one in (3.1) for qubits, the control parameters λ refer to the parameters B_j^μ and J_{ij}^μ. However, the Hamiltonian for geometric quantum computation is usually much more general and does not contains elements manifesting a peculiar form of qubits. The logical qubits of a geometric scheme are often implicitly encoded into the subspace undergoing the cyclic evolution.

Let the states $|\alpha_j(t)\rangle$ $(j = 1, 2, \ldots)$ form an *instantaneous basis* of the Hilbert space $\mathcal{H}$. These states are often *chosen* to be the instantaneous eigenstates of the Hamiltonian $\hat{H}(t)$,

$$\hat{H}(t) |\alpha_j(t)\rangle = E_j(t) |\alpha_j(t)\rangle, \tag{3.50}$$

with time-dependent eigenenergies $E_j(t)$. However, the choice is completely arbitrary as long as they are *cyclic*, $|\alpha_j(\tau)\rangle = |\alpha_j(0)\rangle$, and *smooth* enough to be differentiable with respect to time (and hence with respect to the parameters λ_μ). As the basis varies with time, the bases at different instants t and t' must be related to each other

$$|\alpha_j(t')\rangle = \sum_i |\alpha_i(t)\rangle \, U_{ij}(t, t') \tag{3.51}$$

by a unitary matrix $U_{ij}(t, t') := \langle \alpha_i(t) | \alpha_j(t') \rangle$ as further shown in Eqs. (A.12) and (A.13). Choosing different bases at different times is analogous to choosing different reference frames at different positions as also illustrated in Fig. 3.5. It is important to perceive the relation between different reference frames. Naturally, the unitary matrix $U(t)$ plays an important role when describing the dynamics in a time-dependent problem.

Given that the choice is arbitrary, the instantaneous basis states $|\alpha_j(t)\rangle$ are not related to the physical state directly. A physical state should satisfy the Schrödinger equation

$$i\frac{d}{dt} |\psi(t)\rangle = \hat{H}(t) |\psi(t)\rangle. \tag{3.52}$$

Suppose that a physical state $|\psi(t')\rangle$ at a given time t' is expanded in the instantaneous basis $|\alpha_j(t')\rangle$ as

$$|\psi(t')\rangle = \sum_j |\alpha_j(t')\rangle \langle \alpha_j(t') | \psi(t')\rangle. \tag{3.53}$$

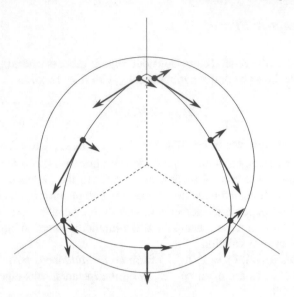

Fig. 3.5 Illustration of a parallel transport on the unit sphere

At later time $t > t'$, it must be expanded in another instantaneous basis $|\alpha_i(t)\rangle$ in the form

$$|\psi(t)\rangle = \sum_{ij} |\alpha_i(t)\rangle \, U_{ij}(t,t') \langle \alpha_j(t')|\psi(t')\rangle , \qquad (3.54)$$

where $U_{ij}(t,t')$ is the unitary matrix in (3.51) that describes the basis change to $|\alpha_i(t)\rangle$ from $|\alpha_j(t')\rangle$. Putting (3.54) into the time-dependent Schrödinger equation (3.52), one can see that the unitary matrix $U(t,t')$ satisfies the dynamical equation

$$i\frac{\partial}{\partial t}U(t,t') = [H(t) - iA(t)]U(t,t') , \qquad (3.55)$$

where $H(t)$ is the matrix representation of the Hamiltonian $\hat{H}(t)$ in the instantaneous basis $|\alpha_j(t)\rangle$, $H_{ij}(t) := \langle \alpha_i(t)| \hat{H}(t) |\alpha_j(t)\rangle$, and the matrix $A(t)$ has been defined by

$$A_{ij}(t) := \langle \alpha_i(t)| \frac{d}{dt} |\alpha_j(t)\rangle . \qquad (3.56)$$

The additional term $-iA(t)$ in (3.55) is a non-inertial effect similar to the one we have observed as in Eq. (3.10) in the rotating frame in the Rabi oscillation. The solution to Eq. (3.55) is formally given by

$$U(t,t') = \text{Texp}\left[-\int_{t'}^{t} ds \, \{A(s) + iH(s)\} \right] , \qquad (3.57)$$

where T denotes the time ordering.[2]

So far, everything has been completely general. For a geometric quantum computation, it is convenient to identify a dynamical subspace $\mathcal{K}(t) \subset \mathcal{H}$ such that

(a) $\mathcal{K}(t)$ undergoes a *cyclic evolution*, $\mathcal{K}(\tau) = \mathcal{K}(0)$;
(b) $\hat{H}(t)$ vanishes within $\mathcal{K}(t)$ for any time $0 < t < \tau$.

With these two conditions satisfied, the unitary matrix governing the evolution of the state vector $|\psi(t)\rangle$ in (3.54) is determined solely by the matrix $A(t)$,

$$U(t, t') = \mathrm{Texp}\left[-\int_{t'}^{t} ds\, A(s)\right], \tag{3.58}$$

Furthermore, using the chain rule for the total derivative with respect to t, we can rewrite matrix $A(t)$ into the form

$$A_{ij}(t) = \sum_{\mu} \frac{d\lambda_{\mu}}{dt}\left\langle \alpha_i \left| \frac{\partial}{\partial \lambda_{\mu}} \right| \alpha_j \right\rangle = \sum_{\mu} \frac{d\lambda_{\mu}}{dt} A_{ij}^{\mu}, \tag{3.59}$$

where we have put

$$A_{ij}^{\mu} := \left\langle \alpha_i \left| \frac{\partial}{\partial \lambda_{\mu}} \right| \alpha_j \right\rangle. \tag{3.60}$$

Putting $A(t)$ in (3.59) back to (3.58), one can see that the unitary matrix $U(\tau, 0)$ for the whole cycle is completely determined by the closed path in the parameter space traversed by the parameters $\lambda_{\mu}(t)$. That is, for a given loop $\mathcal{C}$ in the parameter space, the unitary matrix $U(\tau, 0) = U(\mathcal{C})$ is given by

$$U(\mathcal{C}) = \mathrm{Pexp}\left[-\sum_{\mu} \oint_{\mathcal{C}} d\lambda_{\mu}\, A^{\mu}\right], \tag{3.61}$$

where P denotes path ordering. Putting it back into the evolution of the physical state in (3.54), we see that even without a dynamical effect ($H = 0$ within subspace $\mathcal{K}$) and after the Hamiltonian returns to the original value, $U(\mathcal{C})$ can make a non-trivial transitions of the states in subspace $\mathcal{K}$.

To further understand $\hat{A}^{\mu}$ and $U(\mathcal{C})$, suppose that one chooses a different basis, say,

$$|\beta_j\rangle := \hat{V}\, |\alpha_j\rangle = \sum_{i} |\alpha_i\rangle\, V_{ij}, \tag{3.62}$$

---

[2] For a given matrix $M(t)$, the time-ordered exponential of $M(t)$ is defined by the series

$$\mathrm{Texp}\left[\int_{0}^{t} ds\, M(s)\right] := I + \int_{0}^{t} ds\, M(s) + \int_{0}^{t} ds_1 \int_{0}^{s_1} ds_2\, M(s_1)M(s_2) + \cdots.$$

Notice the upper limits of the integrals.

where $\hat{V}$ is a unitary operator and V is its matrix representation in the original basis $\{|\alpha_j\rangle\}$. Under the change of basis, matrix A^μ transforms as

$$A^\mu \mapsto V^\dagger A^\mu V + V^\dagger \frac{\partial V}{\partial \lambda_\mu}, \tag{3.63}$$

which is the typical manner in which a vector potential transforms under the gauge transformation in quantum mechanics. In this sense, A^μ is called the *non-Abelian gauge potential*, and it describes the connection between the bases, $|\alpha_i(\lambda_\mu)\rangle$ and $|\alpha_j(\lambda_\mu + \delta\lambda_\mu)\rangle$, at infinitesimally different points along the path C in the parameter space. The gauge connection A^μ and the corresponding non-Abelian geometric phase $U(C)$ are in close analogy with the parallel transport in curved space illustrated in Fig. 3.5.

In short, the geometric quantum computation implements the quantum gate operations by means of the unitary transformation $U(C)$ in (3.61). The computational space is identified by subspace $\mathcal{K}$ undergoing a cyclic evolution. Different choices of the closed path C result in different quantum logic gates.

The *topological scheme* of quantum computation is a peculiar case of the geometric scheme. We will take a glimpse of an example of topological quantum computation later in Sect. 6.5.

3.4 Measurement-Based Scheme

According to the fundamental postulates of quantum mechanics discussed at the very beginning of the book, the quantum state of a system changes as a consequence of two different causes. One is the time-evolution governed by the Schrödinger equation. Both the dynamical and geometric schemes of quantum computation are inherently based on such time evolution. The other cause for the change in quantum states is the collapse of the quantum states after measurement, as dictated by Postulate 3. So can we also use it for quantum computation? At first glance, it may look absurd when considering the uncontrollable and random nature of the collapse of a quantum state upon measurement. Nevertheless, quantum computation has been recently shown to be possible just by employing measurements in different bases, provided the system is prepared in a special quantum state called the *cluster state* or, more generally, the *graph state* (Raussendorf & Briegel, 2001; Raussendorf et al., 2002). Such a scheme of quantum computation is the *measurement-based quantum computation* or *one-way quantum computation*. The reasoning behind the "one-way" nomenclature will become clear below. There are several technical methods to implement measurement-based quantum computation. Here, we introduce elementary ideas, and we kindly refer interested readers to the aforementioned articles and their follow-up literature.

To get an idea of how the measurement-based scheme works, first consider a quantum circuit of the simple form

$$ \tag{3.64} $$

where the input state $|\psi\rangle = |0\rangle c_0 + |1\rangle c_1$ on the first qubit is arbitrary. Recall from Eq. (2.33) that the CNOT gate "copies" the logical basis states of the control qubit to the target qubit. It transforms the input state $|\psi\rangle \otimes |0\rangle$ to $|00\rangle c_0 + |11\rangle c_1$. Then, the additional Hadamard gate on the first qubit leads to the final state

$$ \frac{|0\rangle \otimes |\psi\rangle + |1\rangle \otimes (\hat{Z}|\psi\rangle)}{\sqrt{2}}. \tag{3.65} $$

When the first qubit is measured, the state of the second qubit is set to different states depending on the measurement outcome. When the outcome is 0, the state is identical to the input state of the first qubit; and it is $\hat{Z}|\psi\rangle$ if the outcome is 1. In any case, the state coming out on the second qubit can be set to the (unknown) input state of the first qubit—with an additional operation $\hat{Z}$ if necessary—because we know the measurement outcome.

We can rewrite the quantum circuit in (3.64) using relation (2.44) between the CNOT and CZ gates into the form

$$ \tag{3.66} $$

At this stage, there is no reason why we should prefer the CZ gate to the CNOT gate. However, we will see below that the CZ gate is better suited for the measurement-based scheme. More importantly, the Hadamard gate followed by the measurement in the standard basis on the first qubit can be regarded as a measurement in the Pauli X basis. It is indeed physically achieved by rotating the measurement setup, as we have discussed in Sect. 2.5. In effect, the measurement in the Pauli X basis on the first qubit transfers a quantum state to the second qubit.

Consider the following quantum circuit model.

In[ ]:= `qc = QuantumCircuit[ProductState[S[1] → c@{0, 1}, "Label" → "|ψ)"],`
       `LogicalForm[Ket[], S[2]], S[2, 6], CZ[S[1], S[2]], S[{1, 2}, 6]]`

Out[ ]=

This is the output state.

In[·]:= `out = Elaborate[qc]`

Out[·]= $\dfrac{c_0 \left|\underline{\ }\right\rangle}{\sqrt{2}} - \dfrac{c_0 \left|1_{S_1}\right\rangle}{\sqrt{2}} + \dfrac{c_1 \left|1_{S_1}1_{S_2}\right\rangle}{\sqrt{2}} - \dfrac{c_1 \left|1_{S_2}\right\rangle}{\sqrt{2}}$

As you can see here, when the first qubit is measured and the outcome is 0, then the second qubit is identical to the input state of the first qubit. In case the measurement outcome is 1, the second qubit is set to the initial state of the first qubit with the Pauli Z operated.

In[·]:= `KetFactor[out, S[1]] // LogicalForm`

Out[·]= $\left|c_{\theta_2}\right\rangle \otimes \left(\dfrac{1_c \left|c_{\theta_s}\right\rangle}{\sqrt{S}} - \dfrac{1_2 \left|2_{\theta_s}\right\rangle}{\sqrt{S}} \right) - \left|2_{\theta_2}\right\rangle \otimes \left(\dfrac{1_c \left|c_{\theta_s}\right\rangle}{\sqrt{S}} + \dfrac{1_2 \left|2_{\theta_s}\right\rangle}{\sqrt{S}} \right)$

According to the above observation, we undo the byproduct operator depending on the outcome of the measurement on the first qubit.

In[·]:= `qc2 = QuantumCircuit[qc, Measurement[S[1]], ControlledU[S[1], S[2, 3]]]`

Out[·]=

$\left|\psi\right\rangle$ ──●──── H ──◁┈┈┈●┈┈┈

$\left|0\right\rangle$ ── H ──●── H ──────── Z ──

In[·]:= `new = Elaborate[qc2] /. {c[0] * Conjugate[c[0]] + c[1] * Conjugate[c[1]] → 1};`
`KetFactor[new, S[1]] // LogicalForm`

Out[·]= $\left|c_{\theta_2}\right\rangle \otimes \left(1_c \left|c_{\theta_s}\right\rangle + 1_2 \left|2_{\theta_s}\right\rangle \right)$

Next, let us consider a slightly different quantum circuit

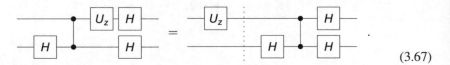

$$\tag{3.67}$$

Compared to the previous quantum circuit, it has a single-qubit rotation $\hat{U}_z$ around the z-axis on the first qubit. In (3.67), the equality holds because $\hat{U}_z$ commutes with the CZ gate. Applying the previous result to the quantum circuit on the right-hand side of (3.67) immediately gives the output state

$$\frac{\left|0\right\rangle \otimes (\hat{U}_z \left|\psi\right\rangle) + \left|1\right\rangle \otimes (\hat{Z}\hat{U}_z \left|\psi\right\rangle)}{\sqrt{2}}. \tag{3.68}$$

Upon the measurement of the first qubit, the output state of the second qubit becomes either $\hat{U}_z \left|\psi\right\rangle$ or $\hat{Z}\hat{U}_z \left|\psi\right\rangle$ depending on the measurement outcome. In either case, the state $\hat{U}_z \left|\psi\right\rangle$ is achieved on the second qubit with an additional operation $\hat{Z}$ if necessary. Again, the key point here is that the unitary operation $\hat{U}_z$ followed by a measurement in the standard basis on the first qubit can be regarded as a measurement in the rotated basis (see Sect. 2.5). In the end, we have achieved the state with unitary gate $\hat{U}_z$ applied on it by means of a (rotated) measurement.

Consider the following quantum circuit model.

```
In[ ]:= Let=Real, φ[
    qc { QuantumCircuit=LogicalForm=Ket=[, S=2[[,
      S=2, 6[, CZ=S=1[, S=2[[, Rotation=φ, S=1, 3[[, S=}1, 2], 6[[
```

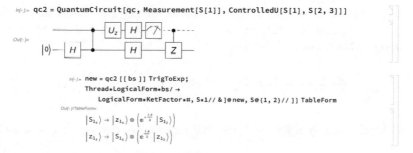

This shows how the logical basis states in the first qubit is transformed through the above quantum circuit model.

```
In[ ]:= bs = Basis[S[1]];
    out = qc ** bs // TrigToExp // Garner;
    Thread[LogicalForm[bs] →
        LogicalForm[KetFactor[#, S[1]] & /@ out, S@{1, 2}]] // TableForm
```

$$\left|S_{1_z}\right\rangle \rightarrow \left|S_{1_z}\right\rangle \otimes \left(\frac{e^{-\frac{i\phi}{2}}}{\sqrt{2}}\right) + \left|z_{1_z}\right\rangle \otimes \left(\frac{e^{-\frac{i\phi}{2}}}{\sqrt{2}}\right)$$

$$\left|z_{1_z}\right\rangle \rightarrow \left|S_{1_z}\right\rangle \otimes \left(\frac{e^{\frac{i\phi}{2}}\left|z_{1_z}\right\rangle}{\sqrt{2}}\right) + \left|z_{1_z}\right\rangle \otimes \left(-\frac{e^{\frac{i\phi}{2}}\left|z_{1_z}\right\rangle}{\sqrt{2}}\right)$$

When the first qubit is measured, the output state of the second qubit depends on the measurement outcome. In either case, however, the input state of the first qubit with the unitary gate operated can be recovered in the second qubit with an operation of the Pauli Z if the measurement outcome is 1.

In accordance with the above argument, the byproduct operator can be fixed as follows.

```
In[ ]:= qc2 = QuantumCircuit[qc, Measurement[S[1]], ControlledU[S[1], S[2, 3]]]
```

```
In[ ]:= new = qc2 [[ bs ]] TrigToExp;
    Thread*LogicalForm*bs/ →
        LogicalForm*KetFactor*#, S*1// & ]@new, S@{1, 2}// ]] TableForm
```

$$\left|S_{1_z}\right\rangle \rightarrow \left|z_{1_z}\right\rangle \otimes \left(e^{-\frac{i\phi}{2}}\left|S_{1_z}\right\rangle\right)$$

$$\left|z_{1_z}\right\rangle \rightarrow \left|S_{1_z}\right\rangle \otimes \left(e^{\frac{i\phi}{2}}\left|z_{1_z}\right\rangle\right)$$

Since the inverse of the Hadamard gate equals to itself, $\hat{H}^{-1} = \hat{H}$, one can remove the second Hadamard gate on the second qubit by applying $\hat{H}$. This leads to the quantum circuit

$$\tag{3.69}$$

Accounting for the extra Hadamard gate that we applied, we obtain the output state

$$\frac{1}{\sqrt{2}} \sum_{x_1=0,1} |x_1\rangle \otimes (\hat{H}\hat{Z}^{x_1}\hat{U}_z |\psi\rangle) \tag{3.70}$$

for the above quantum circuit. Now, the quantum circuit requires no quantum gates once an entanglement has been created by the CZ gate, and only measurement is needed. The cost is that the Hadamard gate appears as a byproduct on the output qubit. However, we will see below that the trailing Hadamard gate on the second qubit plays an important role to achieve an arbitrary single-qubit rotation in the measurement-based scheme.

3.4.1 Single-Qubit Rotations

Le us now inspect the implementation of arbitrary single-qubit rotations solely based on measurement. Consider the following quantum circuit

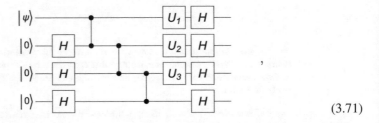

$$\tag{3.71}$$

where the label U_j denotes the single-qubit rotation $\hat{U}_z(\phi_j)$ around the z-axis by angle ϕ_j. The output state from the above quantum circuit can be analyzed by consecutively applying the result in (3.70) from (3.69). It reads as

$$\sum_{x_1,x_2,x_3} |x_1\rangle \otimes |x_2\rangle \otimes |x_3\rangle \otimes \left(\hat{Z}^{x_3} \hat{U}_z(\phi_3) \hat{H} \hat{Z}^{x_2} \hat{U}_z(\phi_2) \hat{H} \hat{Z}^{x_1} \hat{U}_z(\phi_1) |\psi\rangle \right). \tag{3.72}$$

Using the identity $\hat{H}\hat{Z}\hat{H} = \hat{X}$, the final state on the fourth qubit can be written as

$$|\Psi\rangle_4 = \hat{Z}^{x_3} \hat{U}_z(\phi_3) \hat{X}^{x_2} \hat{U}_x(\phi_2) \hat{Z}^{x_1} \hat{U}_z(\phi_3) |\psi\rangle . \tag{3.73}$$

Furthermore, using the identities $\hat{X}\hat{Z}\hat{X} = -\hat{Z}$ and $\hat{Z}\hat{X}\hat{Z} = -\hat{X}$, we arrive at the expression

$$|\Psi\rangle_4 = \hat{Z}^{x_3} \hat{X}^{x_2} \hat{Z}^{x_1} \hat{U}_z((-1)^{x_2}\phi_3)\hat{U}_x((-1)^{x_1}\phi_2)\hat{U}_z(\phi_1) |\psi\rangle . \tag{3.74}$$

The key point to notice here is that upon measurement on the first three qubits, the fourth qubit takes on the state with a single-qubit rotation

$$\hat{U}_z((-1)^{x_2}\phi_3)\hat{U}_x((-1)^{x_1}\phi_2)\hat{U}_z(\phi_3) \tag{3.75}$$

operated on the initial state $|\psi\rangle$ of the first qubit. The rotation consists of three rotations around two perpendicular axes, and these are sufficient to realize any arbitrary single-qubit rotation (see Sect. 2.1.3).

The rotation in (3.75) is not deterministic since the direction—clockwise or anticlockwise—depends on the outcomes x_1, x_2, x_3 of the measurement on the first three qubits. We can thus make it deterministic by delaying the operations on the second qubit after measurement on the first qubit and the operation on the third qubit after measurement on the second qubit. Consider a modified quantum circuit as follows

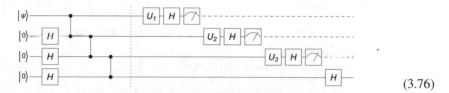

$$(3.76)$$

Suppose that we want to implement a single-qubit Euler rotation[3]

$$\hat{U}(\alpha, \beta, \gamma) := \hat{U}_z(\alpha)\hat{U}_x(\beta)\hat{U}_z(\gamma). \tag{3.77}$$

We first set $\phi_1 = \gamma$. Then, depending on the measurement outcome x_1 on the first qubit, we set $\phi_2 = (-1)^{x_1}\beta$. Similarly, depending on the measurement outcome x_2 on the second qubit, we set $\phi_3 = (-1)^{x_2}\alpha$. The final state on the fourth qubit will become

$$|\Psi\rangle_4 = \hat{Z}^{x_3}\hat{X}^{x_2}\hat{Z}^{x_1}\hat{U}(\alpha, \beta, \gamma)|\psi\rangle \tag{3.78}$$

with the single-qubit rotation fixed *deterministically*. There are still undesired operations, $\hat{Z}^{x_3}\hat{X}^{x_2}\hat{Z}^{x_1}$. However, these operations are irrelevant at the end of quantum computation, and do not have to be corrected. For example, if the final state is measured in the logical basis, $\hat{Z}$ does not affect the readout at all, and the effect of $\hat{X}$ can be handled by a classical post-processing.

Let us construct a quantum circuit model for an arbitrary single-qubit rotation in the measurement-based scheme.

This will simplify the analysis below.

```
Let[Real, c, ϕ]
```

Here is the overall quantum circuit model. It is built in two steps divided by the vertical dashed line. The first step corresponds to the preparation, and the second to the operation.

---

[3] Here we have adopted a different convention for the Euler rotation. One of the most common convention in physics is to use the combination, $\hat{U}_z(\alpha)\hat{U}_y(\beta)\hat{U}_z(\gamma)$.

```
In[*]:= qc = QuantumCircuit[ProductState[S[1] → @c[0], c[1]{, "Label" → Ket["ψ"]],
      LogicalForm[Ket[], S}@2, 3, 4{], S[@2, 3, 4{, 6],
      CZ[S[1], S[2]], CZ[S[2], S[3]], CZ[S[3], S[4]], "Separator",
      Rotation[ϕ[1], S[1, 3], "Label" → "U₁"], S[1, 6], Measurement[S[1]],
      Rotation[ϕ[2], S[2, 3], "Label" → "U₂"], S[2, 6], Measurement[S[2]],
      Rotation[ϕ[3], S[3, 3], "Label" → "U₃"], S[3, 6], Measurement[S[3]], S[@4{, 6]]
```

Out[*]:=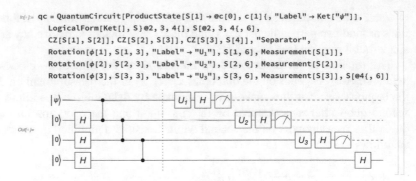

For the sake of analysis, we build the first part of the above quantum circuit model separately.

```
In[*]:= qc0 = QuantumCircuit[
      ProductState[S[1] → @c[0], c[1]{, "Label" → Ket["ψ"]],
      LogicalForm[Ket[], S}@2, 3, 4{],
      S[@2, 3, 4{, 6], CZ[S[1], S[2]], CZ[S[2], S[3]], CZ[S[3], S[4]]]
```

Out[*]:=

Perform a measurement on the first qubit in a properly rotated basis.

```
In[*]:= out1 [ QuantumCircuit]qc0, Rotation]ϕ]1=, S]1, 3=, "Label" → "U₁"=,
      S]1, 6=, Measurement]S]1=== // Elaborate;
      x1 [ Readout]out1, S]1==
```

Out[*]:= 0

Depending on the outcome x1 out of the measurement on the first qubit, rotate suitable the measurement setup for the second qubit and perform the measurement.

```
In[*]:= out2 = QuantumCircuit[out1, Rotation[-(1/^x1) ϕ[2*, S[2, 3*, "Label" → "U₂"*,
      S[2, 6*, Measurement[S[2*** ]] Elaborate;
      x2 = Readout[out2, S[2**
```

Out[*]:= 0

Similarly, depending on the measurement outcome x2 on the second qubit, adjust the measurement setup for the third qubit and take the measurement.

```
In[*]:= out3 = QuantumCircuit[out2, Rotation[ (-1) ^x2 * ϕ[3], S[3, 3], "Label" → "U₃"],
      S[3, 6], Measurement[S[3]], S[4, 6]] // Elaborate;
      out3 = out3 /. {c[0] ^2 + c[1] ^2 → 1};
      x3 = Readout[out3, S[3]]
```

Out[*]:= 0

Check the result.

```
In[ ]:= expr = MultiplyPower[S[4, 3], x3] **
           MultiplyPower[S[4, 1], x2] ** MultiplyPower[S[4, 3], x1] **
           Rotation[ϕ[3], S[4, 3]] ** Rotation[ϕ[2], S[4, 1]] **
           Rotation[ϕ[1], S[4, 3]] ** (Ket[S[1] → x1, S[2] → x2, S[3] → x3] **
             ProductState[S[4] → {c[0], c[1]}]) // Elaborate
```

$$Out[ ]= |-\rangle \left(Cos\left[\frac{\phi_1}{2}\right]\left(c_\theta Cos\left[\frac{\phi_2}{2}\right] - i\,c_1 Sin\left[\frac{\phi_2}{2}\right]\right) + Sin\left[\frac{\phi_1}{2}\right]\left(-i\,c_\theta Cos\left[\frac{\phi_2}{2}\right] + c_1 Sin\left[\frac{\phi_2}{2}\right]\right)\right) \times$$

$$\left(Cos\left[\frac{\phi_3}{2}\right] - i\,Sin\left[\frac{\phi_3}{2}\right]\right) +$$

$$|1_{S_4}\rangle \left(Sin\left[\frac{\phi_1}{2}\right]\left(i\,c_1 Cos\left[\frac{\phi_2}{2}\right] - c_\theta Sin\left[\frac{\phi_2}{2}\right]\right) + Cos\left[\frac{\phi_1}{2}\right]\left(c_1 Cos\left[\frac{\phi_2}{2}\right] - i\,c_\theta Sin\left[\frac{\phi_2}{2}\right]\right)\right) \times$$

$$\left(Cos\left[\frac{\phi_3}{2}\right] + i\,Sin\left[\frac{\phi_3}{2}\right]\right)$$

```
In[ ]:= out3 - expr
```

$$Out[ ]= 0$$

3.4.2   CNOT Gate

Le us now construct a quantum circuit to implement a CNOT gate in the measurement-based scheme. There are several ways to do so, and here we adopt the example described in the quantum circuit

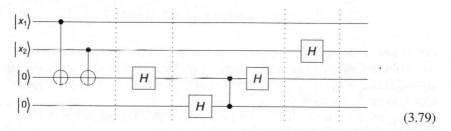

$$(3.79)$$

The first two CNOT gates brings the input state—recall Eq. (2.56) and in (3.24)—to

$$|x_1\rangle \otimes |x_2\rangle \otimes |x_1 \oplus x_2\rangle \otimes |0\rangle \tag{3.80}$$

at the instant indicated by the first vertical dotted line. At the second vertical dotted line, the state becomes

$$|x_1\rangle \otimes |x_2\rangle \otimes \left(\hat{H}\,|x_1 \oplus x_2\rangle\right) \otimes |0\rangle . \tag{3.81}$$

Now we use (3.69), in effect, to transfer the state stored in the third qubit to the fourth qubit. It gives the state

$$\sum_{y_3} |x_1\rangle \otimes |x_2\rangle \otimes |y_3\rangle \otimes \left(\hat{H}\hat{Z}^{y_3}\hat{H}\,|x_1 \oplus x_2\rangle\right)$$

$$= \sum_{y_3} |x_1\rangle \otimes |x_2\rangle \otimes |y_3\rangle \otimes \left(\hat{X}^{y_3}\,|x_1 \oplus x_2\rangle\right).$$

$$(3.82)$$

Finally, applying the Hadamard gate on the second qubit, as shown in Eq. (2.19), leads to

$$\sum_{y_2 y_3} |x_1\rangle \otimes |y_2\rangle \otimes |y_3\rangle \otimes \left(\hat{X}^{y_3}\,|x_1 \oplus x_2\rangle\right)(-1)^{y_2 x_2}. \qquad (3.83)$$

The phase factor -1 occurs when $x_2 = y_2 = 1$. By inspection, one can see that the same factor arises when one applies $\hat{Z}$ on both the first and fourth qubits. This leads to the final state out of the quantum circuit as follows

$$|\Psi\rangle_{\text{out}} = \sum_{y_2, y_3} \left(\hat{Z}^{y_2}\,|x_1\rangle\right) \otimes |y_2\rangle \otimes |y_3\rangle \otimes \left(\hat{X}^{y_3}\hat{Z}^{y_2}\,|x_1 \oplus x_2\rangle\right). \qquad (3.84)$$

When one performs measurement on the second and third qubits, depending on the outcomes y_2 and y_3, the final state stored in the first and fourth qubits is given by

$$|\Psi\rangle_{14} = \left(\hat{Z}^{y_2}\,|x_1\rangle\right) \otimes \left(\hat{X}^{y_3}\hat{Z}^{y_2}\,|x_1 \oplus x_2\rangle\right). \qquad (3.85)$$

Let us ignore the byproduct operators, $\hat{X}$ and/or $\hat{Z}$, for the moment. We then see that in this quantum circuit, the first qubit plays the role of control qubit. The input state of the target qubit is put in the second qubit, and the output state is supposed to appear in the fourth qubit.

Using the relation (2.44) between the CZ and CNOT gates, we can rewrite the quantum circuit in (3.79) into the form

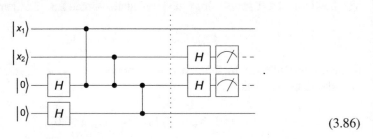

$$(3.86)$$

It is clear that once the entanglement is generated among the four qubits, then it is sufficient to perform a measurement on the second and third qubits to implement the CNOT gate. One has to apply $\hat{Z}$ on both the first and fourth qubit when $y_2 = 1$ and $\hat{X}$ as well if $y_3 = 1$. But again, in many cases, the additional step is not necessary or can be done with simple classical post-processing.

Here we demonstrate the CNOT gate based on the measurement-based scheme.

Let us consider the following quantum circuit model. The first qubit plays a role of the control qubit. The input state of the target qubit enters the second qubit and comes out on the forth qubit.

In[ ]:= qc1 = QuantumCircuit[LogicalForm[Ket[], S@{3, 4}],
   S[{3, 4}, 6], CZ[S[1], S[3]], CZ[S[2], S[3]], CZ[S[3], S[4]],
   "Separator", S[{2, 3}, 6], Measurement[S@{2, 3}]]

Out[ ]=

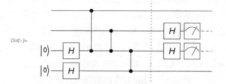

This shows how the logical basis states on the first two qubits are transformed to the first and forth qubit. The result is not exactly what we expect from the CNOT gate since we have not corrected the byproduct operators.

In[ ]:= bs = Basis[S@{1, 2}];
   out = qc1 ** bs;
   Thread[LogicalForm[bs] → LogicalForm[out, S@{1, 2, 3, 4}]] // TableForm
Out[ ]//TableForm=
$$\left|0_{s_1}0_{s_2}\right\rangle \rightarrow \left|0_{s_1}1_{s_2}0_{s_3}0_{s_4}\right\rangle$$
$$\left|0_{s_1}1_{s_2}\right\rangle \rightarrow \left|0_{s_1}0_{s_2}1_{s_3}0_{s_4}\right\rangle$$
$$\left|1_{s_1}0_{s_2}\right\rangle \rightarrow \left|1_{s_1}1_{s_2}0_{s_3}1_{s_4}\right\rangle$$
$$\left|1_{s_1}1_{s_2}\right\rangle \rightarrow \left|1_{s_1}0_{s_2}1_{s_3}1_{s_4}\right\rangle$$

The following quantum circuit model includes the correction of the byproduct operators.

In[ ]:= qc2 = QuantumCircuit[qc1, ControlledU[S[3], S[4, 1]],
   ControlledU[S[2], S[4, 3]], ControlledU[S[2], S[1, 3]]]

Out[ ]=

We see that the output state stored on the first and forth qubit is the state we expect.

In[ ]:= bs = Basis[S@{1, 2}];
   out = qc2 ** bs;
   Thread[LogicalForm[bs] → LogicalForm[out, S@{1, 2, 3, 4}]] // TableForm
Out[ ]//TableForm=
$$\left|0_{s_1}0_{s_2}\right\rangle \rightarrow \left|0_{s_1}0_{s_2}1_{s_3}0_{s_4}\right\rangle$$
$$\left|0_{s_1}1_{s_2}\right\rangle \rightarrow \left|0_{s_1}1_{s_2}1_{s_3}1_{s_4}\right\rangle$$
$$\left|1_{s_1}0_{s_2}\right\rangle \rightarrow \left|1_{s_1}0_{s_2}1_{s_3}1_{s_4}\right\rangle$$
$$\left|1_{s_1}1_{s_2}\right\rangle \rightarrow \left|1_{s_1}1_{s_2}1_{s_3}0_{s_4}\right\rangle$$

In the above example, both the input and output states of the control qubit are stored in the same qubit whereas the target state is transferred from one to another qubit. In many practical cases, especially in the middle of quantum computation, it is more convenient to transfer the target qubit to another qubit as well. It can be achieved by using similar methods with 10 or 15 qubits (Raussendorf & Briegel, 2001; Raussendorf et al., 2003).

3.4.3 Graph States

For measurement-based quantum computation, the crucial resource is the state prepared by the quantum circuit elements on the left of the vertical dashed line in (3.76). Once the state has been prepared, from then on, quantum logic gates can be implemented solely by measurements in rotated bases. Such a state is called a *graph state*. Specifically, consider a graph $\mathcal{G}$ where a set of vertices are connected by edges. A qubit is located at each vertex. Let $\hat{U}_{ab}$ denotes the CZ gate on the two qubits located at the vertex a and b. To generate the graph state $|\mathcal{G}\rangle$ associated with the graph $\mathcal{G}$, prepare each qubit first in the state $|+\rangle := (|0\rangle + |1\rangle)/\sqrt{2}$. Then for each pair $\langle ab \rangle$ of vertices a and b connected by an edge, apply the CZ gate $\hat{U}_{ab}$ on the corresponding qubits. In short,

$$|\mathcal{G}\rangle := \prod_{\langle ab \rangle} \hat{U}_{ab} \, |+\rangle^{\otimes n} \, , \tag{3.87}$$

where the product is over all edges in the graph.

A graph state is created by applying the CZ gate on the two qubits linked by the graph.

For example, consider a linear graph.

In[ ]:= `gr = Graph[{S[1] ↔ S[2], S[2] ↔ S[3], S[3] ↔ S[4]}, VertexLabels → "Index"}`

Out[ ]=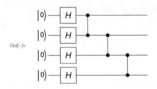

The follewi ongqnu ulai h wtate qwfleateg mc the rodbs qnu byantyp f qfyq p oged

In[ ]:= `qc = QuantumCircuit[LogicalForm[Ket[], S@{1, 2, 3, 4}], gr]`

Out[ ]=

```
|0⟩—[H]—•
|0⟩—[H]—•———•
|0⟩—[H]——————•———•
|0⟩—[H]——————————•
```

In[ ]:= `vec [ Elaborate]qc=;`
`vec // LogicalForm`

Out[ ]= $\frac{\theta}{S}\big|1_{2_a}1_{2_3}1_{2_4}1_{2_5}\big\rangle - \frac{\theta}{S}\big|1_{2_a}1_{2_3}1_{2_4}0_{2_5}\big\rangle - \frac{\theta}{S}\big|1_{2_a}1_{2_3}0_{2_4}1_{2_5}\big\rangle + \frac{\theta}{S}\big|1_{2_a}1_{2_3}0_{2_4}0_{2_5}\big\rangle -$

$\frac{\theta}{S}\big|1_{2_a}0_{2_3}1_{2_4}1_{2_5}\big\rangle - \frac{\theta}{S}\big|1_{2_a}0_{2_3}1_{2_4}0_{2_5}\big\rangle + \frac{\theta}{S}\big|1_{2_a}0_{2_3}0_{2_4}1_{2_5}\big\rangle - \frac{\theta}{S}\big|1_{2_a}0_{2_3}0_{2_4}0_{2_5}\big\rangle -$

$\frac{\theta}{S}\big|0_{2_a}1_{2_3}1_{2_4}1_{2_5}\big\rangle - \frac{\theta}{S}\big|0_{2_a}1_{2_3}1_{2_4}0_{2_5}\big\rangle - \frac{\theta}{S}\big|0_{2_a}1_{2_3}0_{2_4}1_{2_5}\big\rangle + \frac{\theta}{S}\big|0_{2_a}1_{2_3}0_{2_4}0_{2_5}\big\rangle +$

$\frac{\theta}{S}\big|0_{2_a}0_{2_3}1_{2_4}1_{2_5}\big\rangle + \frac{\theta}{S}\big|0_{2_a}0_{2_3}1_{2_4}0_{2_5}\big\rangle - \frac{\theta}{S}\big|0_{2_a}0_{2_3}0_{2_4}1_{2_5}\big\rangle + \frac{\theta}{S}\big|0_{2_a}0_{2_3}0_{2_4}0_{2_5}\big\rangle$

The same state can be directly obtained using GraphState.

In[ ]:= **new = GraphState[gr]**

Out[ ]= $\dfrac{\left|\_\right\rangle}{4} - \dfrac{1}{4}\left|1_{s_1}1_{s_2}\right\rangle - \dfrac{1}{4}\left|1_{s_1}1_{s_2}1_{s_3}1_{s_4}\right\rangle + \dfrac{\left|1_{s_1}\right\rangle}{4} + \dfrac{1}{4}\left|1_{s_1}1_{s_2}1_{s_3}\right\rangle -$

$\dfrac{1}{4}\left|1_{s_1}1_{s_2}1_{s_4}\right\rangle + \dfrac{1}{4}\left|1_{s_1}1_{s_3}\right\rangle - \dfrac{1}{4}\left|1_{s_1}1_{s_3}1_{s_4}\right\rangle + \dfrac{1}{4}\left|1_{s_1}1_{s_4}\right\rangle - \dfrac{1}{4}\left|1_{s_1}1_{s_2}\right\rangle +$

$\dfrac{\left|1_{s_2}\right\rangle}{4} + \dfrac{1}{4}\left|1_{s_2}1_{s_3}1_{s_4}\right\rangle + \dfrac{1}{4}\left|1_{s_2}1_{s_4}\right\rangle - \dfrac{1}{4}\left|1_{s_3}1_{s_4}\right\rangle + \dfrac{\left|1_{s_3}\right\rangle}{4} + \dfrac{\left|1_{s_4}\right\rangle}{4}$

In[ ]:= **vec - new // Simplify**

Out[ ]= 0

The graph states are not only useful as a valuable resource for measurement-based quantum computations, but they are also interesting in their own right. Graph states exhibit a high degree of entanglement. They have been studied extensively for the structure of bi-partite and multi-partite entanglement. One can also construct a class of quantum error-correction codes (see Chap. 6) based on graph states. It allows for reliable storage and processing of quantum information in a fault-tolerant way. All these features of graph states can be attributed to the following property: For each vertex a in $\mathcal{G}$, define an operator

$$\hat{C}_a := \hat{X}_a \prod_{b \in \mathcal{N}_a} \hat{Z}_b, \tag{3.88}$$

where the product is over the set $\mathcal{N}_a$ of all vertices b connected to a. $\hat{C}_a$ is often called the *correlation operator* of vertex a. The graph state $|\mathcal{G}\rangle$ associated with $\mathcal{G}$ is the unique simultaneous eigenstate with eigenvalue $+1$ of every $\hat{C}_a$. In other words,

$$\hat{C}_a |\mathcal{G}\rangle = |\mathcal{G}\rangle \tag{3.89}$$

for all vertices a in the graph $\mathcal{G}$. The operators $\hat{C}_a$ are said to *stabilize* the graph state. The stabilizer formalism that we will discuss in Sect. 6.3 provides powerful tools to exploit stabilizing operators.

As another example, consider the following graph and corresponding quantum circuit model.

```
In[-]:= gr = Graph[]S[1@ ↔ S[2@, S[2@ ↔ S[3@, S[1@ ↔ S[3@, S[1@ ↔ S[4@{,
        ImageSize → Small, VertexLabels → "Index"@
     QuantumCircuit[LogicalForm[Ket[@, S}]1, 2, 3, 4{@, gr@
```

Out[-]=

Out[-]=

This is the graph state associated with the above graph.

```
In[-]:= vec = GraphState[gr]
```

$$
Out[-]= \frac{|-\rangle}{4} - \frac{1}{4}\,|1_{S_1}1_{S_2}\rangle - \frac{1}{4}\,|1_{S_1}1_{S_2}1_{S_3}\rangle + \frac{|1_{S_1}\rangle}{4} + \frac{1}{4}\,|1_{S_1}1_{S_2}1_{S_3}1_{S_4}\rangle +
$$
$$
\frac{1}{4}\,|1_{S_1}1_{S_2}1_{S_4}\rangle - \frac{1}{4}\,|1_{S_1}1_{S_3}\rangle + \frac{1}{4}\,|1_{S_1}1_{S_3}1_{S_4}\rangle - \frac{1}{4}\,|1_{S_1}1_{S_4}\rangle - \frac{1}{4}\,|1_{S_2}1_{S_3}\rangle -
$$
$$
\frac{1}{4}\,|1_{S_2}1_{S_3}1_{S_4}\rangle + \frac{|1_{S_2}\rangle}{4} + \frac{1}{4}\,|1_{S_2}1_{S_4}\rangle + \frac{|1_{S_3}\rangle}{4} + \frac{1}{4}\,|1_{S_3}1_{S_4}\rangle + \frac{|1_{S_4}\rangle}{4}
$$

Here are the correlation operators associate with the vertices in the graph.

```
In[-]:= gnr = Stabilizer[gr]
```

$$
Out[-]= \left\{ S_1^x\,S_2^z\,S_3^z\,S_4^z,\ S_1^z\,S_2^x\,S_3^z,\ S_1^z\,S_2^z\,S_3^x,\ S_1^z\,S_4^x \right\}
$$

The graph state is the simultaneous eigenstate with eigenvalue +1 of all correlation operators.

```
In[-]:= (gnr ** vec) / vec
```

Out[-]= {1, 1, 1, 1}

Mathematically, one can associate a graph state with any graph. However, such a state is difficult to generate in practical systems. Doing so requires applying the CZ gate on spatially separated qubits. *Cluster states* are a special subclass of graph states, where the underlying graph is an d-dimensional square grid. The regular connectivity of the underlying graph makes the associated state far more feasible. Indeed, a number of physical methods have been proposed to generate cluster states.

Problems

3.1 The single-parameter controlled Hamiltonian in Eq. (3.16) has two fictitious magnetic fields rotating in opposite senses, so it seems to be a matter of choice to pick either of the two. Argue why it does not work physically to choose the frame rotating in the clockwise sense.

3.2 Consider a system of two qubits interacting with each other through the XY exchange coupling in (3.22). For the given coupling constant J, find a way to physically implement the SWAP gate by tuning the operation time τ. If necessary, you can apply additional single-qubit gates. Ignore a global phase factor (if any). Once the SWAP gate is carried out, you can combine the $\sqrt{\text{SWAP}}$ gate and single qubit gates to construct the CZ gate (see Sect. 2.2.1).

3.3 Consider a toy model similar to the one described in Eq. (3.30) and Fig. 3.2, but with two ground-state levels:

$$\hat{H} = \epsilon \,|\epsilon\rangle \,\langle\epsilon| - \frac{1}{2} \sum_{j=1}^{2} (\Omega_j \,|j\rangle \,\langle\epsilon| + h.c.). \qquad (3.90)$$

Assume that

$$\Omega_1 = \Omega \cos(\theta/2) e^{-i\phi/2}, \quad \Omega_2 = \Omega \sin(\theta/2) e^{+i\phi/2} \quad (\theta, \phi \in \mathbb{R}). \qquad (3.91)$$

(a) Find all eigenstates and the corresponding eigenvalues of the Hamiltonian $\hat{H}$. Show that there exists an eigenstate $|D\rangle$ the eigenvalue of which is always zero regardless of Ω_1 and Ω_2. $|D\rangle$ is the "dark state" of the model.

(b) Calculate the (Abelian) gauge potential $A^\phi(\theta, \phi)$ for fixed θ for the one-dimensional subspace spanned by $|D\rangle$. Plot A^ϕ as a function of θ and ϕ.

(c) Calculate the (Abelian) geometric phase $U(\mathcal{C})$ for the path $\mathcal{C}$ such that θ is fixed and ϕ changes from 0 to 2π.

3.4 Consider the following quantum circuit

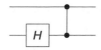

$$(3.92)$$

Show that the output state is given by

$$\frac{|0\rangle \otimes |\psi\rangle + |1\rangle \otimes (\hat{X} \,|\psi\rangle)}{\sqrt{2}}. \qquad (3.93)$$

3.5 Consider the following quantum circuit:

(a) Suppose that the input state is $(|0\rangle \,c_0 + |1\rangle \,c_1) \otimes |0\rangle$. Show that the output state is given by $|0\rangle \otimes |+\rangle \,c_0 + |1\rangle \otimes |-\rangle \,c_1$, where $|\pm\rangle := (|0\rangle \pm |1\rangle)/\sqrt{2}$ is the eigenstates (with eigenvalues ± 1) of the Pauli X operator.

(b) Find the output state for the input state $(|0\rangle c_0 + |1\rangle c_1) \otimes |1\rangle$.

3.6 Let $\hat{U}_\mu(\phi)$ be the single-qubit rotation around the μ-axis by angle ϕ in the Bloch space. Show

   (a) that

$$\hat{X}\hat{U}_z(\phi)\hat{X} = \hat{U}_z(-\phi),\qquad(3.94)$$

   where $\hat{X}$ is the Pauli X operator; and
   (b) that

$$\hat{H}\hat{U}_z(\phi)\hat{H} = \hat{U}_x(\phi),\qquad(3.95)$$

   where $\hat{H}$ is the Hadamard operator.

Chapter 4
Quantum Algorithms

In Chap. 2, we have discussed the use of elementary quantum logic gates to carry out arbitrary quantum computations, and in Chap. 3, we have seen the requirements to physically implement those elementary quantum logic gates. Those discussions are enough to establish physical and realistic models of quantum computation. However, we have disregarded an important issue so far: the efficiency of the implementations. Quantum computers turn out to be technically hard to build, and error rates remain a fundamental concern for quantum computers while classical computers can, in principle, perform the aforementioned calculations anyway. Why should quantum computation be attractive?

Peter Shor's quantum factorization algorithm (Shor, 1994, 1997) brought quantum computation great attention, even from the public, at the turn of the millennium. The factorization of large numbers was the first practically important task that is not feasible on a classical computer but can be performed efficiently on a quantum computer.

In this chapter, we explore several elementary examples of quantum algorithms that efficiently solve problems that are known to be exponentially hard with classical algorithms. Although some of them may be of little use for practical applications, these examples are still interesting to elucidate the ideas and features behind quantum algorithms that distinguish them from classical algorithms.

Quantum teleportation is included in the discussion. It is a quantum communication protocol, rather than a quantum algorithm. Nonetheless, we include it here because it is a simple yet fascinating example demonstrating what one can do with quantum states that is not possible at all with classical information. Quantum teleportation makes key parts of many quantum algorithms as well.

Supplementary Information The online version contains supplementary material available at https://doi.org/10.1007/978-3-030-91214-7_4.

Throughout the chapter, we keep using the same notation as in Chap. 2, Eq. (2.3) in particular.

4.1 Quantum Teleportation

Quantum teleportation is a communication protocol[1] that sends quantum information to a distant party by utilizing a pre-shared pair of entangled particles. The protocol has attracted public interest since it resembles (fictional) teleportation. In science fiction, (hypothetical) teleportation instantly transports matter or energy across space. Quantum teleportation does not transfer physical objects but only quantum information. Quantum teleportation is also conceptually distinguished from telefax. Telefax makes a copy of the document and sends the duplicated copy while the original document remains at its origin. In quantum teleportation, a quantum state is transferred and disappears from the original system.

Of course, transmitting quantum information would be trivial if there were a reliable quantum channel between the parties. However, quantum teleportation assumes that no quantum channel is available. The task would also be straightforward if the two parties were in direct contact with each other. For example, one party can simply use the SWAP gate in Eq. (2.45). Another example would be to use a measurement-based method to transfer a quantum state as in Eq. (3.64). Yet again, in quantum teleportation, the two parties are at such a distance that no joint operation is attainable.

When one has access to both qubits, a SWAP gate for example is enough to send state from one qubit to the other.

```
In[ ]:= Let[Complex, c]
       qc =
         QuantumCircuit[{LogicalForm[Ket[], S[1]], ProductState[S[2] → {c[0], c[1]}]},
           SWAP[S[1], S[2]], {LogicalForm[Ket[], S[2]],
           ProductState[S[1] → {c[0], c[1]}]}, "PortSize" → 2]
```

$$Out[ ]= \quad |0\rangle \longrightarrow\!\times\!\longrightarrow |0\rangle c_0 + |1\rangle c_1$$
$$\qquad |0\rangle c_0 + |1\rangle c_1 \longrightarrow\!\times\!\longrightarrow |0\rangle$$

```
In[ ]:= out = KetFactor@ExpressionFor[qc]
```

$$Out[ ]= \quad \left(c_0 \left|0_{S_1}\right\rangle + c_1 \left|1_{S_1}\right\rangle\right) \otimes \left|0_{S_2}\right\rangle$$

Quantum teleportation is one of the first quantum information protocols that vividly illustrates the significance of quantum entanglement as a valuable resource. In this section, we discuss the physical principle behind quantum teleportation and demonstrate the protocol. Another closely related quantum communication protocol is the so-called *superdense coding*. Since the idea and protocol stand in parallel to

---

[1] In this sense, quantum teleportation may not be classified rigorously as a quantum algorithm. Anyhow, it is included here because it is not only interesting in its own right but also used as part of various quantum algorithms.

the quantum teleportation protocol, we do not discuss it here and refer the readers to the original work (Bennett & Wiesner, 1992).

4.1.1   Nonlocality in Entanglement

Quantum entanglement is a key resource in quantum teleportation as we will see shortly. However, it also reveals an intriguing feature of quantum mechanics: *nonlocality*. Quantum entanglement and the accompanying nonlocality are largely unexpected consequences of the superposition principle of quantum states, first pointed out by Einstein et al. (1935). Before we discuss an implementation of quantum teleportation, we take a brief look at the nonlocal property buried in quantum entanglement.

Suppose that Alice and Bob share a pair of qubits in an entangled state

$$|\Psi\rangle = \frac{|0\rangle \otimes |0\rangle + |1\rangle \otimes |1\rangle}{\sqrt{2}}. \tag{4.1}$$

Recall that it can be generated using the quantum entangler circuit in Sect. 2.2.1. When Bob measures his qubit, the measurement readout is just random and can be either 0 or 1. However, if Alice performs a measurement on her qubit anytime before Bob's measurement, Bob's result is completely fixed by the result of Alice's measurement.

This is the quantum entangler circuit. It generates a maximally entangled state.

In[ ]:= qc = QuantumCircuit[LogicalForm[Ket[], S@{1, 2}], S[1, 6], CNOT[S[1], S[2]]]

Out[ ]=

$|0\rangle$ —[H]—●—
$|0\rangle$ ————⊕—

Here is the output state. It is indeed a maximally entangled state.

In[ ]:= vec = ExpressionFor[qc];
vec // LogicalForm

Out[ ]= $\dfrac{\left|0_{S_1} 0_{S_2}\right\rangle}{\sqrt{2}} + \dfrac{\left|1_{S_1} 1_{S_2}\right\rangle}{\sqrt{2}}$

Now Bob measure on his qubit.

In[ ]:= qc2 = QuantumCircuit[qc, "Spacer", Measurement@S[2]]

Out[ ]=

$|0\rangle$ —[H]—●———
$|0\rangle$ ————⊕—[◿]---

This shows that the measurement outcome is just random.

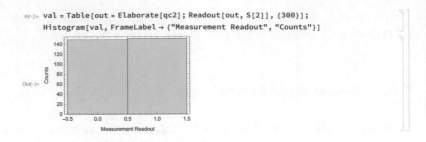

```
In[ ]:= val = Table[out = Elaborate[qc2]; Readout[out, S[2]], {300}];
        Histogram[val, FrameLabel → {"Measurement Readout", "Counts"}]
```

Here Alice, possessing the first qubit, measures her qubit before Bob measures his qubit.

```
In[ ]:= qc3 = QuantumCircuit[qc, "Spacer",
        Measurement@S[1], "Separator", Measurement@S[2]]
```

This shows that the measurement outcomes of Alice's and Bobs' are perfectly correlated. This illustrates that Bob's measurement results have affected by Alice measurement.

```
In[ ]:= val = Boole@Table[out = Elaborate[qc2];
        Equal @@ Readout[out, S@{1, 2}], {300}];
        Histogram[val, {-0.5, 1.5, 1},
        FrameLabel → {"Bob's readout equal to Alice's", "Counts"}]
```

The above conclusion holds however far Alice and Bob are separated and however soon Bob measures after Alice does. Somehow Alice's measurement affects Bob's measurement "instantaneously". It seemingly violates Einstein's special theory of relativity, which dictates that nothing can travel faster than light. This has forced many people to hesitate to believe in quantum mechanics, that is until John Bell proposed an experimental test of an inequality that was later named after him (Bell, 1966). An experimental result violating Bell's inequality would mean that quantum mechanics is not a *local realistic* theory. Aspect et al. (1981) reported results of actual experiments that supported the nonlocality of quantum mechanics. Since Bell's pioneering work, many other inequalities have been proposed as tests of the nonlocality of quantum mechanics. One of the best known is the Clauser-Horne-Shimony-Holt inequality (Clauser et al., 1969).

Nonlocality tests have attracted renewed interest when Greenberger et al. (1990) proposed a new test *without an inequality* using three spin-1/2 particles (i.e., three

qubits). It is now known as the GHZ test after the authors. Unlike Bell-type nonlocality tests, it does not use an inequality but is based on an equality. Later, Hardy
(1992, 1993) proposed a similar test without an inequality for two qubits. Hardy's
test is interesting since it works for almost all entangled states while the GHZ test
uses a particular entangled state.

Here we introduce Hardy's test. We then follow the arguments by Goldstein
(1994), rather than Hardy's original work.

Alice and Bob are spatially separated. Suppose that they share a pair of qubits in
an entangled state of the form[2]

$$|\Psi\rangle = \frac{|0\rangle \otimes |0\rangle + |0\rangle \otimes |1\rangle + |1\rangle \otimes |0\rangle}{\sqrt{3}}. \tag{4.2}$$

Each person chooses either $\hat{Z}$ or $\hat{X}$ and measures the selected observable on her/his
qubit.

(a) If both Alice and Bob choose $\hat{Z}$, then there is no chance for both to get outcome
    -1 because $|\Psi\rangle$ does not include the state $|1\rangle \otimes |1\rangle$.
(b) If Alice measures $\hat{Z}$ and Bob measures $\hat{X}$, no event occurs for which Alice gets
    1 and Bob gets -1. This becomes clear when we rewrite $|\Psi\rangle$ in the form

$$|\Psi\rangle = \frac{\sqrt{2}}{\sqrt{3}} |0\rangle \otimes |+\rangle + \frac{1}{\sqrt{3}} |1\rangle \otimes |0\rangle , \tag{4.3}$$

where $|\pm\rangle = (|0\rangle \pm |1\rangle)/\sqrt{2}$ are the eigenstates of $\hat{X}$ belonging to the eigenval-
ues ± 1. If Alice's qubit is in $|0\rangle$, then Bob's qubit is definitely in $|+\rangle$, and they
will never get the outcomes 1 and -1, respectively.
(c) If Alice measures $\hat{X}$ while Bob measures $\hat{Z}$, they will never get the respective
    outcomes -1 and 1 for a similar reason. This becomes clear by putting $|\psi\rangle$ in
    the form

$$|\Psi\rangle = \frac{1}{\sqrt{3}} |0\rangle \otimes |1\rangle + \frac{\sqrt{2}}{\sqrt{3}} |+\rangle \otimes |0\rangle . \tag{4.4}$$

The expected measurement results are summarized in Table 4.1.

Now, what if both Alice and Bob measure $\hat{X}$ on their qubits? What would a *local
realistic* argument predict given the known facts (a)–(c)? Einstein et al. (1935) argues,
"If, without in any way disturbing a system, we can predict with certainty the value of
a physical quantity, then there exists an element of physical reality corresponding to
this physical quantity." This is called the *reality* requirement. They also argue, "Since
at the time of measurement the two systems no longer interact, no real change can
take place in the second system in consequence of anything that may be done to the
first system." This is called the *locality* requirement.

---

[2] Here we assume a particular entangled state for a simple introduction, but Hardy's test has almost
no restriction.

Table 4.1 Hardy's test of nonlocality without inequality. The first row lists the combinations of observables to measure. The first column lists possible measurement outcomes. The three zeros indicate known facts about the state of the given system, meaning that a measurement of the combination never gives the corresponding outcome. Then, a local realistic argument would predict that the probability P for measurement $(\hat{X}, \hat{X})$ to give $(-1, -1)$ is zero

	$(\hat{Z}, \hat{Z})$	$(\hat{Z}, \hat{X})$	$(\hat{X}, \hat{Z})$	$(\hat{X}, \hat{X})$
$(+1, +1)$				
$(+1, -1)$		0		
$(-1, +1)$			0	
$(-1, -1)$	0			P

Let us deduce the result of the measurement combination $(\hat{X}, \hat{X})$ respecting the two requirements above. Fact (a) implies that either Alice's or Bob's qubit must be in the state $|0\rangle$. Suppose that Alice's qubit is in $|0\rangle$. Then, Fact (b) implies that Bob's qubit must be in $|+\rangle$. In turn, this means that when measuring $(\hat{X}, \hat{X})$ they will never observe the outcome $(-1, -1)$. Next consider the case where Bob's qubit is in $|0\rangle$. Then, Fact (c) implies that Alice's qubit must be in $|+\rangle$. In turn, this means that measurement of $(\hat{X}, \hat{X})$ will never give $(-1, -1)$. Therefore, in any case, the probability for Alice and Bob to get the outcome $(-1, -1)$ from a measurement of $(\hat{X}, \hat{X})$ is zero.

What does quantum mechanics predict? The probability for the outcome $(-1, -1)$ from a measurement of $(\hat{X}, \hat{X})$ is given by

$$P = |\langle -- |\Psi\rangle|^2 = \frac{1}{12}. \tag{4.5}$$

It contradicts the above local realistic argument! This implies that quantum theory is nonlocal.

4.1.2 Implementation of Quantum Teleportation

Now let us turn back to the quantum teleportation protocol. Suppose that Bob wants to send one bit of quantum information, say, $|\psi\rangle_C = |0\rangle_C \psi_0 + |1\rangle_C \psi_1$ stored in Charlie's qubit, to Alice residing in a place far away from Bob (and Charlie). It is important to note that the quantum state $|\psi\rangle$ is unknown. The no-cloning theorem (see Sect. 1.3.1) prevents a naive approach such as copying and transmitting the quantum state.

To teleport a quantum state, Alice and Bob need to share an entangled pair of qubits in advance. Suppose that the shared pair of qubits is in the state

$$|\beta_0\rangle_{AB} = \frac{|00\rangle_{AB} + |11\rangle_{AB}}{\sqrt{2}}, \tag{4.6}$$

which is one of the four Bell states. Recall that the entanglement is generated by a combination of the Hadamard and CNOT gate (see Sect. 2.2.1). Thus, the initial state $|\Psi\rangle$ of the three qubits at the beginning of the procedure is given by

$$|\Psi\rangle = |\beta_0\rangle_{AB} \otimes |\psi\rangle_C = \frac{|000\rangle\psi_0 + |110\rangle\psi_0 + |001\rangle\psi_1 + |111\rangle\psi_1}{\sqrt{2}} \quad (4.7)$$

Rewriting the parts consisting of qubits B and C in the Bell basis—the basis consisting of the four Bell states in Eq. (2.88)—leads to

$$|\Psi\rangle = (|0\rangle\psi_0 + |1\rangle\psi_1)_A \otimes |\beta_0\rangle_{BC} + (|0\rangle\psi_1 + |1\rangle\psi_0)_A \otimes |\beta_1\rangle_{BC}$$
$$+ (|0\rangle\psi_1 - |1\rangle\psi_0)_A \otimes |\beta_2\rangle_{BC} + (|0\rangle\psi_0 - |1\rangle\psi_1)_A \otimes |\beta_3\rangle_{BC}. \quad (4.8)$$

The crucial point here is that the state of A in the first term is identical to the quantum state $|\psi\rangle$, and those in the rest are also closely related to $|\psi\rangle$ by the Pauli operators. More explicitly, one can rewrite the total state vector as

$$|\Psi\rangle = |\psi\rangle_A \otimes |\beta_0\rangle_{BC} + \left(\hat{S}_A^x |\psi\rangle_A\right) \otimes |\beta_1\rangle_{BC}$$
$$+ \left(i\hat{S}_A^y |\psi\rangle_A\right) \otimes |\beta_2\rangle_{BC} + \left(\hat{S}_A^z |\psi\rangle_A\right) \otimes |\beta_3\rangle_{BC} \quad (4.9)$$

Now Bob performs a measurement on the two qubits B and C, owned by Bob, *in the Bell basis*. The measurement will yield outcomes $\mu = 0, 1, 2, 3$ and will collapse the total state vector into the corresponding term, $\hat{S}_A^\mu |\psi\rangle_A \otimes |\beta_\mu\rangle$ (there is an additional factor of i for $\mu = 2$ but the global phase factor is physically irrelevant). The remaining task is for Bob to inform Alice of the outcome μ so that Alice recover the desired state $|\psi\rangle$ by operating the inverse operator $\hat{S}_A^\mu$ on her qubit. The information regarding the measurement outcome amounts to two bits, and it requires only a classical channel for transmission. In short, the quantum teleportation protocol consists of the following steps:

1. Alice and Bob generate an entangled pair of qubits, A and B, and share the pair between them. This can be done anytime before the procedure actually starts. Bob prepares a quantum state to send in a separate qubit C.
2. Bob makes a Bell measurement, i.e., the measurement in the Bell basis, on his two qubits B and C.
3. Bob sends the two-bit information regarding the measurement outcome to Alice through a classical communication channel.
4. Alice operates a proper inverse operator to recover the desired quantum state.

Here is a simulation of the quantum teleportation protocol using Q3.

The initial state of the total system at the beginning of the protocol.

```
In[ ]:= Let[Complex, ψ]
        vec = Ket[] × ψ[0] + Ket[S[3] → 1] × ψ[1];
        vec // LogicalForm
```
$$Out[ ]= \left| 0_{S_3} \right\rangle \psi_0 + \left| 1_{S_3} \right\rangle \psi_1$$

```
In[ ]:= tot = BellState[S@{0, 2}, 0] ** vec;
        tot // LogicalForm
```
$$Out[ ]= \frac{\left| 0_{S_0} 0_{S_2} 0_{S_3} \right\rangle \psi_0}{\sqrt{2}} + \frac{\left| 1_{S_0} 1_{S_2} 0_{S_3} \right\rangle \psi_0}{\sqrt{2}} + \frac{\left| 0_{S_0} 0_{S_2} 1_{S_3} \right\rangle \psi_1}{\sqrt{2}} + \frac{\left| 1_{S_0} 1_{S_2} 1_{S_3} \right\rangle \psi_1}{\sqrt{2}}$$

The total state is rewritten in the Bell basis for qubits S[2,None] and S[3,None].

```
In[ ]:= bs = BellState[S@{2, 3}];
        prj = Dyad[#, #] & /@ bs;
        KetFactor /@ (prj ** tot) // TableForm
```
$$Out[ ]//TableForm=$$
$$\frac{\left(\left| 0_{S_2} 0_{S_3} \right\rangle + \left| 1_{S_2} 1_{S_3} \right\rangle \right) \otimes \left(\left| 0_{S_0} \right\rangle \psi_0 + \left| 1_{S_0} \right\rangle \psi_1 \right)}{2\sqrt{2}}$$
$$\frac{\left(\left| 0_{S_2} 1_{S_3} \right\rangle + \left| 1_{S_2} 0_{S_3} \right\rangle \right) \otimes \left(\left| 1_{S_0} \right\rangle \psi_0 + \left| 0_{S_0} \right\rangle \psi_1 \right)}{2\sqrt{2}}$$
$$\frac{\left(-\left| 0_{S_2} 1_{S_3} \right\rangle + \left| 1_{S_2} 0_{S_3} \right\rangle \right) \otimes \left(\left| 1_{S_0} \right\rangle \psi_0 - \left| 0_{S_0} \right\rangle \psi_1 \right)}{2\sqrt{2}}$$
$$\frac{\left(\left| 0_{S_2} 0_{S_3} \right\rangle - \left| 1_{S_2} 1_{S_3} \right\rangle \right) \otimes \left(\left| 0_{S_0} \right\rangle \psi_0 - \left| 1_{S_0} \right\rangle \psi_1 \right)}{2\sqrt{2}}$$

Step 1. Alice and Bob generates an entangled pair. Bob prepares a quantum state in a separate qubit.

```
In[ ]:= qc1 = QuantumCircuit[LogicalForm[Ket[], S@{1, 2}],
          ProductState[S[3] → {ψ[0], ψ[1]}, "Label" → Ket[ψ]], S[1, 6],
          CNOT[S[1], S[2]], "Separator", "Invisible" → S[1.5]]
```

Step 2. Bob performs a Bell measurement on his qubits. This can be done by reversing the entangler circuit.

```
In[ ]:= qc2a = QuantumCircuit[CNOT[S[2], S[3]], S[2, 6]];
        op = ExpressionFor[qc2a];
        bs = BellState[S@{2, 3}];
        ls = op ** bs;
        LogicalForm@Transpose@Thread[bs → ls] // TableForm
```
$$Out[ ]//TableForm=$$
$$\frac{\left| 0_{S_2} 0_{S_3} \right\rangle + \left| 1_{S_2} 1_{S_3} \right\rangle}{\sqrt{2}} \to \left| 0_{S_2} 0_{S_3} \right\rangle$$
$$\frac{\left| 0_{S_2} 1_{S_3} \right\rangle + \left| 1_{S_2} 0_{S_3} \right\rangle}{\sqrt{2}} \to \left| 0_{S_2} 1_{S_3} \right\rangle$$
$$\frac{\left| 0_{S_2} 1_{S_3} \right\rangle - \left| 1_{S_2} 0_{S_3} \right\rangle}{\sqrt{2}} \to \left| 1_{S_2} 1_{S_3} \right\rangle$$
$$\frac{\left| 0_{S_2} 0_{S_3} \right\rangle - \left| 1_{S_2} 1_{S_3} \right\rangle}{\sqrt{2}} \to \left| 1_{S_2} 0_{S_3} \right\rangle$$

This is the corresponding quantum circuit model.

In[ ]:= `qc2 = QuantumCircuit["Separator", qc2a,`
`    Measurement@S@{2, 3}, "Visible" → S[1], "Invisible" → S[1.5]]`

Out[ ]:=

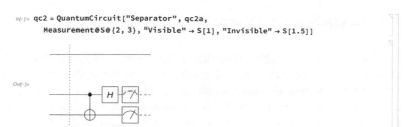

Step 3. Bob sends the measurement outcome to Alice through a classical channel.
Step 4. Alice applies a proper unitary gate on her qubit in accordance with Bob's message.
These two steps are simulated by a feedback control.

In[ ]:= `qc2 = QuantumCircuit[Measurement@S@{2, 3}, "Separator",`
`    ControlledU[S[2], S[1, 3]], ControlledU[S[3], S[1, 1]], "Invisible" → S[1.5]]`

Out[ ]:=

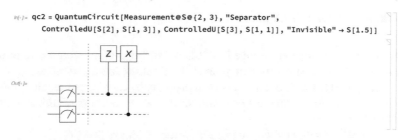

Combining all the steps, one can transmit a quantum state to a remote party without a quantum channel.

In[ ]:= `qc = QuantumCircuit[in = ProductState[S[3] → {ψ[0], ψ[1]}, "Label" → Ket[ψ]],`
`    LogicalForm[Ket[], S@{1, 2}], S[1, 6], CNOT[S[1], S[2]], "Separator",`
`    "Spacer", CNOT[S[2], S[3]], S[2, 6], Measurement[S@{2, 3}],`
`    "Separator", ControlledU[S[2], S[1, 3]], ControlledU[S[3], S[1, 1]],`
`    ProductState[S[1] → {ψ[0], ψ[1]}, "Label" → Ket[ψ]], "Invisible" → S[1.5]]`

Out[ ]:=

Check the result.

In[ ]:= `out = ExpressionFor[qc] /. {Conjugate[ψ[0]] * ψ[0] + Conjugate[ψ[1]] * ψ[1] → 1};`
`    LogicalForm@KetFactor[out, S@{2, 3}]`

Out[ ]:= $\left|0_{S_2} 0_{S_3}\right\rangle \otimes \left(\left|0_{S_1}\right\rangle \psi_0 + \left|1_{S_1}\right\rangle \psi_1\right)$

Try with an explicit numbers as well.

In[ ]:= `{ψ[0], ψ[1]} = Normalize@RandomVector[];`
`    out = ExpressionFor[qc];`
`    in → LogicalForm@KetFactor[out, S@{2, 3}]`

Out[ ]:= $\left((-0.378488 + 0.186883\ i)\ |0\rangle + (0.859853 + 0.287185\ i)\ |1\rangle\right)_{S_3} →$
$\left|1_{S_2} 1_{S_3}\right\rangle \otimes \left((-0.378488 + 0.186883\ i) \times \left((1.+0.\ i)\ |0_{S_1}\rangle - (1.52529 + 1.5119\ i)\ |1_{S_1}\rangle\right)\right)$

4.2 Deutsch-Jozsa Algorithm & Variants

The Deutsch-Jozsa algorithm (Deutsch, 1985; Deutsch & Jozsa, 1992) is known as
the first quantum algorithm that is faster than the best classical counterpart. Although
it is not useful for practical applications, it is still interesting since it illustrates some
aspects of *quantum parallelism* and provides a hint for why quantum computers are
faster than classical computers. It has inspired two variants: the Bernstein-Vazirani
algorithm and Simon's algorithm.

4.2.1 *Quantum Oracle*

An *oracle* in computer science is a "black box" operation with a certain unknown
property. In a decision problem, such as the Deutsch-Jozsa and related problems, we
are supposed to figure out the unknown property by running the oracle. Before going
further, let us first examine a *quantum oracle*—a quantum mechanical implementation of an oracle.

Typically, a classical oracle is described by a binary function

$$f : \{0, 1\}^m \to \{0, 1\}^n \,, \tag{4.10}$$

that maps an m-bit input to an n-bit output. The function $f(x)$ may not be invertible
in general.

Given a classical oracle f, a proper quantum implementation should operate on
qubits and allow superposition in the input states. One naive approach is to define
an operator $\hat{O}$ by $\hat{O} |x\rangle = |f(x)\rangle$ for any state in the logical basis of the m-qubit
register. This would not work because, as mentioned above, the function $f(x)$ is not
invertible in general, and hence operator $\hat{O}$ cannot be unitary.

To overcome such an issue, first extend the mapping f by adding an auxiliary
register of n bits to the input and keeping the original input value so that both the
input and output registers have $(m + n)$ bits. An extended mapping, $\{0, 1\}^{m+n} \to$
$\{0, 1\}^{m+n}$, is then defined by the association

$$(x, y) \mapsto (x, f(x) \oplus y) \,, \tag{4.11}$$

where $x \in \{0, 1\}^m$ and $y \in \{0, 1\}^n$ are bit strings of the m-bit native register and
the n-bit auxiliary register, respectively. Although the function $f(x)$ itself may not
be invertible, the extended mapping in (4.11) is always one-to-one regardless of
the function $f(x)$ (see Problem 4.1). Due to this property, the extended mapping
in (4.11) is widely used to convert a classical code to a form that is suitable for
(classical) *reversible computation*.[3]

---

[3] For a complete reversible computation, there is another important step required to remove the
"garbage" bits.

The *quantum oracle* corresponding to the classical oracle f is simply an implementation of the extended mapping (4.11) on quantum registers. It is a quantum gate operation defined by the association

$$\hat{U}_f : |x\rangle \otimes |y\rangle \mapsto |x\rangle \otimes |f(x) \oplus y\rangle \, , \qquad (4.12)$$

where $|x\rangle$ and $|y\rangle$ are the logical basis states belonging to the native and auxiliary register of m and n qubits, respectively. Since the extended mapping (4.11) is one-to-one and the logical basis states are orthonormal, the operator $\hat{U}_f$ is unitary, as given in Problem 4.1. Recall that $\hat{U}_f$ is a linear operator and can act on any arbitrary superposition states. Then, in quantum circuit model, a quantum oracle is depicted diagrammatically as follows

$$(4.13)$$

In this particular example, the first three qubits are from the native register and the last two qubits belong to the auxiliary register. The bit values of the auxiliary qubits are flipped conditionally depending on the value $f(x)$ as a function of the bit values x of the native qubits.

Here is a simple example, which is used in the Deutsch-Jozsa algorithm.

Consider a balanced function f as a classical oracle.

```
Clear[f];
f[0, 0] = f[1, 1] = 0;
f[0, 1] = f[1, 0] = 1;
```

Here is the corresponding quantum oracle.

```
In[ ]:= cc = S@{1, 2};
        tt = S[3];
        qc = QuantumCircuit[Oracle[f, cc, tt]]
```

Out[ ]=

```
In[ ]:= op = ExpressionFor[qc]
```

$$Out[\ ]= \frac{1}{2} + \frac{1}{2} S_1^z S_2^z - \frac{1}{2} S_1^z S_2^z S_3^x + \frac{S_3^x}{2}$$

```
In[ ]:= bs = Basis@Join[cc, {tt}];
        bs // LogicalForm
```

$$Out[ ]= \left\{ \left| 0_{S_1} 0_{S_2} 0_{S_3} \right\rangle, \; \left| 0_{S_1} 0_{S_2} 1_{S_3} \right\rangle, \; \left| 0_{S_1} 1_{S_2} 0_{S_3} \right\rangle, \right.$$
$$\left. \left| 0_{S_1} 1_{S_2} 1_{S_3} \right\rangle, \; \left| 1_{S_1} 0_{S_2} 0_{S_3} \right\rangle, \; \left| 1_{S_1} 0_{S_2} 1_{S_3} \right\rangle, \; \left| 1_{S_1} 1_{S_2} 0_{S_3} \right\rangle, \; \left| 1_{S_1} 1_{S_2} 1_{S_3} \right\rangle \right\}$$

```
In[ ]:= out = op ** bs;
        out // LogicalForm
```

$$Out[ ]= \left\{ \left| 0_{S_1} 0_{S_2} 0_{S_3} \right\rangle, \; \left| 0_{S_1} 0_{S_2} 1_{S_3} \right\rangle, \; \left| 0_{S_1} 1_{S_2} 1_{S_3} \right\rangle, \right.$$
$$\left. \left| 0_{S_1} 1_{S_2} 0_{S_3} \right\rangle, \; \left| 1_{S_1} 0_{S_2} 1_{S_3} \right\rangle, \; \left| 1_{S_1} 0_{S_2} 0_{S_3} \right\rangle, \; \left| 1_{S_1} 1_{S_2} 1_{S_3} \right\rangle, \; \left| 1_{S_1} 1_{S_2} 0_{S_3} \right\rangle \right\}$$

```
In[ ]:= Thread[bs → out] // LogicalForm // TableForm
```
Out[ ]//TableForm=

$$\left| 0_{S_1} 0_{S_2} 0_{S_3} \right\rangle \rightarrow \left| 0_{S_1} 0_{S_2} 0_{S_3} \right\rangle$$
$$\left| 0_{S_1} 0_{S_2} 1_{S_3} \right\rangle \rightarrow \left| 0_{S_1} 0_{S_2} 1_{S_3} \right\rangle$$
$$\left| 0_{S_1} 1_{S_2} 0_{S_3} \right\rangle \rightarrow \left| 0_{S_1} 1_{S_2} 1_{S_3} \right\rangle$$
$$\left| 0_{S_1} 1_{S_2} 1_{S_3} \right\rangle \rightarrow \left| 0_{S_1} 1_{S_2} 0_{S_3} \right\rangle$$
$$\left| 1_{S_1} 0_{S_2} 0_{S_3} \right\rangle \rightarrow \left| 1_{S_1} 0_{S_2} 1_{S_3} \right\rangle$$
$$\left| 1_{S_1} 0_{S_2} 1_{S_3} \right\rangle \rightarrow \left| 1_{S_1} 0_{S_2} 0_{S_3} \right\rangle$$
$$\left| 1_{S_1} 1_{S_2} 0_{S_3} \right\rangle \rightarrow \left| 1_{S_1} 1_{S_2} 0_{S_3} \right\rangle$$
$$\left| 1_{S_1} 1_{S_2} 1_{S_3} \right\rangle \rightarrow \left| 1_{S_1} 1_{S_2} 1_{S_3} \right\rangle$$

Let us consider another example, which is used in Simon's algorithm.

Consider a two-to-one function as a classical oracle.

```
Clear[f];
f[0, 0] = f[1, 1] = {0, 1};
f[0, 1] = f[1, 0] = {1, 1};
```

Here is an implementation of the corresponding quantum oracle.

```
In[ ]:= cc = S@{1, 2};
        tt = S@{3, 4};
        qc = QuantumCircuit[Oracle[f, cc, tt]]
```

Out[ ]=

```
In[ ]:= op = Elaborate@Oracle[f, cc, tt]
```

$$Out[ ]= \frac{1}{2} S_3^x S_4^x + \frac{1}{2} S_1^z S_2^z S_4^x - \frac{1}{2} S_1^z S_2^z S_3^x S_4^x + \frac{S_4^x}{2}$$

```
In[ ]:= bs = Basis@Join[cc, tt];
        bs // LogicalForm
```

$$Out[ ]= \left\{ \left| 0_{S_1} 0_{S_2} 0_{S_3} 0_{S_4} \right\rangle, \; \left| 0_{S_1} 0_{S_2} 0_{S_3} 1_{S_4} \right\rangle, \; \left| 0_{S_1} 0_{S_2} 1_{S_3} 0_{S_4} \right\rangle, \; \left| 0_{S_1} 0_{S_2} 1_{S_3} 1_{S_4} \right\rangle, \right.$$
$$\left| 0_{S_1} 1_{S_2} 0_{S_3} 0_{S_4} \right\rangle, \; \left| 0_{S_1} 1_{S_2} 0_{S_3} 1_{S_4} \right\rangle, \; \left| 0_{S_1} 1_{S_2} 1_{S_3} 0_{S_4} \right\rangle, \; \left| 0_{S_1} 1_{S_2} 1_{S_3} 1_{S_4} \right\rangle,$$
$$\left| 1_{S_1} 0_{S_2} 0_{S_3} 0_{S_4} \right\rangle, \; \left| 1_{S_1} 0_{S_2} 0_{S_3} 1_{S_4} \right\rangle, \; \left| 1_{S_1} 0_{S_2} 1_{S_3} 0_{S_4} \right\rangle, \; \left| 1_{S_1} 0_{S_2} 1_{S_3} 1_{S_4} \right\rangle,$$
$$\left. \left| 1_{S_1} 1_{S_2} 0_{S_3} 0_{S_4} \right\rangle, \; \left| 1_{S_1} 1_{S_2} 0_{S_3} 1_{S_4} \right\rangle, \; \left| 1_{S_1} 1_{S_2} 1_{S_3} 0_{S_4} \right\rangle, \; \left| 1_{S_1} 1_{S_2} 1_{S_3} 1_{S_4} \right\rangle \right\}$$

```
In[ ]:= out = op ** bs;
       out // LogicalForm
```

$$
\begin{aligned}
\textit{Out[ ]=} \{ &|0_{s_1}0_{s_2}0_{s_3}1_{s_4}\rangle, \ |0_{s_1}0_{s_2}0_{s_3}0_{s_4}\rangle, \ |0_{s_1}0_{s_2}1_{s_3}1_{s_4}\rangle, \ |0_{s_1}0_{s_2}1_{s_3}0_{s_4}\rangle, \\
&|0_{s_1}1_{s_2}1_{s_3}1_{s_4}\rangle, \ |0_{s_1}1_{s_2}1_{s_3}0_{s_4}\rangle, \ |0_{s_1}1_{s_2}0_{s_3}1_{s_4}\rangle, \ |0_{s_1}1_{s_2}0_{s_3}0_{s_4}\rangle, \\
&|1_{s_1}0_{s_2}1_{s_3}1_{s_4}\rangle, \ |1_{s_1}0_{s_2}1_{s_3}0_{s_4}\rangle, \ |1_{s_1}0_{s_2}0_{s_3}1_{s_4}\rangle, \ |1_{s_1}0_{s_2}0_{s_3}0_{s_4}\rangle, \\
&|1_{s_1}1_{s_2}0_{s_3}1_{s_4}\rangle, \ |1_{s_1}1_{s_2}0_{s_3}0_{s_4}\rangle, \ |1_{s_1}1_{s_2}1_{s_3}1_{s_4}\rangle, \ |1_{s_1}1_{s_2}1_{s_3}0_{s_4}\rangle \}
\end{aligned}
$$

```
In[ ]:= tbl = Thread[Rule[bs, out]] // LogicalForm;
       tbl[[ ;; 8]] // TableForm
```

$$
\begin{aligned}
\textit{Out[ ]//TableForm=} \\
|0_{s_1}0_{s_2}0_{s_3}0_{s_4}\rangle &\rightarrow |0_{s_1}0_{s_2}0_{s_3}1_{s_4}\rangle \\
|0_{s_1}0_{s_2}0_{s_3}1_{s_4}\rangle &\rightarrow |0_{s_1}0_{s_2}0_{s_3}0_{s_4}\rangle \\
|0_{s_1}0_{s_2}1_{s_3}0_{s_4}\rangle &\rightarrow |0_{s_1}0_{s_2}1_{s_3}1_{s_4}\rangle \\
|0_{s_1}0_{s_2}1_{s_3}1_{s_4}\rangle &\rightarrow |0_{s_1}0_{s_2}1_{s_3}0_{s_4}\rangle \\
|0_{s_1}1_{s_2}0_{s_3}0_{s_4}\rangle &\rightarrow |0_{s_1}1_{s_2}1_{s_3}1_{s_4}\rangle \\
|0_{s_1}1_{s_2}0_{s_3}1_{s_4}\rangle &\rightarrow |0_{s_1}1_{s_2}1_{s_3}0_{s_4}\rangle \\
|0_{s_1}1_{s_2}1_{s_3}0_{s_4}\rangle &\rightarrow |0_{s_1}1_{s_2}0_{s_3}1_{s_4}\rangle \\
|0_{s_1}1_{s_2}1_{s_3}1_{s_4}\rangle &\rightarrow |0_{s_1}1_{s_2}0_{s_3}0_{s_4}\rangle
\end{aligned}
$$

There are several interesting features of a quantum oracle that can be immediately noticed from the definition in (4.12). In Sect. 2.2.1, we noted that the CNOT gate makes a copy of the logical state of the control register to the target register when the latter has been prepared initially in the state $|0\rangle$, as in Eq. (2.36). This property is exploited in many applications, especially to generate entanglement as in Eqs. (2.33) and (2.37). The quantum oracle has a similar property, but it makes a copy of the image $|f(x)\rangle$ rather than the state $|x\rangle$ itself of the native register to the ancillary register,

$$
|x\rangle \otimes |0\rangle \mapsto |x\rangle \otimes |f(x)\rangle . \tag{4.14}
$$

Suppose that a native quantum register is in the superposition $\frac{1}{2^{m/2}} \sum_{x=0}^{2^m-1} |x\rangle$ and the ancillary quantum register in in the state $|0\rangle \equiv |0\rangle^{\otimes n}$. The quantum oracle transforms the state as

$$
\frac{1}{2^{m/2}} \sum_{x=0}^{2^m-1} |x\rangle \otimes |0\rangle \mapsto \frac{1}{2^{m/2}} \sum_{x=0}^{2^m-1} |x\rangle \otimes |f(x)\rangle . \tag{4.15}
$$

Just like the state in (2.33) from the CNOT gate, the state (4.15) from the quantum oracle is also entangled unless f is constant. In this case, the entanglement is controlled by the (classical) oracle f.

Here is demonstrate the feature of quantum oracle that makes a copy of the image f(x) of the native register to the ancillary qubit.

In this particular example, we consider a one-to-one function, but it can be arbitrary (but not constant if an entanglement is necessary).

```
Clear[f];
f[0, 0] = {1, 1};
f[0, 1] = {1, 0};
f[1, 0] = {0, 1};
f[1, 1] = {0, 0};
```

```
In[ ]:=  cc = {1, 2};
         tt = {3, 4};
         qc = QuantumCircuit[
           LogicalForm[Ket[], Join[S@cc, S@tt]], S[cc, 6], Oracle[f, S@cc, S@tt]]
```

The output state is an entangled state (unless the classical oracle f is constant).

```
In[ ]:=  out = ExpressionFor[qc];
         LogicalForm[out, Join[S@cc, S@tt]]
```

$$Out[ ]= \frac{1}{2}\left|0_{s_1}0_{s_2}1_{s_3}1_{s_4}\right\rangle + \frac{1}{2}\left|0_{s_1}1_{s_2}1_{s_3}0_{s_4}\right\rangle + \frac{1}{2}\left|1_{s_1}0_{s_2}0_{s_3}1_{s_4}\right\rangle + \frac{1}{2}\left|1_{s_1}1_{s_2}0_{s_3}0_{s_4}\right\rangle$$

To make clearer the copies made tot the ancillary register, it may be useful to rewrite the state vector in a form that distinguishes the native and ancillary register.

```
In[ ]:=  KetFactor[out, S@cc] // LogicalForm
```

$$Out[ ]= \left|0_{s_1}0_{s_2}\right\rangle \otimes \left(\frac{1}{2}\left|1_{s_3}1_{s_4}\right\rangle\right) + \left|0_{s_1}1_{s_2}\right\rangle \otimes \left(\frac{1}{2}\left|1_{s_3}0_{s_4}\right\rangle\right) +$$
$$\left|1_{s_1}0_{s_2}\right\rangle \otimes \left(\frac{1}{2}\left|0_{s_3}1_{s_4}\right\rangle\right) + \left|1_{s_1}1_{s_2}\right\rangle \otimes \left(\frac{1}{2}\left|0_{s_3}0_{s_4}\right\rangle\right)$$

In Sect. 2.2.2, we have seen that the controlled-unitary gate induces a phase shift on the control register—rather than on the target register—when the target register is in an eigenstate of the unitary operator. A similar method can be used to induce a phase shift conditionally on every term that satisfies a certain condition. For example, let $f : \{0, 1\}^n \to \{0, 1\}$ be a classical oracle, and suppose that we are given a state

$$|\psi\rangle = \sum_{x=0}^{2^n-1} |x\rangle\, \psi_x \tag{4.16}$$

on an n-qubit register. We want to get a new state where every term $|x\rangle$ satisfying $f(x) = 1$ flips its amplitude ψ_x to $-\psi_x$

$$\left|\psi'\right\rangle = \sum_{x=0}^{2^n-1} |x\rangle\, (-1)^{f(x)}\psi_x \tag{4.17}$$

In other words, we want an *effective* quantum gate that maps the logical basis states as

$$|x\rangle \mapsto |x\rangle\, (-1)^{f(x)}, \quad x = 0, 1, 2, \ldots, 2^n - 1\,. \tag{4.18}$$

This can be achieved using the quantum oracle corresponding to the function f and preparing an auxiliary qubit in the state $|-\rangle := (|0\rangle - |1\rangle)/\sqrt{2}$, the eigenstate of the Pauli X operator belonging to the eigenvalue -1, as depicted in the quantum circuit

$$(4.19)$$

Indeed, for a logical state $|x\rangle$ with $f(x) = 1$,

$$|x\rangle \otimes |-\rangle = \frac{|x\rangle \otimes |0\rangle - |x\rangle \otimes |1\rangle}{\sqrt{2}} \mapsto \frac{|x\rangle \otimes |1\rangle - |x\rangle \otimes |0\rangle}{\sqrt{2}} = -|x\rangle \otimes |-\rangle \;,$$

$$(4.20)$$

while nothing happens for $|x\rangle$ with $f(x) = 0$. One can find an example of a more general conditional phase shift in Problem 4.2.

This is an interesting twist of the quantum oracle. It is useful in Grover's search algorithm.

The classical oracle mark the solutions, {0,0}=0 and {1,1}=1 in this particular case.

```
Clear[f];
f[0, 0] = 1; f[0, 1] = 0; f[1, 0] = 0; f[1, 1] = 1;
```

This quantum circuit model implements the desired mapping.

```
In[·]:= cc = {1, 2};
tt = {3};
ct = Join[cc, tt];
qc = QuantumCircuit[LogicalForm[Ket[S[tt] → 1], S[ct]],
    S[ct, 6], Oracle[f, S@cc, S@tt], S[tt, 6]]
```

```
In[·]:= out = ExpressionFor[qc];
KetFactor[out, S[tt]] // LogicalForm
```

$$Out[·]= -\frac{1}{2}\left(\left|0_{S_1}\right\rangle - \left|1_{S_1}\right\rangle\right) \otimes \left(\left|0_{S_2}\right\rangle - \left|1_{S_2}\right\rangle\right) \otimes \left|1_{S_3}\right\rangle$$

Check the result.

```
In[·]:= bs = Basis[S@cc];
ff = f @@@ IntegerDigits[Range[0, 2^2 - 1], 2, 2];
Thread[bs → Power[-1, ff]] // LogicalForm
```

$$Out[·]= \left\{\left|0_{S_1}0_{S_2}\right\rangle \to -1,\; \left|0_{S_1}1_{S_2}\right\rangle \to 1,\; \left|1_{S_1}0_{S_2}\right\rangle \to 1,\; \left|1_{S_1}1_{S_2}\right\rangle \to -1\right\}$$

```
In[·]:= new = bs.Power[-1, ff] / 2;
new // LogicalForm
```

$$Out[·]= \frac{1}{2}\left(-\left|0_{S_1}0_{S_2}\right\rangle + \left|0_{S_1}1_{S_2}\right\rangle + \left|1_{S_1}0_{S_2}\right\rangle - \left|1_{S_1}1_{S_2}\right\rangle\right)$$

4.2.2 Deutsch-Jozsa Algorithm

Let us now discuss the Deutsch-Jozsa algorithm. Proposed first by Deutsch & Jozsa (1992), the algorithm determines whether a given function is balanced or constant. It is of little use in practice, but it is one of the first examples of a quantum algorithm that is exponentially faster than any possible classical algorithms. Furthermore, it is so simple that it clearly reveals some aspects of *quantum parallelism* and the operational principle of the quantum oracle.

The Deutsch-Jozsa problem is defined as follows: Consider a classical function $f : \{0, 1\}^n \to \{0, 1\}$. It is a classical oracle taking n-bit inputs and returning 0 or 1. The function is known to be either constant—either 0 or 1 for all inputs—or balanced—0 for one half of the possible inputs and 1 for the other half. The task is to determine whether f is constant or balanced by using the oracle the least number of times.

In classical algorithms, one has to evaluate the oracle $(2^{n-1} + 1) \approx 2^n/2$ times in the worst case. That is, one has to try almost half of the possible inputs before reliably determining the unknown property of the oracle. As we will see now, the Deutsch-Jozsa algorithm determines the property from a single query to the *quantum oracle*.

The Deutsch-Jozsa algorithm is summarized in the quantum circuit

$$(4.21)$$

The last qubit is an ancillary qubit that induces the conditional phase shifts on the first n qubits. The Hadamard gate on it ensures that the ancillary qubit is in the eigenstate $|-\rangle$ of the Pauli X operator just before the quantum oracle operates. According to Eq. (2.18), the Hadamard gates before the quantum oracle create a linear superposition of all the logical basis states of the n-qubit native register

$$|0\rangle \xrightarrow{\hat{H}^{\otimes n}} \frac{1}{2^{n/2}} \sum_{x=0}^{2^n-1} |x\rangle \tag{4.22}$$

Next, as seen in Eq. (4.19), the quantum oracle induces the conditional phase shifts

$$\sum_{x=0}^{2^n-1} |x\rangle \to \sum_{x} |x\rangle (-1)^{f(x)}. \tag{4.23}$$

Another operation of the Hadamard gates shuffles and causes additional phase factors in accordance with (2.19), and it leads to the output state to be measured

$$\frac{1}{2^{n/2}} \sum_x |x\rangle (-1)^{f(x)} \xrightarrow{\hat{H}^{\otimes n}} \frac{1}{2^n} \sum_{y=0}^{2^n-1} |y\rangle \sum_{x=0}^{2^n-1} (-1)^{f(x)+x\cdot y} \tag{4.24}$$

To see the effect of the function f on the final state, suppose that f is a constant function. Then, the output state is given by

$$\frac{(-1)^{f(0)}}{2^n} \sum_{y=0}^{2^n-1} |y\rangle \sum_{x=0}^{2^n-1} (-1)^{x\cdot y} = (-1)^{f(0)} |0\rangle \tag{4.25}$$

Every measurement on each of the n qubits should yield zero with unit probability. To make the analysis more explicit, consider the probability to find the n-qubit register in the state $|0\rangle \equiv |0\rangle \otimes |0\rangle \otimes \cdots \otimes |0\rangle$

$$P_0 = \frac{1}{2^n} \left| \sum_{x=0}^{2^n-1} (-1)^{f(x)} \right|^2 . \tag{4.26}$$

This assures that $P_0 = 1$ for a constant function f. When the function f is balanced, there are as many terms with -1 as 1, and the sum is always zero. There must be at least one qubit in the state $|1\rangle$ if the function f is balanced. Therefore, to determine whether a given function f is constant or balanced, one needs to run the quantum oracle just once, and then check whether the measurement outcome is 0. If the outcome is 0, then the function must be constant; and balanced otherwise.

Consider a balanced function as an example.

```
Clear[f];
f[0, 0] = f[1, 1] = 0;
f[0, 1] = f[1, 0] = 1;
```

Here is a quantum circuit model of the Deutsch-Jozsa algorithm. The final Hadamard gate on the third qubit is not necessary, but we put it here to make the output state more readable.

```
In[·]:= cc = {1, 2};
        tt = {3};
        all = {1, 2, 3};
        qc = QuantumCircuit[LogicalForm[Ket[S[3] → 1], S@all],
          S[all, 6], Oracle[f, S@cc, S@tt], S[all, 6]]
```

In[·]:= `out = ExpressionFor[qc]`

Out[·]= $\left| 1_{s_1} 1_{s_2} 1_{s_3} \right\rangle$

4.2.3   Bernstein-Vazirani Algorithm

Suppose that we are given a binary function $f : \{0, 1\}^n \rightarrow \{0, 1\}$, the value of which is determined by a secret string s of n bits

$$f : x \mapsto x \cdot s \quad (\text{mod } 2). \tag{4.27}$$

The Bernstein-Vazirani problem finds the secret bit string s by making queries to the oracle f. Classically, one needs n queries to the function f to infer the secret string s.

The Bernstein-Vazirani algorithm (Bernstein & Vazirani, 1993, 1997) is implemented in the same quantum circuit (4.21) as the Deutsch-Jozsa algorithm, only that the analysis of the final readout is different. The first set of the Hadamard gates and the quantum oracle in (4.21) leads the n-qubit register to the state

$$\sum_{x=0}^{2^n-1} |x\rangle \, (-1)^{x \cdot s}. \tag{4.28}$$

Taking the inverse of the mapping in (2.19), one can observe that the second set of the Hadamard gates converts the above state to the final state $|s\rangle$ that is set by the secret bit string. Therefore, a simple measurement of the final state in the logical basis will just reveal the secret string.

Consider a secrete string of bits. The task is to find the secrete string.

```
string = {0, 1};
```

In the Bernstein-Vazirani algorithm, the value of the classical oracle f is determined by the given secrete string.

```
Clear[f];
f[x__] := Mod[{x}.string, 2]
```

For example, $f[1,1] = 0 \cdot 1 + 1 \cdot 1 = 1$.

In[·]:= `f[1, 1]`

Out[·]= `1`

Here is a quantum circuit model of the Bernstein-Vazirani algorithm. The final Hadamard gate on the third qubit is not necessary and we put it here to make the output state more readable.

```
In[ ]:= cc = {1, 2};
       tt = {3};
       all = {1, 2, 3};
       qc = QuantumCircuit[LogicalForm[Ket[S[3] → 1], S@all],
         S[all, 6], Oracle[f, S@cc, S@tt], S[all, 6]]
```

```
In[ ]:= out = ExpressionFor[qc];
       LogicalForm[out, S@cc]
```

$$Out[ ]= \left| 0_{S_1} 1_{S_2} 1_{S_3} \right\rangle$$

Here is the secrete string successfully retrieved.

```
In[ ]:= answer = out[S@cc]
```

$$Out[ ]= \{0, 1\}$$

4.2.4 Simon's Algorithm

Finally, let us turn to Simon's algorithm (Simon, 1997). It was the first quantum algorithm featuring an exponential speed-up over the known best classical algorithm for a specific problem. The algorithm is known to have inspired the quantum factorization algorithm. Furthermore, it has recently been shown that Simon's algorithm can be used to break symmetric-key cryptosystems.

In Simon's problem, we are given a function $f : \{0, 1\}^n \to \{0, 1\}^n$ and a secret string s of n bits. For all $x, y \in \{0, 1\}^n$, $f(x) = f(y)$ if and only if $y = x \oplus s$. Note that f is either one-to-one ($s = 0$) or two-to-one ($s \neq 0$). The task is to find the secret string s with as few queries to function f as possible. Classically, one needs queries to $f(x)$ with up to $2^{n-1} + 1$ different inputs.[4] Unlike the Deutsch-Jozsa problem, Simon's problem is known to be hard to solve, even probabilistically.

Simon's algorithm is summarized in Fig. 4.1 in two slightly different quantum circuits. The measurement on the ancillary (second) register in the quantum circuit in Fig. 4.1b can be delayed until the measurement is made on the native (first) register, or even dropped off completely as in the quantum circuit in Fig. 4.1a. That is, it is not essential, but depending on your taste, it may simplify an analysis of the algorithm. Here, we adopt the quantum circuit in Fig. 4.1a. The first Hadamard gate on the native register transforms the input state of the whole system—again, see Eq. (2.18)—as

---

[4] More precisely, one needs order of $\sqrt{2^n}$ queries to encounter a pair of two bit strings leading to the same result with probability greater than 1/2. The number being $\sqrt{2^n}$ rather than 2^n is related to the so-called "birthday paradox".

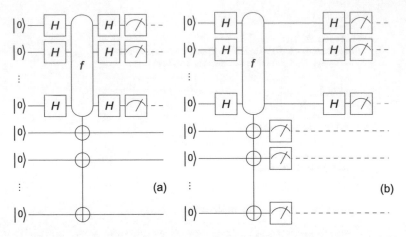

Fig. 4.1 Quantum circuits for Simon's algorithm. Measurement is not essential on the ancillary register in (**b**). It can be performed later than the second Hadamard gate on the native register, or even dropped off completely as in (**a**)

$$|0\rangle \otimes |0\rangle \xrightarrow{\hat{H}^{\otimes n}} \frac{1}{2^{n/2}} \sum_{x=0}^{2^n-1} |x\rangle \otimes |0\rangle \ . \tag{4.29}$$

As we noted in (4.15), the quantum oracle makes a copy of the image $|f(x)\rangle$ of the state $|x\rangle$ of the native register to the ancillary register, and leads to

$$\frac{1}{2^{n/2}} \sum_{x=0}^{2^n-1} |x\rangle \otimes |f(x)\rangle \ . \tag{4.30}$$

Finally, the second set of Hadamard gates on the native register maps the above state into (see Eq. (2.19))

$$\sum_{y=0}^{2^n-1} |y\rangle \otimes \frac{1}{2^n} \sum_{x=0}^{2^n-1} |f(x)\rangle \, (-1)^{x \cdot y}. \tag{4.31}$$

The measurement on the native register yields an n-bit string y. The probability for a particular string y is determined by the squared norm, $P_y = \langle \psi_y | \psi_y \rangle$, of the y-dependent state $|\psi_y\rangle$ of the ancillary register

$$|\psi_y\rangle := \frac{1}{2^n} \sum_{x=0}^{2^n-1} |f(x)\rangle \, (-1)^{x \cdot y} \ . \tag{4.32}$$

Let us then examine different cases. First, suppose that the secret string $s = 0$. In this case, the function f is one-to-one, and the state $|\psi_y\rangle$ in (4.32) simply consists of terms that are a rearrangement of the logical basis states

$$|\psi_y\rangle := \frac{1}{2^n} \sum_{z=0}^{2^n-1} |z\rangle (-1)^{y \cdot f^{-1}(z)}, \qquad (4.33)$$

where f^{-1} is the inverse of f, and hence $P_y = 2^{-n}$ for all y. In other words, for $s = 0$, the measurement produces a random bit string y with uniform probability. Now, suppose that $s \neq 0$. In this case, the function f is two-to-one such that $f(x) = f(x \oplus s)$. The terms in the summation in (4.32) appear in pairs giving the same state, $|f(x)\rangle = |f(x \oplus s)\rangle$. To be more explicit, let $\mathcal{B} := f(\{0, 1\}^n)$ be the image of f. Then,

$$|\psi_y\rangle := \frac{1}{2^n} \sum_{z \in \mathcal{B}} |z\rangle \left\{ (-1)^{y \cdot a_z} + (-1)^{y \cdot b_z} \right\}, \qquad (4.34)$$

where a_z and b_z are the elements in the preimage (or inverse image) of z under f, $f(a_z) = f(b_z) = z$. Since $b_z = a_z \oplus s$ and $(a_z \oplus s) \cdot y = (a_z \cdot y) \oplus (s \cdot y)$, it follows that

$$\left\{ (-1)^{a_z \cdot y} + (-1)^{b_z \cdot y} \right\} = (-1)^{y \cdot a_z} \left\{ 1 + (-1)^{y \cdot s} \right\}. \qquad (4.35)$$

If $y \cdot s$ is odd, $y \cdot s = 1 \pmod 2$, then $|\psi_y\rangle$ is a null vector, and $P_y = 0$ for such a bit string y. On the other hand, if $y \cdot s$ is even, $y \cdot s = 0 \pmod 2$, then $|\psi_y\rangle$ is a finite vector and is independent of y. That is, $P_y = 2^{1-n}$ regardless of y as long as $y \cdot s$ is even. In short, the bit string y resulting from the measurement on the native (first) register always satisfies $y \cdot s = 0 \pmod 2$ for any secret bit string s. To find $s \equiv (s_1 s_2 \ldots s_n)_2$, one needs to run the algorithms repeated to get $(n - 1)$ linearly independent bit strings $y^{(i)} \equiv (y_1^{(i)} y_2^{(i)} \ldots y_n^{(i)})_2$ $(i = 1, 2, \ldots, n - 1)$, and then solve the set of equations

$$\sum_{j=1}^{n} y_j^{(i)} s_j = 0 \pmod 2. \qquad (4.36)$$

It can be argued that the probability to get $n - 1$ linearly independent bit strings out of n runs is slightly larger than $1/4$. Therefore, the required number of queries to the quantum oracle is in the order of n, exponentially smaller than for the classical algorithm.

Consider again a secrete bit string.

```
string = {1, 1};
```

Consider a two-to-one function obeying the rule (specified in Simon's problem).

```
Clear[f];
f[0, 0] = f[1, 1] = {0, 1};
f[0, 1] = f[1, 0] = {1, 1};
```

Here is an implementation of the corresponding quantum oracle.

```
In[ ]:= cc = {1, 2};
tt = {3, 4};
all = Join[cc, tt];
qc = QuantumCircuit[LogicalForm[Ket[], S@all],
     S[cc, 6], Oracle[f, S@cc, S@tt], S[cc, 6], Measurement[S@cc]]
```

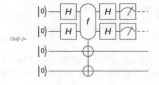

Out[ ]=

```
In[ ]:= out = ExpressionFor[qc];
LogicalForm[out, S@all]
result = Readout[out, S@cc]
```

$$Out[ ]= \frac{\left|0_{s_1}0_{s_2}0_{s_3}1_{s_4}\right\rangle}{\sqrt{2}} + \frac{\left|0_{s_1}0_{s_2}1_{s_3}1_{s_4}\right\rangle}{\sqrt{2}}$$

Out[ ]= {0, 0}

```
In[ ]:= eqs = Table[out = ExpressionFor[Matrix[qc], S@all];
     result = Readout[out, S@cc], {2}]
```

Out[ ]= {{0, 0}, {1, 1}}

Now let us examine a lager system. Suppose that we are given a secrete bit string.

```
string = {1, 1, 0};
```

This is a function consistent with the above secrete bit string.

```
Clear[f];
f[0, 0, 0] = f[1, 1, 0] = {0, 1, 1};
f[0, 0, 1] = f[1, 1, 1] = {1, 1, 1};
f[0, 1, 0] = f[1, 0, 0] = {1, 0, 0};
f[0, 1, 1] = f[1, 0, 1] = {0, 0, 1};
```

Here is an implementation of the corresponding quantum oracle.

```
In[ ]:= cc = {1, 2, 3};
     tt = {4, 5, 6};
     all = Join[cc, tt];
     qc1 = QuantumCircuit[LogicalForm[Ket[], S@all],
         S[cc, 6], Oracle[f, S@cc, S@tt], S[cc, 6]];
     qc2 = QuantumCircuit[qc1, Measurement[S@cc]]
```

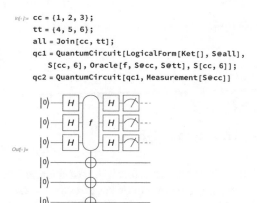

This is one way to get the measurement outcome.

```
In[ ]:= out = ExpressionFor[qc2];
     LogicalForm[out, S@all]
     result = Readout[out, S@cc]
```

$$Out[ ]= \frac{1}{2} \left| 1_{S_1}1_{S_2}1_{S_3}0_{S_4}0_{S_5}1_{S_6} \right\rangle + \frac{1}{2} \left| 1_{S_1}1_{S_2}1_{S_3}0_{S_4}1_{S_5}1_{S_6} \right\rangle - \frac{1}{2} \left| 1_{S_1}1_{S_2}1_{S_3}1_{S_4}0_{S_5}0_{S_6} \right\rangle - \frac{1}{2} \left| 1_{S_1}1_{S_2}1_{S_3}1_{S_4}1_{S_5}1_{S_6} \right\rangle$$

Out[ ]= {1, 1, 1}

To make repeated measurements, it is more efficient to first compute the state just before the measurement.

```
new = ExpressionFor[qc1];
```

Now we perform the measurement repeatedly.

```
In[ ]:= data = Table[out = Measurement[new, S@cc];
           Readout[out, S@cc], {12}];
        data // TableForm
```

```
Out[ ]//TableForm=
        1    1    0
        0    0    0
        1    0    0
        1    0    0
        0    0    0
        0    1    0
        1    0    0
        0    1    0
        1    1    0
        0    0    0
        0    0    0
        0    0    0
```

As two linearly independent vectors (bit strings), we choose these:

```
In[ ]:= mat = {{1, 1, 0}, {0, 0, 1}}
Out[ ]= {{1, 1, 0}, {0, 0, 1}}
```

Then, the linear equation, mat.ss=0 (mod 2), for the Boolean variables ss:={s1,s2,s3} is given by the following, which agrees with the given secrete bit string.

In[ ]:= ss = {1, 1, 0}

Out[ ]= {1, 1, 0}

In[ ]:= Mod[mat.ss, 2]

Out[ ]= {0, 0}

4.3   Quantum Fourier Transform (QFT)

The quantum Fourier transform is a unitary transformation of quantum states. It is analogous to the *discrete Fourier transform* of a finite set of numbers. In the case of the quantum Fourier transform, the numbers are replaced with quantum states.

The quantum Fourier transform can be performed efficiently on a quantum computer consisting of n qubits with only $\mathcal{O}(n^2)$ elementary quantum gate operations, compared to $\mathcal{O}(n2^n)$ gates for the best known classical algorithm for a discrete Fourier transform. This exponential speedup in the quantum Fourier transform algorithm relative to the classical counterpart enabled the celebrated quantum factorization algorithm to achieve a similar efficiency. Apart from the quantum factorization algorithm, the quantum Fourier transformation is a key part of many other quantum algorithms, such as the quantum phase estimation (Sect. 4.4), the order-finding problem, the discrete logarithm, and most importantly the hidden subgroup problem. Such a wide range of applications of the quantum Fourier transform stems from the fact that all known quantum algorithms featuring an exponential speedup over classical algorithms are variations of the hidden subgroup problem and, as first realized by Kitaev (1996), the key step to solve the latter problem is the quantum Fourier transform.

4.3.1   Definition and Physical Meaning

To make the physical meaning underlying the quantum Fourier transform clearer, in this section, we use a slightly different notation for the logical basis states, and denote them by

$$|X_x\rangle := |x_1\rangle \otimes |x_2\rangle \otimes \cdots \otimes |x_n\rangle , \quad x = 0, 1, 2, \ldots, 2^n - 1, \tag{4.37}$$

where as usual, x_j are the binary digits of the index $x = (x_1 x_2 \ldots x_n)_2$. The quantum Fourier transform (QFT for short) is a unitary transformation defined by the association

$$\hat{U}_{\text{QFT}} |X_y\rangle := \frac{1}{2^{n/2}} \sum_{x=0}^{2^n-1} |X_x\rangle \, e^{ixp_y} \quad (y = 0, 1, 2, \ldots, 2^n - 1), \tag{4.38}$$

where p_y is the yth "wave number" defined by

$$p_y := \frac{2\pi y}{2^n} = 2\pi (0.y_1 y_2 \ldots y_n)_2 . \qquad (4.39)$$

The expression in the right-hand side of (4.38) is of a familiar form of the discrete Fourier transform except that instead of numerical coefficients for the exponential factor e^{ixp_y}, there appear state vectors. Therefore, one may think of the quantum Fourier transform as a *vector-valued* discrete Fourier transformation.

However, in many physical applications, there is a more important aspect of the quantum Fourier transform than its formal similarity to the discrete Fourier transform. It turns out to be useful and inspiring to regard the QFT as a basis change, a unitary transformation from the logical basis

$$\left\{ |X_x\rangle : x = 0, 1, \ldots, 2^n - 1 \right\} \qquad (4.40)$$

to the so-called "conjugate basis"

$$\left\{ |P_y\rangle := \hat{U}_{\mathrm{QFT}} |X_y\rangle : y = 0, 1, \ldots, 2^n - 1 \right\} . \qquad (4.41)$$

The key observation is that in the "continuum" limit ($L := 2^n \to \infty$), the logical basis corresponds to the eigenbasis of the "position" and the conjugate basis to that of the "momentum".[5] Indeed, one can see that the following two observables

$$\hat{X} := \sum_x |X_x\rangle x \langle X_x| , \quad \hat{P} := \sum_y |P_y\rangle p_y \langle P_y| \qquad (4.42)$$

bearing eigenvalues x and p_y, respectively, satisfy the relation

$$e^{-ia\hat{P}} \hat{X} e^{+ia\hat{P}} = \hat{X} - a\hat{I} . \qquad (4.43)$$

This implies that $e^{-ia\hat{P}}$ is a translation operator, and $\hat{P}$ is the generator of the translation, i.e., the momentum.

The quantum Fourier transform appears frequently in many quantum algorithms (notably the quantum phase estimation algorithm in Sect. 4.4), either in an explicit or in a disguised form. The relation between the logical and conjugate basis defined above is useful to understand the principle behind the particular algorithms.

---

[5] This notion can be made rigorous even in a finite dimensional Hilbert space using the unitary operators $\exp(-ip\hat{X})$ and $\exp(-ia\hat{P})$, which generate the so-called *Weyl-Heisenberg group*. See (Weyl, 1931, Sect. IV.14) for example.

4.3.2   Quantum Implementation

The quantum Fourier transform (QFT) algorithm is to efficiently implement the unitary operator (4.37) in terms of the elementary gates.

Getting back to the definition, the quantum Fourier transform maps the quantum states by

$$\hat{U}_{\text{QFT}} \left| X_y \right\rangle = \frac{1}{2^{n/2}} \sum_{x=0}^{2^n-1} \left| X_x \right\rangle e^{ixp_y} \tag{4.44}$$

with wave number

$$p_y := \frac{2\pi y}{2^n} = 2\pi (0.y_1 y_2 \ldots y_n)_2 . \tag{4.45}$$

Since the characteristic properties of the Fourier transform of any kind are attributed to an oscillatory phase factor e^{ixp_y}, we start by elaborating it. The product xp_y in the exponent is given by

$$xp_y = \frac{2\pi xy}{2^n} = 2\pi (0.x_1 x_2 \ldots x_n)_2 y = 2\pi y \sum_{k=1}^{n} x_k 2^{-k} , \tag{4.46}$$

which recasts the phase factor into the form

$$e^{ixp_y} = \prod_{k=1}^{n} e^{2\pi iyx_k 2^{-k}} . \tag{4.47}$$

We put it back in (4.44) and rewrite the sum over the integer values x with the sum over the bit values $x_j \in \{0, 1\}$, which lead to

$$\hat{U}_{\text{QFT}} \left| X_y \right\rangle = \bigotimes_{k=1}^{n} \sum_{x_k} \left| x_k \right\rangle e^{2\pi ix_k y/2^k} . \tag{4.48}$$

For a fixed k, the ratio $y/2^k$ is represented in the binary digits as

$$y/2^k = (y_1 y_2 \cdots y_{n-k}.y_{n-k+1} \cdots y_{n-1} y_n)_2 . \tag{4.49}$$

The integer part does not affect the sum because $e^{2\pi im} = 1$ for any integer m, and only the fractional part $(0.y_{n-k+1} \cdots y_{n-1} y_n)_2$ does. Depriving the phase factors $e^{ix_k p_y}$ of the integer parts, we arrive at the expression for the result of the transformation

$$\hat{U}_{\text{QFT}} \left| X_y \right\rangle = \bigotimes_{k=1}^{n} \sum_{x_k} \left| x_k \right\rangle e^{2\pi ix_k (0.y_{n-k+1} \cdots y_{n-1} y_n)_2} . \tag{4.50}$$

Explicitly writing the sums over bit values x_j, it reads as

$$\hat{U}_{QFT} |X_y\rangle = \left(\frac{|0\rangle + |1\rangle \, e^{2\pi i 0.y_n}}{\sqrt{2}} \right) \otimes \left(\frac{|0\rangle + |1\rangle \, e^{2\pi i 0.y_{n-1}y_n}}{\sqrt{2}} \right) \otimes$$
$$\cdots \otimes \left(\frac{|0\rangle + |1\rangle \, e^{2\pi i 0.y_1 y_2 \cdots y_n}}{\sqrt{2}} \right). \quad (4.51)$$

Interestingly, the state resulting from the quantum Fourier transform is clearly a product state. The first factor is stored in the first qubit of the quantum register, and consecutive factors are stored in the corresponding qubits in the same order. For later analysis, it is useful to reverse the order. This is equivalent to perform the qubit-reversing unitary transformation

$$\hat{U}_{QBR} |y_1\rangle \otimes |y_2\rangle \otimes \cdots \otimes |y_n\rangle = |y_n\rangle \otimes \cdots \otimes |y_2\rangle \otimes |y_1\rangle. \quad (4.52)$$

Applying first the qubit-reversing transformation before the quantum Fourier transform, we arrive at the final expression that we analyze subsequently

$$\hat{U}_{QFT}\hat{U}_{QBR} |X_y\rangle = \left(\frac{|0\rangle + |1\rangle \, e^{2\pi i \, 0.y_1}}{\sqrt{2}} \right) \otimes \left(\frac{|0\rangle + |1\rangle \, e^{2\pi i \, 0.y_2 y_1}}{\sqrt{2}} \right) \otimes$$
$$\cdots \otimes \left(\frac{|0\rangle + |1\rangle \, e^{2\pi i \, 0.y_n \cdots y_2 y_1}}{\sqrt{2}} \right). \quad (4.53)$$

So far, we have done nothing but re-express the defining Eq. (4.44) of the quantum Fourier transform. Now, we identify the elementary quantum logic gates that compose the transform. The product in (4.53) is particularly useful to analyze. Let us examine the state stored on the kth qubit,

$$\frac{|0\rangle + |1\rangle \, e^{2\pi i \, (0.y_k \cdots y_2 y_1)_2}}{\sqrt{2}}. \quad (4.54)$$

Noting that

$$2\pi \, (0.y_k \ldots y_2 y_1)_2 = \sum_{j=1}^{k} y_j 2^{j-k} \pi, \quad (4.55)$$

we see that qubits $1, 2, \ldots k$ makes relative phase shifts $2^{1-k}\pi, 2^{2-k}\pi, \cdots, \pi$, respectively, on the state in (4.54). Among these, the phase shift π depends on the state $|y_k\rangle$ of the kth qubit itself, and it is equivalent to the Hadamard gate. Therefore, we can regard that the state (4.54) stored on the kth qubit is a result of the transformation

$$|y_1\rangle \otimes \cdots \otimes |y_{k-1}\rangle \otimes |y_k\rangle \mapsto |y_1\rangle \otimes \cdots \otimes |y_{k-1}\rangle \otimes \prod_{j=1}^{k-1} \hat{T}^{y_j}(2^{j-k}\pi)\hat{H}|y_k\rangle \,,$$
(4.56)

where $\hat{T}(\phi)$ denotes the relative phase shift

$$\hat{T}(\phi) = |0\rangle\langle 0| + |1\rangle\langle 1| e^{i\phi} \doteq \begin{bmatrix} 1 & 0 \\ 0 & e^{i\phi} \end{bmatrix}.$$
(4.57)

The actual operation of $\hat{T}^{y_j}(2^{j-k}\pi)$ is determined by the logical state $|y_j\rangle$ of the jth qubit, so it is a controlled-unitary gate. For example, the state stored on the third qubit is represented in the quantum circuit as

(4.58)

where we have used the short-hand notation $\hat{T}_l := \hat{T}(\pi/2^l)$. Overall, the quantum Fourier transformation can be represented in the quantum circuit as

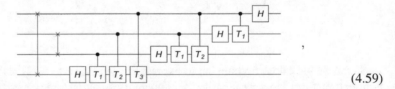

(4.59)

where we have implemented the qubit-reversing transformation $\hat{U}_{\text{QBT}}$ by applying SWAP gates on pairs of qubits from the first and second half of the quantum register. In the above quantum circuit, we have applied $\hat{U}_{\text{QBR}}$ at the beginning. One can apply it at the end as well. Since $\hat{U}_{\text{QBR}}$ corresponds to a simple reversal of the order of qubits, the quantum circuit in (4.59) is identical to the following quantum circuit

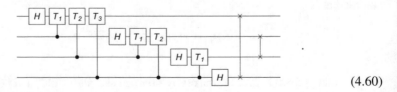

(4.60)

This quantum circuit for the quantum Fourier transformation is found more commonly in the literature.

Here we simulate the quantum Fourier transform on a quantum register of 4 qubits.

```
$n = 4;
```

Here defined are the controlled-phase gates.

```
CP[j_, j_] := S[j, 6]
CP[j_, k_] := ControlledU[S[j],
    Phase[Pi Power[2, k - j], S[k], "Label" -> Subscript["T", j - k]]
    ] /; j > k
CP[j_, k_] := ControlledU[S[j],
    Phase[Pi Power[2, j - k], S[k], "Label" -> Subscript["T", k - j]]
    ] /; j < k
```

We denote by the label T_k the relative phase shift by $\pi/2^k$: $T_k := \begin{pmatrix} 1 & 0 \\ 0 & e^{i\pi/2^k} \end{pmatrix}$.

This is just a short-hand function.

```
SW[j_] := SWAP[S[j], S[$n - j + 1]]
SW[All] := Table[SW[j], {j, 1, $n / 2}]
```

This is the standard implementation, which can be seen commonly in textbooks.

```
In[ ]:= gates = Flatten@Table[CP[j, k], {k, $n, 1, -1}, {j, k, 1, -1}];
        qc1 = QuantumCircuit[Sequence @@ SW[All], Sequence @@ gates]
```

Out[ ]=

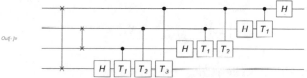

To verify the above quantum circuit model, we compare its matrix representation with the matrix for the discrete Fourier transform.

```
In[ ]:= mat = Matrix[qc1] // TrigToExp;
        new = FourierMatrix[Power[2, $n]];
        DeleteCases[new - mat // Simplify // Flatten, 0]
```

Out[ ]= {}

One can also reverse the order of qubits later to get the following quantum circuit model for the quantum Fourier transform.

```
In[ ]:= gates = Flatten@Table[CP[j, k], {k, 1, $n}, {j, k, $n}];
        qc2 = QuantumCircuit[Sequence @@ gates, Sequence @@ SW[All]]
```

Out[ ]=

Again, verify it by comparing the matrix representation with the matrix for the discrete Fourier transform.

```
In[ ]:= mat = Matrix[qc2] // TrigToExp;
       new = FourierMatrix[Power[2, $n]];
       DeleteCases[new - mat // Simplify // Flatten, 0]

Out[ ]= {}
```

4.3.3 Semiclassical Implementation

The semiclassical implementation is not only interesting from a physical point of view, but it is also extremely interesting given the current level of technology (Griffiths & Niu, 1996). Since it does not require any entanglement, a quantum information resource that is the most fragile and hence difficult to maintain, many experiments are using this implementation whenever the QFT and/or the quantum phase estimation (see Sect. 4.4) is in need (Chiaverini, 2005; Higgins *et al.*, 2007).

The line of argument in the original work (Griffiths & Niu, 1996) to derive the semiclassical implementation of the quantum Fourier transformation is interesting in its own right, and it offers another physically-motivated derivation of the quantum Fourier transform algorithm. Here, however, we will not repeat the original arguments. Instead, we simply make use of the relation between the quantum Fourier transform and its inverse. The inverse quantum Fourier transform is given (as in Problem 4.4) by

$$\hat{U}_{\text{QFT}}^{\dagger} |X_y\rangle = \frac{1}{2^{n/2}} \sum_x |X_x\rangle \, e^{-ixp_y} \, , \qquad (4.61)$$

which follows directly from the orthogonality relation

$$\frac{1}{2^n} \sum_y e^{i(x-x')p_y} = \delta_{xx'} \, . \qquad (4.62)$$

The expression (4.61) suggests that the inverse quantum Fourier transform is just another form of the quantum Fourier transform, and indeed, this is a characteristic property common to any kind of Fourier transform. This implies that one can recover the original quantum Fourier transform from the inverse quantum Fourier transform, and vice versa, in several ways. For example, note that $\hat{U}_{\text{QFT}}^{\dagger} |X_y\rangle$ is an element of the conjugate basis because $e^{-ixp_y} = e^{ix(2\pi - p_y)}$ for any x and y. In fact, $\hat{U}_{\text{QFT}}^{\dagger} |X_y\rangle = |P_{2^n - y}\rangle$. More useful in the context of the semiclassical implementation of the quantum Fourier transform is to note that the inverse quantum Fourier transform $\hat{U}_{\text{QFT}}^{\dagger}$ is nothing but the quantum Fourier transform with the phase factor e^{ixp_y} replaced by e^{-ixp_y}. It corresponds to replacing the relative phase shifts $\hat{T}(\phi)$ in (4.59) by $T^{\dagger}(\phi)$. Conversely, if we start with the quantum circuit in (4.59), take its Hermitian conjugate (i.e., inverse), and replace $\hat{T}^{\dagger}(\phi) \equiv \hat{T}(-\phi)$ with $\hat{T}(\phi)$, then we should recover the original quantum Fourier transform. Through this procedure, we get the following quantum circuit

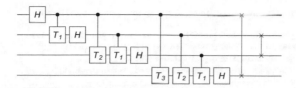

for the quantum Fourier transform. Compared to the quantum circuit in (4.59), the above quantum circuit has its Hadamard gate on a fixed qubit coming before the qubit "controls" the operation of the relative phase shift $\hat{T}(\phi)$ on other qubits. This difference brings a significant consequence. To see it, we put explicit measurements on the qubits,

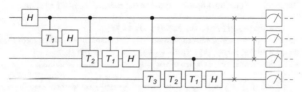

According to the *principle of deferred measurement*, the measurement can be performed before the controlled unitary gates

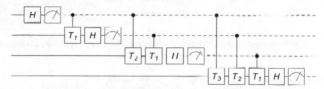

This form is exactly the quantum circuit derived in Griffiths & Niu (1996). The important point is that after the measurement, the controlled-unitary gates become just classical feedback controls that require no entanglement.

Here we simulate the semiclassical implementation of the quantum Fourier transform. Again, we consider a quantum register of four qubits.

```
$n = 4;
SS = S[Range[$n], None];
```

Here defined are the controlled-phase gates.

```
CP[j_, j_] := S[j, 6]
CP[j_, k_] := ControlledU[S[j],
    Phase[Pi Power[2, k-j], S[k], "Label" → Subscript["T", j-k]]
    ] /; j > k
CP[j_, k_] := ControlledU[S[j],
    Phase[Pi Power[2, j-k], S[k], "Label" → Subscript["T", k-j]]
    ] /; j < k
```

```
In[ ]:= in = Ket @@ Thread[SS → RandomChoice[{0, 1}, $n]];
        LogicalForm[in, SS]
```

Out[ ]= $\left| 1_{S_1} 1_{S_2} 0_{S_3} 1_{S_4} \right\rangle$

This is a quantum circuit model for the quantum Fourier transformation.

```
In[·]:= gates = Flatten@Table[CP[j, k], {k, $n, 1, -1}, {j, k, 1, -1}];
       swaps = Table[SWAP[S[k], S[$n - k + 1]], {k, 1, $n / 2}];
       qc1 = QuantumCircuit[LogicalForm[in, SS],
           Sequence @@ swaps, Sequence @@ gates, Measurement[SS]]
```

Out[·]=
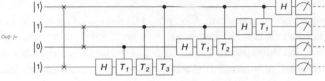

This is the semiclassical implementation of the quantum Fourier transform.

```
In[·]:= gates = Table[CP[j, k], {k, 1, $n}, {j, 1, k}];
       gates = Flatten@MapIndexed[Append[#1, Measurement@S[#2]] &, gates];
       swaps = Table[SWAP[S[k], S[$n - k + 1]], {k, 1, $n / 2}];
       qc2 = QuantumCircuit[LogicalForm[in, SS], Sequence @@ gates]
```

Out[·]=
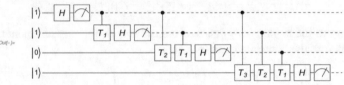

To verify the equivalence of the two quantum circuit models, we compare the probability distributions.

```
In[·]:= Timing[data1 = Table[out = ExpressionFor[qc1];
           FromDigits[Readout[out, SS], 2], {500}];]
```

Out[·]= {73.093, Null}

```
In[·]:= Histogram[data1, FrameLabel → {"Measurement Outcome", "Counts"}]
```

Out[·]=
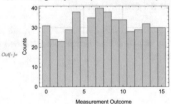

Note that the bit strings of the measurement outcomes from the semiclassical model must be reversed .

```
In[·]:= Timing[data2 = Table[out = ExpressionFor[qc2];
           FromDigits[Reverse@Readout[out, SS], 2], {500}];]
```

Out[·]= {18.3252, Null}

In[ ]:= `Histogram[data2, FrameLabel → {"Measurement Outcome", "Counts"}]`

Out[ ]=

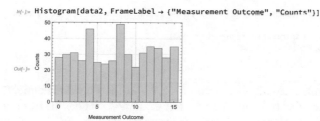

The discrepancy between the two distributions is due to the finite sample size.

Another way to test the equivalence of the fully quantum and semiclassical implementation of the quantum Fourier transformation is to take as the input state one that ends up in a product state.

In[ ]:= `vecs = FourierMatrix[Power[2, $n], FourierParameters → {0, -1}].Basis[SS];`
`in = RandomChoice[vecs];`
`in // LogicalForm`

Out[ ]= $\frac{1}{4} \left|0_{S_1}0_{S_2}0_{S_3}0_{S_4}\right\rangle + \frac{1}{4} i \left|0_{S_1}0_{S_2}0_{S_3}1_{S_4}\right\rangle - \frac{1}{4} \left|0_{S_1}0_{S_2}1_{S_3}0_{S_4}\right\rangle - \frac{1}{4} i \left|0_{S_1}0_{S_2}1_{S_3}1_{S_4}\right\rangle +$

$\frac{1}{4} \left|0_{S_1}1_{S_2}0_{S_3}0_{S_4}\right\rangle + \frac{1}{4} i \left|0_{S_1}1_{S_2}0_{S_3}1_{S_4}\right\rangle - \frac{1}{4} \left|0_{S_1}1_{S_2}1_{S_3}0_{S_4}\right\rangle - \frac{1}{4} i \left|0_{S_1}1_{S_2}1_{S_3}1_{S_4}\right\rangle +$

$\frac{1}{4} \left|1_{S_1}0_{S_2}0_{S_3}0_{S_4}\right\rangle + \frac{1}{4} i \left|1_{S_1}0_{S_2}0_{S_3}1_{S_4}\right\rangle - \frac{1}{4} \left|1_{S_1}0_{S_2}1_{S_3}0_{S_4}\right\rangle - \frac{1}{4} i \left|1_{S_1}0_{S_2}1_{S_3}1_{S_4}\right\rangle +$

$\frac{1}{4} \left|1_{S_1}1_{S_2}0_{S_3}0_{S_4}\right\rangle + \frac{1}{4} i \left|1_{S_1}1_{S_2}0_{S_3}1_{S_4}\right\rangle - \frac{1}{4} \left|1_{S_1}1_{S_2}1_{S_3}0_{S_4}\right\rangle - \frac{1}{4} i \left|1_{S_1}1_{S_2}1_{S_3}1_{S_4}\right\rangle$

In[ ]:= `gates = Flatten@Table[CP[j, k], {k, $n, 1, -1}, {j, k, 1, -1}];`
`swaps = Table[SWAP[S[k], S[$n - k + 1]], {k, 1, $n / 2}];`
`qc1 = QuantumCircuit[{in, "Label" → None},`
`  Sequence @@ swaps, Sequence @@ gates, Measurement[SS]]`

Out[ ]=

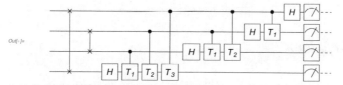

In[ ]:= `gates = Table[CP[j, k], {k, 1, $n}, {j, 1, k}];`
`gates = Flatten@MapIndexed[Append[#1, Measurement@S[#2]] &, gates];`
`swaps = Table[SWAP[S[k], S[$n - k + 1]], {k, 1, $n / 2}];`
`qc2 = QuantumCircuit[{in, "Label" → None}, Sequence @@ gates]`

Out[ ]=

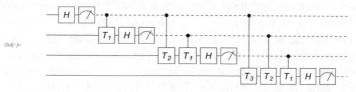

In[ ]:= `out1 = ExpressionFor[qc1];`
`LogicalForm[out1, SS]`

Out[ ]= $\left|1_{S_1}1_{S_2}0_{S_3}0_{S_4}\right\rangle$

Again, recall that the order of qubits are reversed in the semiclassical implementation.

```
In[*]:= out2 = ExpressionFor[N@qc2] // Chop;
        LogicalForm[out2, SS]
Out[*]= 1. |0_{S_1} 0_{S_2} 1_{S_3} 1_{S_4}⟩
```

From the quantum circuits discussed above, we see that quantum Fourier transform requires an order of n^2 elementary gates. For example, consider the quantum circuit in (4.59). There are n Hadamard gates, one for each qubit. The number of controlled-unitary gates is given by $\sum_{k=1}^{n}(k-1) = n(n-1)/2$. There are $n/2$ SWAP gates as well, each of which requires three CNOT gates. Overall, we need $n(n+4)/2 \sim \mathcal{O}(n^2)$ gates. This computational load is compared to $\mathcal{O}(n2^n)$ gates for the fast Fourier transform (FFT), the best classical algorithm for the discrete Fourier transform.

4.4   Quantum Phase Estimation (QPE)

Quantum phase estimation is a procedure to estimate the phase of the unknown eigenvalue of an eigenstate of a unitary operator. Like the quantum Fourier transformation, the quantum algorithm for the quantum phase estimation is one of the key elements of many quantum algorithms, including the quantum factorization algorithm. In fact, the quantum Fourier transform is a part of the quantum phase estimation algorithm, and it almost always appears in that way rather than independently. In this sense, the quantum phase estimation is a key subroutine for most quantum algorithms. Indeed, all quantum algorithms known so far can be regarded as a quantum phase estimation in one form or another (Cleve et al., 1998).[6]

4.4.1   Definition

Let $\hat{U}$ be a unitary operator on the Hilbert space associated with a register consisting of n qubits. Suppose that the quantum register is known to be prepared in an eigenstate state $|\phi\rangle$ of $\hat{U}$. Its eigenvalue $e^{-2\pi i\phi}$ is desired but *unknown*. The quantum phase estimation (QPE) procedure gets the phase value $2\pi\phi$ of the eigenvalue.

Note that the quantum phase estimation is directly related to the measurement. Recall that finding the eigenvalue $e^{-2\pi i\phi}$ reveals the corresponding eigenstate $|\phi\rangle$. A unitary operator can be written as $\hat{U} = e^{-2\pi i\hat{Q}}$ for some observable $\hat{Q}$, so an estimation of phase corresponds to the measurement of the quantity $\hat{Q}$. In this sense, one can regard the quantum phase estimation procedure as a realization of the quantum measurement on quantum computers. We will discuss this aspect in more detail in Sect. 4.4.4.

---

[6] Mathematically speaking, they belong to the class called the hidden subgroup problem.

To get a general idea behind the quantum phase estimation procedure, take an auxiliary register of m qubits. We call it the "probe" register, for a reason that will be clear later in this section. We prepare the probe register in the superposition $\propto \sum_x |x\rangle$ of all elements in the logical basis. The superposition state can be generated by applying the Hadamard gate, as seen Eq. (2.18). The state vector of the total system becomes

$$\left(\hat{H}^{\otimes m} |0\rangle\right) \otimes |\phi\rangle = \frac{1}{2^{m/2}} \sum_{x=0}^{2^m - 1} |x\rangle \otimes |\phi\rangle \tag{4.63}$$

Next, perform a unitary transformation on the total system defined by

$$\hat{U}_{\text{QPE}} : |x\rangle \otimes |y\rangle \mapsto |x\rangle \otimes \left(\hat{U}^x |y\rangle\right) , \tag{4.64}$$

where $|x\rangle$ $(x = 0, 1, \cdots, 2^m - 1)$ belongs to the probe register and $|y\rangle$ $(y = 0, 1, \ldots, 2^n - 1)$ to the native register. The transformation is a controlled-unitary gate, performing the transformation $\hat{U}$ repeatedly depending on value of x on the native register. Upon the controlled unitary transformation, the state vector becomes

$$\frac{1}{2^{m/2}} \sum_x |x\rangle \otimes \left(\hat{U}^x |\phi\rangle\right) = \frac{1}{2^{m/2}} \sum_x |x\rangle \otimes \left(|\phi\rangle \, e^{-2\pi i x \phi}\right)$$

$$= \left(\frac{1}{2^{m/2}} \sum_{x=0}^{2^m - 1} |x\rangle \, e^{-2\pi i x \phi}\right) \otimes |\phi\rangle \tag{4.65}$$

Note that the original register still remains in the same state $|\phi\rangle$, while the probe register has picked up a relative phase shift proportional to x, $e^{-2\pi i x \phi}$. The state in (4.65) stored on the probe register is thus formally the same as the (inverse) quantum Fourier transform (see (4.38) and (4.61)). For the moment, let us assume that ϕ $(0 \le \phi < 1)$ takes one of the discrete values

$$\frac{0}{2^m}, \frac{1}{2^m}, \frac{2}{2^m}, \ldots, \frac{2^m - 1}{2^m}. \tag{4.66}$$

In this case, performing a quantum Fourier transform $\hat{U}_{\text{QFT}}$ on the probe register brings it to the logical basis state $|2^m \phi\rangle$. A subsequent measurement on the probe register in the logical basis reveals the value ϕ. The procedure described above is then summarized in the following diagram

$$\tag{4.67}$$

The normalized phase ϕ is a continuous parameter. In general, it takes values other than the discrete values listed in (4.66). In such a case, the number m of the qubits in the probe register is not sufficient to accurately estimate the phase ϕ. Thus, the measurement randomly produces different values. Fortunately, the probability for a particular value is distributed sharply (with a larger probe register resulting in a sharper probability distribution) around the value $2^m\phi$. We will come back to this point below.

4.4.2 Implementation

Let us now discuss an actual implementation of the individual steps. The preparation of the probe register in the overall superposition state is achieved by using the Hadamard gate (see Sect. 2.1.2)

$$|0\rangle^{\otimes m} \to \hat{H}^{\otimes m} |0\rangle^{\otimes m} = \frac{1}{2^{m/2}} \sum_{x=0}^{2^m-1} |x\rangle . \tag{4.68}$$

Explicitly, the controlled-$\hat{U}^x$ transformation in Eq. (4.64) reads as

$$|x\rangle \otimes \hat{U}^x |y\rangle = |x_1\rangle \otimes |x_2\rangle \otimes \cdots \otimes |x_m\rangle \otimes \hat{U}^{x_1 2^{m-1}} \hat{U}^{x_2 2^{m-2}} \cdots \hat{U}^{x_m 2^0} |y\rangle . \tag{4.69}$$

It is a product of controlled-$\hat{U}^{2^j}$ gates depending on the value x_j of the jth qubit in the probe register. Thus it is implemented by using the following quantum circuit

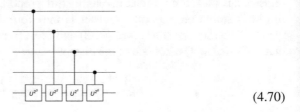

$$\tag{4.70}$$

Here, it is worth noting that an efficient quantum phase estimation algorithm requires an efficient implementation of the controlled-U^{2^j} gates. One of the most common unitary transformations that allows for efficient implementation is *modular multiplication*, which is used in the order-finding problem that is discussed in Sect. 4.5.2.

The controlled-$\hat{U}^x$ gate transforms the state of the total system as

$$\frac{1}{2^{m/2}} \sum_{x=0}^{2^m-1} |x\rangle \otimes |\phi\rangle \to \left(\frac{1}{2^{m/2}} \sum_{x=0}^{2^m-1} |x\rangle e^{-2\pi i x \phi} \right) \otimes |\phi\rangle \tag{4.71}$$

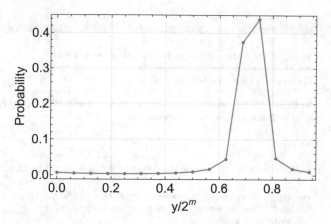

Fig. 4.2 The probability P_y to estimate the phase ϕ as $y/2^m$ by using $m = 4$ "probe" qubits when the true phase value is $\phi = 18/25$

The final quantum Fourier transformation on the probe register transforms the state as

$$\frac{1}{2^{m/2}} \sum_x |x\rangle\, e^{-2\pi i x\phi} \rightarrow \sum_y |y\rangle \left(\frac{1}{2^m} \sum_x |x\rangle\, e^{2\pi i(y-2^m\phi)x/2^m} \right) \qquad (4.72)$$

When $2^m\phi$ is one of the discrete values $0, 1, 2, \ldots, 2^m - 1$, the quantum Fourier transform brings the probe to its single logical basis state $|2^m\phi\rangle$. From the orthogonality relation,

$$\frac{1}{2^m} \sum_{k=0}^{2^m-1} e^{2\pi i(x-y)k/2^m} = \delta_{xy} . \qquad (4.73)$$

In general, though, it is not the case because the state is a superposition of different logical states. Upon the measurement on the probe register in the logical basis, the probability to get the outcome y is given by

$$P_y = \left| \frac{1}{2^m} \sum_{x=0}^{2^m-1} e^{i2\pi(y-2^m\phi)x/2^m} \right|^2 = \frac{1}{2^{2m}} \left| \frac{1 - e^{2\pi i(y-2^m\phi)}}{1 - e^{2\pi i(y-2^m\phi)/2^m}} \right|^2 . \qquad (4.74)$$

For a reasonably large m, probability P_y sharply peaks around $2^m\phi$ (see Fig. 4.2). It is precisely 1 only when $2^m\phi$ is one of the integers $0, 1, \ldots, 2^m - 1$, as mentioned above.

We take an ancillary register consisting of four qubits, which we call the "probe" register.

```
$m = 4; (* the number of qubits in the "probe" register *)
prb = Range[$m]; (* the "probe" register *)
sys = $m + 1; (* the "system" register *)
```

We consider a rotation around the z-axis for the unitary operator. In this case, the logical basis consists of the eigenstates of the unitary operator.

```
φ = 18 / 25;
```

Here is the controlled $- U^{2^{m-j}}$ operator to be used in the phase estimation algorithm.

```
ctrlU[j_] := ControlledU[S[j], Phase[-2 Pi φ * Power[2, $m - j], S[sys]],
   "Label" → Superscript["U", Superscript["2", ToString[$m - j]]],
   "LabelSize" → 0.65]
```

In[·]:= `qc = QuantumCircuit[LogicalForm[Ket[], S@prb],`
     `Ket[S[sys] → 1], {S[prb, 6], "LabelSize" → 0.8}, ctrlU[1],`
     `ctrlU[2], ctrlU[3], ctrlU[4], QuantumFourierTransform[S@prb]]`

Out[·]=
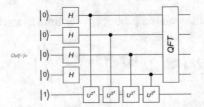

In this particular example, we already know that the "system" register remains in its initial state and what the initial state is. To make the process general, however, we take the partial trace over the system register to focus on the "probe" register.

```
out = Matrix[N@qc];
rho = PartialTrace[out, {sys}];
```

From the density matrix of the probe register, we can get the probability to read out the possible values of the normalized phase.

In[·]:= `probability = Transpose@{Range[0, 2^$m - 1] / 2^$m, Normal@Chop@Diagonal@rho};`
     `g = ListLinePlot[probability,`
     `   PlotRange → {{0, 1}, All}, PlotMarkers → Automatic,`
     `   FrameLabel → {"φ", "Probability"}]`

Out[·]=

4.4.3 Accuracy

We now estimate the error bound of the phase estimation.[7] We split $2^m\phi$ into the integer part p and the fractional part δ so that $2^m\phi = p + \delta$ with $0 \leq \delta < 1$. The fractional part δ cannot be stored in the m-qubit probe register, and it causes an error in the estimated phase. For a given resource with finite size, we cannot avoid it.

One can get the best estimate when measurement outcome y points correctly to p or $p + 1$. The error $\varepsilon := |y/2^m - \phi|$ in the estimated phase would be less than 2^{-m}. The probability P_y is distributed sharply (for a large m) around p and $p + 1$. The probabilities at $y = p$ and $y = p + 1$ are bigger than those at any other values of y. Therefore, it is useful to bound the probability for the error in the estimated phase to be less than 2^{-m},

$$P(\varepsilon < 2^{-m}) = P_p + P_{p+1}. \tag{4.75}$$

Unfortunately, the probability distribution P_y has finite width, and outcome y does not always result in the desired value. Nevertheless, P_y for $y = p$ and $y = p + 1$ are bigger than for any other values of y.

It follows from (4.74) that

$$P(\varepsilon < 2^{-m}) = \frac{1}{2^{2m}} \left| \frac{\sin(\pi\delta)}{\sin(\pi\delta/2^m)} \right|^2 + \frac{1}{2^{2m}} \left| \frac{\sin(\pi(1-\delta))}{\sin(\pi(1-\delta)/2^m)} \right|^2. \tag{4.76}$$

As a function of δ, it is symmetric around $\delta = 1/2$. That is, $1/2 - \delta \to \delta$ does not change the expression. Furthermore, it has a maximum of 1 for $\delta = 0$ because in this case $P_p = 1$, and no error can arise. It has minimum at $\delta = 1/2$ because $P_p = P_{p+1}$ in this case. We thus have the inequality

$$P(\varepsilon < 2^{-m}) \geq \frac{1}{2^{2m-1}} \left| \frac{\sin(\pi/2)}{\sin(\pi/2^{m+1})} \right|^2 = \frac{1}{2^{2m-1}\sin^2(\pi/2^{m+1})}. \tag{4.77}$$

Note that $(\pi/2^{m+1})/\sin(\pi/2^{m+1})$ monotonically decreases and converges to 1 as m increases to infinity. Therefore, we have

$$P(\varepsilon < 2^{-m}) \geq \frac{8}{\pi^2} \approx 0.81. \tag{4.78}$$

As m increases, we can improve the accuracy with error less than 2^{-m} with success probability larger than 0.81.

---

[7] A different method and probability bound are presented in Nielsen & Chuang (2011).

4.4.4 Simulation of the von Neumann Measurement

The von Neumann scheme of measurement (see Sect. 1.3.1) is an idealistic mechanism that implements a projection measurement. As already mentioned, the quantum phase estimation procedure is, in spirit, equivalent to the von Neumann scheme of measurement. In fact, the quantum phase estimation was put forward to simulate the von Neumann scheme on quantum computers (Kitaev, 1996, 1997). The equivalence of the von Neumann scheme of measurement and the quantum phase estimation procedure has also inspired direct tomography of wave functions and its variants (Lundeen *et al.*, 2011; Vallone & Dequal, 2016). Therefore, it will be interesting for scientific as well as heuristic purposes to examine the equivalence in more detail.

We consider a "system" consisting of n qubits and pick a "probe" of m qubits. To maintain the physical nature of the von Neumann scheme, let us consider two bases for the probe, $\{|X_x\rangle\}$ and $\{|P_x\rangle\}$ ($x = 0, 1, 2, \ldots, 2^m - 1$), that are conjugate to each other and related by the quantum Fourier transform $|P_x\rangle = \hat{U}_{\mathrm{QFT}} |X_x\rangle$. We call them the "position" and "momentum" basis, respectively (see Sect. 4.3.1). To simplify the argument, we assume the system to be in a definite yet unknown eigenstate $|a\rangle$ of an observable $\hat{A}$. We want to find eigenvalue a by conducting a measurement procedure.

Following the von Neumann scheme, we prepare the probe in the state $|X_0\rangle$, which refers to a definite position. Then, we let the system and probe interact with each other by coupling the "momentum" operator $\hat{P}$ of the probe to the observable $\hat{A}$ of the system with strength g. After an interaction for a finite duration of time, the probe is measured in the basis $|X_y\rangle$, as illustrated in the following schematic diagram

We rewrite the interaction unitary operator [to be compared with (1.53)] as

$$\hat{U}_{\mathrm{int}} = \exp\left(-ig\hat{P} \otimes \hat{A}\right) = \sum_x |P_x\rangle \langle P_x| \otimes e^{-igp_x\hat{A}} = \sum_x |P_x\rangle \langle P_x| \otimes \hat{U}^x,$$

(4.79)

where the operator $\hat{U}$ acting on the system has been defined by

$$\hat{U} = \exp\left[-i\left(\frac{2\pi g}{2^n}\right)\hat{A}\right].$$

(4.80)

In this new interpretation, operator $\hat{U}$ is repeatedly operated x times on the system depending on the wave number p_x (eigenvalue of $\hat{P}$) in the probe. This interpretation is seemingly opposite to the original idea in the von Neumann scheme, where the translation operator $\hat{T}_a$ is operated on the probe depending on eigenvalue a of $\hat{A}$ in the system. This reinterpretation is depicted schematically in the following diagram [to be compared to the diagram in (1.54)]

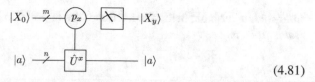

$$(4.81)$$

Note that the probe register controls the operation of $\hat{U}$ on the system in the "momentum" basis $\{|P_y\rangle\}$, and hence the momentum basis is the computational basis (logical basis). On the other hand, the von Neumann scheme requires the measurement to be performed in the "position" basis $\{|X_y\rangle\}$. As discussed in Sect. 2.5, any measurement in a basis different from the logical basis can be achieved by applying a proper unitary transformation corresponding to the basis change, which in this case is the quantum Fourier transform. Similarly, one can obtain the input state $|X_0\rangle$ by acting $\hat{U}_{\text{QFT}}^{\dagger}$ on $|P_0\rangle$. These changes lead to the schematic diagram

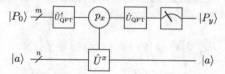

Finally, with all bases changed to the logical basis $\{|P_x\rangle\}$, we can drop the addition labels 'P' and 'p' from the states and eigenvalues. Also recall that the action of the quantum Fourier transform on $|P_0\rangle$ can be replaced by Hadamard gates, $\hat{U}_{\text{QFT}}^{\dagger}|P_0\rangle = \hat{H}^{\otimes m}|P_0\rangle$ (see Problem 4.5). Therefore, the von Neumann measurement can be depicted as in the following diagram, which is identical to diagram (4.67) corresponding to the quantum phase estimation procedure.

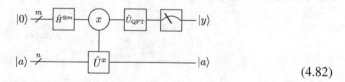

$$(4.82)$$

4.5 Applications

Before concluding this chapter, let us examine some examples where the quantum Fourier transform and the quantum phase estimation can be applied. A Fourier transform is particularly useful for periodic effects. Thus, it is natural to use the quantum Fourier transform to find the unknown period of a given function. The order-finding problem is a specific example, where the function is the *modular exponentiation*. However, as mentioned earlier, the quantum Fourier transform cannot be used independently even though it is the key part. One needs a procedure to induce the relative phase shifts to be extracted with the quantum Fourier transform. The procedure is a type of quantum phase estimation. Mathematically, the period-finding and order-finding problem belong to a wider class of problem known as the *hidden sub-group problem*.

4.5.1 The Period-Finding Algorithm

Let $f : \{0, 1\}^m \to \{0, 1\}^n$ be a periodic function, and

$$f(x + \lambda) = f(x) \tag{4.83}$$

for all x and fixed λ. The period-finding problem is to find the unknown (primary) period λ with the least queries to the function f. For simplicity, we assume that λ is a divider of 2^m. Here we also assume that the quantum oracle $\hat{U}_f$ corresponding to the function f is implemented efficiently for the mapping

$$\hat{U}_f : |x\rangle \otimes |y\rangle \mapsto |x\rangle \otimes |y \oplus f(x)\rangle . \tag{4.84}$$

Given the sequence of the function values $f(0), f(1), \ldots, f(\lambda - 1)$, one can perform a discrete Fourier transform of the sequence to obtain

$$F_k := \frac{1}{\sqrt{\lambda}} \sum_{x=0}^{\lambda-1} f(x) e^{-2\pi i k x/\lambda}. \tag{4.85}$$

Analogously, it is also possible to define new states[8]

$$|F_k\rangle = \frac{1}{\sqrt{\lambda}} \sum_{x=0}^{\lambda-1} |f(x)\rangle \, e^{+2\pi i k x/\lambda} \tag{4.86}$$

for $k = 0, 1, 2, \ldots, \lambda - 1$ using the discrete Fourier transform of the sequence of states $|f(x)\rangle$ with $x = 0, 1, 2, \ldots, \lambda - 1$. In general, the new states $|F_k\rangle$ may not be linearly independent with each other because $|f(x)\rangle = |f(y)\rangle$ for some $0 \le x, y < \lambda$.[9]

The period-finding algorithm is thus summarized in the quantum circuit in Fig. 4.3. We prepare the control register in the state $2^{-m/2} \sum_x |x\rangle$. The quantum oracle (4.84) then makes the transformation

---

[8] Here we take the opposite sign in the exponent of the phase factor $e^{2\pi i k x/\lambda}$ to be consistent with the quantum Fourier transform shown in (4.38).

[9] In this respect, the period-finding problem in the present form is slightly different from the hidden subgroup problem. In the hidden subgroup problem, the function $f : \mathcal{G} \to \mathcal{S}$ from a group $\mathcal{G}$ to a set $\mathcal{S}$ is assumed to separate the cosets. In other words, $f(x) = f(y)$ *if and only if* x and y belong to the same coset of the given subgroup $\mathcal{H} \subset \mathcal{G}$.

Fig. 4.3 Quantum circuit
model for the period-finding
algorithm

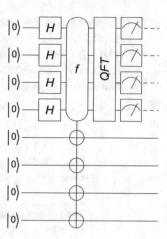

$$\frac{1}{2^{m/2}} \sum_{x=0}^{2^m-1} |x\rangle \otimes |0\rangle \rightarrow \frac{1}{2^{m/2}} \sum_{x=0}^{2^m-1} |x\rangle \otimes |f(x)\rangle \, . \tag{4.87}$$

We now replace the states $|f(x)\rangle$ with $|F_k\rangle$ in (4.86) using the inverse relation

$$|f(x)\rangle = \frac{1}{\sqrt{\lambda}} \sum_{k=0}^{\lambda-1} |F_k\rangle \, e^{-2\pi i k x/\lambda} \, , \tag{4.88}$$

which leads to the expression for the state of the total system

$$\frac{1}{\sqrt{\lambda}} \sum_{k=0}^{\lambda-1} \left(\frac{1}{2^{m/2}} \sum_{x=0}^{2^m-1} |x\rangle \, e^{-2\pi i x k/\lambda} \right) \otimes |F_k\rangle \, . \tag{4.89}$$

This reveals that the quantum oracle has induced relative phase shifts in the control
register that are proportional to the logical value x. As in the quantum phase estimation procedure discussed previously in Sect. 4.4, one can extract the relative phase
shifts by employing the quantum Fourier transform (see Eqs. (4.71) and (4.72)).
Doing so corresponds to estimating the (normalized) "phase" k/λ, from which one
can find the period λ (see below). In this sense, one can regard the period-finding
problem to fall in the quantum phase estimation category.

Of course, one apparent difference is that the relative phase shifts are induced
by the quantum oracle in this case. Furthermore, unlike the standard quantum phase
estimation procedure, the phase k/λ to be estimated is randomly selected from the
values corresponding to $k = 0, 1, \ldots, \lambda - 1$. It is thus possible that the observed k
may have a common divider with λ so that $k/\lambda = k'/\lambda'$ for another pair of integers

k' and λ'. For this reason, one cannot determine the period λ completely from the recorded value k/λ. The denominator (and the integer multiples of it) is only a candidate for the period λ. However, it does not pose a serious obstacle given than it is easy to check if a candidate is a true period or not by evaluating the function for a few input values. In passing, note that the probability P_k for k to take a particular value,

$$P_k = \frac{\langle F_k | F_k \rangle}{\lambda}, \tag{4.90}$$

is not uniformly distributed and varies with k.[10] Again, this is because $f(x) = f(y)$ may hold for some $0 \le x, y < \lambda$.

We demonstrate the basic ideas in the period-finding algorithm. More complete example follows below.

```
$n = 4;
$N = Power[2, $n];
```

We take an example of the modular exponential function, $f(x) = a^x \pmod{L}$ for a fixed a and L. Here a and L are assumed to be co-primes.

```
$a = 3;
Clear[f];
f[k_] := PowerMod[$a, k, $N]
```

The function is periodic.

In[·]:= `ListPlot[Table[{k, f[k]}, {k, 0, $N - 1}], Joined → True,`
`    PlotMarkers → Automatic, FrameLabel → {"x", HoldForm@f[x]}]`

Out[·]=

The classical oracle (in the binary form) corresponding to the function f (expressed in usual integers).

```
Clear[g]
g[x__] := IntegerDigits[f[FromDigits[{x}, 2]], 2, $n]
```

Here is the quantum circuit model of the algorithm.

---

[10] This is another difference from the hidden subgroup problem.

```
In[ ]:= cc = Range[$n];
       tt = Range[$n + 1, $n + $n];
       qc0 = QuantumCircuit[
           LogicalForm[Ket[], Join[S@cc, S@tt]], S[cc, 6], Oracle[g, S@cc, S@tt],
           QuantumFourierTransform[S@cc]];
       qc = QuantumCircuit[qc0, Measurement[S@cc]]
```

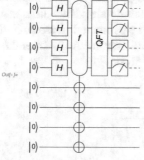

```
In[ ]:= out = ExpressionFor[qc];
       LogicalForm[out, Join[S@cc, S@tt]]
       result = Readout[out, S@cc]
```

$$Out[ ]= \frac{1}{2} \left| 1_{s_1} 0_{s_2} 0_{s_3} 0_{s_4} 0_{s_5} 0_{s_6} 0_{s_7} 1_{s_8} \right\rangle - \frac{1}{2} \left| 1_{s_1} 0_{s_2} 0_{s_3} 0_{s_4} 0_{s_5} 0_{s_6} 1_{s_7} 1_{s_8} \right\rangle +$$
$$\frac{1}{2} \left| 1_{s_1} 0_{s_2} 0_{s_3} 0_{s_4} 1_{s_5} 0_{s_6} 0_{s_7} 1_{s_8} \right\rangle - \frac{1}{2} \left| 1_{s_1} 0_{s_2} 0_{s_3} 0_{s_4} 1_{s_5} 0_{s_6} 1_{s_7} 1_{s_8} \right\rangle$$

$$Out[ ]= \{1, 0, 0, 0\}$$

From the measurement outcome, one can find the period. In this case, the period is 4.

```
In[ ]:= frac = FromDigits[result, 2] / $N
```

$$Out[ ]= \frac{1}{2}$$

For a more complete overview, let us examine all possible outcomes.

```
In[ ]:= new = ExpressionFor[qc0];
       LogicalForm[new, Join[S@cc, S@tt]]
```

$$Out[ ]= \frac{1}{4} \left| 0_{s_1} 0_{s_2} 0_{s_3} 0_{s_4} 0_{s_5} 0_{s_6} 0_{s_7} 1_{s_8} \right\rangle + \frac{1}{4} \left| 0_{s_1} 0_{s_2} 0_{s_3} 0_{s_4} 0_{s_5} 0_{s_6} 1_{s_7} 1_{s_8} \right\rangle +$$
$$\frac{1}{4} \left| 0_{s_1} 0_{s_2} 0_{s_3} 0_{s_4} 1_{s_5} 0_{s_6} 0_{s_7} 1_{s_8} \right\rangle + \frac{1}{4} \left| 0_{s_1} 0_{s_2} 0_{s_3} 0_{s_4} 1_{s_5} 0_{s_6} 1_{s_7} 1_{s_8} \right\rangle + \frac{1}{4} \left| 0_{s_1} 1_{s_2} 0_{s_3} 0_{s_4} 0_{s_5} 0_{s_6} 0_{s_7} 1_{s_8} \right\rangle +$$
$$\frac{1}{4} i \left| 0_{s_1} 1_{s_2} 0_{s_3} 0_{s_4} 0_{s_5} 0_{s_6} 1_{s_7} 1_{s_8} \right\rangle - \frac{1}{4} \left| 0_{s_1} 1_{s_2} 0_{s_3} 0_{s_4} 1_{s_5} 0_{s_6} 0_{s_7} 1_{s_8} \right\rangle -$$
$$\frac{1}{4} i \left| 0_{s_1} 1_{s_2} 0_{s_3} 0_{s_4} 1_{s_5} 0_{s_6} 1_{s_7} 1_{s_8} \right\rangle + \frac{1}{4} \left| 1_{s_1} 0_{s_2} 0_{s_3} 0_{s_4} 0_{s_5} 0_{s_6} 0_{s_7} 1_{s_8} \right\rangle - \frac{1}{4} \left| 1_{s_1} 0_{s_2} 0_{s_3} 0_{s_4} 0_{s_5} 0_{s_6} 1_{s_7} 1_{s_8} \right\rangle +$$
$$\frac{1}{4} \left| 1_{s_1} 0_{s_2} 0_{s_3} 0_{s_4} 1_{s_5} 0_{s_6} 0_{s_7} 1_{s_8} \right\rangle - \frac{1}{4} \left| 1_{s_1} 0_{s_2} 0_{s_3} 0_{s_4} 1_{s_5} 0_{s_6} 1_{s_7} 1_{s_8} \right\rangle + \frac{1}{4} \left| 1_{s_1} 1_{s_2} 0_{s_3} 0_{s_4} 0_{s_5} 0_{s_6} 0_{s_7} 1_{s_8} \right\rangle -$$
$$\frac{1}{4} i \left| 1_{s_1} 1_{s_2} 0_{s_3} 0_{s_4} 0_{s_5} 0_{s_6} 1_{s_7} 1_{s_8} \right\rangle - \frac{1}{4} \left| 1_{s_1} 1_{s_2} 0_{s_3} 0_{s_4} 1_{s_5} 0_{s_6} 0_{s_7} 1_{s_8} \right\rangle + \frac{1}{4} i \left| 1_{s_1} 1_{s_2} 0_{s_3} 0_{s_4} 1_{s_5} 0_{s_6} 1_{s_7} 1_{s_8} \right\rangle$$

```
In[ ]:= data = Table[out = Measurement[new, S@cc];
         FromDigits[Readout[out, S@cc], 2], {15}]
```

$$Out[ ]= \{12, 8, 0, 12, 4, 4, 4, 0, 0, 8, 0, 8, 8, 4, 8\}$$

$In[\cdot]:=$ **sec = Union[data / $N]**

$Out[\cdot]=$ $\left\{ 0, \dfrac{1}{4}, \dfrac{1}{2}, \dfrac{3}{4} \right\}$

The period must be 4, which agrees with the true value.

A more serious hurdle is that the estimated phase ϕ is only approximate to the precise value k/λ, $\phi \approx k/\lambda$, because the register is of finite size. This difference makes it more difficult to determine the period λ from the estimated phase ϕ. Fortunately, a continued fraction representation provides an efficient method to find the period.

A *continued fraction* is an expression of the form

$$a_0 + \cfrac{1}{a_1 + \cfrac{1}{a_2 + \cfrac{1}{\ddots + \cfrac{1}{a_n}}}} , \tag{4.91}$$

where $a_0, a_1, \ldots, a_n$ are positive integers.[11] It is often denoted by

$$[a_0; a_1, a_2, \ldots, a_n]. \tag{4.92}$$

One can represent any real number with a unique continued fraction.[12] For a given number, the procedure to find a continued fraction is simple. For example, consider $13/5$. We first split it into its integer part which is 2 and fractional part $3/5$, so $13/5 = 2 + 3/5$. Next, we take the reciprocal of $3/5$ and split it into its integer and fractional parts, $5/3 = 1 + 2/3$. By repeating the procedure, we find that

$$\frac{13}{5} = 2 + \cfrac{1}{1 + \cfrac{1}{1 + \frac{1}{2}}} , \tag{4.93}$$

or equivalently,

$$\frac{13}{5} = [2; 1, 1, 2]. \tag{4.94}$$

The well-known Euclidean algorithm, dating back to Euclid's book *Elements*, efficient in finding a continued fraction representation of a number. A continued fraction may be terminated by a finite n, and in such case is said to be finite. If a continued fraction continues the iteration without termination, then it is called an infinite continued fraction. The continued fraction $[a_0; a_1, a_2, \ldots, a_m]$ is called the mth *convergent* of $[a_0; a_1, a_2, \ldots, a_m, a_{m+1}, \ldots, a_n]$ $(0 \le m \le n)$.

Although not so common in daily life, continued fraction representations are, in many respects, more natural mathematically than the decimal representations or

---

[11] There are several other conventions for continued fractions. Here we only use the so-called *simple* or *canonical* continued fractions.

[12] For a finite continued fraction, there is a trivial ambiguity such as $[a_0; a_1, a_2, \ldots, a_n] = [a_0; a_1, a_2, \ldots, (a_n - 1), 1]$. We regard such continued fractions to be equivalent.

other similar representations with a fixed base. To see this, consider the following properties: The continued fraction for any rational number is finite, which stands in contrast with decimal representations. For example, $5/3 = [1; 1, 2]$ whereas $5/3 = 1.666\ldots$. On the other hand, continued fractions for irrational numbers are infinite. For example, the golden ratio is represented by the continued fraction, $(1 + \sqrt{5})/2 = [1; 1, 1, 1, \ldots]$. Furthermore, continued fractions for a number and its reciprocal are identical except for a shift by one place. In other words, $[a_0; a_1, a_2, \ldots, a_n]$ and $[0; a_0, a_1, a_2, \ldots, a_n]$ are reciprocal of each other.

The method to determine the period λ from the estimated phase $\phi \approx k/\lambda$ is based on a theorem from number theory. It states[13] that if

$$\left| \frac{k}{\lambda} - \phi \right| \le \frac{1}{2\lambda^2} \tag{4.95}$$

for any integers k and λ, then k/λ is a convergent of the continued fraction for ϕ. Note that ϕ is accurate to m bits. The error is less than 2^{-m} with probability higher than 0.81 (Sect. 4.4.3). Since $\lambda \ll 2^m$, the condition in (4.95) is satisfied. Therefore, the continued fraction of the estimated phase ϕ gives coprimes k'/λ' such that $k'/\lambda' = k/\lambda$. λ' and integer multiples of λ' are candidates for the period. We then evaluate (classically) the function to check if λ (or any integer multiple of λ') is the true period λ of the function.

Now we consider a full period-finding problem. We consider a four-qubit system.

```
$n = 4;
$N = Power[2, $n];
```

We define an arbitrary function with period 3.

```
Clear[f];
f[0] = 1; f[1] = 5; f[2] = 4;
f[x_] := f[Mod[x, 3]]
```

This plot illustrates that the function is indeed periodic with period 3.

```
In[·]:= ListPlot[Table[{k, f[k]}, {k, 0, $N - 1}], Joined → True,
          PlotMarkers → Automatic, FrameLabel → {"x", HoldForm@f[x]}]
```

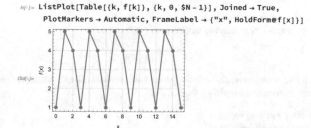

The classical oracle (in the binary form) corresponding to the function f (expressed in usual integers).

---

[13] A proof is found in Nielsen & Chuang (2011, Appendix 4).

```
Clear[g]
g[x__] := IntegerDigits[f[FromDigits[{x}, 2]], 2, $n]
```

Here is the quantum circuit model of the algorithm.

```
In[ ]:= cc = Range[$n];
tt = Range[$n + 1, $n + $n];
qc0 = QuantumCircuit[
    LogicalForm[Ket[], Join[S@cc, S@tt]], S[cc, 6], Oracle[g, S@cc, S@tt],
    QuantumFourierTransform[S@cc]];
qc = QuantumCircuit[qc0, Measurement[S@cc]]
```

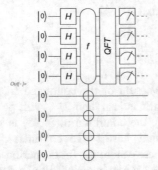

This is a full run of the above quantum circuit model.

```
In[ ]:= Timing[out = ExpressionFor[qc];]
LogicalForm[out, Join[S@cc, S@tt]];
result = Readout[out, S@cc]
```

```
Out[ ]= {6.88555, Null}
```

```
Out[ ]= {0, 1, 0, 0}
```

In order to check more random results, it is more efficient to first calculate the output state right before the measurement.

```
In[ ]:= Timing[new = ExpressionFor[qc0];]
```

```
Out[ ]= {3.20231, Null}
```

From the above output state, we perform measurements (as many times as we like) to get random results.

```
In[ ]:= out = Measurement[new, S@cc];
phi = FromDigits[Readout[out, S@{cc}], 2] / $N
```

$$Out[ ]= \frac{11}{16}$$

From the above random result, we determine the period of the function by utilizing the continued fraction representation of the estimated phase.

```
In[ ]:= cnt = ContinuedFraction[phi]
candidates = Convergents[cnt]
```

```
Out[ ]= {0, 1, 2, 5}
```

$$Out[ ]= \left\{0, 1, \frac{2}{3}, \frac{11}{16}\right\}$$

From the above convergents, we have two candidates 3 (and its multiples 6, 9, 12, 15 are also candidates) and 16. Evaluation of the function for some arguments confirms that 3 is the period of the function.

4.5.2 The Order-Finding Algorithm

In the previous section, the function f was not specified explicitly, but the corresponding quantum oracle (4.84) was assumed to be implemented efficiently. Here we consider a specific function, the *modular exponentiation*

$$f(x) = a^x \quad (\text{mod } N) \tag{4.96}$$

for fixed positive integers a and N. The two integers a and N are assumed to be coprimes. The function is a periodic function,

$$a^r = 1 \quad (\text{mod } N) \tag{4.97}$$

for some integer r. The primary period (the smallest period) r is called the order of a modulo N. The order-fining problem is to determine the order for given a and N. It is known to be hard to solve classically. The order-finding algorithm is the central part of the quantum factorization algorithm to be discussed in Sect. 4.5.3.

Unlike the period-finding algorithm, the order-finding algorithm does not resort to quantum oracle, and implement the function with elementary gates. More specifically, it uses the quantum phase estimation algorithm to estimate the phase of the unitary operator

$$\hat{U} : |x\rangle \mapsto |ax \,(\text{mod } N)\rangle \,, \tag{4.98}$$

the quantum mechanical extension of the *modular multiplication*.

A repeated application of the modular multiplication gives the modular exponentiation

$$\hat{U}^y : |x\rangle \mapsto |a^y x \,(\text{mod } N)\rangle. \tag{4.99}$$

As such, applying the modular multiplication r times should bring the system back to the original state. Therefore, the states

$$|1\rangle, \hat{U}\,|1\rangle, \hat{U}^2\,|1\rangle, \cdots, \hat{U}^{r-1}\,|1\rangle \tag{4.100}$$

form a closed cycle. In physics, it is common to construct the eigenstates of $\hat{U}$ that generates the cycle with the discrete Fourier transform of the states in the list.[14] By definition, the latter is the quantum Fourier transform [see Eq. (4.38)]. Indeed, define

---

[14] In condensed matter physics, the resulting states are called *Bloch states*.

$$|\phi_k\rangle := \frac{1}{\sqrt{r}} \sum_{x=0}^{r-1} \hat{U}^r |1\rangle e^{2\pi i x k/r} = \frac{1}{\sqrt{r}} \sum_{x=0}^{r-1} |a^x \; (\mathrm{mod}\, N)\rangle e^{2\pi i x k/r} \qquad (4.101)$$

for $k = 0, 1, \ldots, r - 1$. We see that

$$\hat{U} |\phi_k\rangle = \frac{1}{\sqrt{r}} \sum_{x=0}^{r-1} |a^{x+1} \; (\mathrm{mod}\, N)\rangle e^{2\pi i x k/r} = e^{-2\pi i k/r} |\phi_k\rangle . \qquad (4.102)$$

In other words, $|\phi_k\rangle$ is an eigenstate of $\hat{U}$ belonging to the eigenvalue $e^{-2\pi i k/r}$. Therefore, by preparing the target register in one of the eigenstates $|\phi_k\rangle$ and applying the standard quantum phase estimation procedure, one would be able to estimate the "phase" $2\pi \phi_k := 2\pi k/r$. One would then find the order r from it.

Unfortunately, it is highly non-trivial to prepare an register in one of those eigenstates. Instead, we prepare the target register in a superposition of *all* eigenstates. It is also a common dictum in physics that a localized distribution in real space is unlocalized in the Fourier space and *vice versa*. Indeed, we find that

$$\sum_{k=0}^{r-1} |\phi_k\rangle = |1\rangle \equiv |0\rangle \otimes \cdots \otimes |0\rangle \otimes |1\rangle . \qquad (4.103)$$

The quantum phase estimation procedure randomly selects one from the possible values $0, 1/r, \ldots, (r-1)/r$. Furthermore, the estimated phase ϕ is only approximate to one of the precise values $\phi_k = k/r$. We have already seen this situation in the period-finding algorithm. The (classical) post-processing is exactly the same as in the period-finding algorithm.

There still remains a question when it comes to the performance of the order-finding algorithm. The performance of the algorithm relies on how efficiently one can perform the controlled-$\hat{U}^{2^j}$ gate (see Sect. 4.4.2). As mentioned above in (4.99), $\hat{U}^y$ is nothing but the modular exponentiation. Fortunately, it is known that the modular exponentiation can be implemented efficiently on a quantum computer (Vedral et al., 1996). Let $\hat{V}$ be the modular multiplication on *two* registers (to be compared with $\hat{U}$ with fixed a in (4.98)),

$$\hat{V} : |x\rangle \otimes |y\rangle \mapsto |x\rangle \otimes |xy \; (\mathrm{mod}\, N)\rangle . \qquad (4.104)$$

The basic idea is to duplicate the logical state, $|x\rangle \otimes |0\rangle \to |x\rangle \otimes |x\rangle$ by employing the CNOT gate, and then to apply the modular multiplication on the two registers, $|x\rangle \otimes |x\rangle \to |x^2 \; (\mathrm{mod}\, N)\rangle$. Repeating the procedure j times produces the desired result. For example, the following quantum circuit effectively performs $\hat{U}^{2^3}$,

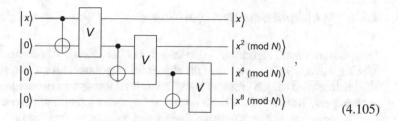

$$(4.105)$$

where each horizontal line denotes a *register* of n qubits, not a single qubit, and $x = 0, 1, \ldots, 2^n - 1$.

4.5.3 Quantum Factorization Algorithm

The quantum factorization algorithm is composed of two parts, one quantum mechanical and the other classical. The quantum mechanical part is just the order-finding algorithm. The classical part determines the factors from the order based on number theory.

Let N be an odd[15] composite integer. We want to factorize it. Suppose that $N = p_1^{v_1} \cdots p_m^{v_m}$, where p_j are prime numbers and v_j are integers, is the prime factorization.

We start by choosing an integer x uniformly at random from the set

$$\{x : 1 \leq x < N, \ \gcd(x, N) = 1\}. \tag{4.106}$$

We calculate the order r of x modulo N such that $x^r = 1 \ (\mathrm{mod} \ N)$. This is the step where the quantum order-finding algorithm is required. It is likely that the order r is even and $x^{r/2} \neq -1 \ (\mathrm{mod} \ N)$. According to number theory, the actually probability for this condition to be satisfied is larger than $1 - 2^{-m}$.

With x and r at hand, we put $y = x^{r/2}$. It is a non-trivial solution to

$$y^2 = 1 \quad (\mathrm{mod} \ N). \tag{4.107}$$

It is non-trivial in the sense that $y \neq \pm 1 \ (\mathrm{mod} \ N)$. A non-trivial solution to (4.107) is useful because number theory guarantees that either or both of $\gcd(y \pm 1, N)$ is a non-trivial factor of N. When N is an L-bit composite number, it takes an order of $\mathcal{O}(L^3)$ operations to calculate $\gcd(y \pm 1, N)$.

---

[15] If N is even, 2 is one of the factors, start with $N/2$.

4.5.4 *Quantum Search Algorithm*

The quantum search algorithm—also known as *Grover's algorithm* after its inventor (Grover, 1996, 1997)—is a quantum algorithm to conduct an *unstructured* search. The unstructured search problem is to find a small number of elements with a specified property from within a huge *unsorted* (or, more generally, *unstructured*) data set. The quantum search algorithm finds a desired element with a high probability with an order of $\sqrt{N}$ queries to the function that defines the property, where N is the size of the data set. Classical algorithms needs N queries. Interestingly, the quantum search algorithm is known to be already optimal in the sense that any quantum algorithm for unstructured search needs an order of $\sqrt{N}$ queries. In any case, the superiority of the quantum algorithm over its classical counterparts is merely quadratic (i.e., polynomial) rather than exponential. However, the quadratic speedup cannot be underestimated. The efficacy is considerable for large N. Furthermore, the quantum search algorithm can apply to a wide range of problems. For example, it enables us to determine the existence and number of solutions of so-called NP-complete problems and to search exhaustively for the solutions.

To understand and analyze the quantum search algorithm, it is convenient to introduce two closely related unitary operators. One is a *Householder transformation*. Let $|v\rangle$ be a normalized vector in the Hilbert space. The Householder transformation associated with $|v\rangle$ is a unitary transformation defined by

$$\hat{U}_v := \hat{I} - 2 |v\rangle \langle v|. \tag{4.108}$$

Geometrically, it is a reflection in the hyperplane perpendicular to the vector $|v\rangle$ and containing the origin (null vector) as illustrated in Fig. 4.4a. For this reason, it is also known as *Householder reflection*. The other is Grover's diffusion operator. Grover's diffusion operator associated with $|v\rangle$ is defined by

$$\hat{V}_v := 2 |v\rangle \langle v| - \hat{I} = -\hat{U}_v. \tag{4.109}$$

It is a reflection in the line along the vector $|v\rangle$ as shown in Fig. 4.4b. It can also be regarded as the Householder reflection $\hat{U}_v$ associated with the same vector $|v\rangle$ followed by another reflection through the origin [see Fig. 4.4b].

One can directly implement the Householder reflection $\hat{U}_v$ for a known state $|v\rangle$. Suppose that $|v\rangle$ is achieved from $|0\rangle \equiv |0\rangle^{\otimes n}$ by a unitary transformation $\hat{W}$, $|v\rangle = \hat{W} |0\rangle$.[16] Then it follows that

$$\hat{U}_v = \hat{W} \left(\hat{I} - 2 |0\rangle \langle 0| \right) \hat{W}^\dagger = \hat{W} \hat{U}_0 \hat{W}^\dagger. \tag{4.110}$$

Note that $\hat{U}_0$ flips the phase of $|0\rangle$ to $- |0\rangle$ but leaves other states unaffected. It is equivalent to writing

---

[16] Recall that there always exists such a unitary operator $\hat{W}$. See Theorem A.16.

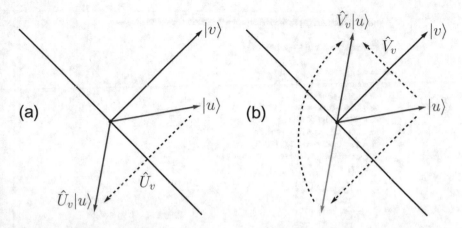

Fig. 4.4 Geometric interpretation of **a** Householder transformation and **b** Grover's diffusion operator. **a** The Householder transformation $\hat{U}_v$ associated with a normalized vector $|v\rangle$ is a reflection in the hyperplane containing the origin and perpendicular to $|v\rangle$. **b** Grover's diffusion operator $\hat{V}_v$ with respect to $|v\rangle$ is a reflection in the line along the vector $|v\rangle$. It can also be regarded as the Householder transformation $\hat{U}_v$ associated with the same vector $|v\rangle$ followed by another reflection through the origin

$$\hat{U}_0 : |x_1 \cdots x_{n-1}\rangle \otimes |x_n\rangle \mapsto \begin{cases} |x_1 \cdots x_{n-1}\rangle \otimes \left(-\hat{Z}\,|x_n\rangle\right) & x_1 = \cdots = x_{n-1} = 0; \\ |x_1 \cdots x_{n-1}\rangle \otimes |x_n\rangle & \text{(otherwise)}. \end{cases}$$

(4.111)

It is a variant of multi-qubit controlled-$\hat{Z}$ gate. More explicitly, $\hat{U}_0$ for four qubits is described by the following quantum circuit

$$\hat{U}_0 \doteq$$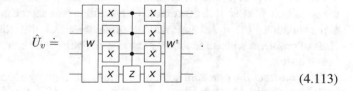

(4.112)

Overall, the Householder transformation associated with $|v\rangle$ can be implemented by a quantum circuit of the form

$$\hat{U}_v \doteq$$

(4.113)

One can achieve Grover's diffusion operator $\hat{V}_v$ for a known vector $|v\rangle$ in the same manner (up to a global phase factor -1).

Here is a quantum circuit model for Grover's diffusion operator.

We first implement $\left(1 - 2 \left|0\right\rangle \left\langle 0\right|\right)$.

```
In[ ]:= $n = 3;
jj = Range[$n];
ss = S[jj, None];
cz = CZ @@@ Subsets[ss, {2}];
cz = Successive[CZ, ss];
cz = CZ[S[1], #] & /@ Rest@ss;
qc = QuantumCircuit[S[jj, 1],
  ControlledU[S[Most@jj], S[$n, 3]],
  S[jj, 1]]
```

Out[ ]=

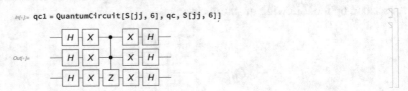

```
In[ ]:= bs = Basis[S[jj]];
out = qc ** bs;
Thread[bs → out] // LogicalForm // TableForm
```

Out[ ]//TableForm=

$$\left|0_{S_1}0_{S_2}0_{S_3}\right\rangle \to -\left|0_{S_1}0_{S_2}0_{S_3}\right\rangle$$
$$\left|0_{S_1}0_{S_2}1_{S_3}\right\rangle \to \left|0_{S_1}0_{S_2}1_{S_3}\right\rangle$$
$$\left|0_{S_1}1_{S_2}0_{S_3}\right\rangle \to \left|0_{S_1}1_{S_2}0_{S_3}\right\rangle$$
$$\left|0_{S_1}1_{S_2}1_{S_3}\right\rangle \to \left|0_{S_1}1_{S_2}1_{S_3}\right\rangle$$
$$\left|1_{S_1}0_{S_2}0_{S_3}\right\rangle \to \left|1_{S_1}0_{S_2}0_{S_3}\right\rangle$$
$$\left|1_{S_1}0_{S_2}1_{S_3}\right\rangle \to \left|1_{S_1}0_{S_2}1_{S_3}\right\rangle$$
$$\left|1_{S_1}1_{S_2}0_{S_3}\right\rangle \to \left|1_{S_1}1_{S_2}0_{S_3}\right\rangle$$
$$\left|1_{S_1}1_{S_2}1_{S_3}\right\rangle \to \left|1_{S_1}1_{S_2}1_{S_3}\right\rangle$$

This is Grover's diffusion operator (up to an irrelevant global phase factor of -1).

```
In[ ]:= qc1 = QuantumCircuit[S[jj, 6], qc, S[jj, 6]]
```

Out[ ]=

Let us now describe the unstructured search problem more precisely. Suppose that we have a large list of items labeled by $x = 0, 1, 2, \ldots, N - 1$. Among the items, there are a small number M $(1 \le M \ll N)$ of items with a unique property. The problem is to seek those items. The property is usually described by a classical oracle, $f : \{0, 1\}^n \to \{0, 1\}$, where we assume $N \equiv 2^n$. The desired items are designated by the condition $f(x) = 1$. In other words, an unstructured search is to find the solutions to the equation $f(x) = 1$. This is why the quantum search algorithm applies to a broad class of problems where a function $f(x)$ is well defined and can be evaluated without heavy cost.

We construct the quantum search algorithm combining Householder reflections and Grover's diffusion operators. Let $\mathcal{M}$ be the set of solutions,

$$\mathcal{M} := \left\{x : f(x) = 1, \quad x = 0, 1, \ldots, 2^n - 1\right\}. \tag{4.114}$$

Define a state of superposition consisting of the solutions

$$|\omega\rangle := \frac{1}{\sqrt{M}} \sum_{x \in \mathcal{M}} |x\rangle, \tag{4.115}$$

and another consisting of items that are not solutions

$$|\omega_\perp\rangle := \frac{1}{\sqrt{N-M}} \sum_{x \notin \mathcal{M}} |x\rangle. \tag{4.116}$$

Obviously, they are orthogonal to each other, $\langle \omega | \omega_\perp \rangle = 0$. Neither of the two states are known and can be prepared. Rather, $|\omega\rangle$ is the state we want to produce through the algorithm. Now consider the overall-superposition state, consisting all logical basis states

$$|\alpha\rangle := \frac{1}{\sqrt{N}} \sum_{x=0}^{N-1} |x\rangle. \tag{4.117}$$

As we have seen several times, this state can be prepared by applying the Hadamard gates $\hat{H}^{\otimes n}$ on $|0\rangle$. Since $|\alpha\rangle$ involves both solutions and non-solutions, one can rewrite it into the form

$$|\alpha\rangle = \sqrt{\frac{M}{N}} |\omega\rangle + \sqrt{\frac{N-M}{N}} |\omega_\perp\rangle. \tag{4.118}$$

The advantage of the overall-superposition state $|\alpha\rangle$ is that one can make use of the linearity of quantum mechanics—or quantum parallelism—to evaluate an operator on every state all at once. Indeed, applying the Householder reflection $\hat{U}_\omega$ associated with the state vector $|\omega\rangle$ on $|\alpha\rangle$ produces

$$\hat{U}_\omega |\alpha\rangle = -\sqrt{\frac{M}{N}} |\omega\rangle + \sqrt{\frac{N-M}{N}} |\omega_\perp\rangle = \frac{1}{\sqrt{N}} \sum_{x=0}^{N-1} |x\rangle (-1)^{f(x)}. \tag{4.119}$$

With a single operation of $\hat{U}_\omega$, we have flagged the states $|x\rangle$ associated with the solutions by the reversed phase factor -1. Unfortunately, however, we cannot directly single out those states. Instead, the tactic is to amplify the component along the state $|\omega\rangle$. This can be achieved by applying Grover's diffusion operator $\hat{V}_\alpha$ associated with $|\alpha\rangle$, $\hat{U}_\omega |\alpha\rangle \to \hat{V}_\alpha \hat{U}_\omega |\alpha\rangle$. As illustrated geometrically in Fig. 4.5a, it brings the state closer to the state $|\omega\rangle$.

At this stage, it is convenient to define the *Grover rotation* operator

$$\hat{G} := \hat{V}_\alpha \hat{U}_\omega. \tag{4.120}$$

$\hat{G}$ describes a rotation within the subspace spanned by $|\omega\rangle$ and $|\alpha\rangle$ towards the desired state $|\omega\rangle$. For a more quantitative analysis, we define the angle θ by

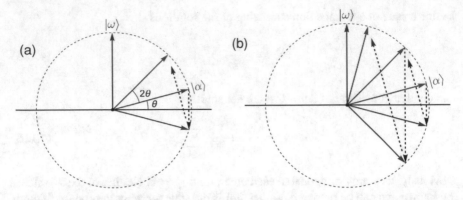

Fig. 4.5 Geometric interpretation of the quantum search algorithm. **a** The application of the Householder reflection $\hat{U}_\omega$ followed by Grover's diffusion operator $\hat{V}_\alpha$ leads to the Grover rotation, $\hat{G} := \hat{V}_\alpha \hat{U}_\omega$, a rotation by angle 2θ. θ is the angle between $|\alpha\rangle$ and the horizontal axis. **b** Alternating applications of $\hat{U}_\omega$ and $\hat{V}_\alpha$, or equivalently repeated applications of the Grover rotation $\hat{G}$, bring $|\alpha\rangle$ increasingly closer to $|\omega\rangle$

$$\sin(\theta/2) = \sqrt{\frac{M}{N}}, \quad \cos(\theta/2) = \sqrt{\frac{N-M}{N}} \tag{4.121}$$

so that

$$|\alpha\rangle = |\omega\rangle \sin(\theta/2) + |\omega_\perp\rangle \cos(\theta/2). \tag{4.122}$$

It corresponds to the state rotated from $|\omega_\perp\rangle$ by angle θ around the axis perpendicular to both $|\omega\rangle$ and $|\omega_\perp\rangle$ in the Bloch space corresponding to the two-dimensional subspace spanned by $|\omega\rangle$ and $|\omega_\perp\rangle$. Accordingly, $\hat{G}$ corresponds to a rotation by angle 2θ around the same axis. This is illustrated in Fig. 4.5a. The k applications of the Grover rotation $\hat{G}$ give

$$|\psi_k\rangle = \hat{G}^k |\alpha\rangle = |\omega\rangle \sin((k+1/2)\theta) + |\omega_\perp\rangle \cos((k+1/2)\theta). \tag{4.123}$$

Starting from $|\alpha\rangle$ in (4.122), the resulting state $|\psi_k\rangle$ gets closest to the desired state $|\omega\rangle$ when $(k+1/2)\theta \approx \pi/2$. Therefore, the optimal number K of the Grover rotations to apply is given by

$$K = \text{round}(\pi/2\theta - 1/2). \tag{4.124}$$

Assuming that $M \ll N$, we observe from (4.121) that

$$\theta \approx 2\sin^{-1}\left(\sqrt{M/N}\right) \approx 2\sqrt{M/N} \ll 1. \tag{4.125}$$

It follows that

$$K \approx \frac{\pi}{4}\sqrt{\frac{N}{M}}. \tag{4.126}$$

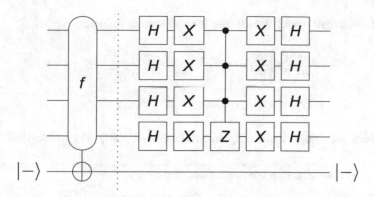

Fig. 4.6 A quantum circuit model for the implementation of the Grover rotation. The first part before the vertical dashed line effectively implements the Householder transformation $\hat{U}_\omega$. The second part corresponds to Grover's diffusion operator $\hat{V}_\alpha$

This means that we can find the solution by applying the Grover rotations $\hat{G}$ an order of $\sqrt{N}$ times.

We can make the estimation of the performance more rigorous and get the upper bound. We note that

$$K \leq \text{ceil}(\pi/2\theta). \tag{4.127}$$

In accordance with (4.121), we note that

$$\theta/2 \geq \sin(\theta/2) = \sqrt{M/N}, \tag{4.128}$$

where we have assumed that $M \leq N/2$. Putting it to (4.127), we find that

$$K \leq \frac{\pi}{4}\sqrt{\frac{N}{M}}. \tag{4.129}$$

In short, the estimation in (4.126) turns out to be the worst case estimation.

In the above, we have described the quantum search algorithm in terms of the Grover rotation consisting of the Householder reflection $\hat{U}_\omega$ and Grover's diffusion operator $\hat{V}_\alpha$. As we have noted in (4.113), $\hat{V}_\alpha$ can be implemented in terms of elementary gates because the state $|\alpha\rangle$ is known. However, $|\omega\rangle$ is unknown, and one cannot implement $\hat{U}_\omega$ in a similar way. Fortunately, we have noted in (4.119) that the effect of $\hat{U}_\omega$ is flagging the states $|x\rangle$ corresponding to the solutions with the reversed phase factor -1. One can achieve exactly the same effect by employing the quantum oracle (Sect. 4.2.1) corresponding to the classical oracle $f(x)$. In particular, with the ancillary qubit prepared in the state $|-\rangle := (|0\rangle - |1\rangle)/\sqrt{2}$, the quantum oracle induces a conditional phase shift as follows [see Eqs. (4.19) and (4.20)]

$$|x\rangle \otimes |-\rangle \mapsto \left(|x\rangle (-1)^{f(x)}\right) \otimes |-\rangle . \tag{4.130}$$

In quantum circuit model, the identity is expressed as follows

$$|x\rangle\Big\{\boxed{f}\Big\}|x\rangle\,(-1)^{f(x)} \quad |x\rangle\Big\{\boxed{f}\Big\}\hat{U}_\omega\,|x\rangle$$
$$=$$
$$|-\rangle \;\longrightarrow\!\!\oplus\!\!\longrightarrow\; |-\rangle \qquad\quad |-\rangle \;\longrightarrow\!\!\oplus\!\!\longrightarrow\; |-\rangle \tag{4.131}$$

The overall quantum circuit to implement the Grover rotation is depicted in Fig. 4.6.

Here we demonstrate the quantum search algorithm.

We work with 4 (system) qubits and one ancillary qubit. The ancillary qubit is for the quantum oracle.

```
$n = 4;
cc = Range[$n - 1];
tt = $n;
jj = Range[$n];
aa = $n + 1;
```

Here we specify a classical oracle. The function f is more human readable. The function g will be used in the actual quantum oracle.

```
Clear[f]
f[3] = 1; f[_] = 0;
g[x__] := f@FromDigits[{x}, 2]
```

The ancilla qubit is prepared in the following state.

```
In[·]:= in = ProductState[S[aa] → {1, -1} / Sqrt[2]];
        in // Elaborate // LogicalForm
```

$$Out[\cdot]= \frac{|0_{S_5}\rangle}{\sqrt{2}} - \frac{|1_{S_5}\rangle}{\sqrt{2}}$$

This is a quantum circuit for the Grover rotation.

```
In[·]:= grv = QuantumCircuit[
          Oracle[g, S[jj], S[aa]], "Separator",
          S[jj, 6], S[jj, 1], ControlledU[S[cc], S[tt, 3]], S[jj, 1], S[jj, 6],
          "PortSize" → {0.5, 0.2}
        ]
```

Out[·]=

We apply the Grover rotation once on the input state.

In[ ]:= `qc1 = QuantumCircuit[LogicalForm[Ket[], S[jj]],`
`{in, "Label" → "|-)"}, S[jj, 6], "Separator", grv]`

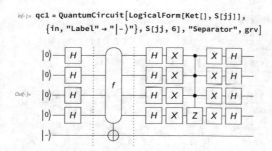

Out[ ]=

Here is the result of it.

In[ ]:= `out1 = Elaborate[qc1]`

Out[ ]=
$$
-\frac{3\left|_-\right)}{16\sqrt{2}} - \frac{3\left|1_{S_1}\right)}{16\sqrt{2}} - \frac{3\left|1_{S_1}1_{S_2}\right)}{16\sqrt{2}} - \frac{3\left|1_{S_1}1_{S_2}1_{S_3}\right)}{16\sqrt{2}} - \frac{3\left|1_{S_1}1_{S_2}1_{S_3}1_{S_4}\right)}{16\sqrt{2}} +
$$
$$
\frac{3\left|1_{S_1}1_{S_2}1_{S_3}1_{S_4}1_{S_5}\right)}{16\sqrt{2}} + \frac{3\left|1_{S_1}1_{S_2}1_{S_3}1_{S_5}\right)}{16\sqrt{2}} - \frac{3\left|1_{S_1}1_{S_2}1_{S_4}\right)}{16\sqrt{2}} + \frac{3\left|1_{S_1}1_{S_2}1_{S_4}1_{S_5}\right)}{16\sqrt{2}} +
$$
$$
\frac{3\left|1_{S_1}1_{S_2}1_{S_5}\right)}{16\sqrt{2}} - \frac{3\left|1_{S_1}1_{S_3}\right)}{16\sqrt{2}} - \frac{3\left|1_{S_1}1_{S_3}1_{S_4}\right)}{16\sqrt{2}} + \frac{3\left|1_{S_1}1_{S_3}1_{S_4}1_{S_5}\right)}{16\sqrt{2}} + \frac{3\left|1_{S_1}1_{S_3}1_{S_5}\right)}{16\sqrt{2}} -
$$
$$
\frac{3\left|1_{S_1}1_{S_4}\right)}{16\sqrt{2}} + \frac{3\left|1_{S_1}1_{S_4}1_{S_5}\right)}{16\sqrt{2}} + \frac{3\left|1_{S_1}1_{S_5}\right)}{16\sqrt{2}} + \frac{3\left|1_{S_2}\right)}{16\sqrt{2}} - \frac{3\left|1_{S_2}1_{S_3}\right)}{16\sqrt{2}} - \frac{3\left|1_{S_2}1_{S_3}1_{S_4}\right)}{16\sqrt{2}} +
$$
$$
\frac{3\left|1_{S_2}1_{S_3}1_{S_4}1_{S_5}\right)}{16\sqrt{2}} + \frac{3\left|1_{S_3}1_{S_4}1_{S_5}\right)}{16\sqrt{2}} - \frac{3\left|1_{S_4}1_{S_4}\right)}{16\sqrt{2}} + \frac{3\left|1_{S_2}1_{S_4}1_{S_5}\right)}{16\sqrt{2}} + \frac{3\left|1_{S_2}1_{S_5}\right)}{16\sqrt{2}} -
$$
$$
\frac{11\left|1_{S_3}1_{S_4}\right)}{16\sqrt{2}} - \frac{3\left|1_{S_3}\right)}{16\sqrt{2}} + \frac{11\left|1_{S_3}1_{S_4}1_{S_5}\right)}{16\sqrt{2}} + \frac{3\left|1_{S_3}1_{S_5}\right)}{16\sqrt{2}} - \frac{3\left|1_{S_4}\right)}{16\sqrt{2}} + \frac{3\left|1_{S_4}1_{S_5}\right)}{16\sqrt{2}} + \frac{3\left|1_{S_5}\right)}{16\sqrt{2}}
$$

We apply the Grover rotation again on the above output state.

In[ ]:= `qc2 = QuantumCircuit[{out1, "Label" → "|ψ)"}, grv]`

Out[ ]= $|\psi\rangle$

In[ ]:= **out2 = Elaborate[qc2]**

Out[ ]=
$$\frac{5\,|\underline{\ }\rangle}{64\,\sqrt{2}} - \frac{5\,|1_{s_1}1_{s_2}1_{s_3}1_{s_4}1_{s_5}\rangle}{64\,\sqrt{2}} + \frac{5\,|1_{s_1}\rangle}{64\,\sqrt{2}} + \frac{5\,|1_{s_1}1_{s_2}\rangle}{64\,\sqrt{2}} + \frac{5\,|1_{s_1}1_{s_2}1_{s_3}\rangle}{64\,\sqrt{2}} +$$

$$\frac{5\,|1_{s_1}1_{s_2}1_{s_3}1_{s_4}\rangle}{64\,\sqrt{2}} - \frac{5\,|1_{s_1}1_{s_2}1_{s_3}1_{s_5}\rangle}{64\,\sqrt{2}} - \frac{5\,|1_{s_1}1_{s_2}1_{s_4}1_{s_5}\rangle}{64\,\sqrt{2}} + \frac{5\,|1_{s_1}1_{s_2}1_{s_4}\rangle}{64\,\sqrt{2}} -$$

$$\frac{5\,|1_{s_1}1_{s_2}1_{s_5}\rangle}{64\,\sqrt{2}} + \frac{5\,|1_{s_1}1_{s_3}\rangle}{64\,\sqrt{2}} + \frac{5\,|1_{s_1}1_{s_3}1_{s_4}\rangle}{64\,\sqrt{2}} - \frac{5\,|1_{s_1}1_{s_3}1_{s_4}1_{s_5}\rangle}{64\,\sqrt{2}} - \frac{5\,|1_{s_1}1_{s_3}1_{s_5}\rangle}{64\,\sqrt{2}} +$$

$$\frac{5\,|1_{s_1}1_{s_4}\rangle}{64\,\sqrt{2}} - \frac{5\,|1_{s_1}1_{s_4}1_{s_5}\rangle}{64\,\sqrt{2}} - \frac{5\,|1_{s_1}1_{s_5}\rangle}{64\,\sqrt{2}} - \frac{5\,|1_{s_2}1_{s_3}1_{s_4}1_{s_5}\rangle}{64\,\sqrt{2}} + \frac{5\,|1_{s_2}\rangle}{64\,\sqrt{2}} + \frac{5\,|1_{s_2}1_{s_3}\rangle}{64\,\sqrt{2}} +$$

$$\frac{5\,|1_{s_2}1_{s_3}1_{s_4}\rangle}{64\,\sqrt{2}} - \frac{5\,|1_{s_2}1_{s_3}1_{s_5}\rangle}{64\,\sqrt{2}} + \frac{5\,|1_{s_2}1_{s_4}\rangle}{64\,\sqrt{2}} - \frac{5\,|1_{s_2}1_{s_4}1_{s_5}\rangle}{64\,\sqrt{2}} - \frac{5\,|1_{s_2}1_{s_5}\rangle}{64\,\sqrt{2}} -$$

$$\frac{61\,|1_{s_3}1_{s_4}1_{s_5}\rangle}{64\,\sqrt{2}} - \frac{5\,|1_{s_3}1_{s_5}\rangle}{64\,\sqrt{2}} + \frac{5\,|1_{s_3}\rangle}{64\,\sqrt{2}} + \frac{61\,|1_{s_3}1_{s_4}\rangle}{64\,\sqrt{2}} - \frac{5\,|1_{s_4}1_{s_5}\rangle}{64\,\sqrt{2}} + \frac{5\,|1_{s_4}\rangle}{64\,\sqrt{2}} - \frac{5\,|1_{s_5}\rangle}{64\,\sqrt{2}}$$

We repeat the same procedure.

In[ ]:= **qc3 = QuantumCircuit[{out2, "Label" → "|ψ)"}, grv]**

Out[ ]= $|\psi\rangle$
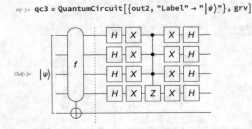

In[ ]:= **out3 = Elaborate[qc3]**

Out[ ]=
$$\frac{13\,|\underline{\ }\rangle}{256\,\sqrt{2}} - \frac{13\,|1_{s_1}1_{s_2}1_{s_3}1_{s_4}1_{s_5}\rangle}{256\,\sqrt{2}} + \frac{13\,|1_{s_1}\rangle}{256\,\sqrt{2}} + \frac{13\,|1_{s_1}1_{s_2}\rangle}{256\,\sqrt{2}} + \frac{13\,|1_{s_1}1_{s_2}1_{s_3}\rangle}{256\,\sqrt{2}} +$$

$$\frac{13\,|1_{s_1}1_{s_2}1_{s_3}1_{s_4}\rangle}{256\,\sqrt{2}} - \frac{13\,|1_{s_1}1_{s_2}1_{s_3}1_{s_5}\rangle}{256\,\sqrt{2}} - \frac{13\,|1_{s_1}1_{s_2}1_{s_4}1_{s_5}\rangle}{256\,\sqrt{2}} + \frac{13\,|1_{s_1}1_{s_2}1_{s_4}\rangle}{256\,\sqrt{2}} -$$

$$\frac{13\,|1_{s_1}1_{s_2}1_{s_5}\rangle}{256\,\sqrt{2}} + \frac{13\,|1_{s_1}1_{s_3}\rangle}{256\,\sqrt{2}} + \frac{13\,|1_{s_1}1_{s_3}1_{s_4}\rangle}{256\,\sqrt{2}} - \frac{13\,|1_{s_1}1_{s_3}1_{s_4}1_{s_5}\rangle}{256\,\sqrt{2}} - \frac{13\,|1_{s_1}1_{s_3}1_{s_5}\rangle}{256\,\sqrt{2}} +$$

$$\frac{13\,|1_{s_1}1_{s_4}\rangle}{256\,\sqrt{2}} - \frac{13\,|1_{s_1}1_{s_4}1_{s_5}\rangle}{256\,\sqrt{2}} - \frac{13\,|1_{s_1}1_{s_5}\rangle}{256\,\sqrt{2}} - \frac{13\,|1_{s_2}1_{s_3}1_{s_4}1_{s_5}\rangle}{256\,\sqrt{2}} + \frac{13\,|1_{s_2}\rangle}{256\,\sqrt{2}} + \frac{13\,|1_{s_2}1_{s_3}\rangle}{256\,\sqrt{2}} +$$

$$\frac{13\,|1_{s_2}1_{s_3}1_{s_4}\rangle}{256\,\sqrt{2}} - \frac{13\,|1_{s_2}1_{s_3}1_{s_5}\rangle}{256\,\sqrt{2}} + \frac{13\,|1_{s_2}1_{s_4}\rangle}{256\,\sqrt{2}} - \frac{13\,|1_{s_2}1_{s_4}1_{s_5}\rangle}{256\,\sqrt{2}} - \frac{13\,|1_{s_2}1_{s_5}\rangle}{256\,\sqrt{2}} -$$

$$\frac{251\,|1_{s_3}1_{s_4}\rangle}{256\,\sqrt{2}} + \frac{13\,|1_{s_3}\rangle}{256\,\sqrt{2}} + \frac{251\,|1_{s_3}1_{s_4}1_{s_5}\rangle}{256\,\sqrt{2}} - \frac{13\,|1_{s_3}1_{s_5}\rangle}{256\,\sqrt{2}} - \frac{13\,|1_{s_4}1_{s_5}\rangle}{256\,\sqrt{2}} + \frac{13\,|1_{s_4}\rangle}{256\,\sqrt{2}} - \frac{13\,|1_{s_5}\rangle}{256\,\sqrt{2}}$$

In this case, the optimal number of applications of the Grover algorithm is four. So, this is supposed to be the last step.

In[ ]:= `qc4 = QuantumCircuit[{out3, "Label" → "|ψ)"}, grv]`

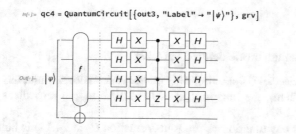

Out[ ]= $|\psi\rangle$

This is the final state.

In[ ]:= **out4 = Elaborate[qc4]**

Out[ ]= $-\dfrac{171\,|{\ldots}\rangle}{1024\,\sqrt{2}} - \dfrac{171\,|1_{s_1}\rangle}{1024\,\sqrt{2}} - \dfrac{171\,|1_{s_1}1_{s_2}\rangle}{1024\,\sqrt{2}} - \dfrac{171\,|1_{s_1}1_{s_2}1_{s_3}\rangle}{1024\,\sqrt{2}} - \dfrac{171\,|1_{s_1}1_{s_2}1_{s_3}1_{s_4}\rangle}{1024\,\sqrt{2}} +$

$\dfrac{171\,|1_{s_1}1_{s_2}1_{s_3}1_{s_4}1_{s_5}\rangle}{1024\,\sqrt{2}} + \dfrac{171\,|1_{s_1}1_{s_2}1_{s_3}1_{s_5}\rangle}{1024\,\sqrt{2}} - \dfrac{171\,|1_{s_1}1_{s_2}1_{s_4}\rangle}{1024\,\sqrt{2}} + \dfrac{171\,|1_{s_1}1_{s_2}1_{s_4}1_{s_5}\rangle}{1024\,\sqrt{2}} +$

$\dfrac{171\,|1_{s_1}1_{s_2}1_{s_5}\rangle}{1024\,\sqrt{2}} - \dfrac{171\,|1_{s_1}1_{s_3}\rangle}{1024\,\sqrt{2}} - \dfrac{171\,|1_{s_1}1_{s_3}1_{s_4}\rangle}{1024\,\sqrt{2}} + \dfrac{171\,|1_{s_1}1_{s_3}1_{s_4}1_{s_5}\rangle}{1024\,\sqrt{2}} +$

$\dfrac{171\,|1_{s_1}1_{s_3}1_{s_5}\rangle}{1024\,\sqrt{2}} - \dfrac{171\,|1_{s_1}1_{s_4}\rangle}{1024\,\sqrt{2}} + \dfrac{171\,|1_{s_1}1_{s_4}1_{s_5}\rangle}{1024\,\sqrt{2}} + \dfrac{171\,|1_{s_1}1_{s_5}\rangle}{1024\,\sqrt{2}} - \dfrac{171\,|1_{s_2}\rangle}{1024\,\sqrt{2}} -$

$\dfrac{171\,|1_{s_2}1_{s_3}\rangle}{1024\,\sqrt{2}} - \dfrac{171\,|1_{s_2}1_{s_3}1_{s_4}\rangle}{1024\,\sqrt{2}} + \dfrac{171\,|1_{s_2}1_{s_3}1_{s_4}1_{s_5}\rangle}{1024\,\sqrt{2}} + \dfrac{171\,|1_{s_2}1_{s_3}1_{s_5}\rangle}{1024\,\sqrt{2}} -$

$\dfrac{171\,|1_{s_2}1_{s_4}\rangle}{1024\,\sqrt{2}} + \dfrac{171\,|1_{s_2}1_{s_4}1_{s_5}\rangle}{1024\,\sqrt{2}} + \dfrac{171\,|1_{s_2}1_{s_5}\rangle}{1024\,\sqrt{2}} - \dfrac{781\,|1_{s_3}1_{s_4}1_{s_5}\rangle}{1024\,\sqrt{2}} - \dfrac{171\,|1_{s_3}\rangle}{1024\,\sqrt{2}} +$

$\dfrac{781\,|1_{s_3}1_{s_4}\rangle}{1024\,\sqrt{2}} + \dfrac{171\,|1_{s_3}1_{s_5}\rangle}{1024\,\sqrt{2}} - \dfrac{171\,|1_{s_4}\rangle}{1024\,\sqrt{2}} + \dfrac{171\,|1_{s_4}1_{s_5}\rangle}{1024\,\sqrt{2}} + \dfrac{171\,|1_{s_5}\rangle}{1024\,\sqrt{2}}$

To analyze the result, we ignore the ancillary qubit. Note that the result is a density matrix.

`new = Matrix@PartialTrace[out4, S[aa]];`

The diagonal elements of the density matrix are the probability in the logical basis.

`prb = Diagonal[new];`

As you can see below, the probability for x=3 is the highest. We conclude that the quantum search algorithm works reasonably well.

In[ ]:= `ListLinePlot[Transpose@{Range[0, 2^$n - 1], prb},`
     `PlotMarkers → Automatic,`
     `FrameLabel → {"x", "Probability"}]`

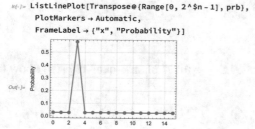

Problems

4.1. Consider the quantum oracle defined in (4.12).

(a) Show that classically (operating only on the basis states without any super-position of them), the mapping in (4.11) is one-to-one regardless of the function f.

(b) Show that for any function f, the transformation $\hat{U}_f$ in (4.12) is unitary.

4.2. **(conditional phase shift)** Let $f : \{0, 1\}^n \to \{0, 1\}$ be a (classical) oracle. Suppose that we have a quantum computer consisting of an n-qubit "control register" and a single-qubit "target register". Using a quantum oracle, construct a quantum circuit that shifts the phase by the factor $e^{i\phi}$ of every term in $|x\rangle$ satisfying $f(x) = 1$ of the n-qubit register, while keeping the second single-qubit register intact. The quantum circuit effectively transforms the states of the control qubit as

$$\sum_x |x\rangle \mapsto \sum_x |x\rangle \, e^{i\phi f(x)}. \tag{4.132}$$

A simple application of quantum oracle as in Eq. (4.19) corresponds to $\phi = \pi$. Hint: See the implementation of the controlled-unitary gate in Sect. 2.2.2.

4.3. Prove the identity

$$e^{-ia\hat{P}} \hat{X} e^{+ia\hat{P}} = \hat{X} - a \tag{4.133}$$

in Eq. (4.43).
Hint: See Appendix A.4.2, and recall that $e^{-ia\hat{P}} = \sum_y |P_y\rangle e^{-iap_y} \langle P_y|$.

4.4. Using the orthogonality relation

$$\sum_{z=0}^{2^n-1} e^{i(x-y)p_z} = 2^n \delta_{xy} \tag{4.134}$$

for all $x, y = 0, 1, \ldots, 2^n - 1$, prove the identity

$$\hat{U}_{\text{QFT}}^{\dagger} |X_y\rangle = \frac{1}{2^{n/2}} \sum_x |X_x\rangle \, e^{-ixp_y} \tag{4.135}$$

in Eq. (4.61).

4.5. Consider the logical state $|X_0\rangle \equiv |0\rangle^{\otimes n}$ of an n-qubit register. Show that

$$\hat{U}_{\text{QFT}} |X_0\rangle = \hat{U}_{\text{QFT}}^{\dagger} |X_0\rangle = \hat{H}^{\otimes n} |0\rangle^{\otimes n} = \frac{1}{2^{n/2}} \sum_{x=0}^{2^n-1} |X_x\rangle = |P_0\rangle. \tag{4.136}$$

Chapter 5
Quantum Decoherence

In the previous chapters, our discussion and arguments have been mainly based on the principles of quantum physics for closed systems. However, no practical system is realistically closed. A system is naturally subjected to its interactions with the surrounding system, which is commonly called the *environment*. There is also a more fundamental reason for the notion of an *open quantum system* in quantum mechanics. The theory of quantum mechanics is intrinsically probabilistic, meaning that the verification of any quantum principle should be tested statistically by taking repeated measurements and incorporating the resulting data. The measurement process inevitably requires coupling the system to a measuring device. Moreover, for quantum computation and more generally for quantum information processing, we desire preparation, manipulation, and measurement of quantum states. All those procedures require the system to be coupled to external equipment.

In principle, one can regard the combined system enclosing both the system and the environment as a closed system, and thus apply the quantum mechanical principles to the total system. However, the environment is typically large—and since perfect isolation is impossible, the total system is eventually the whole universe—with a huge number of degrees of freedom. A complete microscopic description incorporating the environmental degrees of freedom is not only impractical but also of little use. First of all, such a description is tremendously complicated and hard to solve. A solution, if any, would lead to an intractable amount of information, the vast majority of which would be irrelevant to the physical effects exhibited by the system itself.

A more reasonable and practical approach is thus to seek an effective description of open quantum systems in terms of only the system's degrees of freedom. An effective theory is achieved in two stages: First, ignorance of the environmental degrees of freedom brings about a statistical mixture of pure states for the system. The state

Supplementary Information The online version contains supplementary material available at https://doi.org/10.1007/978-3-030-91214-7_5.

M.-S. Choi, *A Quantum Computation Workbook*, https://doi.org/10.1007/978-3-030-91214-7_5

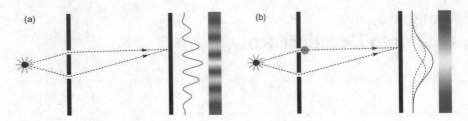

Fig. 5.1 Double-slit experiments with single electrons. **a** The number density of electrons accumulated on the screen exhibits a clear interference pattern. **b** A particle detector (gray disk) is located at the upper slit. In this case, no interference pattern is observed

of the system is no longer a pure state and is described by the density operator. We have already introduced this description in Sect. 1.1.2. Second, the influence of the environment should be reflected on the (effective) dynamical evolution of the density operator in a way that does not depend on the details of the environment and of the system-environment coupling. A powerful mathematical tool is provided by the formalism of quantum operations.

In this chapter, we first take toy models to examine the decoherence process on the elementary and qualitative level. We then introduce quantum operations formalism. The two common and complementary representations of quantum operations are discussed together with simple examples. Quantum operations are used not only for the dynamical processes of open quantum systems but also for the quantum theory of generalized measurement. Next, we turn to the quantum master equation approach to open quantum systems. This is an approximate approach for quantum operations formalism under the Markovian assumption. While quantum operations formalism provides the most general mathematical tool, it is not always possible to find explicit quantum operations for given specific systems. It is far simpler and more insightful to construct the quantum master equation and thus examine the solution to understand the behavior of the open quantum systems in question. In the remaining part of the chapter, we introduce distance measures between quantum states.

5.1   How Does Decoherence Occur?

Before discussing a more rigorous and mathematical description, let us take a toy model and its variants to examine how a quantum state loses phase coherence. As one will see, those models are oversimplifications. Nevertheless, they are sufficient to inspire a qualitative physical picture of decoherence. They provide motivations and elementary examples for the mathematical framework to be introduced in the subsequent sections of the chapter.

The phase coherence of quantum states can be seen best in interference experiments. The most popular interference experiment is the double-slit experiment in Fig. 5.1a. Ranked top in a survey (Crease, 2002) about "the most beautiful

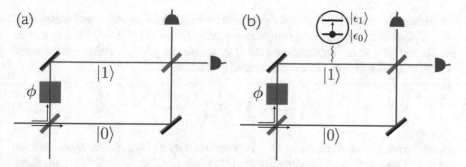

Fig. 5.2 Mach–Zehnder interference experiments. **a** Ideal case without decoherence. The region labeled with ϕ causes a relative phase shift in the photon traveling along the upper arm. **b** The photon traveling along the upper arm interacts with a two-level system and causes it to make a transition from the ground state $|\epsilon_0\rangle$ to the excited state $|\epsilon_1\rangle$

experiment in physics," the double-slit experiment with electrons (Jönsson, 1961) played a decisive role to establish de Broglie's *wave-particle duality* for any particle in quantum physics. Notably, in a later experiment (Tonomura et al., 1989), there was only one electron in the apparatus at any one time. It was a thought experiment that Feynman (Feynman et al., 1963) suggested emphasizing the coherent nature of the quantum state. In Dirac's words (Dirac, 1958),[1] "each [electron] interferes only with itself."

Feynman's other thought experiment (Feynman et al., 1963) illustrated in Fig. 5.1b is an interesting variant of the double-slit experiment in the present context. In this experiment, a particle detector located at the upper slit can determine which slit each electron passes through. The so-called which-path information breaks the interference pattern and one can observe no interference fringe. A common explanation of the disappearance of interference in this experiment is based on the principle of complementarity. In quantum mechanics, anything has both the wave and particle nature but exhibits only one of them exclusively at a time (in a given situation). Indeed, the experiment is essentially the same as the thought experiment suggested by Einstein in a discussion with Bohr at the Fifth Physical Conference of the Solvay Institute, devoted to "Electrons and Photons" (Bohr, 1949), in 1927. Bohr gave the same argument. However, a closer analysis of the thought experiment also reveals the physical mechanism of decoherence. Here we will focus on this point.

To make the point clearer, it is convenient to consider a slightly different version with the Mach–Zehnder interferometer in Fig. 5.2. In Fig. 5.2a, a photon is incident on the horizontal optical mode (the lower blue arm of the interferometer) and encounters a beam splitter. It can either go through the beam splitter to the other side of the horizontal mode or reflected into the vertical mode (the upper red arm). The photon after the beam splitter is thus in the state of superposition $(|0\rangle + |1\rangle)/\sqrt{2}$, where $|0\rangle$ and $|1\rangle$ describe the photon in the blue and red arms, respectively. In the red arm,

---

[1] Originally he remarked, "Each photon then interferes only with itself."

the photon passes through an optical medium that causes an additional phase shift of ϕ. The state of photon in this stage is given by $(|0\rangle + |1\rangle \, e^{i\phi})/\sqrt{2}$. The two arms are recombined at the second beam splitter, where the photon is either transmitted or reflected. The photon is then described by the state

$$|0\rangle \left(\frac{1 + e^{i\phi}}{2} \right) + |1\rangle \left(\frac{1 - e^{i\phi}}{2} \right), \tag{5.1}$$

where the additional phase factor of -1 (phase shift by π) for a photon reflected from the red to blue arm is responsible for the reflection from an optically hard medium. The photon is finally detected by either detectors at the ends of the optical modes. The probability to detect the photon in the blue arm is given by

$$P_0 = \left| \frac{1 + e^{i\phi}}{2} \right|^2 = \cos^2(\phi/2). \tag{5.2}$$

Similarly, the probability to find the photon in the red arm is given by $P_1 = \sin^2(\phi/2)$. Clearly, the photon detection probability in each output port manifests interference as the relative phase shift ϕ varies. As the analyses of the photon states in respective stages suggest, the Mach–Zehnder interferometer in Fig. 5.2a for a photon is an equivalent of the quantum circuit in Fig. 5.3a for a qubit.

Consider a quantum circuit model equivalent to the Mach-Zender interferometer.

```
In[ ]:= qc = QuantumCircuit[LogicalForm[Ket[], S[1]],
          S[1, 6], Phase[φ, S[1], "Label" → "φ"], S[1, 6]]
```

$Out[ ]=$ $|0\rangle$ —$\boxed{H}$—$\boxed{\phi}$—$\boxed{H}$—

```
In[ ]:= out = Elaborate[qc];
       out // LogicalForm
```

$Out[ ]=$ $\dfrac{1}{2} \left(1 + e^{i\phi} \right) \left| 0_{S_1} \right\rangle + \dfrac{1}{2} \left(1 - e^{i\phi} \right) \left| 1_{S_1} \right\rangle$

```
In[ ]:= Let[Real, φ]
       {P[0, φ_], P[1, φ_]} =
          Abs[Normal@Matrix@out]^2 // Elaborate // ExpToTrig // Simplify
```

$Out[ ]=$ $\left\{ \cos\left[\dfrac{\phi}{2}\right]^2, \sin\left[\dfrac{\phi}{2}\right]^2 \right\}$

```
In[ ]:= Plot[P[0, Pi s], {s, 0, 2}, FrameLabel → {"φ / π", "P₀"}]
```

$Out[ ]=$

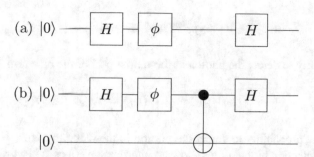

Fig. 5.3 Quantum circuits equivalent to the Mach–Zehnder interference experiments in Fig. 5.2. The Hadamard gate is the equivalent of the beam splitter in the Mach–Zehnder interferometer while the CNOT gate corresponds to the two-level system coupled to the upper optical mode. The unitary gate labeled by ϕ is the phase gate, $|0\rangle \langle 0| + |1\rangle e^{i\phi} \langle 1|$

Let us now consider a variant in Fig. 5.2b. In this experiment, there is a two-level system coupled to the upper optical mode. In other words, if the photon is in the red arm, then it interacts with a two-level system. As a result of the interaction, the two-level system, which is initially prepared in its ground state $|\epsilon_0\rangle$, makes a transition to the excited state $|\epsilon_1\rangle$. The two-level system is the counterpart of the particle detector behind the upper slit in the double-slit experiment in Fig. 5.1b. Here we will regard it as an environment. The crucial point is that unlike the particle detector, one cannot assess the two-level system in this experiment. Nevertheless, through the interaction with the photon traveling in the upper arm, the two-level system destroys the interference. To see this, let us take a closer look at what happens in the experiment. The initial state of the total system consisting of the incident photon and the two-level system is $|0\rangle \otimes |\epsilon_0\rangle$. After the beam splitter and the phase-shifter medium, it becomes

$$\left(\frac{|0\rangle + |1\rangle e^{i\phi}}{\sqrt{2}} \right) \otimes |\epsilon_0\rangle = \frac{|0\rangle \otimes |\epsilon_0\rangle + |1\rangle \otimes |\epsilon_0\rangle e^{i\phi}}{\sqrt{2}}. \tag{5.3}$$

The interaction between the photon and the two-level system causes the transition from $|\epsilon_0\rangle$ to $|\epsilon_1\rangle$ on the condition that the photon is in the upper arm. It leads to

$$\frac{|0\rangle \otimes |\epsilon_0\rangle + |1\rangle \otimes |\epsilon_1\rangle e^{i\phi}}{\sqrt{2}} \tag{5.4}$$

for the state of the total system. The process is analogous to the CNOT gate in quantum computation. Indeed, the Mach–Zehnder experiment in Fig. 5.2b with a two-level system coupled to one of the optical mode is equivalent to the quantum circuit in Fig. 5.3b for two qubits. After the second beam splitter, the quantum state is described by

$$\frac{1}{2} |0\rangle \otimes \left(|\epsilon_0\rangle + |\epsilon_1\rangle \, e^{i\phi}\right) + \frac{1}{2} |1\rangle \otimes \left(|\epsilon_0\rangle - |\epsilon_1\rangle \, e^{i\phi}\right). \tag{5.5}$$

The probability to detect the photon at the output port of the blue arm is thus given by

$$P_0 = \frac{1}{4} \left(\langle\epsilon_0| + e^{-i\phi} \langle\epsilon_1|\right) \left(|\epsilon_0\rangle + |\epsilon_1\rangle \, e^{i\phi}\right) = \frac{1}{2}. \tag{5.6}$$

Likewise, the probability to detect the photon in the other arm is $P_1 = 1/2$. Hence the photon detection probabilities show no interference effect. It is worth noting that the quantum state in (5.4) is a maximally entangled state between the photon and the two-level system—the maximal entanglement is even clearer in (5.4). This means that upon tracing out the two-level system, which we cannot assess, the density operator describing the quantum state of the photon becomes completely random, $\hat{\rho} = \hat{I}/2$, irrespective of the relative phase shift ϕ. That is, due to the interaction and subsequent entanglement with the two-level system, the photon has lost completely the phase coherence in its quantum state.

Consider a quantum circuit model equivalent to the Mach-Zender interference experiment with a two-level system coupled to one of the two arms.

```
In[ ]:= qc = QuantumCircuit[LogicalForm[Ket[], S@{1, 2}],
          S[1, 6], Phase[ϕ, S[1], "Label" → "ϕ"], CNOT[S[1], S[2]], S[1, 6]]
```

```
Out[ ]=
        |0⟩ ─ H ─ ϕ ─────── H ─
        |0⟩ ─────────── ⊕ ──────
```

```
In[ ]:= out = Elaborate[qc];
        out // LogicalForm
```

$$Out[\,]= \frac{1}{2} \left| 0_{S_1} 0_{S_2} \right\rangle + \frac{1}{2} e^{i\phi} \left| 0_{S_1} 1_{S_2} \right\rangle + \frac{1}{2} \left| 1_{S_1} 0_{S_2} \right\rangle - \frac{1}{2} e^{i\phi} \left| 1_{S_1} 1_{S_2} \right\rangle$$

```
In[ ]:= Let[Real, ϕ]
        rho = PartialTrace[out, S[2]]
```

$$Out[\,]= \frac{1}{2}$$

In Feynman's thought experiment in Fig. 5.1b, it has been assumed that the particle detector behind the upper slit is perfect. In the Mach–Zehnder version in Fig. 5.2b, it corresponds to an interaction between the photon and the two-level system such that the quantum state $|\epsilon_1\rangle$ of the two-level system after the interaction is orthogonal to the initial state $|\epsilon_0\rangle$. Recall that in principle one can devise a measurement to discriminate two orthogonal states with certainty. What if the detector is not perfect? For example, suppose that the detector efficiency is so low that it often misses the particle even if it passes through the upper slit. How much would the presence of such a detector affect, if at all, the interference pattern in the screen? Translated to the Mach–Zehnder experiment in Fig. 5.2b, it corresponds to the quantum state $|\epsilon_1'\rangle$ of the two-level system after the interaction with the photon is not orthogonal to $|\epsilon_0\rangle$. It is also analogous to the quantum circuit

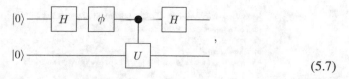

(5.7)

where the unitary gate $\hat{U}$ is responsible for the transition $|\epsilon_0\rangle \rightarrow |\epsilon_1'\rangle$, that is, $|\epsilon_1'\rangle = \hat{U} |\epsilon_0\rangle$. Following the above lines of argument, one can obtain the final state of the combined system of the photon and the two-level system as

$$\frac{1}{2} |0\rangle \otimes \left(|\epsilon_0\rangle + |\epsilon_1'\rangle e^{i\phi}\right) + \frac{1}{2} |1\rangle \otimes \left(|\epsilon_0\rangle - |\epsilon_1'\rangle e^{i\phi}\right), \tag{5.8}$$

which is to be compared to (5.5). The photon detection probabilities are given by

$$P_0 = \frac{1}{4} \left(\langle\epsilon_0| + e^{-i\phi}\langle\epsilon_1'|\right)\left(|\epsilon_0\rangle + |\epsilon_1'\rangle e^{i\phi}\right) = \frac{1}{2}\left[1 + \mathrm{Re}\langle\epsilon_0|\epsilon_1'\rangle e^{i\phi}\right] \tag{5.9a}$$

$$P_1 = \frac{1}{2}\left[1 - \mathrm{Re}\langle\epsilon_0|\epsilon_1'\rangle e^{i\phi}\right] \tag{5.9b}$$

Clearly, when $|\epsilon_1'\rangle$ is orthogonal to $|\epsilon_0\rangle$, $\langle\epsilon_0|\epsilon_1'\rangle = 0$, the previous results in (5.6) are reproduced, and one can observe no interference. Furthermore, when there is no interaction between the photon and the two-level system, the two-level system should remain in the initial state $|\epsilon_0\rangle$, $|\epsilon_1'\rangle = |\epsilon_0\rangle$. In this case, the photon detection probabilities recover interference to the full extent as in (5.2). The results in (5.9) shows that in general, the interference is destroyed only partially depending on how easy $|\epsilon_1'\rangle$ can be, in principle, discriminated from $|\epsilon_0\rangle$. Putting it another way, the better the environment "knows" about the quantum state, the more loss of phase coherence it suffers.

Now consider a quantum circuit model equivalent to the Mach-Zender interference experiment with a two-level system *weakly* coupled to one of the two arms. Here we assume the unitary operator $U = U_y(\theta)$ with $\theta < \pi$ acting conditionally on the two-level system.

```
In[ ]:= qc = QuantumCircuit[LogicalForm[Ket[], S@{1, 2}],
        S[1, 6], Phase[ϕ, S[1], "Label" → "ϕ"],
        ControlledU[S[1], Rotation[θ, S[2, 2]], "Label" → "U"], S[1, 6]]
```

Out[ ]=
```
|0⟩─H─ϕ─●─H─
|0⟩─────U───
```

The final state is an entangled state in general.

```
In[ ]:= out = Elaborate[qc] // Simplify;
        out // LogicalForm
```

Out[ ]= $\frac{1}{2} \times \left(\left(1 + e^{i\phi} \cos\left[\frac{\theta}{2}\right]\right) |0_{S_1} 0_{S_2}\rangle + \left(1 - e^{i\phi} \cos\left[\frac{\theta}{2}\right]\right) |1_{S_1} 0_{S_2}\rangle - e^{i\phi} \left(-|0_{S_1} 1_{S_2}\rangle + |1_{S_1} 1_{S_2}\rangle\right) \sin\left[\frac{\theta}{2}\right] \right)$

However, the entanglement is not a maximally entangled state. The density operator for the photon manifests coherence to certain extent.

```
In[ ]:= Let[Real, ϕ, θ]
       rho = PartialTrace[out, S[2]] // ExpToTrig // Elaborate
```
$$Out[ ]= \frac{1}{2} + \frac{1}{2} \cos\left[\frac{\theta}{2}\right] \times \cos[\phi] \, S_1^z - \frac{1}{2} \cos\left[\frac{\theta}{2}\right] S_1^y \sin[\phi]$$

```
In[ ]:= {P[0, θ_, ϕ_], P[1, θ_, ϕ_]} = Diagonal@Matrix@rho
```
$$Out[ ]= \left\{\frac{1}{2} + \frac{1}{2} \cos\left[\frac{\theta}{2}\right] \times \cos[\phi], \; \frac{1}{2} - \frac{1}{2} \cos\left[\frac{\theta}{2}\right] \times \cos[\phi]\right\}$$

Depending on the strength of the coupling, the interference may survive partially.

```
In[ ]:= tt = Range[0, 4] Pi / 4;
       Plot[Evaluate[P[0, #, Pi s] & /@ tt], {s, 0, 2},
         FrameLabel → {"ϕ / π", "P₀"}, PlotLegends → tt]
```

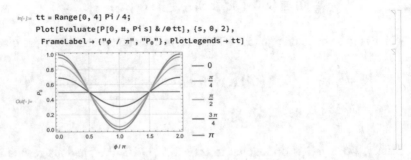

5.2 Quantum Operations

Under a certain physical process, the state of a given system evolves into another state. The time evolution of a closed system is described by unitary operators. What about an open quantum system that interacts with its environment?

The dynamical processes of open quantum systems are described by a special kind of supermaps (Appendix B.2) called *quantum operations*. A supermap transforms density operators to other density operators while preserving their elementary properties. In particular, density operators are positive,[2] so a quantum operation needs to preserve this positivity. However, it turns out that merely preserving positivity is not sufficient and a much stronger condition is required. Imagine that a system has interacted with its surroundings and established an entanglement with them. Now consider an operation acting non-trivially only on the system but trivially on its surroundings. One physically expects the operation to preserve the properties, positivity in particular, of the density operator of the whole containing the system and the surroundings. Essentially, a quantum operation needs to preserve the positivity of not only the density operators of a given system but also all density operators of any extended system including the system itself and its surrounding systems. Mathematically, such a condition is satisfied by *completely positive* supermaps (Definition B.4).

---

[2] Recall that a positive operator is Hermitian by definition.

Now, we define quantum operations more precisely. Let $\mathcal{V}$ and $\mathcal{W}$ be vector spaces. Suppose that $\mathscr{F}$ is a supermap from $\mathcal{L}(\mathcal{V})$ onto $\mathcal{L}(\mathcal{W})$. $\mathscr{F}$ is called a *quantum operation* if it satisfies the following three axioms

(a) $\mathscr{F}$ never increases the trace. That is, $0 \leq \mathrm{Tr}\mathscr{F}(\hat{\rho}) \leq 1$ for any *density operator*[3] $\hat{\rho}$ on $\mathcal{V}$.

(b) $\mathscr{F}$ is *convex linear*. That is, for any probabilities p_j[4] and density operators $\hat{\rho}_j$ on $\mathcal{V}$,

$$\mathscr{F}\left(\sum_j \hat{\rho}_j p_j\right) = \sum_j \mathscr{F}(\hat{\rho}_j) p_j . \tag{5.10}$$

(c) $\mathscr{F}$ is a *completely positive supermap*.[5] That is, not only $\mathscr{F}(\hat{\rho})$ itself is positive for any positive operator $\hat{\rho}$ on $\mathcal{V}$, but $(\mathscr{F} \otimes \mathscr{I})(\hat{\rho})$ is also positive for any positive operator on $\mathcal{V} \otimes \mathcal{E}$ with an arbitrary vector space $\mathcal{E}$.

Most quantum operations preserve the trace. That is, $\mathrm{Tr}\,\mathscr{F}(\hat{\rho}) = 1$ for all density operators $\hat{\rho}$. An important exception is the process associated with a (generalized selective) measurement. When the trace is not preserved, $\mathrm{Tr}\,\mathscr{F}(\hat{\rho})$ gives the probability for the dynamical process $\mathscr{F}$ to occur.

In quantum information theory, the quantum operations preserving trace, i.e., completely positive and trace-preserving supermaps, are called *quantum channels*. Physically, these describe communication channels that can transmit quantum information, as well as classical information.

Another important class of physical phenomena described by quantum operations is *quantum decoherence* or just *decoherence* for short, referring to the loss of quantum coherence. We have briefly introduced this effect in the previous section (Sect. 5.1) through toy models. Quantum operations offers a complete and general description of decoherence.

In this section, we introduce three mathematical methods to describe quantum operations, the Kraus representation in Sect. 5.2.1, the Choi isomorphism in Sect. 5.2.2, and the unitary representation in Sect. 5.2.3. We conclude this section by giving examples of these representations for single-qubit systems in Sect. 5.2.4.

5.2.1   Kraus Representation

A quantum operation is a restricted form of completely positive supermap. Accordingly, for any quantum operation $\mathscr{F}$ from $\mathcal{L}(\mathcal{V})$ onto $\mathcal{L}(\mathcal{W})$, there exist linear maps $\hat{F}_\mu$ from $\mathcal{V}$ onto $\mathcal{W}$ such that

---

[3] That is, $\hat{\rho}$ is positive semi-definite and $\mathrm{Tr}\,\hat{\rho} = 1$. See Sect. 1.1.2 for the precise definition and properties of a density operator.

[4] That is, $0 \leq p_j \leq 1$ and $\sum_j p_j = 1$.

[5] In most literature, it is called a completely positive "map".

$$\mathscr{F}(\hat{\rho}) = \sum_{\mu} \hat{F}_{\mu} \hat{\rho} \hat{F}_{\mu}^{\dagger} \tag{5.11}$$

for all linear operators (not necessarily density operators) $\hat{\rho}$ on $\mathcal{V}$ (Theorem B.5). The linear maps $\hat{F}_{\mu}$ are called the *Kraus elements* or the *Kraus maps* of $\mathscr{F}$. Then, the trace-decreasing condition in axiom (a) imposes the inequalities

$$0 \le \sum_{\mu} \hat{F}_{\mu}^{\dagger} \hat{F}_{\mu} \le 1 . \tag{5.12}$$

One can always choose Kraus elements that are *mutually orthogonal* with respect to the Hilbert–Schmidt inner product, that is,

$$\mathrm{Tr}\, \hat{F}_{\mu}^{\dagger} \hat{F}_{\nu} = 0 \tag{5.13}$$

whenever $\mu \ne \nu$. Through the procedure of choosing orthogonal Kraus elements, one can drastically optimize Kraus elements.

The Kraus representation of a quantum operation as a sum of operators provides powerful tools to analyze the quantum operation. However, at this stage, the Kraus representation follows from a mathematical theorem for a completely positive supermap (Theorem B.5). How does the Kraus representation arise physically? What is the physical meaning of the Kraus elements? These questions are the subject of this subsection, and meanwhile we will also discuss some basic properties of Kraus representation.

The simplest example of the Kraus representation is of the form

$$\mathscr{F}(\hat{\rho}) = \hat{U} \hat{\rho} \hat{U}^{\dagger}, \tag{5.14}$$

involving a single unitary operator. Naturally it describes the unitary dynamics of a closed system. Note that if there are more than one Kraus elements involved in the Kraus representation, the associated dynamics is not unitary even if the Kraus elements are all unitary. For example, consider a supermap

$$\mathscr{F}(\hat{\rho}) = (1 - p)\hat{\rho} + \frac{p}{3} \left(\hat{S}^{x} \hat{\rho} \hat{S}^{x} + \hat{S}^{y} \hat{\rho} \hat{S}^{y} + \hat{S}^{z} \hat{\rho} \hat{S}^{z} \right). \tag{5.15}$$

Here all Kraus elements ($\hat{I}$, $\hat{S}^{x}$, $\hat{S}^{y}$ and $\hat{S}^{z}$ up to normalization factors) are unitary. However, $\mathscr{F}$ causes depolarization in $\hat{\rho}$ for a non-zero value of p.

To see how the Kraus representation arises physically, let us consider a system interacting with its environment. We denote with $\mathcal{V}$ and $\mathcal{E}$ the Hilbert spaces associated with the system and the environment, respectively. For simplicity, we assume that the total system is initially in the product state $\hat{\rho} \otimes \hat{\sigma}$. The total system is a closed system, and the dynamical process afterwards due to the system-environment interaction is described with an overall unitary operator $\hat{U}$ acting on the total system, $\hat{\rho} \otimes \hat{\sigma} \mapsto \hat{U}(\hat{\rho} \otimes \hat{\sigma})\hat{U}^{\dagger}$. Without access to the environment, one has to take a partial trace (Appendix B.3) of the final state over the environment to obtain the state of the system. Putting it all together, the quantum operation $\mathscr{F}$ describing the process is written as

$$\mathscr{F}(\hat{\rho}) = \underset{\mathcal{E}}{\text{Tr}}\, \hat{U}(\hat{\rho} \otimes \hat{\sigma})\hat{U}^\dagger = \sum_\mu \langle \varepsilon_\mu | \hat{U}(\hat{\rho} \otimes \hat{\sigma})\hat{U}^\dagger | \varepsilon_\mu \rangle , \qquad (5.16)$$

where $\{|\varepsilon_\mu\rangle\}$ is an orthonormal basis of $\mathcal{E}$. On the right-hand side of (5.16), the Hermitian product is applied partially and only on $\mathcal{E}$, and the expression still remains as an operator on $\mathcal{V}$. To further investigate the quantum operation, we take the spectral decomposition (Appendix A.4) of $\hat{\sigma}$

$$\hat{\sigma} = \sum_\nu |s_\nu\rangle \langle s_\nu| , \qquad (5.17)$$

where the eigenvectors $|s_\mu\rangle$ have been normalized by their own eigenvalues $s_\mu = \langle s_\mu | s_\mu \rangle$ (see Eq. (A.57)) since $\hat{\sigma}$ is a positive semidefinite operator. Now, define linear maps $\hat{F}_\mu : \mathcal{V} \to \mathcal{W}$ by

$$\hat{F}_\mu := \sum_\nu \underset{\mathcal{E}}{\text{Tr}}\, \hat{U}\left(\hat{I} \otimes |s_\nu\rangle \langle \varepsilon_\mu|\right) = \sum_\nu \langle \varepsilon_\mu | \hat{U} | s_\nu\rangle . \qquad (5.18)$$

Physically, $\hat{F}_\mu$ describes the dynamics of the system under the condition that the environment has made a transition to state $|\varepsilon_\mu\rangle$. They satisfy the closure relation

$$\sum_\mu \hat{F}_\mu^\dagger \hat{F}_\mu = \hat{I} . \qquad (5.19)$$

Putting (5.18) into (5.16), we arrive at the representation

$$\mathscr{F}(\hat{\rho}) = \sum_\mu \hat{F}_\mu \hat{\rho} \hat{F}_\mu^\dagger \qquad (5.20)$$

for the quantum operation.

Let us take a toy model for a specific example. Consider a chain of three qubits. Suppose that the Hamiltonian of the chain is given by

$$\hat{H} = \frac{1}{2} B \hat{S}_1^z + \frac{1}{2} J \left(\hat{S}_1^x \hat{S}_2^x + \hat{S}_2^x \hat{S}_3^x\right) . \qquad (5.21)$$

We regard the first qubit as the "system", and the other two qubits form the "environment". The coupling constant J indicates how strong the system interacts with the environment. Without the coupling J, the system is a closed system, and the evolution of its quantum state should be unitary. This model is overly artificial as the environment is finite and, in fact, very small, just twice larger than the system. However, it is enough to demonstrate the main idea. We suppose that the whole chain is initially in the product state $|\Psi(0)\rangle = |L\rangle \otimes |L\rangle \otimes |L\rangle$, where $|L\rangle := (|0\rangle + i\,|1\rangle)/\sqrt{2}$ is the "left" state that is often used to denote the left-circularly polarized state of a photon. When focused on the system only, the initial state is $|L\rangle$, or equivalently, $\hat{\rho}(0) = |L\rangle \langle L| = \frac{1}{2}\hat{I}_1 + \frac{1}{2}\hat{S}_1^y$. At later time t, the state of the chain is given by

$$|\Psi(t)\rangle = \hat{U}(t)\,|\Psi(0)\rangle,\qquad(5.22)$$

where the unitary operator $\hat{U}(t) = \exp(-it\hat{H})$ governs the evolution of the total system (the chain). Ignoring the two qubits in the environment, we get a (mixed) state of the system

$$\hat{\rho}(t) = \operatorname*{Tr}_{\mathcal{E}}|\Psi(t)\rangle\,\langle\Psi(t)| = \frac{1}{2}\hat{I}_1 - \frac{1}{2}\frac{\sin(\Omega t)}{\Omega/B}\hat{S}_1^x + \frac{1}{2}\cos(\Omega t)\hat{S}_1^y,\qquad(5.23)$$

where $\Omega := \sqrt{B^2 + J^2}$. Now, let us describe the evolution $\hat{\rho}(0) \to \hat{\rho}(t)$ in terms of quantum operation $\mathscr{F}_t$ depending parametrically on time t. More specifically, we want to find Kraus elements $\hat{F}_\mu(t)$ for the quantum operation so that

$$\hat{\rho}(t) = \sum_\mu \hat{F}_\mu(t)\hat{\rho}(0)\hat{F}_\mu^\dagger(t).\qquad(5.24)$$

Following the prescription in (5.18), we can get the Kraus elements

$$\hat{F}_\mu(t) = \operatorname*{Tr}_{\mathcal{E}}\hat{U}(t)\left(\hat{I} \otimes |L\rangle\,\langle\mu_1| \otimes |L\rangle\,\langle\mu_2|\right),\qquad(5.25)$$

where μ_j are the binary digits of $\mu = (\mu_1\mu_2)_2$. More explicitly, they are given by

$$\hat{F}_0(t) = \frac{\cos\left(\frac{\Omega t}{2}\right)\hat{I}_1}{2} - \frac{iB\sin\left(\frac{\Omega t}{2}\right)\hat{S}_1^z}{2\Omega} + \frac{Je^{-iJt}\sin\left(\frac{\Omega t}{2}\right)\hat{S}_1^x}{2\Omega},\qquad(5.26a)$$

$$\hat{F}_1(t) = \frac{\cos\left(\frac{\Omega t}{2}\right)\hat{I}_1}{2} - \frac{iB\sin\left(\frac{\Omega t}{2}\right)\hat{S}_1^z}{2\Omega} + \frac{Je^{+iJt}\sin\left(\frac{\Omega t}{2}\right)\hat{S}_1^x}{2\Omega},\qquad(5.26b)$$

$$\hat{F}_2(t) = \frac{\cos\left(\frac{\Omega t}{2}\right)\hat{I}_1}{2} - \frac{iB\sin\left(\frac{\Omega t}{2}\right)\hat{S}_1^z}{2\Omega} - \frac{Je^{+iJt}\sin\left(\frac{\Omega t}{2}\right)\hat{S}_1^x}{2\Omega},\qquad(5.26c)$$

$$\hat{F}_3(t) = \frac{\cos\left(\frac{\Omega t}{2}\right)\hat{I}_1}{2} - \frac{iB\sin\left(\frac{\Omega t}{2}\right)\hat{S}_1^z}{2\Omega} - \frac{Je^{-iJt}\sin\left(\frac{\Omega t}{2}\right)\hat{S}_1^x}{2\Omega}.\qquad(5.26d)$$

Direct evaluations make evident that these Kraus elements indeed reproduce the dynamical evolution in (5.24) and that $\sum_\mu \hat{F}_\mu^\dagger(t)\hat{F}_\mu(t) = \hat{I}$. The Kraus elements in (5.26) are not mutually orthogonal—and they can be optimized as we will see below. In the absence of the coupling ($J = 0$), all the Kraus elements are identical,

$$\hat{F}_0(t) = \hat{F}_1(t) = \hat{F}_2(t) = \hat{F}_3(t) = \frac{1}{2}\cos\left(\frac{Bt}{2}\right)\hat{I}_1 - \frac{i}{2}\sin\left(\frac{Bt}{2}\right)\hat{S}_1^z.\qquad(5.27)$$

In this case, the quantum operation is described by a single Kraus element

$$\hat{F}(t) := 2\hat{F}_0^\dagger(t)\hat{I}_1 = \cos(Bt/2) - i\sin(Bt/2)\hat{S}_1^z = \exp\left(-i\hat{S}^z Bt/2\right) , \quad (5.28)$$

which is obviously unitary, as it should be.

We consider a chain of three qubits. The first qubit is regarded as the "system", and the other two form the "environment".

```
$L = 3;
jj = Range[$L];
sys = 1;
env = Range[2, $L];
```

Here is the Hamiltonian describing the chain. We are measuring the energy in units of B (B=1).

```
In[·]:= Let[Real, J]
       H = S[sys, 3] / 2 + J / 2 * Total[ChainBy[S[jj, 1], Multiply]]
```
$$Out[·]= \frac{1}{2} J \left(S_1^x S_2^x + S_2^x S_3^x\right) + \frac{S_1^z}{2}$$

The system has a discrete symmetry. It is invariant under the rotation around the z-axis by angle π.

```
In[·]:= V = Rotation[Pi, S[1, 3]] ** Rotation[Pi, S[2, 3]] ** Rotation[Pi, S[3, 3]]
       V ** H ** Dagger[V]
```
$$Out[·]= i \, S_1^z \, S_2^z \, S_3^z$$

$$Out[·]= \frac{1}{2} J S_1^x S_2^x + \frac{1}{2} J S_2^x S_3^x + \frac{S_1^z}{2}$$

The symmetry leads to the degeneracy of the eigenvalues of the Hamiltonian.

```
In[·]:= ProperValues[H]
```
$$Out[·]= \left\{ \frac{1}{2}\left(-J - \sqrt{1+J^2}\right), \frac{1}{2}\left(-J - \sqrt{1+J^2}\right), \frac{1}{2}\left(J - \sqrt{1+J^2}\right), \frac{1}{2}\left(J - \sqrt{1+J^2}\right), \right.$$
$$\left. \frac{1}{2}\left(-J + \sqrt{1+J^2}\right), \frac{1}{2}\left(-J + \sqrt{1+J^2}\right), \frac{1}{2}\left(J + \sqrt{1+J^2}\right), \frac{1}{2}\left(J + \sqrt{1+J^2}\right) \right\}$$

The time-evolution operator of the chain is evaluated.

```
Let[Real, t]
U[t_] = Elaborate@MultiplyExp[-I t H];
```

We suppose that the chain is initially in the state $|L\rangle \otimes |L\rangle \otimes |L\rangle$, where
$$|L\rangle = (|0\rangle + i|1\rangle)/\sqrt{2}.$$

```
In[·]:= Clear[vec]
       vec[0] = ProductState[S@jj → {1, I} / Sqrt[2]]
       vec[t_] = U[t] ** Elaborate[vec[0]];
```
$$Out[·]= \left(\frac{|0\rangle}{\sqrt{2}} + \frac{i\,|1\rangle}{\sqrt{2}}\right)_{S_1} \otimes \left(\frac{|0\rangle}{\sqrt{2}} + \frac{i\,|1\rangle}{\sqrt{2}}\right)_{S_2} \otimes \left(\frac{|0\rangle}{\sqrt{2}} + \frac{i\,|1\rangle}{\sqrt{2}}\right)_{S_3}$$

Here is a set of replacement rules to be used later to simplify expressions.

```
In[·]:= rules = {Sqrt[1 + J^2] → Ω, 1 / Sqrt[1 + J^2] → 1 / Ω}
```

$$Out[·]= \left\{ \sqrt{1 + J^2} \to \Omega, \; \frac{1}{\sqrt{1 + J^2}} \to \frac{1}{\Omega} \right\}$$

```
In[·]:= Clear[rho]
       rho[t_] = PartialTrace[vec[t], S@env] // Elaborate // ExpToTrig // Garner;
       rho[t] /. rules
```

$$Out[·]= \frac{1}{2} + \frac{1}{2} \, Cos[t\,\Omega] \, S_1^y - \frac{S_1^x \, Sin[t\,\Omega]}{2\,\Omega}$$

Take a look at the dynamical evolution of the state of the "system" (the first qubit), after tracing out the "environment" (the other two qubits). On the left, shown is the evolution in the absence of the coupling to the environment. On the right, the evolution is not coherent due to the coupling to the environment. The evolution is still periodic because the environment is finite.

```
In[·]:= bv0 = Block[{J = 0.}, Table[BlochVector[rho[2 Pi * t]], {t, 0, 5, 0.2}]];
       bv = Block[{J = .5}, Table[BlochVector[rho[2 Pi * t]], {t, 0, 5 / J, 0.2}]];
       GraphicsRow@{BlochSphere[{Blue, Bead /@ bv0}], BlochSphere[{Red, Bead /@ bv}]}
```

Out[·]=

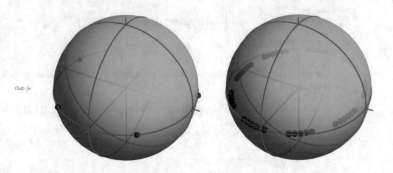

Now, let us examine the evolution in terms of the supermap. This is the initial state of the system.

```
In[·]:= in = Elaborate@ProductState[S[sys] → {1, I} / Sqrt[2]];
       in = Elaborate@Dyad[in, in]
```

$$Out[·]= \frac{1}{2} + \frac{S_1^y}{2}$$

To find the Kraus elements, consider the initial state of the environment.

```
In[·]:= sgm = Elaborate@ProductState[S@env → {1, I} / Sqrt[2]];
       sgm // LogicalForm
```

$$Out[·]= \frac{1}{2} \, \left| 0_{S_2} 0_{S_3} \right\rangle + \frac{1}{2} \, i \, \left| 0_{S_2} 1_{S_3} \right\rangle + \frac{1}{2} \, i \, \left| 1_{S_2} 0_{S_3} \right\rangle - \frac{1}{2} \, \left| 1_{S_2} 1_{S_3} \right\rangle$$

Finally, this is the Kraus elements of the quantum operation.

```
bs = Basis[S@env];
prj = Map[Dyad[sgm, #, S@env] &, bs];
ops = U[t] ** prj // Elaborate;
```

```
In[ ]:= kraus = PartialTrace[#, S@cnv] & /@ ops // ExpToTrig // Elaborate // Garner;
        kraus /. rules
```

$$Out[ ]= \left\{ \frac{1}{2}\cos\left[\frac{t\Omega}{2}\right] \times \left(\cos\left[\frac{Jt}{2}\right] + i\sin\left[\frac{Jt}{2}\right]\right) + \frac{J S_1^x \left(\cos\left[\frac{Jt}{2}\right] - i\sin\left[\frac{Jt}{2}\right]\right) \times \sin\left[\frac{t\Omega}{2}\right]}{2\Omega} + \right.$$

$$\frac{S_1^z \left(-i\cos\left[\frac{Jt}{2}\right] + \sin\left[\frac{Jt}{2}\right]\right)\sin\left[\frac{t\Omega}{2}\right]}{2\Omega}, \frac{1}{2}\cos\left[\frac{t\Omega}{2}\right]\left(i\cos\left[\frac{Jt}{2}\right] + \sin\left[\frac{Jt}{2}\right]\right) +$$

$$\frac{S_1^z \left(\cos\left[\frac{Jt}{2}\right] - i\sin\left[\frac{Jt}{2}\right]\right)\times\sin\left[\frac{t\Omega}{2}\right]}{2\Omega} + \frac{i J S_1^x \left(\cos\left[\frac{Jt}{2}\right] + i\sin\left[\frac{Jt}{2}\right]\right)\times\sin\left[\frac{t\Omega}{2}\right]}{2\Omega},$$

$$\frac{1}{2}\cos\left[\frac{t\Omega}{2}\right]\left(i\cos\left[\frac{Jt}{2}\right] + \sin\left[\frac{Jt}{2}\right]\right) + \frac{S_1^z \left(\cos\left[\frac{Jt}{2}\right] - i\sin\left[\frac{Jt}{2}\right]\right)\times\sin\left[\frac{t\Omega}{2}\right]}{2\Omega} +$$

$$\frac{J S_1^x \left(-i\cos\left[\frac{Jt}{2}\right] + \sin\left[\frac{Jt}{2}\right]\right)\sin\left[\frac{t\Omega}{2}\right]}{2\Omega}, -\frac{1}{2}\cos\left[\frac{t\Omega}{2}\right]\left(\cos\left[\frac{Jt}{2}\right] + i\sin\left[\frac{Jt}{2}\right]\right) +$$

$$\left. \frac{J S_1^x \left(\cos\left[\frac{Jt}{2}\right] - i\sin\left[\frac{Jt}{2}\right]\right)\times\sin\left[\frac{t\Omega}{2}\right]}{2\Omega} + \frac{i S_1^z \left(\cos\left[\frac{Jt}{2}\right] + i\sin\left[\frac{Jt}{2}\right]\right)\times\sin\left[\frac{t\Omega}{2}\right]}{2\Omega}\right\}$$

```
In[ ]:= Clear[new]
        new[t_] = Elaborate@Supermap[kraus][in];
        new[t] /. rules
```

$$Out[ ]= \frac{1}{2} + \frac{1}{2}\cos[t\Omega] S_1^y - \frac{S_1^x \sin[t\Omega]}{2\Omega}$$

```
In[ ]:= new[t] - rho[t] // Elaborate // Garner
Out[ ]= 0
```

In the above arguments, we have derived a Kraus representation for a quantum operation based on a system-plus-environment model. While it provides a useful physical picture of quantum operations, the resulting representation does not look particularly useful at first glance. The Kraus elements $\hat{F}_\mu$ are not orthogonal to each other with respect to the Hilbert–Schmidt inner product in (B.4). Even worse, the number of Kraus elements may be as huge as the dimension of the environmental Hilbert space $\mathcal{E}$ given that the dimension of $\mathcal{E}$ is infinite for any realistic environment. However, neither raises a significant problem. A formal representation in the form of the Kraus representation already facilitates analysis of the quantum operation. Moreover, one can optimize the Kraus elements by reconstructing orthogonal Kraus elements.

Given a set of Kraus elements, how can one actually choose new Kraus elements that are mutually orthogonal.[6] Let $\left\{ \hat{E}_\mu \right\}$ be a basis of the space $\mathcal{L}(\mathcal{V}, \mathcal{W})$ of all linear maps from $\mathcal{V}$ to $\mathcal{W}$. We expand $\hat{F}_\nu$ in the basis

$$\hat{F}_\nu = \sum_\mu \hat{E}_\mu M_{\mu\nu}, \tag{5.29}$$

where M is the matrix of the expansion coefficients. Putting it back to (5.11), we have

$$\mathscr{F}(\hat{\rho}) = \sum_{\mu\nu} \hat{E}_\mu \hat{\rho} \hat{E}_\nu^\dagger \left(M M^\dagger\right)_{\mu\nu}. \tag{5.30}$$

---

[6] It can also be seen in the lines to Theorem B.3. See also Eqs. (5.34) and (5.35).

The square matrix $MM^\dagger$ of size $(\dim \mathcal{V} \dim \mathcal{W})$ is positive semidefinite, and can be decomposed into $MM^\dagger = V\Lambda V^\dagger$, where V is a unitary matrix and Λ is a diagonal matrix with all elements non-negative. We now define the new Kraus elements

$$\hat{F}'_\nu := \sum_\mu \hat{E}_\mu \left(V\sqrt{\Lambda}\right)_{\mu\nu}. \tag{5.31}$$

Then, it is clear that they are mutually orthogonal. Furthermore,

$$\mathscr{F}(\hat{\rho}) = \sum_{\mu=0}^{N-1} \hat{F}'_\mu \hat{\rho} \hat{F}'^\dagger_\mu \tag{5.32}$$

where $N \leq \dim \mathcal{V} \dim \mathcal{W}$. Therefore, it is noted that a set of mutually orthogonal Kraus elements is optimal in the sense that it has no more elements than $(\dim \mathcal{V}) \times (\dim \mathcal{W})$. Applying this method to the previous example, we can get a new set of mutually orthogonal Kraus elements

$$\hat{F}'_0(t) = \cos(\Omega t/2) - i(B/\Omega)\sin(\Omega t/2)\hat{S}_1^z, \tag{5.33a}$$

$$\hat{F}'_1(t) = (J/\Omega)\sin(\Omega t/2)\hat{S}_1^x, \tag{5.33b}$$

from the four Kraus elements in (5.26). The same quantum operation $\hat{\rho}(0) \to \hat{\rho}(t)$ is now specified just by two Kraus elements.

We close this subsection by noting that the Kraus representation is not unique, and we have just seen that given a set of Kraus elements, we could find another set of Kraus elements that are orthogonal to each other. Therefore, there exists unitary freedom for the choice of the Kraus elements. Suppose that the two quantum operations $\mathscr{F}$ and $\mathscr{G}$ are associated with the Kraus elements $\left\{\hat{F}_\mu\right\}$ and $\left\{\hat{G}_\nu\right\}$, respectively.[7] Then, $\mathscr{F} = \mathscr{G}$, that is,

$$\sum_\mu \hat{F}_\mu \hat{\rho} \hat{F}^\dagger_\mu = \sum_\nu \hat{G}_\nu \hat{\rho} \hat{G}^\dagger_\nu \tag{5.34}$$

for all $\hat{\rho} \in \mathcal{L}(\mathcal{V})$ if and only if there exists a unitary matrix U—attaching rows or columns if necessary—such that

$$\hat{G}_\nu = \sum_\mu \hat{F}_\mu U_{\mu\nu}. \tag{5.35}$$

This is analogous to the unitary freedom for the choice of pure states in the specification of a mixed state [see Eqs. (1.14) and (1.15)]. In fact, the underlying mathematical principles are the same. As we have already established the proof of the unitary freedom in the mixed state, let us use it here to prove the unitary freedom in the Kraus representation.

---

[7] The Kraus elements here are not orthogonal, $\mathrm{Tr}\,\hat{F}^\dagger_\mu \hat{F}_\nu \neq 0$ and $\mathrm{Tr}\,\hat{G}^\dagger_\mu \hat{G}_\nu \neq 0$, in general.

If the two sets of Kraus elements satisfy the relation (5.35), it is straightforward to prove the two quantum operations are identical, and this immediately follows from the defining property of a unitary matrix. Let us prove the converse. Suppose that $\mathscr{F} = \mathscr{G}$. Then the corresponding Choi operators (see Appendix B.2.3) should be identical as well, $\hat{C}_{\mathscr{F}} = \hat{C}_{\mathscr{G}}$. Given the Kraus elements, one can evaluate the Choi operators explicitly starting from the maximally entangled state in Eq. (5.42),

$$\hat{C}_{\mathscr{F}} = \sum_{\mu} (\hat{F}_{\mu} \otimes \hat{I}) |\Phi\rangle \langle\Phi| (\hat{F}_{\mu} \otimes \hat{I})^{\dagger} = \sum_{\mu} |F_{\mu}\rangle\langle F_{\mu}|, \tag{5.36a}$$

$$\hat{C}_{\mathscr{G}} = \sum_{\nu} (\hat{G}_{\nu} \otimes \hat{I}) |\Phi\rangle \langle\Phi| (\hat{G}_{\nu} \otimes \hat{I})^{\dagger} = \sum_{\nu} |G_{\nu}\rangle\langle G_{\nu}|, \tag{5.36b}$$

where $|F_{\mu}\rangle, |G_{\nu}\rangle \in \mathcal{W} \otimes \mathcal{V}$ are the Choi vectors (see Appendix B.2.3) corresponding to the linear maps $\hat{F}_{\mu}$ and $\hat{G}_{\nu}$, respectively,

$$|F_{\mu}\rangle := (\hat{F}_{\mu} \otimes \hat{I}) |\Phi\rangle \ , \quad |G_{\nu}\rangle := (\hat{G}_{\nu} \otimes \hat{I}) |\Phi\rangle \ . \tag{5.37}$$

According to Eqs. (1.14) and (1.15),

$$\sum_{\mu} |F_{\mu}\rangle\langle F_{\mu}| = \sum_{\nu} |G_{\nu}\rangle\langle G_{\nu}| \tag{5.38}$$

implies that there exists a unitary matrix U such that

$$|G_{\nu}\rangle = \sum_{\mu} |F_{\mu}\rangle U_{\mu\nu} . \tag{5.39}$$

Finally, we note (Appendix B.2.3) that for arbitrary $|\psi\rangle \in \mathcal{V}$

$$\hat{G}_{\nu} |\psi\rangle = \langle\psi^*|F_{\nu}\rangle = \sum_{\mu} \langle\psi^*|F_{\mu}\rangle U_{\mu\nu} = \sum_{\mu} \hat{F}_{\mu} |\psi\rangle U_{\mu\nu} . \tag{5.40}$$

This asserts the relation (5.35).

In some cases, say, motivated by the (unperturbed) Hamiltonian of the isolated system, there may be a preferred basis. Can we exploit the unitary freedom to change the given set of Kraus elements to another set of Kraus elements that is consistent with the preferred basis? Unfortunately, given two sets of Kraus elements, it is not trivial to check whether they are equivalent or not because the Kraus elements in the relation (5.35) are not normalized. In terms of the normalized Kraus elements, $\hat{F}'_{\mu} = \hat{F}_{\mu}\sqrt{p_{\mu}}$ and $\hat{G}'_{\nu} = \hat{G}_{\nu}\sqrt{q_{\nu}}$, the relation reads as

$$\hat{G}'_{\nu} = \sum_{\mu} \hat{F}'_{\mu} \sqrt{p_{\mu}} U_{\mu\nu} / \sqrt{q_{\nu}}. \tag{5.41}$$

As $\hat{F}'_\mu$ and $\hat{G}'_\nu$ are orthonormal, the matrix $\sqrt{p_\mu}U_{\mu\nu}/\sqrt{q_\nu}$ should also be unitary. In general, it is not trivial to find a unitary matrix U that allows $\sqrt{p_\mu}U_{\mu\nu}/\sqrt{q_\nu}$ to be unitary as well.

5.2.2 Choi Isomorphism

The Choi isomorphism is a one-to-one correspondence between supermaps and (usual) operators (Appendix B.2.3). It allows us to inspect a supermap in terms of the corresponding operator. It can reveal additional properties of the supermap that is not immediately clear from the supermap itself. As such, the Choi isomorphism is not confined to quantum operations and is applicable to the whole space of supermaps. Interestingly though, the Choi isomorphism continues to hold between *completely positive supermaps* and *density operators*.[8] It makes the Choi isomorphism immensely useful for the study of quantum operations.

In this subsection, we introduce the Choi isomorphism and then use it to provide two physically-motivated proofs of the Kraus representation theorem summarized in (5.11).

To exploit the property of $\mathscr{F}$ being completely positive later, take a copy of the original vector space $\mathcal{V}$ as a reference space (or any vector space $\mathcal{R}$ of the same dimension as $\mathcal{V}$) and construct the tensor-product space $\mathcal{V} \otimes \mathcal{V}$ of the original and of the reference space. Then, consider a maximally entangled state

$$|\Phi\rangle := \sum_k |v_k\rangle \otimes |v_k\rangle , \tag{5.42}$$

where $\{|v_k\rangle\}$ is an orthonormal basis of $\mathcal{V}$. Its density operator is given by

$$|\Phi\rangle \langle\Phi| = \sum_{kl} |v_k\rangle \langle v_l| \otimes |v_k\rangle \langle v_l|. \tag{5.43}$$

Now operate an extended supermap $\mathscr{F} \otimes \mathscr{I}$, where $\mathscr{I}$ is the identity superoperator, on $|\Phi\rangle \langle\Phi|$ to get

$$\hat{C}_\mathscr{F} := (\hat{\mathscr{F}} \otimes \mathscr{I})(|\Phi\rangle \langle\Phi|) = \sum_{ij}\sum_{kl} |w_i v_k\rangle \langle w_j v_l| C_{ik;jl} \in \mathcal{L}(\mathcal{W} \otimes \mathcal{V}), \tag{5.44}$$

where C is the *Choi matrix* (see Appendix B.2.3) associated with $\mathscr{F}$,

$$\mathscr{F}(|v_k\rangle \langle v_l|) = \sum_{ij} |w_i\rangle \langle w_j| C_{ik;jl} . \tag{5.45}$$

---

[8] Note that an isomorphism between two spaces does not necessarily hold between their subspaces.

Equation (5.44) implies that $\hat{C}_{\mathscr{F}}$ is an operator (not a superoperator) on $\mathcal{W} \otimes \mathcal{V}$ with its matrix representation given by the Choi matrix C. That is, the operator $\hat{C}_{\mathscr{F}}$ and the matrix C are essentially the same mathematical objects. We call $\hat{C}_{\mathscr{F}}$ the *Choi operator* associated with the supermap $\mathscr{F}$. The association $\mathscr{F} \mapsto \hat{C}_{\mathscr{F}}$ is an isomorphism between supermaps and operators, and it is called the *Choi isomorphism*. The Choi operator $\hat{C}_{\mathscr{F}}$ (and hence the Choi matrix C) completely characterizes the associated supermap $\mathscr{F}$. To see it in a more physically transparent way, it is useful to represent the Choi operator $\hat{C}_{\mathscr{F}}$ in a quantum circuit of the form

$$\hat{C}_F = |\Phi\rangle \left\{ \begin{array}{c} \boxed{\mathscr{F}} \\ \hline \end{array} \right.$$

(5.46)

The Choi operator $\hat{C}_{\mathscr{F}}$ is the result of the evolution of a maximally entangled state $|\Phi\rangle$ in $\mathcal{V} \otimes \mathcal{V}$, composed of the original and reference spaces, under supermap $\mathscr{F}$ acting only on the original space. One way to quantitatively characterize the supermap $\mathscr{F}$ is thus to compare $\hat{C}_{\mathscr{F}}$ with the initial maximally entangled state $|\Phi\rangle$,

$$F_E := \mathrm{Tr} \, |\Phi\rangle \langle \Phi| \, \hat{C}_{\mathscr{F}} = \langle \Phi| \, \hat{C}_{\mathscr{F}} \, |\Phi\rangle. \tag{5.47}$$

The quantitative measure F_E is called the *entanglement fidelity*[9] of the supermap $\mathscr{F}$. It quantifies how well the supermap preserves the initial quantum information. The Choi operator also reflects the properties of the associated supermap. When $\mathscr{F}$ is completely positive, $\hat{C}_{\mathscr{F}}$ is a positive operator. Furthermore, when the supermap $\mathscr{F}$ is a *quantum operation*, satisfying all the axioms (a)–(c), $\hat{C}_{\mathscr{F}}$ is a density operator. Here it is interesting to note that the Choi isomorphism continues to hold between *quantum operations* and *density operators*. This refined isomorphism is called the *channel-state duality*.

Let us now prove the Kraus representation theorem using the Choi isomorphism. It is clear that any supermap in the form (5.11) satisfies the three axioms, and is a quantum operation. The converse is more complicated to prove. There are three common ways to do so, each of which is interesting in its own right. The first method (see Exercise B.5) is based directly on the general operator-sum representation in (B.17). Here, we will discuss the two other methods.

The second method relies on the properties of the Choi operator $\hat{C}_{\mathscr{F}}$. For any pure state $|\psi\rangle = \sum_j |v_j\rangle \psi_j \in \mathcal{V}$, define its conjugate state $|\psi^*\rangle := \sum_j |v_j\rangle \psi_j^*$ and observe that

$$\mathscr{F}(|\psi\rangle \langle \psi|) = \mathrm{Tr}_{\mathcal{V}} \left[\left(\hat{I} \otimes |\psi^*\rangle \langle \psi^*| \right) \hat{C}_{\mathscr{F}} \right] = \langle \psi^*| \, \hat{C}_{\mathscr{F}} \, |\psi^*\rangle. \tag{5.48}$$

As it happens often in this chapter, the Hermitian product in $\langle \psi^*| \, \hat{C}_{\mathscr{F}} \, |\psi^*\rangle$ is merely short-hand notation for the partial trace, and $\langle \psi^*| \, \hat{C}_{\mathscr{F}} \, |\psi^*\rangle$ is an operator on $\mathcal{W}$

---

[9] It is a special case of more general notion of *fidelity*. The fidelity between two pure states $|v\rangle$ and $|w\rangle$ is defined by $|\langle v|w\rangle|$.

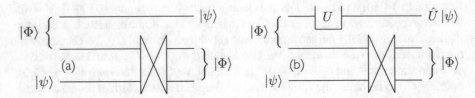

Fig. 5.4 Comparison of quantum teleportation and quantum gate teleportation. **a** Simplified quantum circuit model of quantum teleportation. The Bell measurement is replaced by the projection to $|\Phi\rangle$, and the success probability is $1/d^2$ for $d = dim\mathcal{V}$. **b** A quantum circuit model for quantum gate teleportation. The input state $|\psi\rangle$ on the third qubit results in the unitary-transformed state $\hat{U}|\psi\rangle$ on the first qubit

(not a number). Recalling the quantum circuit representation (5.46) of the Choi isomorphism, the identity (5.48) can be described by the quantum circuit

$$
|\Phi\rangle \left\{ \begin{array}{c} \boxed{\mathscr{F}} \quad\longrightarrow\quad \mathscr{F}(|\psi\rangle\langle\psi|) \\ \longrightarrow\!\!\!\bowtie\!\!\!-\quad |\psi^*\rangle \end{array} \right. \tag{5.49}
$$

where the quantum circuit element $-\!\!\bowtie\!\!-$ represents the projection onto the state specified at the output port. Since $\hat{C}_{\mathscr{F}}$ is a positive operator for a quantum operation $\mathscr{F}$, rewrite it in a spectral decomposition (see Appendix A.4, especially, Eq. (A.57))

$$
\hat{C}_{\mathscr{F}} = \sum_\mu |\varphi_\mu\rangle\langle\varphi_\mu| \, , \tag{5.50}
$$

where each vector $|\varphi_\mu\rangle$ has been normalized so that $\langle\varphi_\mu|\varphi_\mu\rangle$ gives the corresponding (positive) eigenvalue of $\hat{C}_{\mathscr{F}}$, $\hat{C}_{\mathscr{F}}|\varphi_\mu\rangle = |\varphi_\mu\rangle\langle\varphi_\mu|\varphi_\mu\rangle$. We define a linear map $\hat{F}_\mu : \mathcal{V} \to \mathcal{W}$ by the association

$$
\hat{F}_\mu |\psi\rangle = \langle\psi^*|\varphi_\mu\rangle . \tag{5.51}
$$

Note that on the right-hand side of the relation, the Hermitian product is applied partially and only on $\mathcal{V}$ while the remaining part is a vector belonging to $\mathcal{W}$. Putting (5.50) and (5.51) into (5.48), we confirm that

$$
\mathscr{F}(|\psi\rangle\langle\psi|) = \sum_\mu \hat{F}_\mu |\psi\rangle\langle\psi| \hat{F}_\mu^\dagger . \tag{5.52}
$$

$\mathscr{F}$ is linear and $|\psi\rangle$ is arbitrary, proving the statement in the theorem.

Now, turn to the third proof, which is based on the so-called *quantum gate teleportation* protocol. Figure 5.4a shows a simplified quantum circuit model of the quantum teleportation protocol. Compared to the typical quantum teleportation protocol discussed in Sect. 4.1, the Bell measurement has been replaced with the projection onto a single Bell state $|\Phi\rangle$. Due to this modification, the protocol is no longer deterministic.

Nevertheless, with a success probability of $1/d^2$ for $d = dim V$, the input state $|\psi\rangle$ on the third qubit is "teleported" to the first qubit. At the end of the protocol, one can apply any unitary transformation $\hat{U}$ to get $\hat{U}|\psi\rangle$. The result does not change if one applies $\hat{U}$ even before the projection. This variation leads to the quantum circuit depicted in Fig. 5.4b, which is commonly referred to as the quantum gate teleportation protocol.

In the Choi isomorphism, $|U\rangle := \hat{U} \otimes \hat{I}|\Phi\rangle$ is a Choi vector (Appendix B.2.3) corresponding to the unitary operator $\hat{U}$, and hence it completely characterizes $\hat{U}$. The quantum gate teleportation protocol uses $|U\rangle$ as a quantum entanglement resource,[10] and it moves the input state $|\Phi\rangle$ on the third qubit to the unitary-transformed state $\hat{U}|\psi\rangle$ on the first qubit with a success probability of $1/d^2$. The quantum gate teleportation protocol can then be generalized (Problem 5.2) to supermaps for the gate operation and to mixed states for the input state as in the following quantum circuit

$$(5.53)$$

Now consider a state $|\psi\rangle$. In accordance with the quantum gate teleportation protocol in (5.53) and the Choi isomorphism in (5.46), one has

$$\mathscr{F}(|\psi\rangle\langle\psi|)|v_j\rangle = \sum_i |v_i\rangle \left(\langle v_i| \otimes \langle\Phi|\right) \left(\hat{C}_{\mathscr{F}} \otimes |\psi\rangle\langle\psi|\right) \left(|v_j\rangle \otimes |\Phi\rangle\right) \quad (5.54)$$

Again, $\hat{C}_{\mathscr{F}}$ is positive if $\mathscr{F}$ is a quantum operation, so we use the spectral decomposition (5.50) of $\hat{C}_{\mathscr{F}}$. Finally, we define a set of linear operators $\hat{F}_\mu$ by

$$\hat{F}_\mu : |\psi\rangle \mapsto \sum_i |v_i\rangle \left(\langle v_i| \otimes \langle\Phi|\right) \left(|\varphi_\mu\rangle \otimes |\psi\rangle\right). \quad (5.55)$$

Then, we find that

$$\mathscr{F}(|\psi\rangle\langle\psi|) = \sum_\mu \hat{F}_\mu |\psi\rangle\langle\psi| \hat{F}_\mu^\dagger, \quad (5.56)$$

which proves the Kraus representation theorem.

5.2.3 Unitary Representation

A quantum operation can be regarded as a unitary operator on an extended system, which involves an "environment" in addition to the original "system". Although it is not particularly useful for practical applications, the unitary representation

---

[10] As $\hat{U} \otimes \hat{I}$ only operates *locally*, it does not modify the entanglement characteristics of $|\Phi\rangle$.

provides a clear physical insight into the underlying physical processes described by the quantum operation.

Suppose that we are given a quantum operation $\mathscr{F} = \mathcal{L}(\mathcal{V}) \to \mathcal{L}(\mathcal{V})$ represented by the Kraus representation in Eq. (5.11) in terms of the Kraus elements $\hat{F}_\mu$ ($\mu = 0, 1, \ldots, m-1$). As such, we want to construct a system-plus-environment model that produces the same effect as $\mathscr{F}$ when the environment is traced out. Therefore, we need to find a proper vector space $\mathcal{E}$ for the environment and an overall unitary operator $\hat{U}$ acting on the total system $\mathcal{V} \otimes \mathcal{E}$ such that

$$\mathscr{F}(\hat{\rho}) = \mathop{\mathrm{Tr}}_{\mathcal{E}} \hat{U} \left(\hat{\rho} \otimes |\varepsilon_0\rangle \langle \varepsilon_0| \right) \hat{U}^\dagger \tag{5.57}$$

for all $\hat{\rho} \in \mathcal{L}(\mathcal{V})$ and a particular state $|\varepsilon_0\rangle \in \mathcal{E}$.[11] We first construct the vector space $\mathcal{E}$ associated with the environment by choosing an orthonormal basis $\{ |\varepsilon_\mu\rangle : \mu = 0, \ldots, m-1 \}$.[12] We note that the dimension of $\mathcal{E}$ is the same as the number of the Kraus elements $\hat{F}_\mu$ in the Kraus representation (5.11) and is no larger than $(\dim \mathcal{V})^2$. Define a unitary operator $\hat{U}$ on $\mathcal{V} \otimes \epsilon$ by requiring that

$$\hat{U} |\psi\rangle \otimes |\varepsilon_0\rangle = \sum_\mu (\hat{F}_\mu |\psi\rangle) \otimes |\varepsilon_\mu\rangle \tag{5.58}$$

for any $|\psi\rangle \in \mathcal{V}$. Clearly, taking the partial trace over the environment reproduces $\mathscr{F}(|\psi\rangle \langle \psi|)$ as one can see from an explicit evaluation

$$\mathop{\mathrm{Tr}}_{\mathcal{E}} \hat{U} \left(|\psi\rangle \langle \psi| \otimes |\varepsilon_0\rangle \langle \varepsilon_0| \right) \hat{U}^\dagger = \mathop{\mathrm{Tr}}_{\mathcal{E}} \sum_{\mu\nu} \left(\hat{F}_\mu |\psi\rangle \langle \psi| \hat{F}_\nu^\dagger \right) \otimes |\varepsilon_\mu\rangle \langle \varepsilon_\nu|$$

$$= \sum_\mu \hat{F}_\mu |\psi\rangle \langle \psi| \hat{F}_\mu^\dagger \tag{5.59}$$

This relation is linear and holds for an arbitrary vector $|\psi\rangle$, and hence it should holds for any mixed state $\hat{\rho} = \sum_j |\psi_j\rangle p_j \langle \psi_j|$.

5.2.4 Examples

So far, we discussed a general description of quantum noisy processes in terms of the corresponding quantum operations. Let us now present some examples for a single qubit. We consider some limiting cases that nevertheless allow us to easily grasp the physical meaning of the Kraus elements.

---

[11] The choice of $|\varepsilon_0\rangle$ is completely arbitrary, and one can even choose a mixed state.

[12] Here, just for convenience, we have chosen the basis so as for it to include $|\varepsilon_0\rangle$, but it is not necessary.

Phase damping The phase damping process is a decoherence process without involving any relaxation of energy or change in the population over the states. In this sense, it may be regarded as a pure decoherence process without involving any energy relaxation. For this reason, it is also referred to as a *dephasing* process. The unitary representation of a phase damping process in a single-qubit system is given by

$$|0\rangle \otimes |\varepsilon_0\rangle \mapsto |0\rangle \otimes |\varepsilon_0\rangle \tag{5.60a}$$

$$|1\rangle \otimes |\varepsilon_0\rangle \mapsto |1\rangle \otimes |\varepsilon_0\rangle \sqrt{1-p} + |1\rangle \otimes |\varepsilon_1\rangle \sqrt{p}, \tag{5.60b}$$

where $\{|\varepsilon_0\rangle, |\varepsilon_1\rangle\}$ is an orthonormal basis of the vector space $\mathcal{E}$ associated with the environment. It indicates that when and only when the system is in $|1\rangle$, the environment changes its state from the initial state $|\varepsilon_0\rangle$ to another orthogonal state $|\varepsilon_1\rangle$ with probability p. Note that the system remains in the same state in both cases. The key point is that the environment nevertheless "knows" which state the system is in and this knowledge leads to a loss of coherence in the state of the system.

Indeed, the total unitary operator is a controlled-unitary operator with

$$\hat{U} \doteq \begin{bmatrix} \sqrt{1-p} & -\sqrt{p} \\ \sqrt{p} & \sqrt{1-p} \end{bmatrix} \tag{5.61}$$

in the basis $\{|\varepsilon_0\rangle, |\varepsilon_1\rangle\}$. As a result, when the system is prepared in a coherent superposition, the controlled-$\hat{U}$ operation creates an entanglement between the system and the environment (see Sects. 2.2.1 and 2.2.2)

$$(|0\rangle c_0 + |1\rangle c_1) \otimes |\varepsilon_0\rangle \mapsto |0\rangle \otimes |\varepsilon_0\rangle c_0 + |1\rangle \otimes (\hat{U} |\varepsilon_0\rangle)c_1 \tag{5.62}$$

or, more explicitly,

$$(|0\rangle c_0 + |1\rangle c_1) \otimes |\varepsilon_0\rangle \mapsto \left(|0\rangle c_0 + |1\rangle c_1 \sqrt{1-p}\right) \otimes |\varepsilon_0\rangle + |1\rangle \otimes |\varepsilon_1\rangle c_1 \sqrt{p}. \tag{5.63}$$

Due to the entanglement, the final state of the system alone cannot be a pure state (see Sect. 1.1.2) and coherence in the initial state has been lost through the process.

With the prescription in Sect. 5.2.1, the Kraus elements are given by

$$\hat{E}_0 \doteq \begin{bmatrix} 1 & 0 \\ 0 & \sqrt{1-p} \end{bmatrix}, \quad \hat{E}_1 \doteq \begin{bmatrix} 0 & 0 \\ 0 & \sqrt{p} \end{bmatrix} \tag{5.64}$$

and the corresponding quantum operation is given by

$$\mathscr{F}(\hat{\rho}) = \hat{E}_0 \hat{\rho} \hat{E}_0^\dagger + \hat{E}_1 \hat{\rho} \hat{E}_1^\dagger \tag{5.65}$$

Note that they are not orthogonal,

$$\text{Tr}\, \hat{E}_0^\dagger \hat{E}_1 = \sqrt{p(1-p)} \neq 0. \tag{5.66}$$

It is therefore more convenient and efficient to choose mutually orthogonal Kraus elements

$$\hat{F}_0 = \sqrt{\frac{1 + \sqrt{1 - p}}{2}}\,\hat{I}\,, \quad \hat{F}_1 = \sqrt{\frac{1 - \sqrt{1 - p}}{2}}\,\hat{S}^z \qquad (5.67)$$

In short, the quantum operation for the phase damping process is written as

$$\mathscr{F}(\hat{\rho}) = \frac{1 + \sqrt{1 - p}}{2}\,\hat{\rho} + \frac{1 - \sqrt{1 - p}}{2}\,\hat{S}^z \hat{\rho} \hat{S}^z \qquad (5.68)$$

It is also convenient to expand the density operator into

$$\hat{\rho} = \frac{1}{2}\hat{I} + \hat{S}^x \rho_x + \hat{S}_y \rho_y + \hat{S}^z \rho_z\,, \qquad (5.69)$$

where ρ_μ for $\mu = x, y, z$ are real parameters. Then the quantum operation $\mathscr{F}$ gives

$$\mathscr{F}(\hat{\rho}) = \frac{1}{2}\hat{I} + \sqrt{1 - p}\,\hat{S}^x \rho_x + \sqrt{1 - p}\,\hat{S}^y \rho_y + \hat{S}^z \rho_z. \qquad (5.70)$$

Note that the coefficient in $\hat{S}^z$ does not change. It asserts that the populations in $|0\rangle$ and $|1\rangle$, $1/2 + \rho_z$ and $1/2 - \rho_z$, do not change. The phase damping process only causes pure dephasing. In particular, at $p \to 1$, the new density operator approaches,

$$\mathscr{F}(\hat{\rho}) \to \frac{1}{2}\hat{I} + \hat{S}^z \rho_z. \qquad (5.71)$$

The coherence has disappeared completely.

The phase damping process is specified by two Kraus elements.

```
In[ ]:= Let[Real, p]
        ops = {Sqrt[(1 + Sqrt[1 - p]) / 2], Sqrt[(1 - Sqrt[1 - p]) / 2] * S[3]};
        spr = Supermap[ops]
```

$$Out[ ]= \text{Supermap}\left[\left\{\frac{\sqrt{1 + \sqrt{1 - p}}}{\sqrt{2}}, \frac{\sqrt{1 - \sqrt{1 - p}}\ S^z}{\sqrt{2}}\right\}\right]$$

Here p is the probability for the phase to be flipped.

```
In[ ]:= Let[Real, ρ]
        rho = 1 / 2 + ρ[{1, 2, 3}].S[All]
```

$$Out[ ]= \frac{1}{2} + S^x \rho_1 + S^y \rho_2 + S^z \rho_3$$

The supermap transforms the above density operator as follows.

```
In[ ]:= new = spr[rho]
```

$$Out[ ]= \frac{1}{2} + \sqrt{1 - p}\ S^x \rho_1 + \sqrt{1 - p}\ S^y \rho_2 + S^z \rho_3$$

Amplitude damping The amplitude damping process describes the spontaneous decay of the excited state $|1\rangle$ of the system to the ground state $|0\rangle$. The decay is accompanied by the emission of a photon. One can regards that the photon "observes" the system and the information of the system acquired by the photon leads to decoherence. In a unitary representation, the process is described by the overall unitary operator such that

$$|0\rangle \otimes |\varepsilon_0\rangle \mapsto |0\rangle \otimes |\varepsilon_0\rangle \,, \tag{5.72a}$$

$$|1\rangle \otimes |\varepsilon_0\rangle \mapsto |1\rangle \otimes |\varepsilon_0\rangle \sqrt{1-p} + |0\rangle \otimes |\varepsilon_1\rangle \sqrt{p}\,. \tag{5.72b}$$

Thus, the decay occurs with probability p provided that the system is in the state $|1\rangle$. The Kraus elements are given by

$$\hat{F}_0 \doteq \begin{bmatrix} 1 & 0 \\ 0 & \sqrt{1-p} \end{bmatrix}, \quad \hat{F}_1 \doteq \begin{bmatrix} 0 & \sqrt{p} \\ 0 & 0 \end{bmatrix}. \tag{5.73}$$

They are already orthogonal to each other. The quantum operation describing the amplitude damping process reads as

$$\mathscr{F}(\hat{\rho}) = p\hat{S}^+ \hat{\rho} \hat{S}^- +$$
$$\left(\frac{1+\sqrt{1-p}}{2} + \frac{1-\sqrt{1-p}}{2}\hat{S}^z \right) \hat{\rho} \left(\frac{1+\sqrt{1-p}}{2} + \frac{1-\sqrt{1-p}}{2}\hat{S}^z \right)$$
$$\tag{5.74}$$

With the expansion in (5.69), the transformation reads as

$$\mathscr{F}(\hat{\rho}) = \frac{1}{2} + (1-p)\hat{S}^x \rho_x + (1-p)\hat{S}^y \rho_y + \hat{S}^z(1/2 + (1-p)\rho_z). \tag{5.75}$$

In the limit of $p \to 1$, the new density operator approaches the pure state $|0\rangle$,

$$\mathscr{F}(\hat{\rho}) = \frac{1}{2}\hat{I} + \frac{1}{2}\hat{S}^z = |0\rangle \langle 0|. \tag{5.76}$$

regardless of the initial state. This is due to a relaxation from $|1\rangle$ to $|0\rangle$.

The amplitude damping process is specified by two Kraus elements.

```
In[•]:= Let[Real, p]
       ops = {S[10] + Sqrt[1-p] * S[11], Sqrt[p] * S[4]};
       spr = Supermap[ops]
Out[•]= Supermap[{(|0⟩⟨0|)_s + √(1-p) (|1⟩⟨1|)_s, √p S⁺}]
```

Here p is the probability for the phase to be flipped.

```
In[·]:= Let[Real, ρ]
       rho = 1 / 2 + ρ[{1, 2, 3}].S[All]
Out[·]= 1
        ─ + Sˣ ρ₁ + Sʸ ρ₂ + Sᶻ ρ₃
        2
```

The supermap transforms the above density operator as follows.

```
In[·]:= new = spr[rho] // Elaborate
Out[·]= 1                                          ⎛ ⎛ 1     ⎞      ⎞
        ─ + √1-p  Sˣ ρ₁ + √1-p  Sʸ ρ₂ + Sᶻ ⎜p ⎜─ - ρ₃⎟ + ρ₃⎟
        2                                          ⎝ ⎝ 2     ⎠      ⎠
```

Depolarizing In the depolarizing process, the decoherence occurs symmetrically, and there is no distinction of the specific types of actual decoherence processes. The system undergoes an incoherent process with probability p and remains intact with probability $1 - p$. The incoherent process may cause the system to flip the bit value, the phase, or both with equal probability.

The situation can be best described in the unitary representation. Suppose that the system and the environment is in the product state $|\psi\rangle \otimes |\varepsilon_0\rangle$. The decoherence process causes the transition

$$|\psi\rangle \otimes |\varepsilon_0\rangle \mapsto |\psi\rangle \otimes |0\rangle \sqrt{1 - p}$$
$$+ \left(\hat{S}^x |\psi\rangle \otimes |\varepsilon_1\rangle + \hat{S}^y |\psi\rangle \otimes |\varepsilon_2\rangle + \hat{S}^z |\psi\rangle \otimes |\varepsilon_3\rangle \right) \sqrt{\frac{p}{3}}$$
(5.77)

The environment evolves to one of the four mutually orthogonal states. The final states of the environment enables recognizing the process that has occurred (bit flip, phase flip, or both). This causes decoherence on the state of the system.

From the above unitary representation, we can get the Kraus elements

$$\hat{F}_0 = \sqrt{1 - p}\hat{I}, \quad \hat{F}_1 = \sqrt{\frac{p}{3}}\hat{S}^x, \quad \hat{F}_2 = \sqrt{\frac{p}{3}}\hat{S}^y, \quad \hat{F}_3 = \sqrt{\frac{p}{3}}\hat{S}^z.$$
(5.78)

The Kraus elements are already orthogonal to each other. One can also check that they satisfy the completeness relation

$$\sum_{\mu=0}^{3} \hat{F}_\mu^\dagger \hat{F}_\mu = \hat{I}$$
(5.79)

as they should. In the Kraus representation with the above Kraus elements, a density operator $\hat{\rho}$ is transformed under the decoherence process as

$$\mathscr{F}(\hat{\rho}) = (1 - p)\hat{\rho} + \frac{p}{3} \left(\hat{S}^x \hat{\rho} \hat{S}^x + \hat{S}^y \hat{\rho} \hat{S}^y + \hat{S}^z \hat{\rho} \hat{S}^z \right).$$
(5.80)

In terms of the components, it reads as

$$\mathscr{F}(\hat{\rho}) = \frac{1}{2}\hat{I} + \left(1 - \frac{4p}{3}\right)\left(\hat{S}^x \rho_x + \hat{S}^y \rho_y + \hat{S}^z \rho_z\right). \tag{5.81}$$

This implies that under the process, the "spin" polarization (i.e., the Bloch vector corresponding to the resulting density operator)

$$\boldsymbol{P} := (\langle \hat{S}^x \rangle, \langle \hat{S}^y \rangle, \langle \hat{S}^z \rangle) \tag{5.82}$$

shrinks by the factor $(1 - 4p/3)$. Hence the name of the process. The state becomes completely random for $4p/3 = 1$.

The depolarizing process is specified by three Kraus elements.

```
In[·]:= Let[Real, p]
       ops = Prepend[Sqrt[p / 3] * S[All], Sqrt[1 - p]];
       spr = Supermap[ops]
```

$$\text{Out[·]= Supermap}\left[\left\{\sqrt{1-p}, \frac{\sqrt{p}\ S^x}{\sqrt{3}}, \frac{\sqrt{p}\ S^y}{\sqrt{3}}, \frac{\sqrt{p}\ S^z}{\sqrt{3}}\right\}\right]$$

Here p is the probability for the phase to be flipped.

```
In[·]:= Let[Real, ρ]
       rho = 1 / 2 + ρ[{1, 2, 3}].S[All]
```

$$\text{Out[·]= } \frac{1}{2} + S^x \rho_1 + S^y \rho_2 + S^z \rho_3$$

The supermap transforms the above density operator as follows.

```
In[·]:= new = spr[rho]
```

$$\text{Out[·]= } \frac{1}{2} + \left(1 - \frac{4p}{3}\right) S^x \rho_1 + \left(1 - \frac{4p}{3}\right) S^y \rho_2 + \left(1 - \frac{4p}{3}\right) S^z \rho_3$$

5.3 Measurements as Quantum Operations

Generalized measurements (Postulate 3′) can be regarded as a special case of quantum operations. Suppose that a measurement is described by a set of measurement operators $\hat{M}_m$ corresponding to measurement outcomes m. The mapping $\mathscr{F}_m \in \mathcal{L}(V)$ defined by

$$\mathscr{F}_m(\hat{\rho}) = \hat{M}_m \hat{\rho} \hat{M}_m^\dagger \tag{5.83}$$

for each m is obviously a quantum operation. This is natural since the measurement process involves the interaction of the system with the measuring devices. Note that the quantum operation $\mathscr{F}_m$ does not preserve the trace in general,

$$0 \leq \text{Tr}\,\mathscr{F}_m(\rho) \leq 1. \tag{5.84}$$

Physically, $\mathrm{Tr}\,\mathscr{F}_m(\rho)$ gives the probability to get outcome m from the measurement process.

The measurement given above is a *selective measurement*. This physically involves separating an ensemble into subensembles that are distinguished by the measurement outcome. Schwinger (1959) conceived a new notion corresponding to the measurement process prior to the selection stage. It is denominated as a *non-selective measurement*. One can also regard a non-selective measurement as remixing the subensembles after the measurement with the probabilities $\mathrm{Tr}[\mathscr{F}_m(\hat{\rho})]$. A non-selective measurement is thus represented by the quantum operation

$$\mathscr{F}(\hat{\rho}) := \sum_m \mathscr{F}_m(\hat{\rho}) = \sum_m \hat{M}_m \hat{\rho} \hat{M}_m^\dagger . \tag{5.85}$$

In this case, the trace is preserved: $\mathrm{Tr}\,\mathscr{F}(\hat{\rho}) = 1$ for any $\hat{\rho}$. It follows from the completeness relation, $\sum_m \hat{M}_m^\dagger \hat{M}_m = \hat{I}$, satisfied by the measurement operators.

5.4 Quantum Master Equation

Consider an open quantum system interacting with its environment. The system is inevitably subjected to decoherence processes. Then, suppose that the system is in $\hat{\rho}(t)$ at time t. To understand the decoherence processes, we want to examine the state $\hat{\rho}(t')$ at later times $t' > t$. The evolution from $\hat{\rho}(t)$ to $\hat{\rho}(t')$ is described by a quantum operation, which in this case is a completely positive and trace-preserving superoperator. The operator-sum representation (5.11) guarantees the existence of operators $\hat{F}_\mu(t', t)$ such that

$$\hat{\rho}(t') = \sum_\mu \hat{F}_\mu(t', t) \hat{\rho}(t) \hat{F}_\mu^\dagger(t', t) , \tag{5.86a}$$

and satisfying the probability-conserving—trace-preserving—condition

$$\sum_\mu \hat{F}_\mu^\dagger(t', t) \hat{F}_\mu(t', t) = \hat{I} \tag{5.86b}$$

and the orthogonality condition

$$\mathrm{Tr}\,\hat{F}_\mu^\dagger \hat{F}_\nu = 0 \quad (\mu \neq \nu). \tag{5.86c}$$

However, it turns out that under a specific physical situation, it is mostly difficult to determine the relevant operators $\hat{F}_\mu(t', t)$ that properly describe the given situation. This may be because the approach attempts to directly determine $\hat{\rho}(t')$ as a function of time t' given the initial condition set by $\hat{\rho}(t)$. However, it would be more convenient and efficient to express the process in a differential form—a rate equation. After all, both Newton's classical equation of motion and Schrödinger's equation for quantum states are differential equations, describing rates of changes for the state variables.

Can one express an quantum operation with a set of differential equations that is equivalent to the operator-sum representation? Unfortunately, the answer in general is "No". However, under many physically relevant conditions,[13] the operators $\hat{F}_\mu(t', t)$ depend only on the time span $\delta t := t' - t$ but not on the individual instances t' and t. Physically, this implies that the underlying process does not depend on the history, and the assumption is commonly called the *Markov approximation*. Under such conditions, the quantum operation in (5.86) can be reformulated in a differential form and the resulting equation,

$$\frac{d\hat{\rho}}{dt} = \mathscr{L}(\hat{\rho}), \tag{5.87}$$

is called the *Lindblad equation* or *quantum master equation*. Here the superoperator $\mathscr{L}$ defined by

$$\mathscr{L}(\hat{\rho}) := -i[\hat{H}, \hat{\rho}] + \sum_\mu \left(\hat{L}_\mu \hat{\rho} \hat{L}_\mu^\dagger - \frac{1}{2}\hat{L}_\mu^\dagger \hat{L}_\mu \hat{\rho} - \frac{1}{2}\hat{\rho}\hat{L}_\mu^\dagger \hat{L}_\mu \right), \tag{5.88}$$

generates the *quantum Markovian dynamics*, and is called the *Lindblad generator*. The Hermitian operator $\hat{H}$ in (5.88) describes the unitary part of the dynamics. For this reason, $\hat{H}$ is often called the *effective Hamiltonian* of the system, but in general it is not the same as the Hamiltonian when the system is isolated. The operators $\hat{L}_\mu$ in (5.88) are responsible for the non-unitary dynamics and are called the *Lindblad operators* or *quantum jump operators*.

It is also customary to rewrite the Lindblad generator (5.88) into the form

$$\mathscr{L}(\hat{\rho}) = -i[\hat{H}, \hat{\rho}] - \{\hat{G}, \hat{\rho}\} + \sum_\mu \hat{L}_\mu \hat{\rho} \hat{L}_\mu^\dagger, \tag{5.88'}$$

where

$$\hat{G} := \frac{1}{2} \sum_\mu \hat{L}_\mu^\dagger \hat{L}_\mu. \tag{5.89}$$

Interestingly, ignoring the last term in the quantum jump operators in (5.88'), the solution to the Lindblad equation (5.88') is simply given by

$$\hat{\rho}(t) = e^{-it\hat{H}_{\text{non}}} \hat{\rho}(0) e^{it\hat{H}_{\text{non}}^\dagger}, \tag{5.90}$$

which resembles the unitary dynamics in (1.39) with the Hamiltonian $\hat{H}$ replaced with the effective *non-Hermitian Hamiltonian*

---

[13] A notable exception is the case where time-dependent external fields are applied on the system.

$$\hat{H}_{\text{non}} := \hat{H} - i\hat{G}. \tag{5.91}$$

The additional term in $\hat{G}$ of the non-Hermitian Hamiltonian makes a significant difference in the evolution governed by Eq. (5.90) since it causes damping and leads to irreversible population loss in the eigenstates of $\hat{H}$. In this sense, we call $\hat{G}$ the *effective damping operator*. Although the non-Hermitian Hamiltonian approach does not explain all decoherence processes, it lays out an intuitively appealing picture of open systems and has been widely used to describe the effects of a finite life time of (effective) energy levels. The non-Hermitian Hamiltonian also provides a good starting point for various more elaborate methods to investigate decoherence processes. A common example is the so-called *quantum jump approach*. It is an approximate method to solve the Lindblad equation combining the non-unitary evolution in (5.90) due to the non-Hermitian Hamiltonian and the "quantum jumps" due to the quantum jump operators $\hat{L}_\mu$ (Dum et al., 1992; Plenio & Knight, 1998).

The choice of the Lindblad operators $\hat{L}_\mu$ and the effective Hamiltonian $\hat{H}$ is not unique (Breuer & Petruccione, 2002): First, the two sets of Lindblad operators $\left\{\hat{L}_\mu\right\}$ and $\left\{\hat{L}'_\nu\right\}$ give the same Lindblad equation when

$$\hat{L}'_\nu = \sum_\mu \hat{L}_\mu U_{\mu\nu}, \tag{5.92}$$

where U is a unitary matrix. Recall a similar unitary freedom for the choice of the Kraus operators in the specification of quantum operations [see Eqs. (5.34) and (5.35)] as well as for the choice of pure states in the specification of mixed states [see Eqs. (1.14) and (1.15)]. Thanks to the unitary freedom, one can always choose *mutually orthogonal* quantum jump operators,

$$\text{Tr}\,\hat{L}_\mu^\dagger \hat{L}_\nu = 0 \quad (\mu \neq \nu) \tag{5.93}$$

for all μ and ν. Such a choice is optimal in the sense that $(d^2 - 1)$ quantum jump operators, where $d := \dim \mathcal{V}$, is sufficient for any Lindblad equation. The unitary freedom in (5.92) inherits from the unitary freedom for the choice of the Kraus elements in (5.35). The proof is left for an exercise. Second, the Lindblad generator is also invariant under the inhomogeneous transformations

$$\hat{L}_\mu \to \hat{L}'_\mu = \hat{L}_\mu + a_\mu, \tag{5.94a}$$

$$\hat{H} \to \hat{H}' = \hat{H} + \frac{1}{2i}\sum_\mu (a_\mu^* \hat{L}_\mu - a_\mu \hat{L}_\mu^\dagger) + b \tag{5.94b}$$

for any $a_\mu \in \mathbb{C}$ and $b \in \mathbb{R}$. Due to the translational freedom, it is always possible to choose the Lindblad operators to be *traceless*, $\text{Tr}\,\hat{L}_\mu = 0$. Furthermore, for a given Lindblad equation, it is common to impose the condition $\text{Tr}\,\hat{H} = 0$ on the effective

Hamiltonian $\hat{H}$ to make it unique. It is straightforward to prove the translational freedom, and again left as an exercise.

As we have pointed out concerning the unitary freedom in the Kraus representation, the unitary freedom does not necessarily imply that one can exploit it to change a given set of Lindblad operators to any arbitrary choice of Lindblad operators. This is because the Lindblad operators in (5.92) are not normalized. A notable exception is the two-dimensional case (see Sect. 5.4.2).

In the remainder of the section, we derive the quantum master equation (5.88) and discuss methods to solve it.

5.4.1 Derivation

It is straightforward to derive the Lindblad equation (5.88) starting from the Kraus representation (5.86a) under the Markov assumption (see, e.g., Breuer & Petruccione (2002) for details). Here we take a heuristic approach, which is more useful to understand the underlying physics.

As $t' \to t$ ($\delta t \to 0$), it is physically required that $\hat{\rho}(t') \to \hat{\rho}(t)$. This implies that one and only one of $\hat{F}_\mu(\delta t)$ must approach $\hat{I}$. Let us denote it by $\hat{F}_0(\delta t)$. Up to the first order in δt,

$$\hat{F}_0(\delta t) \approx \hat{I} + \hat{L}_0 \, \delta t \,. \tag{5.95}$$

The rest should vanish $\hat{F}_\mu(\delta t) \to 0$ for any $\mu > 0$. Since we physically expect that $\hat{\rho}(t') \approx \hat{\rho}(t) + \mathcal{O}(\delta t)$, $\hat{F}_\mu(\delta t)$ must vanish like $\sqrt{\delta t}$ with δt so that $\hat{F}_\mu(\delta t)\hat{\rho}\hat{F}_\mu^\dagger(\delta) \approx \mathcal{O}(\delta t)$. We put

$$\hat{F}_\mu(\delta t) \approx \hat{L}_\mu \sqrt{\delta t} \quad (\mu > 0)\,. \tag{5.96}$$

As $\hat{L}_\mu$ ($\mu > 0$) directly proportional to $\hat{F}_\mu$, they are all traceless and mutually orthogonal—$\mathrm{Tr}\,\hat{L}_\mu^\dagger\hat{L}_\nu = 0$ for $\mu \neq \nu$. The probability conservation condition, Eq. (5.86b), implies that

$$\hat{L}_0 + \hat{L}_0^\dagger = -\sum_{\mu \neq 0} \hat{L}_\mu^\dagger\hat{L}_\mu\,. \tag{5.97}$$

It suggests that it will be convenient to split $\hat{L}_0$ into the Hermitian and anti-Hermitian part

$$\hat{L}_0 = -\hat{G} - i\hat{H}\,, \tag{5.98}$$

where the Hermitian part $\hat{G}$ is fixed by the operators $\hat{L}_\mu$ with the relation ($\mu > 0$)

$$\hat{G} = \frac{1}{2}\sum_{\mu > 0} \hat{L}_\mu^\dagger\hat{L}_\mu\,. \tag{5.99}$$

whereas $\hat{H}$ remains arbitrary and is determined only by $\hat{F}_0$, which is linearly independent of $\hat{L}_\mu$ ($\mu > 0$). Finally, putting Eqs. (5.95), (5.96), (5.98), and (5.99) into Eq. (5.86) leads to the desired Eq. (5.88) or, equivalently, to (5.88). Note that in this particular derivation, the Lindblad operators $\hat{L}_\mu$ turn out to be traceless and mutually orthogonal automatically without exploiting the unitary freedom in (5.92). Here, the properties inherit from the orthogonality (5.86c) of the Kraus elements.

As mentioned at the beginning of the section, it is difficult in practice to explicitly determine the quantum operations $\hat{\rho}(t) \mapsto \hat{\rho}(t')$ as a function of time. More common approach is to derive the Lindblad equation, often approximately, by reducing the unitary dynamics of the total system consisting of the system plus the environment (see also Sect. 5.2.3). Some examples including perturbative methods are discussed in Breuer & Petruccione (2002).

5.4.2 Examples

Phase damping The Lindblad equation for the phase damping process can be obtained from the Kraus elements in (5.67). We assume that the probability p for the process to occur is proportional to time t, $p = \gamma t$, where γ is the rate of the process per unit time. Expanding the Kraus elements for small t,

$$\hat{F}_0 \approx \hat{I} - \frac{\gamma^t}{8}\hat{I} . \tag{5.100}$$

According to (5.98), we identify the effective Hamiltonian and damping operator with

$$\hat{H} = 0, \quad \hat{G} = \frac{\gamma}{8}\hat{I} , \tag{5.101}$$

respectively. Furthermore,

$$\hat{F}_1 \approx \frac{\sqrt{\gamma t}}{2}\hat{S}^z \tag{5.102}$$

implies that there is one Lindblad operator

$$\hat{L}_1 = \frac{\sqrt{\gamma}}{2}\hat{S}^z . \tag{5.103}$$

Overall, the Lindblad generator for the phase damping process is given by

$$\mathscr{L}(\hat{\rho}) = \frac{\gamma}{8}\hat{\rho} + \frac{\gamma}{4}\hat{S}^z\hat{\rho}\hat{S}^z . \tag{5.104}$$

Amplitude damping The Kraus elements for the amplitude damping process are given in (5.73). In the infinitesimal time t,

$$\hat{F}_0 \approx I - \frac{\gamma}{2}\begin{bmatrix} 0 & 0 \\ 0 & 1 \end{bmatrix}, \quad \hat{F}_1 \approx \sqrt{\gamma}\begin{bmatrix} 0 & 1 \\ 0 & 0 \end{bmatrix} \tag{5.105}$$

Therefore, while the effective Hamiltonian $\hat{H}$ vanishes, the effective damping operator $\hat{G}$ and the Lindblad operator are given by

$$\hat{G} = \frac{\gamma}{4}(1 - \hat{S}^z), \quad \hat{L}_1 = \sqrt{\gamma}\hat{S}^+. \tag{5.106}$$

The Lindblad generator for the amplitude damping is given by

$$\mathscr{L}(\hat{\rho}) = \frac{\gamma}{4}\left[(1 - \hat{S}^z)\hat{\rho} + \hat{\rho}(1 - \hat{S}^z)\right] + \gamma\hat{S}^+\hat{\rho}\hat{S}^-. \tag{5.107}$$

Depolarizing The Kraus elements for the depolarizing process have been obtained in (5.78). Again, assuming $p = \gamma t$ and expanding the Kraus elements for small t give

$$\hat{F}_0 \approx \hat{I} - \frac{\gamma t}{2}\hat{I}, \quad \hat{F}_\mu = \sqrt{\frac{\gamma t}{3}}\hat{S}^\mu \quad (\mu = x, y, z). \tag{5.108}$$

There are three relevant Lindblad operators

$$\hat{L}_\mu = \sqrt{\frac{\gamma}{3}}\hat{S}^\mu \tag{5.109}$$

for $\mu = 1, 2, 3$, and the effective damping operator is given by

$$\hat{G} = \frac{\gamma}{3}\hat{I}. \tag{5.110}$$

Therefore, the Lindblad generator for the depolarizing process is given by

$$\mathscr{L}(\hat{\rho}) = \frac{\gamma}{2}\hat{\rho} + \frac{\gamma}{3}\sum_{\mu=x,y,z}\hat{S}^\mu\hat{\rho}\hat{S}^\mu. \tag{5.111}$$

General damping For a single qubit, any master equation can be put into the form

$$\frac{d\hat{\rho}}{dt} = -i[\hat{H}, \hat{\rho}] - \{\hat{G}, \hat{\rho}\} + \Gamma_+\hat{S}^+\hat{\rho}\hat{S}^- + \Gamma_-\hat{S}^-\hat{\rho}\hat{S}^+ + \Gamma_\phi\hat{S}^z\hat{\rho}\hat{S}^z, \tag{5.112a}$$

where the effective Hamiltonian $\hat{H}$ is arbitrary as long as it is Hermitian, the effective damping operator is given by

$$\hat{G} := \frac{\Gamma_- + \Gamma_+ 2\Gamma_\phi}{4} + \left(\frac{\Gamma_- - \Gamma_+}{4}\right)\hat{S}^z, \tag{5.112b}$$

the real positive parameters $\Gamma_\pm$, Γ_ϕ are the rates at which the decoherence processes associated with the quantum jump operators $\hat{S}^\pm$ and $\hat{S}^z$ occur. In this form, the quantum jump operators, $\hat{S}^\pm$ and $\hat{S}^z$, describe the "simple" transitions—no mixture of different transitions—between the fixed set of states $|0\rangle$ and $|1\rangle$.[14] In general, those states are not the eigenstates of the effective Hamiltonian $\hat{H}$ in the coherent part of the master equation.

To see that the form in Eq. (5.112) is the most general form of the quantum master equation for a single-qubit system, let us start from the general Lindblad equation

$$\frac{d\hat{\rho}}{dt} = -i[\hat{H}, \hat{\rho}] - \{\hat{G}, \hat{\rho}\} + \sum_{\mu=1}^{3} \gamma_\mu \hat{A}_\mu \hat{\rho} \hat{A}_\mu^\dagger, \tag{5.113}$$

where

$$\hat{G} := \frac{1}{2} \sum_\mu \gamma_\mu \hat{A}_\mu^\dagger \hat{A}_\mu. \tag{5.114}$$

The three Lindblad operators are *traceless* and *orthonormal* and $\gamma_\mu \geq 0$. Consider another set of orthonormal operators

$$\hat{L}_1 = \hat{S}^+, \quad \hat{L}_2 = \hat{S}^-, \quad \hat{L}_3 = \frac{1}{\sqrt{2}} \hat{S}^z. \tag{5.115}$$

As both sets $\left\{\hat{A}_\mu\right\}$ and $\left\{\hat{L}_\mu\right\}$ are orthonormal, there exits a unitary matrix U such that

$$\hat{A}_\nu = \sum_\mu \hat{L}_\mu U_{\mu\nu} \tag{5.116}$$

for all $\nu = 1, 2, 3$. In turn, this implies that there exists a unitary operator $\hat{U} \in \mathcal{L}(\mathcal{V})$ such that

$$\hat{A}_\nu = \hat{U} \hat{L}_\nu \hat{U}^\dagger = \sum_\mu \hat{L}_\mu U_{\mu\nu} \tag{5.117}$$

for all $\nu = 1, 2, 3$. Putting (5.117) into the Kraus representation (5.113),

$$\hat{U}^\dagger \frac{d\hat{\rho}}{dt} \hat{U} = -i\hat{U}^\dagger[\hat{H}, \hat{\rho}]\hat{U} - \hat{U}^\dagger\{\hat{G}, \hat{\rho}\}\hat{U} + \sum_\mu \gamma_\mu \hat{L}_\mu^\dagger \hat{U}^\dagger \hat{\rho} \hat{U} \hat{L}_\mu^\dagger. \tag{5.118}$$

Finally, redefining the operators as[15]

$$\hat{\rho}' := \hat{U}^\dagger \hat{\rho} \hat{U}, \quad \hat{H}' := \hat{U}^\dagger \hat{\rho} \hat{U}, \quad \hat{G}' := \hat{U}^\dagger \hat{G} \hat{U} \tag{5.119}$$

---

[14] $\hat{S}^z$ makes a transition to the same state, but it induces different phase factors depending on the states.

[15] Note that $\hat{G}'$ equals to the expression in Eq. (5.112b).

and rescaling the parameters as

$$\Gamma_+ := \gamma_1, \quad \Gamma_- := \gamma_2, \quad \Gamma_- := \frac{1}{2}\gamma_3 \tag{5.120}$$

one arrives at the Lindblad equation of the form in (5.112) for $\hat{\rho}'$. Since $\hat{U}$ is a unitary transformation, it is nothing but a basis change, and $\hat{\rho}$ and $\hat{\rho}'$ are essentially the same. After solving the Lindblad equation, one can easily get $\hat{\rho}$ by applying the inverse transformation. In this sense, the Lindblad equation in (5.112) is the most general form for a single-qubit system.

5.4.3 Solution Methods

The Lindblad equation is a linear equation without an explicit time dependence, and it can always be solved by means of common methods for linear differential equations. More explicitly, in the standard basis, the Lindblad equation can be written as

$$\dot{\rho}_{jk} = \sum_{j'k'} M_{jk;j'k'} \rho_{j'k'}, \tag{5.121}$$

with

$$M_{jk;j'k'} := i(H_{jj'}\delta_{kk'} - \delta_{jj'}H^*_{kk'}) - (G_{jj'}\delta_{kk'} + \delta_{jj'}G^*_{kk'}) + \sum_\mu L_{\mu;jj'}L^*_{\mu;kk'} \tag{5.122}$$

Regarding $\mu := (jk)$ and $\nu := (j'k')$ as collective indices, Eq. (5.121) reads as

$$\dot{\rho}_\mu = \sum_\nu M_{\mu\nu}\rho_\nu, \tag{5.123}$$

which is a typical first-order differential equation for the column vector ρ_μ.

Technically, the differential equation (5.123) is not adequate yet to solve because of the conditions that $\rho_{jk} = \rho^*_{kj}$ and that $\sum_j \rho_{jj} = 1$. The latter condition is reflected in the fact that the determinant of the matrix M is always zero. For example, consider a single-qubit system. Suppose that the Lindblad equation is characterized by the effective Hamiltonian

$$\hat{H} = \frac{1}{2}\Omega\hat{S}^z + \frac{1}{2}\Delta\hat{S}^x \tag{5.124}$$

and the Lindblad operators $\sqrt{\Gamma_\pm}\hat{S}^\pm$. In the matrix form, the Lindblad equation is given by

$$
\frac{d}{dt}
\begin{bmatrix} \rho_{11} \\ \rho_{12} \\ \rho_{21} \\ \rho_{22} \end{bmatrix}
=
\begin{bmatrix}
-\Gamma_- & \frac{i\Delta}{2} & -\frac{i\Delta}{2} & \Gamma_+ \\
\frac{i\Delta}{2} & -\frac{\Gamma_-}{2} - \frac{\Gamma_+}{2} - i\Omega & 0 & -\frac{i\Delta}{2} \\
-\frac{i\Delta}{2} & 0 & -\frac{\Gamma_-}{2} - \frac{\Gamma_+}{2} + i\Omega & \frac{i\Delta}{2} \\
\Gamma_- & -\frac{i\Delta}{2} & \frac{i\Delta}{2} & -\Gamma_+
\end{bmatrix}
\begin{bmatrix} \rho_{11} \\ \rho_{12} \\ \rho_{21} \\ \rho_{22} \end{bmatrix}
\tag{5.125}
$$

Imposing the conditions, $\rho_{11} + \rho_{22} = 1$ and $\rho_{12} = \rho_{21}^*$, the above equation reads as

$$
\frac{d}{dt}
\begin{bmatrix} \rho_{11} \\ \mathrm{Re}\,\rho_{21} \\ \mathrm{Im}\,\rho_{21} \end{bmatrix}
=
\begin{bmatrix}
-\Gamma & 0 & 0 \\
0 & -\Gamma/2 & -i\Gamma/2 \\
-i\Delta & -i\Omega & -\Omega
\end{bmatrix}
\begin{bmatrix} \rho_{11} \\ \mathrm{Re}\,\rho_{21} \\ \mathrm{Im}\,\rho_{21} \end{bmatrix}
+
\begin{bmatrix} \Gamma_+ \\ 0 \\ \Delta/2 \end{bmatrix},
\tag{5.126}
$$

where $\Gamma := \Gamma_+ + \Gamma_-$. It is a typical inhomogeneous first-order differential equation and can be solved using the standard methods. If necessary, various numerical methods can also be applied. This method is extensively discussed in Blum (2012).

In the above discussion, the constraints $\hat{\rho}^\dagger = \hat{\rho}$ and $\mathrm{Tr}\,\hat{\rho} = 1$ have been handled in an ad hoc fashion. They can be dealt with in a systematic way by choosing an appropriate orthonormal basis $\left\{ \hat{B}_\mu : \mu = 0, 1, 2, \ldots, n^2 - 1 \right\}$ for $\mathcal{L}(\mathcal{V})$, where n is the dimension of the vector space $\mathcal{V}$, such that (i) $\hat{B}_0 = \hat{I}/\sqrt{n}$ and (ii) $\hat{B}_\mu^\dagger = \hat{B}_\mu$. Note that the condition (i) implies that the rest elements of the basis are all traceless— $\mathrm{Tr}\,\hat{B}_\mu = 0$ for $\mu = 1, 2, \ldots, n^2 - 1$. We call a basis a *Lindblad basis*. For example, the following operators form a Lindblad basis for a three-level atom:

$$
\frac{1}{\sqrt{3}}\begin{bmatrix} 1 & 0 & 0 \\ 0 & 1 & 0 \\ 0 & 0 & 1 \end{bmatrix},\quad
\frac{1}{\sqrt{6}}\begin{bmatrix} 1 & 0 & 0 \\ 0 & 1 & 0 \\ 0 & 0 & -2 \end{bmatrix},\quad
\frac{1}{\sqrt{2}}\begin{bmatrix} 1 & 0 & 0 \\ 0 & -1 & 0 \\ 0 & 0 & 0 \end{bmatrix},
$$

$$
\frac{1}{\sqrt{2}}\begin{bmatrix} 0 & 1 & 0 \\ 1 & 0 & 0 \\ 0 & 0 & 0 \end{bmatrix},\quad
\frac{1}{\sqrt{2}}\begin{bmatrix} 0 & 0 & 1 \\ 0 & 0 & 0 \\ 1 & 0 & 0 \end{bmatrix},\quad
\frac{1}{\sqrt{2}}\begin{bmatrix} 0 & 0 & 0 \\ 0 & 0 & 1 \\ 0 & 1 & 0 \end{bmatrix},
$$

$$
\frac{1}{\sqrt{2}}\begin{bmatrix} 0 & -i & 0 \\ i & 0 & 0 \\ 0 & 0 & 0 \end{bmatrix},\quad
\frac{1}{\sqrt{2}}\begin{bmatrix} 0 & 0 & -i \\ 0 & 0 & 0 \\ i & 0 & 0 \end{bmatrix},\quad
\frac{1}{\sqrt{2}}\begin{bmatrix} 0 & 0 & 0 \\ 0 & 0 & -i \\ 0 & i & 0 \end{bmatrix}.
\tag{5.127}
$$

Let us examine how various quantities are represented in a Lindblad basis: The components of a density operator $\hat{\rho}$ in the expansion

$$
\hat{\rho} = \sum_{\mu=0}^{n^2-1} \hat{B}_\mu \rho_\mu
\tag{5.128}
$$

are given by $\rho_\mu := \mathrm{Tr}\,\hat{B}_\mu \hat{\rho}$, and they are all real. In particular, $\rho_0 = 1/\sqrt{n}$. More importantly, the Lindblad equation now reads as

$$\dot{\rho}_\mu = \sum_{\nu=1}^{n^2-1} K_{\mu\nu}\rho_\nu + b_\mu \quad (\mu = 1, 2, \ldots, n^2 - 1), \tag{5.129}$$

with the generator matrix K given by

$$K_{\mu\nu} := \mathrm{Tr}\, \hat{B}_\mu^\dagger \mathscr{L}(\hat{B}_\nu), \tag{5.130}$$

where $\mathscr{L}$ is the Lindblad generator in (5.88) or (5.88′), and the inhomogeneous term given by

$$b_\mu := \frac{1}{\sqrt{n}} \mathrm{Tr}\, \hat{B}_\mu^\dagger \mathscr{L}(\hat{B}_0). \tag{5.131}$$

The inhomogeneous differential equation (5.129) has the solution of the form

$$\rho_\mu(t) = \sum_{\nu=1}^{n^2-1} \left[e^{Kt} \right]_{\mu\nu} \rho_\nu(0) + \sum_\nu \left[\int_0^t ds\, e^{Ks} \right]_{\mu\nu} b_\nu. \tag{5.132}$$

Putting this solution for coefficients $\rho_\mu(t)$ back to (5.128) gives the solution $\hat{\rho}(t)$. This method is completely general. Any Lindblad equation can be solved this way, numerically if necessary.

Let us consider a single qubit, and examine a master equation. We assume that the system is initially prepared in a pure state $\left(|0\rangle + |1\rangle\right)/\sqrt{2}$.

```
In[ ]:= init = (1 + S[1]) / 2;
        init // MatrixForm
```
Out[ ]//MatrixForm=
$$\frac{1}{2} \cdot (1 + S^x)$$

This is the effective Hamiltonian.

```
opH = S[3];
```

These are the Lindblad operators.

```
In[ ]:= Let[Real, Γ]
        opL = {Sqrt[Γ["+"]] × S[4], Sqrt[Γ["-"]] × S[5]}
```
Out[ ]= $\{ S^+ \sqrt{\Gamma_+}, S^- \sqrt{\Gamma_-} \}$

This is the Lindblad basis we are going to use. It happens to be equivalent to the basis consisting of the Pauli operators.

```
In[ ]:= lbs = Elaborate@LindbladBasis[S]
        MatrixForm /@ (Matrix[#, S] &) /@ lbs
```
Out[ ]= $\left\{ \dfrac{1}{\sqrt{2}}, \dfrac{S^z}{\sqrt{2}}, \dfrac{S^x}{\sqrt{2}}, \dfrac{S^y}{\sqrt{2}} \right\}$

Out[ ]= $\left\{ \begin{pmatrix} \frac{1}{\sqrt{2}} & 0 \\ 0 & \frac{1}{\sqrt{2}} \end{pmatrix}, \begin{pmatrix} \frac{1}{\sqrt{2}} & 0 \\ 0 & -\frac{1}{\sqrt{2}} \end{pmatrix}, \begin{pmatrix} 0 & \frac{1}{\sqrt{2}} \\ \frac{1}{\sqrt{2}} & 0 \end{pmatrix}, \begin{pmatrix} 0 & -\frac{i}{\sqrt{2}} \\ \frac{i}{\sqrt{2}} & 0 \end{pmatrix} \right\}$

Here are the generator matrix K and the inhomogeneous term when the Lindblad equation is written in the standard form of an inhomogeneous first-order differential equation.

```
In[ ]:= {mat, vec} = LindbladConvert[opH, opL];
        mat // MatrixForm
        vec // MatrixForm
```

Out[ ]//MatrixForm=

$$\begin{pmatrix} -\Gamma_- - \Gamma_+ & 0 & 0 \\ 0 & \frac{1}{2} \times \left(-2\,\dot{\imath} - \frac{\Gamma_-}{2} - \frac{\Gamma_+}{2}\right) + \frac{1}{2} \times \left(2\,\dot{\imath} - \frac{\Gamma_-}{2} - \frac{\Gamma_+}{2}\right) & -\frac{1}{2}\,\dot{\imath}\,\left(-2\,\dot{\imath} - \frac{\Gamma_-}{2} - \frac{\Gamma_+}{2}\right) + \frac{1}{2}\,\dot{\imath}\,\left(2\,\dot{\imath} - \frac{\Gamma_-}{2} - \frac{\Gamma_+}{2}\right) \\ 0 & \frac{1}{2}\,\dot{\imath}\,\left(-2\,\dot{\imath} - \frac{\Gamma_-}{2} - \frac{\Gamma_+}{2}\right) - \frac{1}{2}\,\dot{\imath}\,\left(2\,\dot{\imath} - \frac{\Gamma_-}{2} - \frac{\Gamma_+}{2}\right) & \frac{1}{2} \times \left(-2\,\dot{\imath} - \frac{\Gamma_-}{2} - \frac{\Gamma_+}{2}\right) + \frac{1}{2} \times \left(2\,\dot{\imath} - \frac{\Gamma_-}{2} - \frac{\Gamma_+}{2}\right) \end{pmatrix}$$

Out[ ]//MatrixForm=

$$\begin{pmatrix} \frac{-\Gamma_- + \Gamma_+}{\sqrt{2}} \\ 0 \\ 0 \end{pmatrix}$$

This solves the differential equation based on the generator matrix K and the inhomogeneous term.

```
In[ ]:= Clear[ρ]
        ρ[t_] = Block[
            {Γ, ρ},
            Γ["+"] = .3;
            Γ["-"] = .7;
            Elaborate@LindbladSolve[opH, opL, init, t] // Chop
            ];
        ρ[t] // MatrixForm
```

Out[ ]//MatrixForm=

$$0.5 + 0.5\,e^{-0.5\,t}\,\mathrm{Cos}[2.\,t]\,S^x + \left(-0.2 + 0.2\,e^{-1.\,t}\right)\,S^z + 0.5\,e^{-0.5\,t}\,S^y\,\mathrm{Sin}[2.\,t]$$

To investigate the physical properties of the solution, we calculate the expectation values of the Pauli operators.

```
In[ ]:= {avgX[t_], avgY[t_], avgZ[t_]} = Coefficient[ρ[t], #] & /@ S@{1, 2, 3}
```

Out[ ]= $\left\{0.5\,e^{-0.5\,t}\,\mathrm{Cos}[2.\,t],\ 0.5\,e^{-0.5\,t}\,\mathrm{Sin}[2.\,t],\ -0.2 + 0.2\,e^{-1.\,t}\right\}$

```
In[ ]:= data = Transpose@Table[
            {{t, avgX[t]}, {t, avgY[t]}, {t, avgZ[t]}},
            {t, 0, 10, 0.1}
            ];
        ListLinePlot[data,
         FrameLabel → {"B_z t", "⟨X⟩, ⟨Y⟩, ⟨Z⟩"}
        ]
```

Out[ ]=

Consider a three-level atom with the Λ-type level structure.

```
Let[Qudit, A]
```

In the interaction picture, the Hamiltonian looks like this. We have put the two Rabi transition amplitudes to 1.

```
In[·]:= opH = (1 / 10) A[1 → 1] + (2 / 10) A[2 → 2] + A[2 → 0] + A[0 → 2] + A[2 → 1] + A[1 → 2];
       matH = Matrix[opH];
       matH // MatrixForm
```

$$
\begin{pmatrix} 0 & 0 & 1 \\ 0 & \frac{1}{10} & 1 \\ 1 & 1 & \frac{1}{5} \end{pmatrix}
$$

```
In[·]:= Let[Real, Γ]
       opL = {
         Sqrt[Γ[0, "-"]] * A[2 → 0],
         Sqrt[Γ[0, "+"]] * A[0 → 2],
         Sqrt[Γ[1, "-"]] * A[2 → 1],
         Sqrt[Γ[1, "+"]] * A[1 → 2]}
       matL = Matrix[opL];
       MatrixForm /@ matL
```

$$
Out[·]= \left\{ (|0\rangle\langle2|)\sqrt{\Gamma_{0,-}}, \ (|2\rangle\langle0|)\sqrt{\Gamma_{0,+}}, \ (|1\rangle\langle2|)\sqrt{\Gamma_{1,-}}, \ (|2\rangle\langle1|)\sqrt{\Gamma_{1,+}} \right\}
$$

$$
Out[·]= \left\{ \begin{pmatrix} 0 & 0 & \sqrt{\Gamma_{0,-}} \\ 0 & 0 & 0 \\ 0 & 0 & 0 \end{pmatrix}, \begin{pmatrix} 0 & 0 & 0 \\ 0 & 0 & 0 \\ \sqrt{\Gamma_{0,+}} & 0 & 0 \end{pmatrix}, \begin{pmatrix} 0 & 0 & 0 \\ 0 & 0 & \sqrt{\Gamma_{1,-}} \\ 0 & 0 & 0 \end{pmatrix}, \begin{pmatrix} 0 & 0 & 0 \\ 0 & 0 & 0 \\ 0 & \sqrt{\Gamma_{1,+}} & 0 \end{pmatrix} \right\}
$$

```
In[·]:= Timing[
         ρ[t_] = Block[
           {Γ, init},
           Γ[0, "-"] = 0.05;
           Γ[0, "+"] = 0.01;
           Γ[1, "-"] = 0.025;
           Γ[1, "+"] = 0.005;
           init = DiagonalMatrix[{0, 1, 0}];
           LindbladSolve[matH, matL, init, t]
           ];
         ]
```

```
Out[·]= {3.23239, Null}
```

```
In[·]:= Plot[Evaluate@Diagonal@ρ[t Γi], {t, 0, 10},
         FrameLabel → {"Ω t / π", "Probabilities"},
         PlotRange → All,
         PlotLegends → Automatic]
```

There are two drawbacks in the above approach: First, the size of the generator matrix K increases exponentially with the system size, and in many cases, even numerical methods become intractable. In such cases, the quantum jump approach mentioned earlier—Sect. 5.4—is often used. Second, the resulting solution is given in an explicit matrix representation, and putting the solution in the Kraus representation is tedious or even impractical in many case. This means that for a physical interpretation of the solution, one needs additional analysis specific to the system or situation.

Interestingly, there is a special class of Lindblad equations that allow for a solution directly in the Kraus representation (Nakazato et al., 2006): Let $\{|j\rangle : j = 0, \ldots, n-1\}$ be the eigenbasis from the effective Hamiltonian $\hat{H}$ so that

$$\hat{H} = \sum_j E_j |j\rangle \langle j| . \tag{5.133}$$

We consider a Lindblad equation of the form

$$\frac{d\hat{\rho}}{dt} = -i[\hat{H}, \hat{\rho}] - \{\hat{G}, \hat{\rho}\} + \sum_{jk} \gamma_{jk} \hat{L}_{jk} \hat{\rho} \hat{L}_{jk}^\dagger , \tag{5.134}$$

where every (normalized) quantum jump operator $\hat{L}_{jk}$ corresponds to an incoherent transition between a pair of eigenstates, $\hat{L}_{jk} := |j\rangle \langle k|$, and γ_{jk} characterizes the rate of the process. Note here that the effective damping operator $\hat{G}$,

$$\hat{G} := \frac{1}{2} \sum_{jk} \gamma_{jk} \hat{L}_{jk}^\dagger \hat{L}_{jk} = \frac{1}{2} \sum_k \gamma_k |k\rangle \langle k| \tag{5.135}$$

with

$$\gamma_k := \sum_j \gamma_{jk}, \tag{5.136}$$

commutes with the effective Hamiltonian $\hat{H}$.

To solve the Lindblad equation (5.134), we first recall that the non-unitary evolution in Eq. (5.90) governed by the non-Hermitian Hamiltonian in Eq. (5.91) corresponds to the solution in the absence of the quantum jump operators ($\gamma_{jk} = 0$). It is therefore natural to define the *generalized interaction picture*

$$\hat{\rho}_I(t) := e^{it\hat{H}_{\text{non}}} \hat{\rho}(t) e^{-it\hat{H}_{\text{non}}^\dagger} \tag{5.137}$$

with respect to the non-Hermitian Hamiltonian $\hat{H}_{\text{non}} = \hat{H} - i\hat{G}$. In this interaction picture, the Lindblad equation reads as

$$\frac{d\hat{\rho}_I}{dt} = \sum_{jk} \gamma_{jk} e^{(\gamma_j - \gamma_k)t} \hat{L}_{jk} \hat{\rho}_I \hat{L}_{jk}^\dagger = \sum_{jk} \gamma_{jk} \hat{R}_{jk}(t) , \tag{5.138}$$

where we have defined

$$\hat{R}_{jk}(t) := e^{(\gamma_j - \gamma_k)t} \hat{L}_{jk} \hat{\rho}_I(t) \hat{L}_{jk}^\dagger . \tag{5.139}$$

Since the differential equation (5.138) is equivalent to the integral equation

$$\hat{\rho}_I(t) = \hat{\rho}_I(0) + \sum_{jk} \gamma_{jk} \int_0^t ds \, \hat{R}_{jk}(s) , \tag{5.140}$$

it is now a matter of calculating the new operator $\hat{R}_{jk}(t)$. It follows from (5.138) that the operator $\hat{R}_{jk}(t)$ satisfies the differential equation

$$\frac{d\hat{R}_{jk}}{dt} = \sum_l \Gamma_{kl}^{(j)} \hat{R}_{jl} \tag{5.141}$$

with the matrix $\Gamma^{(j)}$ defined by $\Gamma_{kl}^{(j)} := \delta_{kl}(\gamma_j - \gamma_k) + \gamma_{kl}$. The solution is given by

$$\hat{R}_{jk}(t) = \sum_l \left[e^{t\Gamma^{(j)}} \right]_{kl} \hat{R}_{jl}(0) . \tag{5.142}$$

Putting it back into the integral equation (5.140), we finally obtain

$$\hat{\rho}_I(t) = \hat{\rho}(0) + \sum_{jkl} \gamma_{jk} W_{kl}^{(j)}(t) \hat{L}_{jl} \hat{\rho}(0) \hat{L}_{jl}^\dagger \tag{5.143}$$

with

$$W^{(j)}(s) := \int_0^t ds \, e^{s\Gamma^{(j)}} . \tag{5.144}$$

More explicitly, for the purpose of illustration, the Kraus representation of $\hat{\rho}(t)$ reads as

$$\hat{\rho}(t) = \hat{F}_0(t) \hat{\rho}(0) \hat{F}_0^\dagger(t) + \sum_{jk} \hat{F}_{jk}(t) \hat{\rho}(0) \hat{F}_{jk}^\dagger(t) , \tag{5.145a}$$

where the Kraus elements are given by

$$\hat{F}_0(t) := e^{-it\hat{H}_{\text{non}}} , \quad \hat{F}_{jk}(t) := e^{-it\hat{H}_{\text{non}}} |j\rangle \sqrt{\sum_l \gamma_{jl} W_{lk}^{(j)}(t)} \langle k| . \tag{5.145b}$$

5.4.4 Examples Revisited

Here we consider again the limiting cases discussed in Sects. 5.2.4 and 5.4.2. We will use the method discussed above to solve the Lindblad equations that have been derived in Sect. 5.4.2. From the solutions, we can recover the descriptions in terms of quantum operations in Sect. 5.2.4.

Note that for a single qubit, not surprisingly, the Lindblad basis is given by the Pauli operators. In other words, we expand a density operator $\hat{\rho}$ into

$$\hat{\rho}(t) = \frac{1}{2}\hat{I} + \hat{S}^x \rho_x(t) + \hat{S}^y \rho_y(t) + \hat{S}^z \rho_z(t). \qquad (5.146)$$

Phase damping The Lindblad equation in this case is given by [see also Eq. (5.104)]

$$\frac{d\hat{\rho}}{dt} = \frac{\gamma}{8}\hat{\rho}(t) + \frac{\gamma}{4}\hat{S}^z \hat{\rho}(t)\hat{S}^z . \qquad (5.147)$$

Rewriting the above equation in the Lindblad basis, we have a set of differential equations for the components ρ_μ ($\mu = x, y, z$) of the density operator in Eq. (5.146) as follows

$$\frac{d}{dt}\begin{bmatrix} \rho_x(t) \\ \rho_y(t) \\ \rho_z(t) \end{bmatrix} = -\frac{1}{2}\gamma \begin{bmatrix} 1 & & \\ & 1 & \\ & & 0 \end{bmatrix}\begin{bmatrix} \rho_x(t) \\ \rho_y(t) \\ \rho_z(t) \end{bmatrix}. \qquad (5.148)$$

The equation is diagonal and straight forward to solve it to get

$$\rho_x(t) = e^{-\gamma t/2}\rho_x(0), \quad \rho_y(t) = e^{-\gamma t/2}\rho_y(0), \quad \rho_z(t) = \rho_z(0). \qquad (5.149)$$

Note that the component $\rho_z(t)$ is constant. Again, this is because the phase damping process does not involves any transition between levels. It only causes pure dephasing.

Consider the phase damping process.

It is governed by a single parameter.

```
Let[Real, γ]
```

The Lindblad generator is specified by a single Lindblad operator. Note that in this case, the effective Hamiltonian vanishes.

```
ops = {Sqrt[γ] / 2 × S[3]};
gnr = LindbladGenerator[0, ops]
```

$$\text{LindbladGenerator}\left[\left\{0, \frac{\sqrt{\gamma}\, S^z}{2}\right\}\right]$$

We assume that the system is initially prepared in a pure state.

```
vec = EulerRotation[{0, Pi / 4, 0}, S] ** Ket[]
init = ExpressionFor[Matrix@Dyad[vec, vec], S] // Elaborate;
```

$$\text{Cos}\left[\frac{\pi}{8}\right]|{\downarrow}\rangle + |1_S\rangle \sin\left[\frac{\pi}{8}\right]$$

This is the solution of the Lindblad equation.

```
Clear[rho]
rho[t_] = LindbladSolve[0, ops, init, t]
```

$$\frac{1}{2} + \frac{S^z}{2\sqrt{2}} + \frac{e^{-\frac{t\gamma}{2}} S^+}{2\sqrt{2}} + \frac{e^{-\frac{t\gamma}{2}} S^-}{2\sqrt{2}}$$

As you can see, the time dependence is governed by γt. This indicates that the system has a single time scale, and it is governed by γ. It is thus convenient to use $1/\gamma$ as the unit of time. It is equivalent to put γ=1.

```
γ = 1;
```

```
In[•]:= data = Table[Chop@BlochVector[rho[s Pi]], {s, 0, 3, 0.1}];
       BlochSphere[{Red, Bead /@ data}]
```

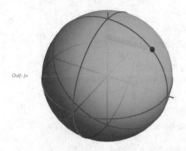

Out[•]=

Amplitude damping Next we consider the amplitude damping process. The Lindblad equation is given by [see (5.107)]

$$\frac{\hat{\rho}}{dt} = \frac{\gamma}{4}\left\{(1 - \hat{S}^z)\hat{\rho}(t) + \hat{\rho}(t)(1 - \hat{S}^z)\right\} + \gamma\hat{S}^+\hat{\rho}(t)\hat{S}^-.\qquad(5.150)$$

In the Lindblad basis, the above equation reads as

$$\frac{d}{dt}\begin{bmatrix} \rho_x(t) \\ \rho_y(t) \\ \rho_z(t) \end{bmatrix} = -\frac{\gamma}{2}\begin{bmatrix} 1 & & \\ & 1 & \\ & & 2 \end{bmatrix}\begin{bmatrix} \rho_x(t) \\ \rho_y(t) \\ \rho_z(t) \end{bmatrix} + \frac{\gamma}{2}\begin{bmatrix} 0 \\ 0 \\ \sqrt{2} \end{bmatrix}.\qquad(5.151)$$

As for the phase damping case, the equation is diagonal. But there is an inhomogeneous term. The solutions are given by

$$\begin{aligned} \rho_x(t) &= e^{-\gamma t/2}\rho_x(0), \\ \rho_y(t) &= e^{-\gamma t/2}\rho_y(0), \\ \rho_z(t) &= e^{-\gamma t/2}\rho_z(0) + \frac{1}{2}(1 - e^{-\gamma t/2}). \end{aligned}\qquad(5.152)$$

We note that

$$\hat{\rho}(t) \to \frac{1}{2}\hat{I} + \frac{1}{2}\hat{S}^z = |0\rangle\langle0|\qquad(5.153)$$

as $t \to \infty$. That is, if we wait long enough, the system definitely goes to the pure state $|0\rangle$. This is due to the relaxation from $|1\rangle$ to $|0\rangle$.

Consider the amplitude damping process.

It is governed by a single parameter.

```
Let[Real, γ]
```

The Lindblad generator is specified by a single Lindblad operator. In this case, the effective Hamiltonian vanishes.

```
In[·]:= ops = {Sqrt[γ] * S[4]};
       gnr = LindbladGenerator[0, ops]
```
$$Out[·]= \text{LindbladGenerator}\left[\left\{0, \sqrt{\gamma}\ S^{+}\right\}\right]$$

We assume that the system is initially prepared in a pure state.

```
In[·]:= vec = EulerRotation[{0, Pi / 4, 0}, S] ** Ket[]
       init = ExpressionFor[Matrix@Dyad[vec, vec], S] // Elaborate;
```
$$Out[·]= \text{Cos}\left[\frac{\pi}{8}\right]\ |{\cdots}\rangle + |1_S\rangle\ \text{Sin}\left[\frac{\pi}{8}\right]$$

This is the solution of the Lindblad equation.

```
In[·]:= rho[t_] = LindbladSolve[0, ops, init, t]
```
$$Out[·]= \frac{1}{2} + \frac{1}{4}\ e^{-t\gamma}\left(-2 + \sqrt{2} + 2\ e^{t\gamma}\right) S^z + \frac{e^{-\frac{t\gamma}{2}}\ S^{+}}{2\sqrt{2}} + \frac{e^{-\frac{t\gamma}{2}}\ S^{-}}{2\sqrt{2}}$$

As you can see, the time dependence is governed by γt. This indicates that the system has a single time scale, and it is governed by γ. It is thus convenient to use $1/\gamma$ as the unit of time. It is equivalent to put γ=1.

```
γ = 1;
```

```
In[·]:= data = Table[Chop@BlochVector[rho[s Pi]], {s, 0, 3, 0.1}];
       BlochSphere[{Red, Bead /@ data}]
```

$Out[·]=$

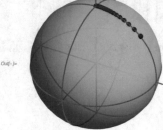

Depolarizing Finally, we examine the depolarizing process. The Lindblad equation is given by [see Eq. (5.111)]

$$\frac{d\hat{\rho}}{dt} = \frac{\gamma}{2}\hat{\rho}(t) + \frac{\gamma}{3}\sum_{\mu=x,y,z}\hat{S}^{\mu}\hat{\rho}(t)\hat{S}^{\mu}. \tag{5.154}$$

The corresponding equation for the components $\rho_{\mu}(t)$ is given by

$$\frac{d}{dt} \begin{bmatrix} \rho_x(t) \\ \rho_y(t) \\ \rho_z(t) \end{bmatrix} = -\frac{4\gamma}{3} \begin{bmatrix} 1 & & \\ & 1 & \\ & & 1 \end{bmatrix} \begin{bmatrix} \rho_x(t) \\ \rho_y(t) \\ \rho_z(t) \end{bmatrix}. \tag{5.155}$$

The solution to the above equation indicates that every component vanishes exponentially

$$\rho_\mu(t) = e^{-4\gamma t/3} \rho_\mu(0). \tag{5.156}$$

In the long time limit ($t \to \infty$), the density operator becomes completely random,

$$\hat{\rho}(t) \to \frac{1}{2}\hat{I}. \tag{5.157}$$

This is the expected characteristic of the depolarizing process.

Consider the depolarizing process.

It is governed by a single parameter.

```
Let[Real, γ]
```

The Lindblad generator is specified by three Lindblad operators. In this case, the effective Hamiltonian vanishes.

```
In[·]:= ops = Sqrt[γ / 3] * S[All];
gnr = LindbladGenerator[0, ops]
```

$$Out[\cdot]= \text{LindbladGenerator}\left[\left\{0, \frac{\sqrt{\gamma}\, S^x}{\sqrt{3}}, \frac{\sqrt{\gamma}\, S^y}{\sqrt{3}}, \frac{\sqrt{\gamma}\, S^z}{\sqrt{3}}\right\}\right]$$

We assume that the system is initially prepared in a pure state.

```
In[·]:= vec = EulerRotation[{0, Pi / 4, 0}, S] ** Ket[]
init = ExpressionFor[Matrix@Dyad[vec, vec], S] // Elaborate;
```

$$Out[\cdot]= \text{Cos}\left[\frac{\pi}{8}\right] |{-}\rangle + |1_s\rangle \, \text{Sin}\left[\frac{\pi}{8}\right]$$

This is the solution of the Lindblad equation.

```
In[·]:= rho[t_] = LindbladSolve[0, ops, init, t]
```

$$Out[\cdot]= \frac{1}{2} + \frac{e^{-\frac{4t\gamma}{3}} S^z}{2\sqrt{2}} + \frac{e^{-\frac{4t\gamma}{3}} S^+}{2\sqrt{2}} + \frac{e^{-\frac{4t\gamma}{3}} S^-}{2\sqrt{2}}$$

As you can see, the time dependence is governed by γt. This indicates that the system has a single time scale, and it is governed by γ. It is thus convenient to use $1/\gamma$ as the unit of time. It is equivalent to put $\gamma{=}1$.

```
γ = 1;
```

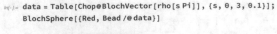

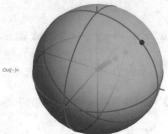

5.5   Distance Between Quantum States

Here we change the subject slightly and bring into focus the question, how close (or different) two quantum states are. We have seen quantum states undergo various operations, unitary or non-unitary, intended or incidental. It is inevitable to quantitatively compare the quantum states before and after those operations to assess their effects. It is easy to imagine more reasons in various other contexts such as quantum communication.

We introduce three methods of quantifying how close two quantum states are. The first two, the Hilbert–Schmidt and trace distances, are *metrics*, a mathematical notion of "distance". It sounds natural to say that two states are closer when the distance between them are smaller. The third, fidelity, is conceptually a relative angle. Since quantum states are normalized, it seems reasonable to regard two quantum states closer when the relative angle is smaller. Fidelity is not a metric in the precise mathematical sense and yet well describes the "closeness" of quantum states in many respects.

In quantum information, the trace distance appears far more frequently than the Hilbert–Schmidt distance. Maybe it is because the classical version of trace distance is popular in statistics. However, there seems to be no profound reason to prefer one to the other in general. Both share many properties in common. One is more convenient in one situation but less in others. It is also heuristic to compare them closely with one another. In this book, we thus discuss them both in parallel.

5.5.1   Norms and Distances

Before we discuss the two metric-based closeness measures, we first recall the relation between Hermitian product, norm, and distance. It will serve as an outline for our discussions on the Hilbert–Schmidt and trace distances.

Any Hermitian product $\langle \cdot, \cdot \rangle$ (Definition A.4) on a vector space gives the notion of *magnitude* or *norm*, defined by

$$\|v\| := \sqrt{\langle v, v \rangle} \tag{5.158}$$

for a vector v. It is called the *canonical norm* associated with the Hermitian product (Appendix A.1.2). Of course, one can define a norm independent of the given Hermitian product as well.

Given a norm, one can then measure the distance $D(v, w)$ between two vectors v and w by the norm (magnitude) of the difference $v - w$. In other words,

$$D(v, w) := \|v - w\|. \tag{5.159}$$

Here we have deliberately avoided to use the bra-ket notation since we will mainly apply these notions to operators (yet regarded as vectors, see Appendix B.1) rather than to state vectors. Why?

For a state vector $|v\rangle$, we have the canonical norm

$$\| |v\rangle \| := \sqrt{\langle v|v \rangle} \tag{5.160}$$

associated with the standard Hermitian product. Then it might be tempting to measure the closeness of $|v\rangle$ and $|w\rangle$ by the distance $\| |v\rangle - |w\rangle \|$. However, there is a serious drawback, and it is not appropriate: Consider two state vectors $|v\rangle$ and $|w\rangle = |v\rangle \, e^{i\phi}$. They represent the same quantum state, whereas the distance between them is finite,

$$\| |v\rangle - |v\rangle \, e^{i\phi} \| = |1 - e^{i\phi}| \neq 0. \tag{5.161}$$

Therefore, $\| |v\rangle - |w\rangle \|$ cannot measure properly how close two quantum states $|v\rangle$ and $|w\rangle$ are. Another problem is that the distance $\| |v\rangle - |w\rangle \|$ is defined only for pure states. No simple modification seems to provide a smooth crossover from pure states to mixed states.

To overcome these problems, we thus seek, from the outset, an operator norm. For a linear operator $\hat{A}$, one can choose the canonical norm

$$\|\hat{A}\| := \sqrt{\left\langle \hat{A}, \hat{A} \right\rangle} \tag{5.162}$$

associated with the given Hermitian product of operators, or any other norm. Then one measures the closeness of two quantum states $\hat{\rho}$ and $\hat{\sigma}$ by $\|\hat{\rho} - \hat{\sigma}\|$. For pure states $|v\rangle$ and $|w\rangle$, the relevant distance is $\| |v\rangle \langle v| - |w\rangle \langle w| \|$. With this measure, the distance between the two states $|v\rangle$ and $|w\rangle = |v\rangle \, e^{i\phi}$ vanishes,

$$\| |v\rangle \langle v| - |v\rangle \, e^{i\phi} e^{-i\phi} \langle v| \| = \| |v\rangle \langle v| - |v\rangle \langle v| \| = 0, \tag{5.163}$$

as it should. Below we introduce two operator-norm-based distance measures.

5.5.2 Hilbert–Schmidt and Trace Norms

Here we introduce two operator norms and discuss their mathematical properties. We will discuss the physical properties and implications of the distances deduced from the corresponding norms in Sect. 5.5.3. If one is more interested in physical aspects, one can skip this subsection and jump to Sect. 5.5.3. Or, one can first study the physical implications there and then come back to investigate the details more thoroughly.

A natural choice of the norm of a linear operator $\hat{A}$ is

$$\|\hat{A}\|_{\mathrm{HS}} := \sqrt{\mathrm{Tr}\,\hat{A}^\dagger \hat{A}}\,. \tag{5.164}$$

The subscript 'HS' indicates that it is the canonical norm associated with the *Hilbert–Schmidt inner product* (Appendix B.1),

$$\left\langle \hat{A}, \hat{B} \right\rangle = \mathrm{Tr}\,\hat{A}^\dagger \hat{B}, \tag{5.165}$$

of two linear operators $\hat{A}$ and $\hat{B}$. If A is the matrix representation of $\hat{A}$ in a given basis, then one can evaluate $\|\hat{A}\|_{\mathrm{HS}}$ in terms of the matrix elements as

$$\|\hat{A}\|_{\mathrm{HS}} = \sqrt{\sum_{ij} A_{ij}^* A_{ij}} = \|(A_{11}, A_{12}, \ldots, A_{21}, A_{22}, \ldots)\|. \tag{5.166}$$

In other words, it equals the norm of the vector obtained by flattening the matrix A into a row (or column) vector.

Another norm we will consider is the *trace norm*. For a linear operator $\hat{A}$, it is defined by

$$\|\hat{A}\|_{\mathrm{tr}} := \mathrm{Tr}\,\sqrt{\hat{A}^\dagger \hat{A}} = \mathrm{Tr}\,\sqrt{\hat{A}\hat{A}^\dagger}. \tag{5.167}$$

Notice the difference between (5.164) and (5.167). Note further that the trace norm of an operator is larger than the magnitude of mere trace,

$$|\,\mathrm{Tr}\,\hat{A}\,| \leq \|\hat{A}\|_{\mathrm{tr}}. \tag{5.168}$$

For example, consider the Pauli operator $\hat{X}$ on a single qubit. Clearly, $\mathrm{Tr}\,\hat{X} = 0$. On the contrary, the trace norm is finite,

$$\|\hat{X}\|_{\mathrm{tr}} = \mathrm{Tr}\,\sqrt{\hat{X}^\dagger \hat{X}} = \mathrm{Tr}\,\sqrt{\hat{I}} = \mathrm{Tr}\,\hat{I} = 2. \tag{5.169}$$

The inequality is even more general, and one can show (Problem 5.7) that

$$|\operatorname{Tr} \hat{A}\hat{U}| \le \|\hat{A}\|_{\mathrm{tr}} \tag{5.170}$$

for any linear operator $\hat{A}$ and unitary operator $\hat{U}$.

For an operator of the form $\hat{A} = |v\rangle\langle v|$, both norms are reduced to the squared norm of the state vector $|v\rangle$,

$$\| \, |v\rangle\langle v| \, \|_{\mathrm{HS}} = \| \, |v\rangle\langle v| \, \|_{\mathrm{tr}} = \langle v|v\rangle = \| \, |v\rangle \, \|^2. \tag{5.171}$$

If $\hat{A}$ is Hermitian, one can show (Problem 5.8) by employing the spectral decomposition (Appendix A.4.1) that the Hilbert–Schmidt norm is the sum of the squared eigenvalues,

$$\|\hat{A}\|_{\mathrm{HS}} = \sqrt{\sum_j a_j^2}, \tag{5.172}$$

while the trace norm is the sum of the absolute values of the eigenvalues,

$$\|\hat{A}\|_{\mathrm{tr}} = \sum_j |a_j|. \tag{5.173}$$

For example, $\hat{X}$ has two eigenvalues ± 1, and hence

$$\|\hat{X}\|_{\mathrm{HS}} = \sqrt{1^2 + (-1)^2} = \sqrt{2}, \quad \|\hat{X}\|_{\mathrm{tr}} = 1 + |-1| = 2. \tag{5.174}$$

More generally, consider a Hermitian operator $\hat{A}$ on a single qubit. It takes the form

$$\hat{A} = a_0 \hat{I} + \boldsymbol{a} \cdot \hat{\boldsymbol{S}}, \tag{5.175}$$

where a_0 is a real number, and $\boldsymbol{a} := (a_x, a_y, a_x)$ a three-dimensional real vector, and $\hat{\boldsymbol{S}} := (\hat{S}^x, \hat{S}^y, \hat{S}^z)$ is the vector of Pauli operators. $\hat{A}$ has two eigenvalues $a_0 \pm \|\boldsymbol{a}\|$. Therefore, we have

$$\|\hat{A}\|_{\mathrm{HS}} = \sqrt{2(a_0^2 + \boldsymbol{a} \cdot \boldsymbol{a})} \tag{5.176}$$

and

$$\|\hat{A}\|_{\mathrm{tr}} = \left|a_0 + \sqrt{\boldsymbol{a} \cdot \boldsymbol{a}}\right| + \left|a_0 - \sqrt{\boldsymbol{a} \cdot \boldsymbol{a}}\right|. \tag{5.177}$$

Interestingly, for traceless Hermitian operators ($a_0 = 0$) on a qubit, they are the same up to a constant numerical factor,

$$\sqrt{2}\|\hat{A}\|_{\mathrm{HS}} = \|\hat{A}\|_{\mathrm{tr}} = 2\|\boldsymbol{a}\|. \tag{5.178}$$

For an arbitrary linear operator $\hat{A}$, singular-value decomposition (Theorem A.22) may be more convenient. We have the relations (Problem 5.8)

$$\|\hat{A}\|_{\mathrm{HS}} = \sqrt{\sum_j s_j^2}, \quad \|\hat{A}\|_{\mathrm{tr}} = \sum_j s_j, \tag{5.179}$$

where s_j are the singular values s_j of $\hat{A}$.

The Hilbert–Schmidt norm cannot exceed the trace norm (Problem 5.9),

$$\|\hat{A}\|_{\mathrm{HS}} \le \|\hat{A}\|_{\mathrm{tr}} \tag{5.180}$$

for all $\hat{A}$. Equality holds if and only if $\hat{A}$ has rank one ($\hat{A}$ takes the form $|v\rangle \langle v|$ for a vector $|v\rangle$). For example, observe that

$$\|\hat{X}\|_{\mathrm{HS}} = \sqrt{2} < \|\hat{X}\|_{\mathrm{tr}} = 2. \tag{5.181}$$

Let us compare the Hilbert-Schmidt and trace norms between quantum states.

```
$d = 3;
data = Table[op = RandomMatrix[$d];
    {HilbertSchmidtNorm[op], TraceNorm[op]}, {10 000}];
```

This illustrates that the Hilbert-Schmidt norm is never larger than the trace norm.

```
In[ ]:= Select[data, First[#] > Last[#] &]
Out[ ]= {}
```

```
In[ ]:= max = Max[data];
ListPlot[data, PlotRange → {{0, 1}, {0, 1}} * max,
  Epilog → Line@{{0, 0}, {1, 1} * max},
  FrameLabel → {"‖A‖_HS", "‖A‖_tr"}, AspectRatio → 1]
```

One striking difference between the Hilbert–Schmidt and trace norms is the effect of quantum operations. Needless to say, under unitary transformations, both are invariant,

$$\|\hat{U}\hat{A}\hat{U}^\dagger\|_{\mathrm{HS}} = \|\hat{A}\|_{\mathrm{HS}}, \quad \|\hat{U}\hat{A}\hat{U}^\dagger\|_{\mathrm{tr}} = \|\hat{A}\|_{\mathrm{tr}}. \tag{5.182}$$

However, the norms vary under general quantum operations. Interestingly, a quantum operation $\mathscr{F}$ never increases the trace norm,[16]

$$\|\mathscr{F}(\hat{A})\|_{\mathrm{tr}} \leq \|\hat{A}\|_{\mathrm{tr}} \tag{5.183}$$

for any linear operator $\hat{A}$. On the contrary, the Hilbert–Schmidt norm sometimes increases under quantum operations.[17] When applied for the distance measure of quantum states, this behavior has a significant physical implication as we will see soon below. This may give a reason to prefer the trace norm (or distance) to the Hilbert–Schmidt norm (or distance).

We examine the effects of quantum operations on the Hilbert-Schmidt and trace norms of operators.

We define a function to construct the Kraus elements from the unitary representation of the quantum operation. Here U is the unitary operator on the total system (system plus environment) and init is the initial (pure) state of the environment. The system and environment are assumed to be initially in a product state.

```
$d = 3;
Clear[kraus]
kraus[U_?MatrixQ, init_?VectorQ] := Map[kraus[U, init, #] &, Range[$d]]
kraus[U_?MatrixQ, init_?VectorQ, k_Integer] := PartialTrace[
   U.CircleTimes[One[$d], Dyad[init, UnitVector[$d, k]]],
   {$d, $d}, {2}];
```

First, we compare the Hilbert-Schmidt norm before and after quantum operations. Sometimes the Hilbert-Schmidt increases after a quantum operation, especially, for operators with large trace.

```
In[·]:= Timing[
   data = Table[
      U = RandomUnitary[$d * $d];
      vec = Normalize[RandomVector[$d]];
      spr = Supermap[kraus[U, vec]];
      mat = RandomReal[{-1, 1}] * One[$d] + 0.4 RandomHermitian[$d];
      {HilbertSchmidtNorm[mat], HilbertSchmidtNorm[spr[mat]]},
      {10 000}
      ];
   ]
Out[·]= {64.4099, Null}
```

---

[16] For a proof when $\hat{A}$ is a traceless Hermitian operator, see Nielsen & Chuang (2011). Pérez-García et al. (2006) provides a more general proof and discusses other norms as well.

[17] At first, it was conjectured that the Hilbert–Schmidt norm for traceless Hermitian operators does not increase under any completely positive trace-preserving supermap. Later, however, it was disproved in Ozawa (2000), where one can find a counter example.

```
In[·]:= max = Max[data];
       ListPlot[data, Epilog → Line@{{0, 0}, {1, 1} * max},
        PlotRange → {{0, 1}, {0, 1}} * max,
        FrameLabel → {"∥A∥_HS", "∥ℱ(A)∥_HS"},
        AspectRatio → 1]
```

Next, we compare the trace distance before and after quantum operations.

```
In[·]:= Timing[
        data = Table[
          U = RandomUnitary[$d * $d];
          vec = Normalize[RandomVector[$d]];
          spr = Supermap[kraus[U, vec]];
          mat = RandomMatrix[$d];
          {TraceNorm[mat], TraceNorm[spr[mat]]},
          {10 000}
         ];
       ]
Out[·]= {63.7983, Null}
```

Unlike the Hilbert-Schmidt norm, the trace norm never increases under a quantum operation.

```
In[·]:= max = Max[data];
       ListPlot[data, Epilog → Line@{{0, 0}, {1, 1} * max},
        PlotRange → {{0, 1}, {0, 1}} * max,
        FrameLabel → {"∥A∥_tr", "∥ℱ(A)∥_tr"},
        AspectRatio → 1]
```

Let us further examine more closely the case for traceless Hermitian operators, which are relevant when investigating the trace distances between mixed states.

```
In[ ]:= Timing[
         data = Table[
           U = RandomUnitary[$d * $d];
           vec = Normalize[RandomVector[$d]];
           spr = Supermap[kraus[U, vec]];
           mat = RandomHermitian[$d];
           mat = mat - Tr[mat] × One[$d];
           {TraceNorm[mat], TraceNorm[spr[mat]]},
           {10 000}
           ];
         ]
Out[ ]= {66.3332, Null}
```

As expected, the trace norm never increases.

```
In[ ]:= max = Max[data];
       ListPlot[data, Epilog → Line@{{0, 0}, {1, 1} * max},
         PlotRange → {{0, 1}, {0, 1}} * max,
         FrameLabel → {"∥A∥_tr", "∥𝓕(A)∥_tr"},
         AspectRatio → 1]
```

5.5.3 Hilbert–Schmidt and Trace Distances

In the previous subsection, we have introduced two frequently used operator norms, the Hilbert–Schmidt and trace norms, and discussed their mathematical properties. They naturally provide distance measures, the *Hilbert–Schmidt distance*

$$D_{\mathrm{HS}}(\hat{A}, \hat{B}) := \|\hat{A} - \hat{B}\|_{\mathrm{HS}} \tag{5.184}$$

and the *trace distance*

$$D_{\mathrm{tr}}(\hat{A}, \hat{B}) := \|\hat{A} - \hat{B}\|_{\mathrm{tr}}, \tag{5.185}$$

between linear operators on a given vector space. Here we are interested the distance measure between quantum states described by density operators, i.e., positive operators with unit trace. Most of the properties of the two distance measures directly follow from those of the corresponding norms. Therefore, here will focus on their physical implications.

Let us first examine the distance measures for qubits. In this case, both the Hilbert–Schmidt and trace distances have a clear geometric meaning. The density operators for single qubits take the form

$$\hat{\rho} = \frac{1}{2} \left(\hat{I} + \boldsymbol{a} \cdot \hat{\boldsymbol{S}} \right), \quad \hat{\sigma} = \frac{1}{2} \left(\hat{I} + \boldsymbol{b} \cdot \hat{\boldsymbol{S}} \right), \tag{5.186}$$

where $\boldsymbol{a} := (a_x, a_y, a_x)$ and $\boldsymbol{b} := (b_x, b_y, b_z)$ are the Bloch vectors (Sect. 1.1) corresponding to $\hat{\rho}$ and $\hat{\sigma}$, respectively. The difference $\hat{\rho} - \hat{\sigma} = (\boldsymbol{a} - \boldsymbol{b}) \cdot \hat{\boldsymbol{S}}/2$ is a traceless Hermitian operator and has two eigenvalues $\pm \|\boldsymbol{a} - \boldsymbol{b}\|/2$. This implies (recall also (5.178)) that

$$\|\hat{\rho} - \hat{\sigma}\|_{\mathrm{HS}} = \frac{\|\boldsymbol{a} - \boldsymbol{b}\|}{\sqrt{2}}, \quad \|\hat{\rho} - \hat{\sigma}\|_{\mathrm{tr}} = \|\boldsymbol{a} - \boldsymbol{b}\|. \tag{5.187}$$

In other words, the trace distance between two quantum states of a single qubit is equal to the Euclidean distance between the corresponding Bloch vectors in the Bloch space. It is also the case for the Hilbert–Schmidt distance (up to a constant factor of $1/\sqrt{2}$).

A stunning property of the trace distance is that it never increases under quantum operations: Suppose that $\mathcal{F}$ is a quantum operation. Then we have (recall Eq. (5.183))

$$\|\mathcal{F}(\hat{\rho}) - \mathcal{F}(\hat{\sigma})\|_{\mathrm{tr}} \leq \|\hat{\rho} - \hat{\sigma}\|_{\mathrm{tr}} \tag{5.188}$$

for any quantum states $\hat{\rho}$ and $\hat{\sigma}$. Physically, one can regard that a quantum operation causes a loss of some piece of information in quantum states. The loss makes it more difficult to distinguish two quantum states. For an analogy, consider two images. If one blurs the images through image processing, then the resulting images become more similar than the original images. The non-increasing feature of the trace distance under quantum operations has many physical implications and applications. One example is that one can construct an entanglement measure based on the trace distance.[18]

---

[18] It turns out that any distance measure non-increasing under quantum operations gives rise to an entanglement measure (Vedral et al., 1997).

Here we demonstrate again that the trace distance never increases under quantum operations, now focusing on quantum states.

```
In[ ]:= Timing[
  data = Table[
    U = RandomUnitary[$d * $d];
    vec = Normalize[RandomVector[$d]];
    spr = Supermap[kraus[U, vec]];
    rho = RandomPositive[3]; rho /= Tr[rho];
    sig = RandomPositive[3]; sig /= Tr[sig];
    {TraceNorm[rho - sig], TraceNorm[spr[rho - sig]]},
    {10 000}
    ];
  ]

Out[ ]= {70.4653, Null}
```

```
In[ ]:= max = Max[data];
  ListPlot[data, Epilog → Line@{{0, 0}, {1, 1} * max},
   PlotRange → {{0, 1}, {0, 1}} * max,
   FrameLabel → {"∥ρ-σ∥_tr", "∥𝓕(ρ)-𝓕(σ)∥_tr"},
   AspectRatio → 1]
```

The distance measures between quantum states are closely related to the distance measure between classical probability distributions. Consider the probability distributions $\{p_m\}$ and $\{q_m\}$ for two random variables with values in a common set $\{m\}$. One way to measure how close they are is to use the Euclidean vector norm

$$D_E(\{p_m\}, \{q_m\}) := \|\boldsymbol{p} - \boldsymbol{q}\| = \sqrt{\sum_m (p_m - q_m)^2} \qquad (5.189)$$

regarding $\boldsymbol{p} := (p_1, p_2, \ldots)$ and $\boldsymbol{q} := (q_1, q_2, \ldots)$ as classical Euclidean vectors. To compare it with the Hilbert–Schmidt distance between quantum states, let us consider the probabilities

$$p_m := \langle m|\hat{\rho}|m\rangle, \quad q_m := \langle m|\hat{\sigma}|m\rangle \qquad (5.190)$$

associated with the projective measurement on two quantum states $\hat{\rho}$ and $\hat{\sigma}$ in an orthonormal basis $\{|m\rangle\}$. More generally, one can consider the probability distributions

$$p_m := \operatorname{Tr} \hat{E}_m \hat{\rho}, \quad q_m := \operatorname{Tr} \hat{E}_m \hat{\sigma} \qquad (5.191)$$

associated with a POVM $\left\{\hat{E}_m\right\}$ measurement on the two quantum states. In either case, one can show (Problem 5.12) that

$$\|\boldsymbol{p} - \boldsymbol{q}\| \le \|\hat{\rho} - \hat{\sigma}\|_{\text{HS}}. \tag{5.192}$$

It is instructive to examine an example for which equality holds in (5.192). Suppose that both $\hat{\rho}$ and $\hat{\sigma}$ are diagonal in $\{|m\rangle\}$,

$$\hat{\rho} = \sum_m |m\rangle\, p_m\, \langle m| \,, \quad \hat{\sigma} = \sum_m |m\rangle\, q_m\, \langle m|\,. \tag{5.193}$$

A simple inspection shows that

$$\|\hat{\rho} - \hat{\sigma}\|_{\text{HS}}^2 = \sum_m (p_m - q_m)^2 = \|\boldsymbol{p} - \boldsymbol{q}\|^2. \tag{5.194}$$

Let us compare the Hilbert-Schmidt distance between quantum states and the vector-norm-based distance between classical probability distributions.

```
data = Table[
    rho = RandomPositive[3]; rho /= Tr[rho];
    sig = RandomPositive[3]; sig /= Tr[sig];
    pp = Chop@Diagonal[rho];
    qq = Chop@Diagonal[sig];
    {HilbertSchmidtNorm[rho - sig], Norm[pp - qq]}, {10 000}];
```

```
In[·]:= max = Max[data];
ListPlot[data, PlotRange → {{0, 1}, {0, 1}} * max,
    Epilog → Line@{{0, 0}, {1, 1} * max},
    FrameLabel → {"∥ρ-σ∥_HS", "∥{p_m-q_m}∥"}, AspectRatio → 1]
```

On the other hand, the trace distance is the quantum generalization of the so-called *Kolmogorov distance*

$$D_K(\{p_m\}, \{q_m\}) := \sum_m |p_m - q_m| \tag{5.195}$$

between two classical probability distributions $\{p_m\}$ and $\{q_m\}$ for two random variables with values in a common set $\{m\}$. To see the relation, consider two mixed states $\hat{\rho}$ and $\hat{\sigma}$. For the moment, let us suppose that both are diagonal in the same orthonormal basis $\{|m\rangle\}$,

$$\hat{\rho} = \sum_m |m\rangle \, p_m \, \langle m| \,, \quad \hat{\sigma} = \sum_m |m\rangle \, q_m \, \langle m| \,. \tag{5.196}$$

Recall that physically, their eigenvalues p_m and q_m are the probabilities to get the outcome m out of the projective measurement associated with the basis $\{|m\rangle\}$. Then, we note that

$$\hat{\rho} - \hat{\sigma} = \sum_m |m\rangle \, (p_m - q_m) \, \langle m| \tag{5.197}$$

is Hermitian with eigenvalues $(p_m - q_m)$. It immediately leads to

$$\|\hat{\rho} - \hat{\sigma}\|_{\text{tr}} = \sum_m |p_m - q_m| = D_K(\{p_m\}, \{q_m\}). \tag{5.198}$$

This is a special case, and in general, the identity does not hold. Yet the classical distance and the trace distance satisfy a definite inequality.

For a given orthonormal basis $\{|m\rangle\}$, consider the two probability distributions

$$p_m = \langle m|\hat{\rho}|m\rangle \,, \quad q_m = \langle m|\hat{\sigma}|m\rangle. \tag{5.199}$$

Or, more generally, consider a POVM $\left\{\hat{E}_m\right\}$ and associated probability distributions

$$p_m = \text{Tr} \, \hat{E}_m \hat{\rho} \,, \quad q_m = \text{Tr} \, \hat{E}_m \hat{\sigma} \,. \tag{5.200}$$

In either case, it turns out (Problem 5.12) that

$$D_K(\{p_m\}, \{q_m\}) \le \|\hat{\rho} - \hat{\sigma}\|_{\text{tr}}. \tag{5.201}$$

The inequality is tight in the sense that equality is attainable by maximizing the distance between probability distributions with an optimal POVM. In other words,

$$\|\hat{\rho} - \hat{\sigma}\|_{\text{tr}} = \max_{\{\hat{E}_m\}} \sum_m |\, \text{Tr} \, \hat{E}_m (\hat{\rho} - \hat{\sigma})|, \tag{5.202}$$

where the maximization is over all possible POVM. Noting that all POVM elements $\hat{E}_m \le \hat{I}$, we can also write that

$$\|\hat{\rho} - \hat{\sigma}\|_{\text{tr}} = \max_{\hat{P}} \text{Tr} \, \hat{P}(\hat{\rho} - \hat{\sigma}), \tag{5.203}$$

where the maximization is over all positive operators $\hat{P}$ such that $\hat{P} \le \hat{I}$.

Let us compare the trace distance between quantum states and the Kolmogorov distance
between classical probability distributions.

```
data = Table[
    rho = RandomPositive[3]; rho /= Tr[rho];
    sig = RandomPositive[3]; sig /= Tr[sig];
    pp = Chop@Diagonal[rho];
    qq = Chop@Diagonal[sig];
    {TraceNorm[rho - sig], Total@Abs[pp - qq]}, {10000}];
```

```
In[·]:= max = Max[data];
    ListPlot[data, PlotRange → {{0, 1}, {0, 1}} * max,
        Epilog → Line@{{0, 0}, {1, 1} * max},
        FrameLabel → {"‖ρ-σ‖_tr", "D_K({p_m},{q_m})"}, AspectRatio → 1]
```

5.5.4   Fidelity

We have discussed two norm-derived distance measures. Fidelity is another frequently used measure of the closeness of quantum states. Unlike the Hermitian and
trace distances, it is not based on a norm. Rather, conceptually it is similar to the relative "angle" between the two quantum states. Since quantum states are normalized,
it seems natural to expect closer quantum states to have a smaller relative angle.
The fidelity is the quantum generalization of the (classical) fidelity between two
probability distributions.

For two quantum states described by density operators $\hat{\rho}$ and $\hat{\sigma}$, the *fidelity* between
them is defined by[19]

$$\hat{F}(\hat{\rho}, \hat{\sigma}) := \text{Tr} \sqrt{\hat{\rho}^{1/2} \hat{\sigma} \hat{\rho}^{1/2}} \tag{5.204}$$

The fidelity can also be computed by means of the trace norm in (5.167),

$$F(\hat{\rho}, \hat{\sigma}) = \left\| \sqrt{\hat{\rho}} \sqrt{\hat{\sigma}} \right\|_{\text{tr}} . \tag{5.205}$$

---

[19] Some people prefers the definition $F(\hat{\rho}, \hat{\sigma}) = \left(\text{Tr} \sqrt{\hat{\rho}^{1/2} \hat{\sigma} \hat{\rho}^{1/2}} \right)^2$.

At first glance, the definition in (5.204) looks asymmetric, but the above identity asserts (see also Problem 5.6) that the fidelity is actually symmetric about $\hat{\rho}$ and $\hat{\sigma}$,

$$F(\hat{\rho}, \hat{\sigma}) = F(\hat{\sigma}, \hat{\rho}). \tag{5.206}$$

Although the fidelity is not derived from a norm, it is invariant under unitary transformations

$$F(\hat{U}\hat{\rho}\hat{U}^\dagger, \hat{U}\hat{\sigma}\hat{U}^\dagger) = F(\hat{\rho}, \hat{\sigma}). \tag{5.207}$$

One can see this by recalling that for any positive operator $\hat{A}$, the square-root function is homomorphic under unitary transformations, $\sqrt{\hat{U}\hat{A}\hat{U}^\dagger} = \hat{U}\sqrt{\hat{A}}\,\hat{U}^\dagger$. When one or both of the two states are pure states, the expression for the fidelity gets simpler,

$$F(|v\rangle \langle v|, \hat{\rho}) = \sqrt{\langle v|\hat{\rho}|v\rangle} \tag{5.208}$$

and

$$F(|v\rangle \langle v|, |w\rangle \langle w|) = |\langle v|w\rangle|. \tag{5.209}$$

The above two limiting cases indicate that geometrically, the fidelity is an extension of the relative "angle". For an analogy, recall that in the three-dimensional Euclidean space, the inner product of two normalized vectors a and b ($\|a\| = \|b\| = 1$) is given by $a \cdot b = \cos\theta$. For quantum states, parallel and anti-parallel arrangements are not distinguished. Consequently, we expect the fidelity to be 1 for identical states, $\hat{\rho} = \hat{\sigma}$, which is indeed true. Below, we will see that the fidelity is bounded between 0 and 1,

$$0 \le F(\hat{\rho}, \hat{\sigma}) \le 1, \tag{5.210}$$

and $F(\hat{\rho}, \hat{\sigma}) = 1$ if and only if $\hat{\rho} = \hat{\sigma}$. The fidelity approaches 1 for closer quantum states and decreases for more different quantum states.

As a simple yet non-trivial example, let us evaluate the fidelity for a single qubit. Consider two states $\hat{\rho}$ and $\hat{\sigma}$, and define $\hat{R} := \hat{\rho}^{1/2}\hat{\sigma}\hat{\rho}^{1/2}$. The operator $\hat{R}$ is positive, and the both eigenvalues r_1 and r_2 of $\hat{R}$ must be positive. One can thus express the fidelity between $\hat{\rho}$ and $\hat{\sigma}$ as

$$F(\hat{\rho}, \hat{\sigma}) = \text{Tr}\sqrt{\hat{R}} = \sqrt{r_1} + \sqrt{r_2}. \tag{5.211}$$

By squaring both sides of the above equation, we get

$$F^2(\hat{\rho}, \hat{\sigma}) = (r_1 + r_2) + 2\sqrt{r_1 r_2} = \text{Tr}\,\hat{R} + 2\sqrt{\det \hat{R}}. \tag{5.212}$$

Using the cyclic invariance of trace,

$$\text{Tr}\,\hat{R} = \text{Tr}\,\hat{\rho}^{1/2}\hat{\sigma}\hat{\rho}^{1/2} = \text{Tr}\,\hat{\sigma}\hat{\rho}^{1/2}\hat{\rho}^{1/2} = \text{Tr}\,\hat{\rho}\hat{\sigma}, \tag{5.213}$$

and the factorization property of determinant,

$$\det \hat{R} = (\det \hat{\rho}^{1/2})(\det \hat{\sigma})(\det \hat{\rho}^{1/2}) = (\det \hat{\rho})(\det \hat{\sigma}), \tag{5.214}$$

we obtain

$$F^2(\hat{\rho}, \hat{\sigma}) = \mathrm{Tr}\, \hat{\rho}\hat{\sigma} + 2\sqrt{(\det \hat{\rho})(\det \hat{\sigma})}. \tag{5.215}$$

Now we put the general form of quantum states in (5.186) into the above expression. In the process, one can use the identity $\mathrm{Tr}\, \hat{S}^\mu \hat{S}^\nu = 2\delta_{\mu\nu}$ for all $\mu, \nu = 0, x, y, z$ to get $\mathrm{Tr}\, \hat{\rho}\hat{\sigma} = (1 + \boldsymbol{a} \cdot \boldsymbol{b})/2$. It is also useful to note that $\hat{\rho}$ has two eigenvalues $(1 \pm \|\boldsymbol{a}\|)/2$ leading to $\det \hat{\rho} = (1 - \boldsymbol{a} \cdot \boldsymbol{a})/4$ and similarly $\det \hat{\sigma} = (1 - \boldsymbol{b} \cdot \boldsymbol{b})/4$. Putting all together, we finally obtain the fidelity in a closed-form expression

$$F(\hat{\rho}, \hat{\sigma}) = \sqrt{\frac{1 + \boldsymbol{a} \cdot \boldsymbol{b} + \sqrt{(1 - \boldsymbol{a} \cdot \boldsymbol{a})(1 - \boldsymbol{b} \cdot \boldsymbol{b})}}{2}} \tag{5.216}$$

If either or both of $\hat{\rho}$ and $\hat{\sigma}$ is a pure state ($\boldsymbol{a} \cdot \boldsymbol{a} = 1$ or $\boldsymbol{b} \cdot \boldsymbol{b} = 1$), then

$$F(\hat{\rho}, \hat{\sigma}) = \sqrt{\frac{1 + \boldsymbol{a} \cdot \boldsymbol{b}}{2}}. \tag{5.217}$$

The most fundamental property of the fidelity is *Uhlmann's theorem*. Suppose that $|\Psi\rangle$ and $|\Phi\rangle$ are the purifications (Sect. 1.1.2) of $\hat{\rho}$ and $\hat{\sigma}$, respectively, into $\mathcal{V} \otimes \mathcal{E}$,

$$\hat{\rho} = \mathrm{Tr}_{\mathcal{E}} |\Psi\rangle \langle \Psi|, \quad \hat{\sigma} = \mathrm{Tr}_{\mathcal{E}} |\Phi\rangle \langle \Phi|. \tag{5.218}$$

Uhlmann's theorem states that the fidelity between $\hat{\rho}$ and $\hat{\sigma}$ cannot be smaller than the fidelity between the purifications $|\Psi\rangle$ and $|\Phi\rangle$,

$$F(\hat{\rho}, \hat{\sigma}) \geq |\langle \Psi|\Phi\rangle|. \tag{5.219}$$

The bound is optimal in the sense that there always exists an optimal purification attaining equality. Putting it another way,

$$F(\hat{\rho}, \hat{\sigma}) = \max_{\{|\Psi\rangle, |\Phi\rangle\}} |\langle \Psi|\Phi\rangle|, \tag{5.220}$$

where the maximization is over all possible purifications (Problems 1.3 and 1.4) of $\hat{\rho}$ and $\hat{\sigma}$. One can prove Uhlmann's formula by following the steps in Problem 5.14.

Uhlmann's theorem is not particularly useful for specific calculations of the fidelity. However, as we will see below, many properties of the fidelity can be assessed more conveniently by employing Uhlmann's theorem than the defining expression (5.204).

As an example of the application of Uhlmann's theorem, we note that the bounds of the fidelity,

$$0 \leq F(\hat{\rho}, \hat{\sigma}) \leq 1, \tag{5.221}$$

follow immediately from (5.220). Let us further examine when the fidelity attains $F(\hat{\rho}, \hat{\sigma}) = 1$. If $\hat{\rho} = \hat{\sigma}$, then it is obvious from (5.204) that $F(\hat{\rho}, \hat{\sigma}) = 1$. To prove the converse, suppose that $\hat{\rho} \neq \hat{\sigma}$. Then, $|\Psi\rangle \neq |\Phi\rangle$ for any purifications $|\Psi\rangle$ and $|\Phi\rangle$ of $\hat{\rho}$ and $\hat{\sigma}$, respectively. Therefore, the fidelity must be strictly less than unity, $F(\hat{\rho}, \hat{\sigma}) < 1$, contradicting the assumption. Thus we have shown that $F(\hat{\rho}, \hat{\sigma}) = 1$ if and only if $\hat{\rho} = \hat{\sigma}$.

Similar to the non-increasing feature of the trace distance, the fidelity never decreases under quantum operations. In other words,

$$F(\mathscr{F}(\hat{\rho}), \mathscr{F}(\hat{\sigma})) \geq F(\hat{\rho}, \hat{\sigma}) \tag{5.222}$$

for any quantum operation $\mathscr{F}$ and quantum states $\hat{\rho}$ and $\hat{\sigma}$ on a vector space $\mathcal{V}$. The interpretation of the inequality (5.222) is the same as for the trace distance. The "damage" due to the quantum operation makes the resulting two states more similar than the original quantum states.

To prove the feature in (5.222), again we use Uhlmann's theorem. Suppose that $|\Psi\rangle$ and $|\Phi\rangle$ be purifications of $\hat{\rho}$ and $\hat{\sigma}$, respectively, into $\mathcal{V} \otimes \mathcal{V}$ such that $F(\hat{\rho}, \hat{\sigma}) = |\langle \Psi | \Phi \rangle|$. We introduce an additional environment $\mathcal{E}$ prepared in a fixed pure state $|\epsilon_0\rangle$ to represent $\mathscr{F}$ with a unitary interaction $\hat{U}$ acting on $\mathcal{V} \otimes \mathcal{V} \otimes \mathcal{E}$ (Sect. 5.2.3),

$$\mathscr{F}(\hat{\rho}) = \underset{\mathcal{V} \otimes \mathcal{E}}{\mathrm{Tr}} \, \hat{U} \, |\Psi\rangle \otimes |\epsilon_0\rangle \tag{5.223}$$

Note that in this model, $\mathcal{V} \otimes \mathcal{E}$ in $\mathcal{V} \otimes (\mathcal{V} \otimes \mathcal{E})$ corresponds to the ultimate "environment". Since $\hat{U} |\Psi\rangle \otimes |\epsilon_0\rangle$ is a purification of $\mathscr{F}(\hat{\rho})$ and similarly, $\hat{U} |\Phi\rangle \otimes |\epsilon_0\rangle$ is a purification of $\mathscr{F}(\hat{\sigma})$, we deduce that

$$F(\mathscr{F}(\hat{\rho}), \mathscr{F}(\hat{\sigma})) \geq \left| (\langle \Psi | \otimes |\epsilon_0\rangle \, \hat{U}\hat{U}^\dagger \, |\Phi\rangle \otimes |\epsilon_0\rangle) \right| = |\langle \Psi | \Phi \rangle| = F(\hat{\rho}, \hat{\sigma}). \tag{5.224}$$

Here we demonstrate the non-decreasing feature of the fidelity under quantum operations.

```
In[·]:= Timing[
        data = Chop@Table[
            U = RandomUnitary[$d * $d];
            vec = Normalize[RandomVector[$d]];
            spr = Supermap[kraus[U, vec]];
            rho = RandomPositive[3]; rho /= Tr[rho];
            sig = RandomPositive[3]; sig /= Tr[sig];
            {Fidelity[rho, sig], Fidelity[spr[rho], spr[sig]]},
            {10 000}
        ];
    ]

Out[·]= {91.0631, Null}
```

```
In[·]:= Select[data, First[#] > Last[#] &]
Out[·]= {}

In[·]:= max = Max[data];
        min = Min[data];
        ListPlot[data, Epilog → Line@{{0, 0}, {1, 1} * max},
         PlotRange → {{min, max}, {min, max}},
         FrameLabel → {"F(ρ,σ)", "F(F(ρ),F(σ))"},
         AspectRatio → 1]
```

The connection of the fidelity of quantum states to the fidelity of classical probability distribution is most clear when the two quantum states $\hat{\rho}$ and $\hat{\sigma}$ are diagonal in the same orthonormal basis $|m\rangle$,

$$\hat{\rho} := \sum_m |m\rangle \, p_m \, \langle m| \, , \quad \hat{\sigma} := \sum_m |m\rangle \, q_m \, \langle m| \, . \tag{5.225}$$

Then, we have

$$F(\hat{\rho}, \hat{\sigma}) = \mathrm{Tr} \sqrt{\hat{\rho}\hat{\sigma}} = \sum_m \sqrt{p_m q_m}. \tag{5.226}$$

The result is identical to the classical fidelity

$$F_{\mathrm{cl}}(\{q_m\}, \{q_m\}) := \sum_m \sqrt{p_m q_m} \tag{5.227}$$

between two probability distributions $\{p_m\}$ and $\{q_m\}$ of random variables with values in $\{m\}$.

To examine the relation in the general case, consider

$$p_m := \langle m|\hat{\rho}|m\rangle \, , \quad q_m := \langle m|\hat{\sigma}|m\rangle \tag{5.228}$$

in the given basis. Alternatively, we can also consider a POVM $\left\{\hat{E}_m\right\}$ with

$$p_m := \mathrm{Tr}\,\hat{E}_m\hat{\rho}, \quad q_m := \mathrm{Tr}\,\hat{E}_m\hat{\sigma}. \tag{5.229}$$

In both cases, one can show (Problem 5.15) that the quantum fidelity is lower than the classical fidelity,

$$F(\hat{\rho}, \hat{\sigma}) \leq F_{cl}(\{p_m\}, \{q_m\}). \tag{5.230}$$

Here we compare the fidelity between quantum states and the fidelity between classical probability distributions.

```
data = Chop@Table[
    rho = RandomPositive[3]; rho /= Tr[rho];
    sig = RandomPositive[3]; sig /= Tr[sig];
    pp = Chop@Diagonal[rho];
    qq = Chop@Diagonal[sig];
    {Fidelity[rho, sig], Total@Sqrt[pp * qq]}, {10000}];
```

This illustrates that the quantum fidelity is lower than the classical fidelity.

```
In[ ]:= min = Min[data];
max = Max[data];
ListPlot[data, PlotRange → {{min, max}, {min, max}},
  Epilog → Line@{{0, 0}, {1, 1} * max},
  FrameLabel → {"F(ρ,σ)", "F_cl({p_m},{q_m})"}, AspectRatio → 1]
```

Problems

5.1. Consider a superoperator with the Kraus representation

$$\mathscr{F}(\hat{\rho}) = \sum_{\mu} \hat{F}_{\mu} \hat{\rho} \hat{F}_{\mu}^{\dagger}. \tag{5.231}$$

Show that $\mathscr{F}$ preserves the trace if and only if

$$\sum_{\mu} \hat{F}_{\mu}^{\dagger} \hat{F}_{\mu} = \hat{I}. \tag{5.232}$$

5.2. By explicitly evaluating the quantum circuit in (5.53), prove the quantum gate teleportation protocol.

5.3. Consider a quantum register of two qubits. Let $\mathscr{F}$ be a quantum operation on the quantum register, specified by the Kraus elements

$$\hat{F}_0 = \sqrt{1 - p_1 - p_2 - p_3}\,\hat{I}\,,$$
$$\hat{F}_1 = \sqrt{p_1}\hat{S}_1^-\,, \quad \hat{F}_2 = \sqrt{p_2}\hat{S}_2^-\,, \quad \hat{F}_3 = \sqrt{p_3}\hat{S}_1^-\hat{S}_2^-\,. \quad (5.233)$$

(a) Construct a quantum circuit to generate the Choi operator $\hat{C}_{\mathscr{F}}$ associated with the quantum operation $\mathscr{F}$. Note that the input state of the system (including an auxiliary quantum register if necessary) must be $|0\rangle \equiv |0\rangle \otimes |0\rangle \otimes \cdots$.
Hint: See the quantum circuit representation in (5.46) to generate the Choi operator. You have to generate the maximally entangled state starting from $|0\rangle$.

(b) Write down the Choi operator $\hat{C}_{\mathscr{F}}$ in terms of the Pauli operators $\hat{S}_1^\mu$ and $\hat{S}_2^\nu$ on the two qubits.

5.4. Consider the following Lindblad equation for a single qubit:

$$\frac{d\hat{\rho}}{dt} = -i[\hat{H}, \hat{\rho}] - i\{\hat{G}, \hat{\rho}\} + \gamma_- |0\rangle \langle 1| \hat{\rho} |1\rangle \langle 0| + \gamma_\phi |1\rangle \langle 1| \hat{\rho} |1\rangle \langle 1|\,,$$
$$(5.234)$$

where the effective Hamiltonian $\hat{H}$ and the damping operator $\hat{G}$ are given by

$$\hat{H} = E_0 |0\rangle \langle 0| + E_1 |1\rangle \langle 1|\,, \quad \hat{G} = \frac{1}{2}(\gamma_- + \gamma_\phi) |1\rangle \langle 1|\,. \quad (5.235)$$

Put $E_0 = 0$ without loss of generality and assume $0 < E_1$. γ_- and γ_ϕ are responsible for the amplitude and phase damping process, respectively. Solve the Lindblad equation and represent the solution $\hat{\rho}(t)$ in the Kraus representation form.

5.5. Consider a three-level atom. Suppose that it is subject to an interaction to its environment, which is described by the following Lindblad equation

$$\frac{d\hat{\rho}}{dt} = -i[\hat{H}, \hat{\rho}] - i\{\hat{G}, \hat{\rho}\} + \eta_0 |0\rangle \langle 2| \hat{\rho} |2\rangle \langle 0| + \eta_1 |1\rangle \langle 2| \hat{\rho} |2\rangle \langle 1|\,, \quad (5.236)$$

where $|0\rangle$ and $|1\rangle$ are the ground-state levels and $|2\rangle$ is the excited state. η_0 is the rates of spontaneous decay of the atomic level $|2\rangle$ to $|0\rangle$. Similarly, η_1 is the spontaneous decay rate from $|2\rangle$ to $|1\rangle$. The effective Hamiltonian is given by

$$\hat{H} = E_1 |1\rangle \langle 1| + E_2 |2\rangle \langle 2| \quad (0 \le E_1 \ll E_2)\,, \quad (5.237)$$

where we have set the ground-state energy to zero $E_0 = 0$. The damping operator is given by

$$\hat{G} = \frac{1}{2}(\eta_0 + \eta_1)\,|2\rangle\,\langle 2|\,. \tag{5.238}$$

Solve the Lindblad equation and put the solution $\hat{\rho}(t)$ in the Kraus representation form. Assume that the atom is initially prepared in the pure state $|1\rangle$.

5.6. Show that

$$\mathrm{Tr}\sqrt{\hat{A}^\dagger\hat{A}} = \mathrm{Tr}\sqrt{\hat{A}\hat{A}^\dagger} \tag{5.239}$$

for any linear operator $\hat{A}$.

5.7. Show that

$$|\,\mathrm{Tr}\,\hat{A}\hat{U}\,| \le \|\hat{A}\|_{\mathrm{tr}} \tag{5.240}$$

for any linear operator $\hat{A}$ and unitary operator $\hat{U}$.
Hint: Use the singular-value decomposition (Theorem A.22) or the polar decomposition (Theorem A.23) of $\hat{A}$.

5.8. Let $\hat{A}$ be a linear operator on a vector space $\mathcal{V}$.

(a) Show that if $\hat{A}$ is Hermitian, then

$$\|\hat{A}\|_{\mathrm{tr}} = \sum_j |a_j|, \tag{5.241}$$

where a_j are the eigenvalues of $\hat{A}$.
Hint: Use the spectral decomposition (Appendix A.4.1) of $\hat{A}$.

(b) Show that

$$\|\hat{A}\|_{\mathrm{tr}} = \sum_j |s_j|, \tag{5.242}$$

where s_j are the singular values of $\hat{A}$.
Hint: Use the singular-value decomposition (Theorem A.22) of $\hat{A}$.

5.9. Show that for any linear operator $\hat{A}$, the Hilbert–Schmidt norm cannot be greater than the trace norm,

$$\|\hat{A}\|_{\mathrm{HS}} \le \|\hat{A}\|_{\mathrm{tr}}, \tag{5.243}$$

where equality holds if and only if $\hat{A} = |v\rangle\,\langle v|$ for a vector $|v\rangle$.
Hint: Use the singular-value decomposition (Theorem A.22) of $\hat{A}$.

5.10. Show that

$$\|\hat{A}\hat{B}\|_{\mathrm{HS}} \le \|\hat{A}\|_{\mathrm{HS}}\|\hat{B}\|_{\mathrm{HS}}, \tag{5.244}$$

and in particular,

$$\|\hat{A}^\dagger\hat{A}\|_{\mathrm{HS}} \le \|\hat{A}\|_{\mathrm{HS}}^2 \tag{5.245}$$

for any linear operators $\hat{A}$ and $\hat{B}$.

Hint: Use the Cauchy–Schwarz inequality (Theorem A.5) and singular-value decomposition (Theorem A.22).

5.11. Let $\hat{F}_\mu$ the Kraus elements of a trace-preserving quantum operation on a vector space $\mathcal{V}$, that is,

$$\mathcal{F}(\hat{\rho}) = \sum_\mu \hat{F}_\mu \hat{\rho} \hat{F}_\mu^\dagger \qquad (5.246)$$

for all linear operators $\hat{\rho}$, and

$$\sum_\mu \hat{F}_\mu^\dagger \hat{F}_\mu = \hat{I}. \qquad (5.247)$$

Show that for the linear operator

$$\hat{G} := \sum_\mu \hat{F}_\mu \hat{F}_\mu^\dagger, \qquad (5.248)$$

$$\|\hat{G}\|_{\mathrm{tr}} = \dim \mathcal{V}. \qquad (5.249)$$

5.12. Let $\hat{A}$ be a linear operator on a vector space $\mathcal{V}$. Show that the following statements are true.

(a) For any orthonormal basis $\{|m\rangle\}$ of $\mathcal{V}$,

$$\sum_m |\langle m|\hat{A}|m\rangle|^2 \leq \|\hat{A}\|_{\mathrm{HS}}, \qquad (5.250a)$$

$$\sum_m |\langle m|\hat{A}|m\rangle| \leq \|\hat{A}\|_{\mathrm{tr}}. \qquad (5.250b)$$

(b) For any POVM $\left\{\hat{E}_m\right\}$ on $\mathcal{V}$,

$$\sum_m (\mathrm{Tr}\, \hat{E}_m \hat{A})^2 \leq \|\hat{A}\|_{\mathrm{HS}}, \qquad (5.251a)$$

$$\sum_m |\mathrm{Tr}\, \hat{E}_m \hat{A}| \leq \|\hat{A}\|_{\mathrm{tr}}. \qquad (5.251b)$$

5.13. Show that for all density operators

$$F(\hat{\rho}, \hat{\sigma}) = F(\hat{\sigma}, \hat{\rho}). \qquad (5.252)$$

Hint: Define $\hat{A} := \sqrt{\hat{\rho}}\sqrt{\hat{\sigma}}$ and recall the identity (5.239).

5.14. Prove Uhlmann's formula in (5.220) by following the steps below.

(a) For an orthonormal basis $\{|v_j\rangle\}$ of $\mathcal{V}$, construct an (unnormalized) maximally entangled state

$$|\Theta\rangle := \sum_j |v_j\rangle \otimes |v_j\rangle. \tag{5.253}$$

(b) Using the theorem in Problem 1.4, write $|\Psi\rangle$ and $|\Phi\rangle$ in the form

$$|\Psi\rangle = (\sqrt{\rho}\hat{U}) \otimes \hat{V} |\Theta\rangle, \quad |\Psi\rangle = (\sqrt{\sigma}\hat{U}') \otimes \hat{V}' |\Theta\rangle, \tag{5.254}$$

where $\hat{U}$, $\hat{U}'$, $\hat{V}$, and $\hat{V}'$ are unitary operators on $\mathcal{V}$.

(c) Show that

$$\langle\Psi|\Phi\rangle = \mathrm{Tr}(\hat{U}^\dagger \sqrt{\rho}\sqrt{\sigma}\,\hat{U}')(\hat{V}^\dagger\hat{V}'). \tag{5.255}$$

(d) Using the theorem in Problem 5.7, show that

$$|\langle\Psi|\Phi\rangle| \leq \mathrm{Tr}\sqrt{\rho}\sqrt{\sigma} = F(\hat{\rho}, \hat{\sigma}). \tag{5.256}$$

(e) Using the polar decomposition of $\sqrt{\hat{\rho}}\sqrt{\hat{\sigma}}$, construct $|\Psi\rangle$ and $|\Phi\rangle$ such that

$$|\langle\Psi|\Phi\rangle| = F(\hat{\rho}, \hat{\sigma}). \tag{5.257}$$

5.15. Let $\hat{\rho}$ and $\hat{\sigma}$ be quantum states, and

$$p_m := \langle m|\hat{\rho}|m\rangle, \quad q_m := \langle m|\hat{\sigma}|m\rangle \tag{5.258}$$

in a given orthonormal basis $\{|m\rangle\}$. Show that

$$F(\hat{\rho}, \hat{\sigma}) \leq F_{\mathrm{cl}}(\{p_m\}, \{q_m\}). \tag{5.259}$$

Hint: Note that

$$F(\hat{\rho}, \hat{\sigma}) = \mathrm{Tr}\sqrt{\hat{\rho}^{1/2}\hat{\sigma}\hat{\sigma}^{1/2}} = \mathrm{Tr}\sqrt{\hat{\rho}}\sqrt{\hat{\sigma}}\,\hat{U}^\dagger \tag{5.260}$$

with the polar decomposition (Theorem A.23)

$$\sqrt{\hat{\rho}}\sqrt{\hat{\sigma}} = \sqrt{\hat{\rho}^{1/2}\hat{\sigma}\hat{\sigma}^{1/2}}\,\hat{U}, \tag{5.261}$$

and use the Cauchy–Schwarz inequality (Theorem A.5).

Chapter 6
Quantum Error-Correction Codes

Nothing is perfect, and everything is prone to errors. But what makes quantum information different from classical information when it comes to error correction?

Any physical system inevitably interacts with its surroundings, which are collectively referred to as the *environment*. These interactions have particularly severe effects on quantum systems. Quantum information is represented through a delicate state of superposition that the environment tends to knock out. This leads to decoherence and the loss of quantum information. Furthermore, quantum gates involved in quantum information processing reside in a continuum of unitary transformations, and an implementation with perfect accuracy is unrealistic for such quantum gates. Even worse, small imperfections may accumulate and result in serious errors in the state undergoing gate operations. On account of such, the errors in quantum information are clearly continuous. Detecting these continuous errors, not to mention correcting them, already seems to be a formidable task.

The principles themselves of quantum mechanics make handling quantum errors a particular challenge. In classical information processing, the basic approach involves creating duplicate copies before processing any information and comparing the output of the different copies to check for any error. For quantum information, this approach is not allowed due to the no-cloning theorem (see Sect. 1.3.1) that prevents copying unknown quantum states. The measurement introduces another obstacle. In a classical case, one can probe the system and correct an error if necessary. However, this tactic does not work in quantum mechanics since the measurement disturbs the quantum states.

In this chapter, we will see that amazingly, despite apparent difficulties, it is possible to successfully correct quantum errors. This is achieved by suitably encoding quantum information. If the quantum information is encoded appropriately, then it

Supplementary Information The online version contains supplementary material available at https://doi.org/10.1007/978-3-030-91214-7_6.

can be recovered by merely correcting a discrete set of errors, as long as the error rate is not too high. We will thus determine the conditions for quantum error-correction codes to protect successfully against probable errors.

6.1   Elementary Examples: Nine-Qubit Code

We start with some elementary examples that require a relatively large use of resources for the given error types, and in this sense, there are more efficient methods. However, their simple structure makes it easy to grasp the fundamental ideas behind the quantum error-correction codes, including how to encode quantum information to protect against errors, how to detect error syndrome, and how to recover quantum information from corrupted data.

6.1.1   Bit-Flip Errors

Suppose that the quantum computer is subjected to quantum noise that flips the basis states of the individual qubits, $|0\rangle \leftrightarrow |1\rangle$, with probability p. The error is called the *bit-flip error*, and we are to examine the code to correct it.

To protect a single-qubit state $|\psi\rangle = |0\rangle c_0 + |1\rangle c_1$ against the bit-flip error, one has to encode it to multiple qubits. For example, one encodes $|\psi\rangle$ to the three-qubit state $|\bar{\psi}\rangle = |000\rangle c_0 + |111\rangle c_1$. In this encoding, the logical basis states are

$$|\bar{0}\rangle := |000\rangle , \quad |\bar{1}\rangle := |111\rangle , \tag{6.1}$$

and these are distinguished from the physical basis states $|0\rangle$ and $|1\rangle$ of each qubit. Therefore, the subspace spanned by the two logical basis states is the *code space*. The projection onto the code space is given by $\hat{P}_0 := |000\rangle \langle 000| + |111\rangle \langle 111|$. The encoding procedure can be achieved through a quantum circuit (recall Fig. 2.5)

$$\tag{6.2a}$$

or (recall Problem 2.5)

$$\tag{6.2b}$$

To see how this encoding protects against the bit-flip error, suppose that the error occurs on the first qubit, whereby the error is described by the operator

$\hat{X} \otimes \hat{I} \otimes \hat{I}$. Then, the encoded state changes to $|\bar{\psi}'\rangle = |100\rangle c_0 + |011\rangle c_1$. This state is orthogonal to the original state $|\bar{\psi}\rangle$, and in principle, in can be discriminated through a proper measurement. Indeed, the measurement associated with the projection $\hat{P}_1 := |100\rangle \langle 100| + |011\rangle \langle 011|$ onto the subspace spanned by $|100\rangle$ and $|011\rangle$ can do it. Similarly, bit flips on the second and third qubit modify the encoded state into $|\bar{\psi}''\rangle = |010\rangle c_0 + |101\rangle c_1$ and $|\bar{\psi}'''\rangle = |001\rangle c_0 + |110\rangle c_1$, respectively. These states reside in the subspaces corresponding to the projection operators $\hat{P}_2 := |010\rangle \langle 010| + |101\rangle \langle 101|$ and $\hat{P}_3 := |001\rangle \langle 001| + |110\rangle \langle 110|$, respectively. As the bit-flip errors on individual qubits bring the encoded state $|\bar{\psi}\rangle$ to mutually orthogonal subspaces, or equivalently, $\hat{P}_i \hat{P}_j = \delta_{ij} \hat{P}_j$, one can set up a projection measurement that is described by the projectors $\hat{P}_0, \hat{P}_1, \hat{P}_2, \hat{P}_3$ to determine the qubit on which the error has actually occurred (if at all). The measurement result is referred to as the *error syndrome*, and its measurement is referred to as the *error-detection* procedure.

It is interesting to note that the states $|\bar{\psi}\rangle$, $|\bar{\psi}'\rangle$, $|\bar{\psi}''\rangle$, and $|\bar{\psi}'''\rangle$ arc all simultaneous eigenstates of the two observables

$$\hat{M}_1 := \hat{Z} \otimes \hat{Z} \otimes \hat{I}, \quad \hat{M}_2 := \hat{I} \otimes \hat{Z} \otimes \hat{Z}. \tag{6.3}$$

Furthermore, since they are mutually orthogonal, they must belong to different pairs of eigenvalues of $\hat{M}_1$ and $\hat{M}_2$. Indeed, we see by inspection that they belong to pairs of eigenvalues $(1, 1)$, $(-1, 1)$, $(-1, -1)$ and $(1, -1)$, respectively. Therefore, the measurements of $\hat{M}_1$ and $\hat{M}_2$ produce exactly the same information regarding the error syndrome as the projection measurement associated with $\hat{P}_j$ outlined above. Although less obvious at this stage, this method of error-detection is easier to extend—and more closely related to the stabilizer formalism to be discussed in Sects. 6.3 and 6.4—than the above measurement in terms of the projectors $\hat{P}_j$, so we will focus mainly on this method from now on.

It is crucial for the error-detection procedure to not reveal the details of the encoding, i.e., the amplitudes c_0 and c_1 in $|\bar{\psi}\rangle$. Otherwise, measuring the error syndrome would destroy the superposition in the quantum states. For example, the measurement of $\hat{M}_1$ just compares the values for the first and second qubit, giving $+1$ if the values are the same and -1 if they are different, without revealing the individual values. Similarly, measuring $\hat{M}_2$ only compares the values of the second and third qubit, but it does not disclose the individual values. Nevertheless, it is still possible to determine where the error occurs. For example, suppose that the quantum noise flips the first qubit. Comparing the values of the first two qubits determines that the error has occurred on either of the first qubits, but not on exact which. Comparing the second and third qubit indicates that the error has occurred on neither (or both) of the qubits. Assuming that the error rate is sufficiently low, the two comparisons can successfully predict that the error occurred on the first qubit.

Once the error-detection procedure identifies the error syndrome, the next step is to correct the error and recover the original encoded state $|\bar{\psi}\rangle$. This step is called the *recovery* procedure. For the bit-flip error, the recovery procedure is quite simple.

Depending on the qubit that has an error, one just has to apply the gate operation $\hat{X} \otimes \hat{I} \otimes \hat{I}$, $\hat{I} \otimes \hat{X} \otimes \hat{I}$, or $\hat{I} \otimes \hat{I} \otimes \hat{X}$; or, of course, just do nothing if the measurement indicates that no error has occurred.

Following the error-correction procedure shown above, the original state is successfully recovered as long as either no error occurs at all or the bit-flip error occurs on only one of the three qubits. The probability for successful recovery is thus given by $(1 - p)^2 + 3p(1 - p)^2 = 1 - 3p^2 + 2p^2$. Compared to a probability of $1 - p$ for a bare state, i.e., a non-encoded state, to remain uncorrupted, the encoding and subsequent error-correction procedure improves the reliability of the storage of the quantum state as long as $p < 1/2$. This rather crude estimation of the quality of the quantum error-correction procedure provides a first impression how the quantum error-correction code works, and a more complete analysis will follow later.

Suppose that we want to encode a single-qubit state to protect it against the bit-flip error.

```
In[ ]:= vec = ProductState[S[1] → {c[0], c[1]}, "Label" → Ket[ψ]]
```
$$Out[ ]= \left(c_0 \left|0\right\rangle + c_1 \left|1\right\rangle\right)_{S_1}$$

The encoding is achieved by a quantum circuit model of the form.

```
In[ ]:= qc = QuantumCircuit[vec,
            LogicalForm[Ket[], S@{2, 3}], CNOT[S[1], S[2]], CNOT[S[1], S[3]]]
```

$$Out[ ]=$$

```
In[ ]:= out = DefaultForm@ExpressionFor[qc];
        out // LogicalForm
```
$$Out[ ]= c_0 \left|0_{S_1} 0_{S_2} 0_{S_3}\right\rangle + c_1 \left|1_{S_1} 1_{S_2} 1_{S_3}\right\rangle$$

Here is another quantum circuit model giving the same encoding.

```
In[ ]:= qc2 = QuantumCircuit[vec,
            LogicalForm[Ket[], S@{2, 3}], CNOT[S[1], S[2]], CNOT[S[2], S[3]]]
```

$$Out[ ]=$$

```
In[ ]:= out2 = DefaultForm@ExpressionFor[qc2];
        out2 // LogicalForm
```
$$Out[ ]= c_0 \left|0_{S_1} 0_{S_2} 0_{S_3}\right\rangle + c_1 \left|1_{S_1} 1_{S_2} 1_{S_3}\right\rangle$$

Consider a bit-flip error occurs randomly

```
In[ ]:= errors = Join[{1}, S[{1, 2, 3}, 1]]
        opE = RandomChoice[errors]
        new = opE ** out;
        new // LogicalForm
```

Out[ ]= $\{1, S_1^x, S_2^x, S_3^x\}$

Out[ ]= S_1^x

Out[ ]= $c_1 \left| 0_{S_1} 1_{S_2} 1_{S_3} \right\rangle + c_0 \left| 1_{S_1} 0_{S_2} 0_{S_3} \right\rangle$

This gives the error syndrome.

```
In[ ]:= {(S[1, 3] ** S[2, 3] ** new) / new,
         (S[2, 3] ** S[3, 3] ** new) / new} // Simplify
```

Out[ ]= $\{-1, 1\}$

6.1.2   Phase-Flip Error

As was already mentioned, a quantum state is a superposition of the logical basis state, and any modulation of the superposition coefficients without altering the logical basis states results in an error. In particular, a mere change in the relative phase in the coefficients leads to a deviation from the original state even though it does not change the occupation probabilities in the logical basis states. Let us consider an extreme case where the relative phase of a single qubit is shifted by π—the relative sign of the superposition coefficients is flipped from $+1$ to -1, and vice versa—with probability p. It is called the *phase-flip error*, and the corresponding error operator is $\hat{Z}$. Such an error is particularly interesting because it has no classical counterpart since classically, the phase does not play any role.

Nevertheless, the phase-flip error is essentially no different from the bit-flip error. This becomes immediately clear when one observes that the phase-flip error switches the states $|\pm\rangle := (|0\rangle \pm |1\rangle)/\sqrt{2}$ to each other. Therefore, a change of basis from $\{|0\rangle, |1\rangle\}$ to $\{|+\rangle, |-\rangle\}$ just converts the phase-flip error to the bit-flip error, which has already been addressed above. Apart from the proper choice of basis, the full error-correction procedure (including encoding, error-detection, and recovery) is the same as protecting against bit-flip errors. It seems ironic that the phase-flip error, which appears to be genuinely a quantum problem, can be countered by a basis change, another feature of quantum mechanics with no classical equivalent. This fact suggests that all types of errors can be handle on an equal footing as the bit-flip error, and it further inspires ideas for how to protect against quantum noise.

Although it is already clear from the above observations, for the sake of later use, we construct the procedure to protect against phase-flip errors. The encoding chooses

$$|\bar{0}\rangle := |+++\rangle , \quad |\bar{1}\rangle := |---\rangle \tag{6.4}$$

as the logical basis states. The unitary transformation to implement the encoding can be described as a quantum circuit of the form

(6.5a)

or

(6.5b)

Compared to the corresponding quantum circuits in (6.2), the above quantum circuits have additional Hadamard gates, $\hat{H}^{\otimes 3}$, to account for the basis change mentioned above. Phase-flip errors on one or fewer qubits of the three can be detected by measuring the observables [recall Eq. (6.3)]

$$\hat{M}'_1 = \hat{H}^{\otimes 3} \hat{M}_1 \hat{H}^{\otimes 3} = \hat{X} \otimes \hat{X} \otimes \hat{I}, \tag{6.6a}$$

$$\hat{M}'_2 = \hat{H}^{\otimes 3} \hat{M}_2 \hat{H}^{\otimes 3} = \hat{I} \otimes \hat{X} \otimes \hat{X}, \tag{6.6b}$$

as $\hat{H}\hat{Z}\hat{H} = \hat{X}$. That is, the phase-flip error on the first, second, and third qubit leads to the measurement results $(-1, 1)$, $(-1, -1)$, and $(1, -1)$, respectively, with an outcome that will be $(1, 1)$ if no error occurs. Once the error syndrome is diagnosed, the error can be fixed by applying the recovery operator $\hat{I} \otimes \hat{I} \otimes \hat{I}$ (doing nothing), $\hat{Z} \otimes \hat{I} \otimes \hat{I}$, $\hat{I} \otimes \hat{Z} \otimes \hat{I}$, or $\hat{I} \otimes \hat{I} \otimes \hat{Z}$ as $\hat{H}\hat{X}\hat{H} = \hat{Z}$.

Suppose that we want to encode a single-qubit state to protect it against the phase-flip error.

$In[\cdot]:=$ `vec = ProductState[S[1] → {c[0], c[1]}, "Label" → Ket[ψ]]`

$Out[\cdot]=$ $\left(c_0 \left|0\right\rangle + c_1 \left|1\right\rangle\right)_{S_1}$

The encoding is achieved by a quantum circuit model of the form.

$In[\cdot]:=$ `qc = QuantumCircuit[vec, LogicalForm[Ket[], S@{2, 3}],`
`    CNOT[S[1], S[2]], CNOT[S[1], S[3]], S[{1, 2, 3}, 6]]`

$Out[\cdot]=$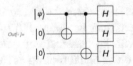

$In[\cdot]:=$ `out = DefaultForm@ExpressionFor[qc];`
`    out // LogicalForm`

$Out[\cdot]=$ $\dfrac{(c_0 + c_1) \left|0_{S_1} 0_{S_2} 0_{S_3}\right\rangle}{2\sqrt{2}} + \dfrac{(c_0 - c_1) \left|0_{S_1} 0_{S_2} 1_{S_3}\right\rangle}{2\sqrt{2}} + \dfrac{(c_0 - c_1) \left|0_{S_1} 1_{S_2} 0_{S_3}\right\rangle}{2\sqrt{2}} + \dfrac{(c_0 + c_1) \left|0_{S_1} 1_{S_2} 1_{S_3}\right\rangle}{2\sqrt{2}} +$

$\dfrac{(c_0 - c_1) \left|1_{S_1} 0_{S_2} 0_{S_3}\right\rangle}{2\sqrt{2}} + \dfrac{(c_0 + c_1) \left|1_{S_1} 0_{S_2} 1_{S_3}\right\rangle}{2\sqrt{2}} + \dfrac{(c_0 + c_1) \left|1_{S_1} 1_{S_2} 0_{S_3}\right\rangle}{2\sqrt{2}} + \dfrac{(c_0 - c_1) \left|1_{S_1} 1_{S_2} 1_{S_3}\right\rangle}{2\sqrt{2}}$

To compare the above result with the desired state, let us construct the encoding explicitly.

In[ ]:= `new = ProductState[S@{1, 2, 3} → {1, 1} / Sqrt[2]] × c[0] +`
        `ProductState[S@{1, 2, 3} → {1, -1} / Sqrt[2]] × c[1]`

$$Out[\ ]= \ c_1 \left(\frac{|0\rangle}{\sqrt{2}} - \frac{|1\rangle}{\sqrt{2}} \right)_{S_1} \otimes \left(\frac{|0\rangle}{\sqrt{2}} - \frac{|1\rangle}{\sqrt{2}} \right)_{S_2} \otimes \left(\frac{|0\rangle}{\sqrt{2}} - \frac{|1\rangle}{\sqrt{2}} \right)_{S_3} +$$

$$c_0 \left(\frac{|0\rangle}{\sqrt{2}} + \frac{|1\rangle}{\sqrt{2}} \right)_{S_1} \otimes \left(\frac{|0\rangle}{\sqrt{2}} + \frac{|1\rangle}{\sqrt{2}} \right)_{S_2} \otimes \left(\frac{|0\rangle}{\sqrt{2}} + \frac{|1\rangle}{\sqrt{2}} \right)_{S_3}$$

In[ ]:= `out - new // Elaborate`

Out[ ]= `0`

Here is another quantum circuit model giving the same encoding.

In[ ]:= `qc2 = QuantumCircuit[vec, LogicalForm[Ket[], S@{2, 3}],`
        `CNOT[S[1], S[2]], CNOT[S[2], S[3]], S[{1, 2, 3}, 6]]`

In[ ]:= `out2 = DefaultForm@ExpressionFor[qc2];`
        `out2 // LogicalForm`

$$Out[\ ]= \ \frac{(c_0 + c_1)\,|0_{S_1}0_{S_2}0_{S_3}\rangle}{2\sqrt{2}} + \frac{(c_0 - c_1)\,|0_{S_1}0_{S_2}1_{S_3}\rangle}{2\sqrt{2}} + \frac{(c_0 - c_1)\,|0_{S_1}1_{S_2}0_{S_3}\rangle}{2\sqrt{2}} + \frac{(c_0 + c_1)\,|0_{S_1}1_{S_2}1_{S_3}\rangle}{2\sqrt{2}} +$$

$$\frac{(c_0 - c_1)\,|1_{S_1}0_{S_2}0_{S_3}\rangle}{2\sqrt{2}} + \frac{(c_0 + c_1)\,|1_{S_1}0_{S_2}1_{S_3}\rangle}{2\sqrt{2}} + \frac{(c_0 + c_1)\,|1_{S_1}1_{S_2}0_{S_3}\rangle}{2\sqrt{2}} + \frac{(c_0 - c_1)\,|1_{S_1}1_{S_2}1_{S_3}\rangle}{2\sqrt{2}}$$

6.1.3 Shor's Nine-Qubit Code

So far we have considered specific types of errors, the bit-flip and phase-flip errors. Obviously, realistic errors are far more general, not only in terms of the above two but also with any continuous change of the complex amplitudes in superposition of quantum states. It may sound rather surprising at first glance, but it turns out that roughly speaking, if a quantum error-correction code can correct the bit-flip and phase-flip errors, then it can correct any arbitrary errors. Before we introduce the precise statement and the corresponding construction of the quantum error-correction code in subsequently sections, we provide an example here: Shor's 9-qubit code. This code just builds upon the bit-flip and phase-flip correction codes discussed above, and yet it corrects completely arbitrary single-qubit errors.

According to the discussion above, protecting against the bit-flip and phase-flip error each requires for three qubits to encode one bit of physical quantum state. Hence, the encoding in Shor's code needs 9 qubits to protect against both types of errors and their possible combinations. More specifically, following the phase-flip correction code, we first encode the physical basis states $|0\rangle$ and $|1\rangle$ in the three-qubit logical basis states $|+++\rangle$ and $|---\rangle$, and we then encode each single-qubit state on three qubits using the bit-flip encoding scheme, that is, $|+\rangle$ by $(|000\rangle + |111\rangle)/\sqrt{2}$

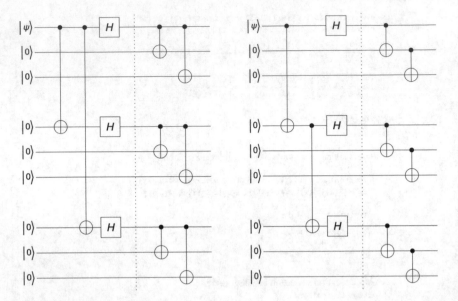

Fig. 6.1 Two quantum circuits for the encoding process in Shor's 9-qubit quantum error-correction code

and $|-\rangle$ by $(|000\rangle - |111\rangle)/\sqrt{2}$. One can regard that the phase-flip correction code is encoded on three blocks, and the bit-flip correction code on each block is encoded on the three qubits. The code words of the overall encoding are given by

$$|\bar{0}\rangle := \frac{(|000\rangle + |111\rangle) \otimes (|000\rangle + |111\rangle) \otimes (|000\rangle + |111\rangle)}{\sqrt{8}}, \tag{6.7a}$$

$$|\bar{1}\rangle := \frac{(|000\rangle - |111\rangle) \otimes (|000\rangle - |111\rangle) \otimes (|000\rangle - |111\rangle)}{\sqrt{8}}. \tag{6.7b}$$

The encoding process can be implemented either of the quantum circuits shown in Fig. 6.1, or their equivalent variants. The first part of each quantum circuit, which is separated from the latter part by the vertical dotted line, is for the phase-flip encoding [recall the quantum circuits in (6.2)] and the second part is to encode each block based on the bit-flip encoding [also recall the quantum circuits in (6.5)].

The syndrome of the bit-flip error on the first block can be diagnosed by measuring the following two observables

$$(\hat{Z} \otimes \hat{Z} \otimes \hat{I}) \otimes (\hat{I} \otimes \hat{I} \otimes \hat{I}) \otimes (\hat{I} \otimes \hat{I} \otimes \hat{I}), \tag{6.8a}$$

$$(\hat{I} \otimes \hat{Z} \otimes \hat{Z}) \otimes (\hat{I} \otimes \hat{I} \otimes \hat{I}) \otimes (\hat{I} \otimes \hat{I} \otimes \hat{I}). \tag{6.8b}$$

This is the way in which the bit-flip error is detected for a state encoded in three qubits. Similarly, the bit-flip errors on the rest of the blocks are diagnosed by measurement of the observables

$$(\hat{I} \otimes \hat{I} \otimes \hat{I}) \otimes (\hat{Z} \otimes \hat{Z} \otimes \hat{I}) \otimes (\hat{I} \otimes \hat{I} \otimes \hat{I}), \tag{6.8c}$$

$$(\hat{I} \otimes \hat{I} \otimes \hat{I}) \otimes (\hat{I} \otimes \hat{Z} \otimes \hat{Z}) \otimes (\hat{I} \otimes \hat{I} \otimes \hat{I}), \tag{6.8d}$$

$$(\hat{I} \otimes \hat{I} \otimes \hat{I}) \otimes (\hat{I} \otimes \hat{I} \otimes \hat{I}) \otimes (\hat{Z} \otimes \hat{Z} \otimes \hat{I}), \tag{6.8e}$$

$$(\hat{I} \otimes \hat{I} \otimes \hat{I}) \otimes (\hat{I} \otimes \hat{I} \otimes \hat{I}) \otimes (\hat{I} \otimes \hat{Z} \otimes \hat{Z}). \tag{6.8f}$$

Once the qubit with the bit-flip error has been identified, the error can be corrected by applying $\hat{X}$ on the qubit.

The phase-flip error on any single qubit switches the relative sign in the encoded state of the whole block that it belongs to. Following the phase-flip correction code, it is thus sufficient to compare the signs of different blocks. The observable to compare signs of the first two blocks is

$$(\hat{X} \otimes \hat{X} \otimes \hat{X}) \otimes (\hat{X} \otimes \hat{X} \otimes \hat{X}) \otimes (\hat{I} \otimes \hat{I} \otimes \hat{I}). \tag{6.9a}$$

Similarly, the signs of the second and third blocks can be compared through the measurement of the observable

$$(\hat{I} \otimes \hat{I} \otimes \hat{I}) \otimes (\hat{X} \otimes \hat{X} \otimes \hat{X}) \otimes (\hat{X} \otimes \hat{X} \otimes \hat{X}). \tag{6.9b}$$

Once the block affected by a phase-flip error has been identified, applying the operator $\hat{Z} \otimes \hat{Z} \otimes \hat{Z}$ on the block recovers the original state.

Here is a widely known quantum circuit model for Shor's 9-qubit quantum error correction code.

```
In[ ]:= qc = QuantumCircuit[
        LogicalForm[Ket[], S@Range[2, 9]],
        CNOT[S[1], S[4]], CNOT[S[1], S[7]],
        S[{1, 4, 7}, 6],
        {CNOT[S[1], S[2]], CNOT[S[4], S[5]], CNOT[S[7], S[8]]},
        {CNOT[S[1], S[3]], CNOT[S[4], S[6]], CNOT[S[7], S[9]]},
        "Invisible" → S@{3.5, 6.5}
        ]
```

This is the logical basis state $\left|0\right\rangle$ of the 9-qubit code.

```
In[ ]:= out0 = ExpressionFor@QuantumCircuit[Ket[S[1] → 0], qc];
        KetFactor@LogicalForm[out0, S@Range[9]]
```

$$Out[ ]= \frac{\left(\left|0_{S_1}0_{S_2}0_{S_3}\right\rangle + \left|1_{S_1}1_{S_2}1_{S_3}\right\rangle\right) \otimes \left(\left|0_{S_4}0_{S_5}0_{S_6}\right\rangle + \left|1_{S_4}1_{S_5}1_{S_6}\right\rangle\right) \otimes \left(\left|0_{S_7}0_{S_8}0_{S_9}\right\rangle + \left|1_{S_7}1_{S_8}1_{S_9}\right\rangle\right)}{2\sqrt{2}}$$

This is the logical basis state $\left|1\right\rangle$ of the 9-qubit code.

```
In[ ]:= out1 = ExpressionFor@QuantumCircuit[Ket[S[1] → 1], qc];
        KetFactor@LogicalForm[out1, S@Range[9]]
```

$$Out[ ]= \frac{\left(\left|0_{S_1}0_{S_2}0_{S_3}\right\rangle - \left|1_{S_1}1_{S_2}1_{S_3}\right\rangle\right) \otimes \left(\left|0_{S_4}0_{S_5}0_{S_6}\right\rangle - \left|1_{S_4}1_{S_5}1_{S_6}\right\rangle\right) \otimes \left(\left|0_{S_7}0_{S_8}0_{S_9}\right\rangle - \left|1_{S_7}1_{S_8}1_{S_9}\right\rangle\right)}{2\sqrt{2}}$$

Here is another equivalent implementation of Shor's 9-qubit quantum error correction code.

```
In[·]:= qc = QuantumCircuit[
        LogicalForm[Ket[], S@Range[2, 9]],
        CNOT[S[1], S[4]], CNOT[S[4], S[7]],
        S[{1, 4, 7}, 6],
        {CNOT[S[1], S[2]], CNOT[S[4], S[5]], CNOT[S[7], S[8]]},
        {CNOT[S[2], S[3]], CNOT[S[5], S[6]], CNOT[S[8], S[9]]},
        "Invisible" → S@{3.5, 6.5}
        ]
```

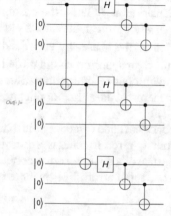

Out[·]=

This is the logical basis state $|0\rangle$ of the 9-qubit code.

```
In[·]:= out0 = ExpressionFor@QuantumCircuit[Ket[S[1] → 0], qc];
        KetFactor@LogicalForm[out0, S@Range[9]]
```

$$Out[·]= \frac{\left(\left|0_{S_1}0_{S_2}0_{S_3}\right\rangle + \left|1_{S_1}1_{S_2}1_{S_3}\right\rangle\right) \otimes \left(\left|0_{S_4}0_{S_5}0_{S_6}\right\rangle + \left|1_{S_4}1_{S_5}1_{S_6}\right\rangle\right) \otimes \left(\left|0_{S_7}0_{S_8}0_{S_9}\right\rangle + \left|1_{S_7}1_{S_8}1_{S_9}\right\rangle\right)}{2\sqrt{2}}$$

This is the logical basis state $|1\rangle$ of the 9-qubit code.

```
In[·]:= out1 = ExpressionFor@QuantumCircuit[Ket[S[1] → 1], qc];
        KetFactor@LogicalForm[out1, S@Range[9]]
```

$$Out[·]= \frac{\left(\left|0_{S_1}0_{S_2}0_{S_3}\right\rangle - \left|1_{S_1}1_{S_2}1_{S_3}\right\rangle\right) \otimes \left(\left|0_{S_4}0_{S_5}0_{S_6}\right\rangle - \left|1_{S_4}1_{S_5}1_{S_6}\right\rangle\right) \otimes \left(\left|0_{S_7}0_{S_8}0_{S_9}\right\rangle - \left|1_{S_7}1_{S_8}1_{S_9}\right\rangle\right)}{2\sqrt{2}}$$

So far, we have seen that the bit-flip and phase-flip errors on single qubits can be successfully detected and corrected as expected by constructing the code. Now the crucial point is that the same code can correct completely arbitrary errors as long as the error occurs in single qubits. Let $|\psi\rangle = |\bar{0}\rangle c_0 + |\bar{1}\rangle c_1$ be the initial state of the encoded qubits. Suppose that an error occurs in one of the qubits. The error can be efficiently described by a quantum operation $\mathscr{F}$. Suppose that $\mathscr{F}$ is specified by the Kraus elements $\hat{F}_\mu$. Then, the state corrupted by the error is given by

$$\mathscr{F}(|\psi\rangle\langle\psi|) = \sum_\mu \hat{F}_\mu |\psi\rangle\langle\psi| \hat{F}_\mu^\dagger. \tag{6.10}$$

It is a statistical mixture of various pure states. Let us focus on one of them, say, $\hat{F}_\mu |\psi\rangle$. As the Kraus element $\hat{F}_\mu$ is an operator on a single qubit j, it can be expanded in the Pauli operators,

$$\hat{F}_\mu = \hat{I}_j C_{0\mu} + \hat{X}_j C_{1\mu} + \hat{Y}_j C_{2\mu} + \hat{Z}_j C_{3\mu}, \tag{6.11}$$

where $\hat{C}_{\lambda\mu}$ are complex coefficients and the subscript j attached to the Pauli operators indicate that the operators act on the particular qubit j. Thus, the corruption state $\hat{F}_\mu |\psi\rangle$ is a superposition of the initial state $|\psi\rangle$ without error, the state $\hat{X}_j |\psi\rangle$ affected by the bit-flip error, the state $\hat{Z}_j |\psi\rangle$ affected by the phase-flip error, and the state $\hat{Y}_j |\psi\rangle = i\hat{X}\hat{Z} |\psi\rangle$ affected by both. When the error syndrome is diagnosed, the state collapses to the corresponding state, and the proper recovery operation discussed above brings back the initial state. The same procedure then applies to all other Kraus elements of $\mathscr{F}$.

This example clearly demonstrates that once the code can correct just bit-flip and phase-flip errors, then it can correct any arbitrary error. In other words, even though the errors in the qubits form a continuum, the fate of a quantum error-correction code is determined by its ability to correct a discrete set of errors. This is more rigorously established in the next section.

6.2   Quantum Error Correction

The characteristic property of the quantum information that distinguishes it from classical information is that it can take an arbitrary *continuous* superposition of logical states. Naturally, errors in quantum information form a continuum given that any slightest deviation in the superposition results in an error. At first glance, it may cast immense difficulties even in detecting errors. Ironically, "continuous" quantum information bears a striking difference from the information stored in classical *analog* systems, and errors in the quantum information can be "discretized" so to speak. In this section, we will show this in two stages. We first discuss the conditions for the errors to be corrected. We then proceed to show that correcting merely a discrete set of errors suffices to correct the continuum of errors in quantum information.

6.2.1   Quantum Error-Correction Conditions

We first present the *quantum error-correction conditions*. These conditions allow to determine, through simple tests, whether a quantum code protects against a certain class of quantum noise. Consider a quantum code associated with the code space $\mathcal{V}$. We describe the error with a quantum operation $\mathscr{E}$ specified by Kraus elements $\{\hat{E}_\mu\}$

(see Sect. 5.2.1). We want to show[1] that there exists an error-correction or recovery operation $\mathscr{R}$ that corrects $\mathscr{E}$ on $\mathcal{V}$ if and only if every pair of the Kraus elements satisfies

$$\hat{P}\hat{E}_\mu^\dagger \hat{E}_\nu \hat{P} = A_{\mu\nu}\hat{P} \tag{6.12}$$

for some Hermitian matrix $A_{\mu\nu}$, where $\hat{P}$ is the projector onto $\mathcal{V}$. In the course, physical meaning of condition (6.12) will become clear.

Supposing that condition in (6.12) is satisfied, we can construct the recovery operation $\mathscr{R}$. The matrix A is Hermitian by assumption, so it can be diagonalized into the form

$$A = U\Lambda U^\dagger \tag{6.13}$$

for some unitary matrix U and diagonal matrix Λ. Define new Kraus elements $\hat{F}_\nu = \sum_\mu \hat{E}_\mu U_{\mu\nu}$, then the condition (6.12) reads as

$$\hat{P}\hat{F}_\mu^\dagger \hat{F}_\nu \hat{P} = \delta_{\mu\nu}\lambda_\nu \hat{P}, \tag{6.14}$$

where λ_μ are the diagonal elements of Λ. Physically, this means that different Kraus elements $\hat{F}_\mu$ bring the code space $\mathcal{V}$ to mutually orthogonal subspaces, $\hat{F}_\mu \mathcal{V} \perp \hat{F}_\nu \mathcal{V}$ for $\mu \neq \nu$. Therefore, measuring the error syndrome corresponds to identifying one of the subspaces, and the subsequent recovery procedure corresponds to bringing the subspace back to the code space. Mathematically, each subspace can be identified by finding the projection operator onto it. To find the projection operators, we note that the polar decomposition (Theorem A.23) of $\hat{F}_\mu \hat{P}$ leads to

$$\hat{F}_\mu \hat{P} = \hat{V}_\mu \sqrt{\hat{P}\hat{F}_\mu^\dagger \hat{F}_\mu \hat{P}} = \hat{V}_\mu \sqrt{\lambda_\mu \hat{P}} = \sqrt{\lambda_\mu}\hat{V}_\mu \hat{P}, \tag{6.15}$$

where $\hat{V}_\mu$ is a unitary operator. We define

$$\hat{P}_\mu := \hat{V}_\mu \hat{P}\hat{V}_\mu^\dagger, \tag{6.16}$$

and assert that they are indeed the desired projection operators

$$\hat{P}_\mu \hat{P}_\nu = \hat{P}_\mu^\dagger \hat{P}_\nu = \frac{\hat{V}_\mu \hat{P}\hat{F}_\mu^\dagger \hat{F}_\nu \hat{P}\hat{V}_\mu^\dagger}{\sqrt{\lambda_\mu \lambda_\nu}} = \delta_{\mu\nu}\hat{P}_\nu. \tag{6.17}$$

Now, the error syndrome can be diagnosed by a projection measurement (Postulate 3′) associated with projectors $\hat{P}_\mu$. After a diagnosis of the syndrome, one can simply apply the unitary operator $\hat{V}_\mu^\dagger$ on the resulting state to recover the initial state.

We have constructed a recovery procedure starting from condition (6.12) assumed for the error process. To show that the condition is most general, suppose that the

---

[1] We follow the proof in Nielsen & Chuang (2011, Sect. 10.3).

error described by $\mathscr{E}$ is correctable. This means that there exists a quantum operation $\mathscr{R}$ that describes the recovery process. Mathematically, it can be specified by the Kraus elements $\{\hat{R}_\nu\}$ such that

$$\mathscr{R}(\mathscr{E}(\hat{P}\hat{\rho}\hat{P})) = c\hat{P}\hat{\rho}\hat{P} \tag{6.18}$$

for some constant $0 < c \leq 1$. In terms of the Kraus elements of $\mathscr{E}$ and $\mathscr{R}$, it reads as

$$\sum_{\mu\nu} \hat{R}_\mu \hat{E}_\nu \hat{P}\hat{\rho}(\hat{R}_\mu \hat{E}_\nu \hat{P})^\dagger = c\hat{P}\hat{\rho}\hat{P} \tag{6.19}$$

for all $\hat{\rho}$. The set $\{\hat{R}_\mu \hat{E}_\nu \hat{P}\}$ of Kraus elements defines the composite quantum operation $\mathscr{R} \circ \mathscr{E}$ describing the recovery after the error. Equation (6.19) indicates that the same quantum operation $\mathscr{R} \circ \mathscr{E}$ is described by a single Kraus element $\sqrt{c}\hat{P}$. According to the unitary freedom for Kraus elements discussed in Sect. 5.2.1, particularly described in Eqs. (5.34) and (5.35), there must exist complex numbers $W_{\mu\nu}$ such that

$$\hat{R}_\mu \hat{E}_\nu = \sqrt{c}\hat{P}W_{\mu\nu} \tag{6.20}$$

and

$$\sum_{\mu\nu} W_{\mu\nu}^* W_{\mu\nu} = 1. \tag{6.21}$$

Therefore, it follows that

$$\sum_\lambda (\hat{R}_\lambda \hat{E}_\mu \hat{P})^\dagger (\hat{R}_\lambda \hat{E}_\nu \hat{P}) = \sum_\lambda \hat{P}\hat{E}_\mu^\dagger \hat{R}_\lambda^\dagger \hat{R}_\lambda \hat{E}_\nu \hat{P} = \hat{P}\hat{E}_\mu^\dagger \hat{E}_\nu \hat{P} = c(W^\dagger W)_{\mu\nu}\hat{P}, \tag{6.22}$$

where we have used the fact that $\sum_\lambda \hat{R}_\lambda^\dagger \hat{R}_\lambda = \hat{I}$ because $\mathscr{R}$ is trace-preserving. Since the matrix $A := cW^\dagger W$ is certainly Hermitian by construction, the last equality in (6.22) asserts that the Kraus elements $\{\hat{E}_\mu\}$ satisfy condition (6.12).

6.2.2 Discretization of Errors

Once a quantum code is proven to protect against a particular type of quantum noise in accordance with the quantum error-correction conditions, it is robust against a surprisingly wide class of types of quantum noise. In fact, it turns out that if a quantum noise operation $\mathscr{E}$ with error operators $\{\hat{E}_\mu\}$ can be corrected by a quantum code $\mathcal{V}$, then the code also protects against any quantum noise $\mathscr{F}$ with error operators $\{\hat{F}_\nu\}$ that are linear superpositions of $\{\hat{E}_\mu\}$,

$$\hat{F}_\nu = \sum_\mu \hat{E}_\mu M_{\mu\nu}, \tag{6.23}$$

where $M_{\mu\nu}$ is a matrix of complex numbers.

The proof is simple. Given that the quantum noise $\mathcal{E}$ is correctable on code space $\mathcal{V}$, it must satisfy the quantum error-correction conditions in (6.12). Multiply the equations with the Matrix $M^\dagger$ and M,

$$\sum_{\mu'\nu'} \hat{P} M^\dagger_{\mu\mu'} \hat{E}^\dagger_{\mu'} \hat{E}_{\nu'} M_{\nu'\nu} \hat{P} = \sum_{\mu'\nu'} M^\dagger_{\mu\mu'} A_{\mu'\nu'} M_{\nu'\nu} \hat{P} \qquad (6.24)$$

This leads to

$$\hat{P} \hat{F}^\dagger_\mu \hat{F}_\nu \hat{P} = (M^\dagger A M)_{\mu\nu} \hat{P}, \qquad (6.25)$$

which is formally the same as the quantum error-correction conditions. Therefore, the quantum noise $\mathcal{F}$ must be correctable on $\mathcal{V}$ as well.

Although quite simple to prove, the implication of the above statement is huge. Consider a class of errors on single qubits. Suppose that it is described by a quantum operation $\mathcal{E}$ specified by Kraus elements $\{\hat{E}_{j\mu}\}$, where the index j indicates the qubit subject to the error and μ denotes different error processes. Since $\hat{E}_{j\mu}$ are operators on the single qubit j, they can be expanded in terms of the Pauli operators $\hat{S}^0_j$, $\hat{S}^x_j$, $\hat{S}^y_j$, and $\hat{S}^z_j$. In order to check if a given quantum error-correction code protects against arbitrary single-qubit errors, one has only to inspect the condition

$$\hat{P} \hat{S}^\mu_j \hat{S}^\nu_j \hat{P} = A_{\mu\nu} \hat{P} \qquad (6.26)$$

for all j and a Hermitian matrix A. In other words, when one constructs a quantum error-correction code, it is sufficient to check a finite (and hence discrete) set of conditions to ensure that the code protects against arbitrary single-qubit errors.

6.3 Stabilizer Formalism

There are several ways to construct a quantum error-correction code. One of the most popular ways is to make an analogy to classical linear codes with the Calderbank–Shor–Steane (CSS) codes as a typical example. In this book, we skip the approach and discuss stabilizer codes. Stabilizer codes do not allow for a direct counterpart in classical codes, not to speak of its elegance and simplicity. Before we discuss stabilizer codes in the next section, we introduce the basic framework of the stabilizer formalism. Stabilizer formalism was first put forward by Gottesman (1996). This formalism is explained in great detail in Gottesman (1997) and is also presented in Nielsen & Chuang (2011). Recently, stabilizer formalism has been extended to the measurement-based quantum computation scheme (Browne & Briegel, 2016).

A quantum error-correction code encodes quantum states of a fixed number of qubits in a subspace of the Hilbert space of more qubits. This subspace is called the *code space* of the particular code. The code space is specified by the logical basis states that span it, and we have seen some examples in Eqs. (6.1), (6.4), and (6.7) from Sect. 6.1. A natural way to analyze the code is to inspect how the quantum states in the code space are affected by the errors. Mathematically, this is achieved by examining the evolution of the quantum states under the quantum operation describing the error process as in Sect. 6.2. This corresponds to the usual Schrödinger picture of dynamical processes in quantum mechanics.

It may sound odd at first glance, but a code space can also be specified by operators that preserve it, and errors can be described by tracking those operators. This is the task for which the stabilizer formalism is. In a broad sense, it is analogous to the Heisenberg picture of dynamical processes in quantum mechanics.

Let $\mathcal{G}$ be a group of operators acting on the quantum states of a quantum system (see Appendix C for a brief introduction to group theory). In quantum computation, the most relevant group is the unitary group $U(2^n)$ of quantum logic gate operators on n qubits. Then, let $|\psi\rangle$ be a quantum state. When it remains unchanged under an operator $\hat{G}$ in the group $\mathcal{G}$,

$$\hat{G}|\psi\rangle = |\psi\rangle, \tag{6.27}$$

the state is said to be *stabilized* by $\hat{G}$, or equivalently, $\hat{G}$ is said to *stabilize* $|\psi\rangle$. For example, the Bell state $|\Phi\rangle := (|00\rangle + |11\rangle)/\sqrt{2}$ is stabilized by $\hat{X} \otimes \hat{X}$. The set of operators that stabilize a quantum state $|\psi\rangle$ forms a subgroup of $\mathcal{G}$. This subgroup is called the *stabilizer subgroup* or simply *stabilizer* of the quantum state $|\psi\rangle$. For example, the Bell state $|\Phi\rangle$ is also stabilized by $\hat{Z} \otimes \hat{Z}$ and $-\hat{Y} \otimes \hat{Y}$ as well as, trivially, by $\hat{I} \otimes \hat{I}$. These operators form a subgroup of $U(2^2)$. In other words, the subgroup

$$\{\hat{I} \otimes \hat{I}, \ \hat{X} \otimes \hat{X}, \ \hat{Z} \otimes \hat{Z}, \ -\hat{Y} \otimes \hat{Y}\} \tag{6.28}$$

is the stabilizer of the Bell state $|\Phi\rangle$.

Consider one of the Bell states.

```
In[ ]:= ket = BellState[S@{1, 2}, 3];
       ket // LogicalForm
```

$$Out[ ]= \frac{|0_{S_1} 0_{S_2}\rangle - |1_{S_1} 1_{S_2}\rangle}{\sqrt{2}}$$

```
In[ ]:= grp = {1, -S[1, 1] ** S[2, 1], S[1, 3] ** S[2, 3], S[1, 2] ** S[2, 2]};
       PauliForm[grp]
```

$$Out[ ]= \{I \otimes I, \ -(X \otimes X), \ Z \otimes Z, \ Y \otimes Y\}$$

```
In[ ]:= new = grp ** ket;
       new // LogicalForm
```

$$Out[ ]= \left\{ \frac{|0_{S_1} 0_{S_2}\rangle}{\sqrt{2}} - \frac{|1_{S_1} 1_{S_2}\rangle}{\sqrt{2}}, \ \frac{|0_{S_1} 0_{S_2}\rangle}{\sqrt{2}} - \frac{|1_{S_1} 1_{S_2}\rangle}{\sqrt{2}}, \ \frac{|0_{S_1} 0_{S_2}\rangle}{\sqrt{2}} - \frac{|1_{S_1} 1_{S_2}\rangle}{\sqrt{2}}, \ \frac{|0_{S_1} 0_{S_2}\rangle}{\sqrt{2}} - \frac{|1_{S_1} 1_{S_2}\rangle}{\sqrt{2}} \right\}$$

The notion of stabilizer is extended to subspaces. Let $\mathcal{V}$ be a subspace of the Hilbert space associated with a quantum system. The stabilizer $\mathcal{S}$ of $\mathcal{V}$ is the set of all operators $\hat{G}$ in $\mathcal{G}$ that stabilize all quantum states in $\mathcal{V}$. For example, consider the code space $\mathcal{V}$ of the bit-flip correction code discussed in Sect. 6.1.1. $\mathcal{V}$ is spanned by the two logical basis states $|000\rangle$ and $|111\rangle$. The stabilizer $\mathcal{S}$ of $\mathcal{V}$ is given by

$$\mathcal{S} = \{\hat{I} \otimes \hat{I} \otimes \hat{I}, \ \hat{Z} \otimes \hat{Z} \otimes \hat{I}, \ \hat{I} \otimes \hat{Z} \otimes \hat{Z}, \ \hat{Z} \otimes \hat{I} \otimes \hat{Z}\}. \qquad (6.29)$$

Consider the code space $\mathcal{V}$ of the bit-flip correction code. It is spanned by the following two logical basis states.

In[-]:= `ss = S[{1, 2, 3}, None];`
`bs = {Ket[], Ket[S@{1, 2, 3} → 1]};`
`bs // LogicalForm`

Out[-]:= $\left\{ \left|0_{S_1}0_{S_2}0_{S_3}\right\rangle, \ \left|1_{S_1}1_{S_2}1_{S_3}\right\rangle \right\}$

The stabilizer $\mathcal{S}$ of $\mathcal{V}$ is given as follows.

In[-]:= `grp = {1, S[1, 3] ** S[2, 3], S[2, 3] ** S[3, 3], S[1, 3] ** S[3, 3]};`
`grp // PauliForm`

Out[-]:= `{I⊗I⊗I, Z⊗Z⊗I, I⊗Z⊗Z, Z⊗I⊗Z}`

This checks the above group indeed stabilizes $\mathcal{V}$.

In[-]:= `grp ** bs⟦1⟧ // LogicalForm[#, ss] &`
`grp ** bs⟦2⟧ // LogicalForm[#, ss] &`

Out[-]:= $\left\{ \left|0_{S_1}0_{S_2}0_{S_3}\right\rangle, \ \left|0_{S_1}0_{S_2}0_{S_3}\right\rangle, \ \left|0_{S_1}0_{S_2}0_{S_3}\right\rangle, \ \left|0_{S_1}0_{S_2}0_{S_3}\right\rangle \right\}$

Out[-]:= $\left\{ \left|1_{S_1}1_{S_2}1_{S_3}\right\rangle, \ \left|1_{S_1}1_{S_2}1_{S_3}\right\rangle, \ \left|1_{S_1}1_{S_2}1_{S_3}\right\rangle, \ \left|1_{S_1}1_{S_2}1_{S_3}\right\rangle \right\}$

Consider the code space of the phase-flip correction code. It is spanned by the following two logical basis states.

In[-]:= `ss = S[{1, 2, 3}, None];`
`bs =`
`   {ProductState[ss → {1, 1} / Sqrt[2]], ProductState[ss → {1, -1} / Sqrt[2]]};`
`bs // LogicalForm`

Out[-]:= $\left\{ \left(\frac{|0\rangle}{\sqrt{2}} + \frac{|1\rangle}{\sqrt{2}}\right)_{S_1} \otimes \left(\frac{|0\rangle}{\sqrt{2}} + \frac{|1\rangle}{\sqrt{2}}\right)_{S_2} \otimes \left(\frac{|0\rangle}{\sqrt{2}} + \frac{|1\rangle}{\sqrt{2}}\right)_{S_3}, \right.$
$\left. \left(\frac{|0\rangle}{\sqrt{2}} - \frac{|1\rangle}{\sqrt{2}}\right)_{S_1} \otimes \left(\frac{|0\rangle}{\sqrt{2}} - \frac{|1\rangle}{\sqrt{2}}\right)_{S_2} \otimes \left(\frac{|0\rangle}{\sqrt{2}} - \frac{|1\rangle}{\sqrt{2}}\right)_{S_3} \right\}$

The stabilizer $\mathcal{S}$ of $\mathcal{V}$ is given as follows.

In[-]:= `grp = {1, S[1, 1] ** S[2, 1], S[2, 1] ** S[3, 1], S[1, 1] ** S[3, 1]};`
`grp // PauliForm`

Out[-]:= `{I⊗I⊗I, X⊗X⊗I, I⊗X⊗X, X⊗I⊗X}`

This checks the above group indeed stabilizes $\mathcal{V}$.

In[·]:= `grp ** bs〚1〛 // Elaborate // KetFactor // LogicalForm[#, ss] &`

$$
Out[·]= \left\{ \frac{(|0_{s_1}\rangle + |1_{s_1}\rangle) \otimes (|0_{s_2}\rangle + |1_{s_2}\rangle) \otimes (|0_{s_3}\rangle + |1_{s_3}\rangle)}{2\sqrt{2}}, \right.
$$

$$
\frac{(|0_{s_1}\rangle + |1_{s_1}\rangle) \otimes (|0_{s_2}\rangle + |1_{s_2}\rangle) \otimes (|0_{s_3}\rangle + |1_{s_3}\rangle)}{2\sqrt{2}},
$$

$$
\frac{(|0_{s_1}\rangle + |1_{s_1}\rangle) \otimes (|0_{s_2}\rangle + |1_{s_2}\rangle) \otimes (|0_{s_3}\rangle + |1_{s_3}\rangle)}{2\sqrt{2}},
$$

$$
\left. \frac{(|0_{s_1}\rangle + |1_{s_1}\rangle) \otimes (|0_{s_2}\rangle + |1_{s_2}\rangle) \otimes (|0_{s_3}\rangle + |1_{s_3}\rangle)}{2\sqrt{2}} \right\}
$$

In[·]:= `grp ** bs〚2〛 // Elaborate // KetFactor // LogicalForm[#, ss] &`

$$
Out[·]= \left\{ \frac{(|0_{s_1}\rangle - |1_{s_1}\rangle) \otimes (|0_{s_2}\rangle - |1_{s_2}\rangle) \otimes (|0_{s_3}\rangle - |1_{s_3}\rangle)}{2\sqrt{2}}, \right.
$$

$$
\frac{(|0_{s_1}\rangle - |1_{s_1}\rangle) \otimes (|0_{s_2}\rangle - |1_{s_2}\rangle) \otimes (|0_{s_3}\rangle - |1_{s_3}\rangle)}{2\sqrt{2}},
$$

$$
\frac{(|0_{s_1}\rangle - |1_{s_1}\rangle) \otimes (|0_{s_2}\rangle - |1_{s_2}\rangle) \otimes (|0_{s_3}\rangle - |1_{s_3}\rangle)}{2\sqrt{2}},
$$

$$
\left. \frac{(|0_{s_1}\rangle - |1_{s_1}\rangle) \otimes (|0_{s_2}\rangle - |1_{s_2}\rangle) \otimes (|0_{s_3}\rangle - |1_{s_3}\rangle)}{2\sqrt{2}} \right\}
$$

Under a quantum operation, a given subspace evolves to a different subspace, and hence the stabilizer of the subspace changes accordingly. In stabilizer formalism, one describes the quantum operation by tracking the stabilizer rather than the subspace or quantum states that are contained in it. For a general parent group $\mathcal{G}$ such as the unitary group $U(2^n)$, the stabilizer formalism may involve unnecessary mathematical complications. However, for many quantum information processing procedures including quantum error correction, the Pauli operators and their tensor products are sufficient. In this case, stabilizer formalism turns out to be extremely simple and insightful.

6.3.1   Pauli Group

In general, finding the stabilizer of a given subspace and vice versa may be non-trivial. Fortunately, when operators are restricted to the Pauli operators or their tensor products on a fixed number n of qubits, the stabilizers have particularly simple structures and useful properties. Formally, such operators form a group (Appendix C), called the *Pauli group* $\mathcal{P}(n)$ on n qubits. Moreover, in accordance with the discretization of errors discussed in Sect. 6.2, such operators are enough to describe any quantum noise operation. Therefore, stabilizer subgroups of the Pauli group provide convenient and powerful mathematical tools to construct and implement a wide class of quantum error-correction codes. In the remaining of the book, we assume that the parent group of any stabilizer is a Pauli group.

The Pauli group on a single qubit consists of the single-qubit Pauli operators. Enlisting the elements, it is given by

$$\mathcal{P}(1) := \left\{ \pm\hat{I}, \pm i\hat{I}, \pm\hat{X}, \pm i\hat{X}, \pm\hat{Y}, \pm i\hat{Y}, \pm\hat{Z}, \pm i\hat{Z} \right\}. \tag{6.30}$$

The additional phase factors ± 1 and $\pm i$ are required because the product of two Pauli operators gives another Pauli operator accompanied by factor $\pm i$. Otherwise, the underlying set is not closed under the group multiplication. The Pauli group on n qubits is given by

$$\mathcal{P}(n) := \left\{ \pm 1, \pm i \right\} \times \left\{ \hat{I}, \hat{X}, \hat{Y}, \hat{Z} \right\}^{\otimes n}. \tag{6.31}$$

We call elements of the Pauli group on n qubits n-qubit *Pauli operators*.

As the number of elements in the Pauli group grows fast with qubits, 4^{n+1}, it is convenient to describe the group in terms of the *generators* (Definition C.2). For example, the single-qubit Pauli group is generated by $\hat{X}$, $\hat{Y}$, and $\hat{Z}$ (or by $i\hat{I}$, $\hat{X}$, and $\hat{Z}$). We denote the generators by enclosing them in $\langle\!\langle \cdots \rangle\!\rangle$. Hence, one can write

$$\mathcal{P}(1) = \left\langle\!\left\langle \{\hat{X}, \hat{Y}, \hat{Z}\} \right\rangle\!\right\rangle. \tag{6.32}$$

The Pauli group on n qubits is generated by tensor products of the Pauli operators acting non-trivially on one and only one qubit. For example, we observe that

$$\mathcal{P}(2) = \left\langle\!\left\langle \{\hat{X} \otimes \hat{I}, \hat{Y} \otimes \hat{I}, \hat{Z} \otimes \hat{I}, \hat{I} \otimes \hat{X}, \hat{I} \otimes \hat{Y}, \hat{I} \otimes \hat{Z}\} \right\rangle\!\right\rangle \tag{6.33}$$

for the Pauli group on two qubits. We often omit the number n of qubits from the Pauli group specification $\mathcal{P}(n)$ and write just $\mathcal{P}$ when there is no risk of confusion.

Here is the Pauli group on a single qubit.

```
In[·]:= grp = PauliGroup[1]
Out[·]:= PauliGroup[1]

In[·]:= elm = GroupElements[grp];
        PauliForm[elm]
Out[·]:= {I, X, Y, Z, -I, -X, -Y, -Z, i I, i X, i Y, i Z, -i I, -i X, -i Y, -i Z}
```

The number of elements quickly increases with the number of qubits.

```
In[·]:= grp = PauliGroup[2]
Out[·]:= PauliGroup[2]

In[·]:= elm = GroupElements[grp];
        PauliForm[elm]
Out[·]:= {I⊗I, I⊗X, I⊗Y, I⊗Z, X⊗I, X⊗X, X⊗Y, X⊗Z, Y⊗I, Y⊗X, Y⊗Y, Y⊗Z,
         Z⊗I, Z⊗X, Z⊗Y, Z⊗Z, -(I⊗I), -(I⊗X), -(I⊗Y), -(I⊗Z), -(X⊗I),
         -(X⊗X), -(X⊗Y), -(X⊗Z), -(Y⊗I), -(Y⊗X), -(Y⊗Y), -(Y⊗Z), -(Z⊗I),
         -(Z⊗X), -(Z⊗Y), -(Z⊗Z), i I⊗I, i I⊗X, i I⊗Y, i I⊗Z, i X⊗I, i X⊗X,
         i X⊗Y, i X⊗Z, i Y⊗I, i Y⊗X, i Y⊗Y, i Y⊗Z, i Z⊗I, i Z⊗X, i Z⊗Y, i Z⊗Z,
         -i I⊗I, -i I⊗X, -i I⊗Y, -i I⊗Z, -i X⊗I, -i X⊗X, -i X⊗Y, -i X⊗Z,
         -i Y⊗I, -i Y⊗X, -i Y⊗Y, -i Y⊗Z, -i Z⊗I, -i Z⊗X, -i Z⊗Y, -i Z⊗Z}
```

```
In[ ]:= GroupOrder@grp
Out[ ]= 64
```

Groups can be conveniently described by the use of generators.

```
In[ ]:= grp = PauliGroup[1]
Out[ ]= PauliGroup[1]
```

For example, the single-qubit Pauli group is generated by three Pauli operators.

```
In[ ]:= gnr = GroupGenerators[grp];
       PauliForm[gnr]
Out[ ]= {X, Y, Z}
```

Here is a generating set of the Pauli group on two qubits. Six generators are enough to generate it.

```
In[ ]:= gnr = GroupGenerators[PauliGroup[2]];
       PauliForm[gnr]
Out[ ]= {I⊗X, I⊗Y, I⊗Z, X⊗I, Y⊗I, Z⊗I}
```

Operators in Pauli groups have two extremely useful properties that eventually determine the group-theoretical structure of the stabilizers and stabilizer codes. First, any two elements of a Pauli group either commute or anti-commute with each other. Second, every element of the Pauli group squares to $\pm\hat{I}$. The elements that square to $\hat{I}$ have eigenvalues ± 1 whereas those that square to $-\hat{I}$ have eigenvalues $\pm i$.

Pauli group has two extremely useful properties. These are illustrated here.

To avoid the irrelevant phase factors ±1 and ±i, we play with simple tensor products of single-qubit Pauli operators rather than the full Pauli group.

```
In[ ]:= ops = Flatten@Outer[Multiply, S[1, Full], S[2, Full]]
Out[ ]= {S_2^0, S_2^x, S_2^y, S_2^z, S_1^x, S_1^x S_2^x, S_1^x S_2^y, S_1^x S_2^z, S_1^y, S_1^y S_2^x, S_1^y S_2^y, S_1^y S_2^z, S_1^z, S_1^z S_2^x, S_1^z S_2^y, S_1^z S_2^z}
```

```
In[ ]:= PauliForm[ops]
Out[ ]= {I⊗I, I⊗X, I⊗Y, I⊗Z, X⊗I, X⊗X, X⊗Y,
        X⊗Z, Y⊗I, Y⊗X, Y⊗Y, Y⊗Z, Z⊗I, Z⊗X, Z⊗Y, Z⊗Z}
```

This table shows that the elements of the Pauli group either commute or anti-commute with each other.

```
In[ ]:= mat = Outer[GottesmanTest, ops, ops];
       TableForm[mat[-6 ;;, -6 ;;],
         TableHeadings → {PauliForm@ops[-6 ;;], PauliForm@ops[-6 ;;]},
         TableAlignments → Right]
Out[ ]//TableForm=
```

	Y⊗Y	Y⊗Z	Z⊗I	Z⊗X	Z⊗Y	Z⊗Z
Y⊗Y	1	-1	-1	1	-1	1
Y⊗Z	-1	1	-1	1	1	-1
Z⊗I	-1	-1	1	1	1	1
Z⊗X	1	1	1	1	-1	-1
Z⊗Y	-1	1	1	-1	1	-1
Z⊗Z	1	-1	1	-1	-1	1

The rule is also simple that allows to determine which elements commute and which elements anti-commute .

Any element of the Pauli group squares to ±1. Elements with the phase factor ±i squares to -1.

```
In[·]:= elm = GroupElements[PauliGroup[S@{1, 2}]];
       sqr = Map[# ** # &, elm];
       TableForm[{PauliForm[elm][[-8 ;;]], sqr[[-8 ;;]]}]
Out[·]//TableForm=
       - i Y⊗I    - i Y⊗X    - i Y⊗Y    - i Y⊗Z    - i Z⊗I    - i Z⊗X    i Z⊗Y    - i Z⊗Z
       -1         -1         -1         -1         -1         -1         -1       -1
```

The above property implies that any element of the Pauli group has eigenvalues ±1 or ±i.

In most applications of the Pauli group, an explicit multiplication table is not necessary, commutation (or anti-commutation) of the elements is sufficient. As long as the commutation of the elements is concerned, one can ignore the phase factors ± 1 and $\pm i$ in the elements of the Pauli group. Mathematically, this can be achieved by considering the *quotient group*

$$\mathcal{P}'(n) := \mathcal{P}(n)/\mathcal{Z}(n), \tag{6.34}$$

where

$$\mathcal{Z}(n) := \left\{ \pm \hat{I}^{\otimes n}, \pm i \hat{I}^{\otimes n} \right\} \tag{6.35}$$

is an invariant subgroup of $\mathcal{P}(n)$.[2] For example, in the quotient group

$$\mathcal{P}' = \left\{ \{\pm \hat{I}, \pm i \hat{I}\}, \{\pm \hat{X}, \pm i \hat{X}\}, \{\pm \hat{Y}, \pm i \hat{Y}\}, \{\pm \hat{Z}, \pm i \hat{Z}\} \right\} \tag{6.36}$$

on a single qubit, the elements $\pm \hat{X}$ and $\pm i \hat{X}$ are not distinguished any longer. They are regarded as equivalent and are denoted collectively by the *coset*

$$\hat{X}\mathcal{Z} = \left\{ \pm \hat{X}, \pm i \hat{X} \right\} \tag{6.37}$$

or, equivalently, $\mathcal{Z}\hat{X}$. This leads to a drastically simple structure of $\mathcal{P}'$ compared to $\mathcal{P}$. For example, $\hat{X}\hat{Y}$ is now equivalent to $\hat{Y}\hat{X}$, that is,

$$\hat{X}\hat{Z}\mathcal{Z} = \hat{Z}\hat{X}\mathcal{Z}. \tag{6.38}$$

This means that unlike the genuine Pauli group $\mathcal{P}(n)$, the quotient group $\mathcal{P}'(n)$ is Abelian and is isomorphic to $(\mathbb{Z}_2)^{2n}$.

For later use, let us establish an isomorphism (i.e., one-to-one correspondence) explicitly between $\mathcal{P}'(n)$ and $(\mathbb{Z}_2)^{2n}$. It suffices to say that on each qubit, we associate each Pauli operator with an ordered pair in $\mathbb{Z}_2 \times \mathbb{Z}_2$ as follows

$$\hat{I} \leftrightarrow (0,0), \quad \hat{X} \leftrightarrow (1,0), \quad \hat{Z} \leftrightarrow (0,1), \quad \hat{Y} \leftrightarrow (1,1). \tag{6.39}$$

For example, given a tensor product $\hat{G} = i\hat{X} \otimes \hat{Z} \otimes \hat{I} \otimes \hat{Y}$, on a system of four qubits, the coset $\hat{G}\mathcal{Z}$ corresponds to $(1, 0; 0, 1; 0, 0; 1, 1)$, that is,

---

[2] In fact, $\mathcal{Z}(n)$ is the center of $\mathcal{P}(n)$, the largest Abelian invariant subgroup of $\mathcal{P}(n)$ (see Appendix C).

$$\hat{G}\mathcal{Z} \leftrightarrow (1, 0; 0, 1; 0, 0; 1, 1). \tag{6.40}$$

Recall that the Abelian group $(\mathbb{Z}_2)^{2n}$ with bit-wise addition modulo 2 as group multiplication can also be regarded as a vector space over the field $\mathbb{Z}_2$ of the binary numbers (see Appendix C.5). For this reason, it is convenient to denote the correspondence in (6.40) by

$$\left| \hat{G} \right\rangle := (1, 0; 0, 1; 0, 0; 1, 1)^T, \quad \left\langle \hat{G} \right| := (1, 0; 0, 1; 0, 0; 1, 1). \tag{6.41}$$

We call the vector $\left| \hat{G} \right\rangle$ in $(\mathbb{Z}_2)^{2n}$ the *Gottesman vector* of the operator $\hat{G}$ (more precisely, of the coset $\hat{G}\mathcal{Z}$).

Ignoring the phase factors in the operators of Pauli group, we are dealing with the factor group of the Pauli group with respect to the center {I, -I, iI, -iI}.

```
In[·]:= op = -I * S[1, 2] ** S[3, 1] ** S[4, 3]
        PauliForm[op, ss = S@{1, 2, 3, 4}]
```

```
Out[·]= - i S₁ʸ S₃ˣ S₄ᶻ
```

```
Out[·]= - i Y ⊗ I ⊗ X ⊗ Z
```

This gives the Gottesman vector of the above operator (more precisely, the coset represented by the operator).

```
In[·]:= vec = GottesmanVector[op, ss]
```

```
Out[·]= {1, 1, 0, 0, 1, 0, 0, 1}
```

An operator representing the coset is recovered.

```
In[·]:= new = FromGottesmanVector[vec, ss]
```

```
Out[·]= S₁ʸ S₃ˣ S₄ᶻ
```

The correspondence between the Pauli operators (more precisely, cosets in $\mathcal{P}'$) and the Gottesman vectors is a group isomorphism that preserves group multiplication. Specifically, the multiplication of Pauli operators corresponds to addition (modulo 2) of the corresponding Gottesman vectors. This provides a convenient way to check for the independence of Pauli operators. Suppose that the elements $\hat{G}_1, \hat{G}_2, \ldots, \hat{G}_k$ of $\mathcal{P}(n)$ are *independent*, so that none of them can be obtained (up to a phase factor) by a multiplication of the others. Then, it follows (see Problem 6.2) that the corresponding Gottesman vectors $\left| \hat{G}_1 \right\rangle, \left| \hat{G}_2 \right\rangle, \ldots, \left| \hat{G}_k \right\rangle$ are *linearly independent* of each other in $(\mathbb{Z}_2)^{2n}$. The converse is also true in that if the Gottesman vectors $\left| \hat{G}_1 \right\rangle, \left| \hat{G}_2 \right\rangle, \ldots, \left| \hat{G}_k \right\rangle$ are linearly independent of each other in $(\mathbb{Z}_2)^{2n}$, then the operators $\hat{G}_1, \hat{G}_2, \ldots, \hat{G}_k$ are independent (up to a phase factor) of each other in $\mathcal{P}(n)$. This method is especially useful when one wants to pick up independent generators from an over-generating set of a stabilizer.

This correspondence also carries the commutation relation of Pauli operators to a certain type of relation of the corresponding Gottesman vectors. Let $\hat{G}_1$ and $\hat{G}_2$

be n-qubit Pauli operators. It is straightforward to show by inspection that if they commute with each other, then

$$\left\langle \hat{G}_1 \middle| \hat{J} \middle| \hat{G}_2 \right\rangle = 0, \tag{6.42}$$

where the operator $\hat{J}$ on Gottesman vectors is defined by

$$\hat{J} := \bigoplus_{j=1}^{n} \begin{bmatrix} 0 & 1 \\ 1 & 0 \end{bmatrix} = \begin{bmatrix} 0 & 1 & & \\ 1 & 0 & & \\ & & \ddots & \\ & & & 0 & 1 \\ & & & 1 & 0 \end{bmatrix}. \tag{6.43}$$

If $\hat{G}_1$ and $\hat{G}_2$ anti-commute with each other, then

$$\left\langle \hat{G}_1 \middle| \hat{J} \middle| \hat{G}_2 \right\rangle = 1. \tag{6.44}$$

The converse of each of the above statements also holds. The operator $\hat{J}$ equips the space $(\mathbb{Z}_2)^{2n}$ of Gottesman vectors with an inner product.

One can use Gottesman vectors to prove another interesting property of Pauli groups: Consider again a set of independent Pauli operators $\hat{G}_1, \hat{G}_2, \ldots, \hat{G}_k$. Then, there exits an operator $\hat{G}$ in $\mathcal{P}(n)$ such that $\hat{G}$ anti-commutes with one and only one of $\hat{G}_1, \hat{G}_2, \ldots, \hat{G}_k$ but commutes with the rest (Problem 6.3). For example, consider the two independent operators $\hat{G}_1 := \hat{Z} \otimes \hat{Z} \otimes \hat{I}$ and $\hat{G}_2 := \hat{I} \otimes \hat{Z} \otimes \hat{Z}$. The operator $\hat{G} := \hat{X} \otimes \hat{I} \otimes \hat{I}$ anti-commutes with $\hat{G}_1$ but commutes with $\hat{G}_2$. On the other hand, the operator $\hat{G}' := \hat{I} \otimes \hat{Z} \otimes \hat{Y}$ commutes with $\hat{G}_1$ and anti-commutes with $\hat{G}_2$.

6.3.2 Properties of the Stabilizers

The simple structure and useful properties of the Pauli groups enable us to inspect many aspects of the stabilizers without peculiar specifications. Here we discuss and summarize the general properties of the stabilizers that are to be exploited repeatedly later.

An immediate consequence of the properties of the Pauli groups in Sect. 6.3.1, in particular that any two elements of a Pauli group either commutes or anti-commutes, is that all elements of a stabilizer must commute with each other (Problem 6.1). That is, any stabilizer is an *Abelian* subgroup of the Pauli group. Also recall that every element of a Pauli group squares to $\pm\hat{I}$. If an operator in a Pauli group squares to $-\hat{I}$, then its eigenvalue must be $\pm i$ and it can stabilize no state. Putting these observations together, we can set down the necessary conditions for a subgroup S of the Pauli group to stabilize a non-trivial subspace $\mathcal{V}$ in a compact form:

(a) all elements of S commute, and

(b) S does not include $-\hat{I}$ as an element.

Let us now inspect the relation between a stabilizer and the subspace stabilized by it more closely: Suppose that S is a stabilizer on n qubits generated by k independent operators $\hat{G}_1, \ldots, \hat{G}_k$. Since all elements in the stabilizer commute each other and square to $\hat{I}$, S is essentially the same as $(\mathbb{Z}_2)^k$. Mathematically, S is isomorphic to the Abelian group $(\mathbb{Z}_2)^k$ with the group multiplication given by the addition modulo 2; see Appendix C.5.

The states stabilized by S are simultaneous eigenvectors of the generators of S. The more generators S has, the more constraints the states have to satisfy. Naturally, a bigger number of generators of S results in a smaller dimension of the subspace $\mathcal{V}$ stabilized by S. More explicitly, the proposition is that the dimension of $\mathcal{V}$ is equal to 2^{n-k}. In other words, the subspace $\mathcal{V}$ can encode $(n - k)$ logical qubits. It is one of the most basic properties that make stabilizer codes so simple and versatile as we will see in Sect. 6.4 below. To see this, note that any n-qubit Pauli operator, say $\hat{G}_1$, without a phase factor $\pm i$ divides the Hilbert space $\mathcal{H}$ into two orthogonal subspaces respectively belonging to eigenvalues ± 1. Consequently, the subspace $\mathcal{W}_1$ belonging to eigenvalue 1 is the subspace stabilized by $\hat{G}_1$ and has dimension 2^{n-1}. An additional Pauli operator $\hat{G}_2$ that commutes $\hat{G}_1$ divides $\mathcal{W}_1$ again into two orthogonal subspaces. One of them belonging to eigenvalue 1, $\mathcal{W}_2$, is stabilized by both $\hat{G}_1$ and $\hat{G}_2$. Obviously, $\mathcal{W}_2$ is 2^{n-2}-dimensional. One can repeat the argument to confirm the proposition. For example, the operator $\hat{Z} \otimes \hat{Z}$ stabilizes any combination of $|00\rangle$ and $|11\rangle$ or, equivalently, any combination of

$$\frac{|00\rangle + |11\rangle}{\sqrt{2}}, \quad \frac{|00\rangle - |11\rangle}{\sqrt{2}}. \tag{6.45}$$

The operator $\hat{X} \otimes \hat{X}$ stabilizes any combination of

$$\frac{|00\rangle + |11\rangle}{\sqrt{2}}, \quad \frac{|01\rangle + |10\rangle}{\sqrt{2}}. \tag{6.46}$$

The vector space stabilized by each of the operators is thus two dimensional. On the other hand, only the states proportional to $(|00\rangle + |11\rangle)/\sqrt{2}$ are stabilized simultaneously by both operators. Therefore, the subspace stabilized by the stabilizer $\langle\{\hat{X} \otimes \hat{X}, \hat{Z} \otimes \hat{Z}\}\rangle$ is one dimensional. We observe that the operator $\hat{X} \otimes \hat{X}$ divides the subspace spanned by the vectors in (6.45) into two subspace, one spanned by $(|00\rangle + |11\rangle)/\sqrt{2}$ and the other by $(|00\rangle - |11\rangle)/\sqrt{2}$. The operator $\hat{Z} \otimes \hat{Z}$ divides the subspace spanned by the vector in (6.46) in a similar manner.

One can prove the above proposition more rigorously. Note that the operators $(\hat{I} \pm \hat{G}_j)/2$ for each j are the projectors onto the eigensubspaces of $\hat{G}_j$ belonging to the eigenvalues ± 1, respectively. We generalize them and define

$$\hat{P}_x := \prod_{j=1}^{k} \frac{\hat{I} + (-1)^{x_j}\hat{G}_j}{2} \tag{6.47}$$

for each binary string $x := (x_1 x_2 \ldots x_k)_2$. $\hat{P}_x$ is the projector onto the simultaneous eigensubspace of the generators $\hat{G}_1, \ldots, \hat{G}_k$ belonging to the eigenvalue $(-1)^{x_1 + \cdots + x_k}$. As such, they are orthogonal, that is,

$$\hat{P}_x \hat{P}_y = \delta_{xy} \hat{P}_x \tag{6.48}$$

for all x, y in $(\mathbb{Z}_2)^k$. In particular, $\hat{P}_0$ is the projector onto the subspace $\mathcal{V}$ stabilized by S. Previously (in Sect. 6.3.2), we noted that given a set of independent Pauli operators, there exists an operator that anti-commutes with only one operator and commutes with the other. Accordingly, for each x, one can find an operator $\hat{G}_x$ such that

$$\hat{G}_x \hat{P}_{(0,\ldots,0)} \hat{G}_x = \hat{P}_x. \tag{6.49}$$

This means that the subspaces $\hat{P}_x \mathcal{H}$ for all x have the same dimension as $\mathcal{V}$. Finally, note that

$$\sum_{x=0}^{2^k-1} \hat{P}_x = \hat{I} \tag{6.50}$$

because

$$\sum_{x_1, x_2, \ldots} (-1)^{x_1 + x_2 + \cdots} = \sum_{x_1} (-1)^{x_1} \sum_{x_2} (-1)^{x_2} \cdots = 0. \tag{6.51}$$

Therefore, the 2^k projectors $\hat{P}_x$ divide the 2^n-dimensional Hilbert space $\mathcal{H}$ into orthogonal subspaces of the same dimension, and the dimension of each subspace must be $2^n / 2^k = 2^{n-k}$.

We have seen above that a subspace $\mathcal{V}$ can be identified by its stabilizer S. For quantum information processing within the subspace $\mathcal{V}$, one also needs to specify a logical basis of $\mathcal{V}$. In stabilizer formalism, logical basis states are specified indirectly through *logical operators*. Let us first take an example. Consider again the subspace $\mathcal{V}$ stabilized by $S = \langle\{\hat{Z} \otimes \hat{Z}\}\rangle$. Suppose that we choose the two states

$$|\bar{0}\rangle = \frac{|00\rangle + |11\rangle}{\sqrt{2}}, \quad |\bar{1}\rangle = \frac{|00\rangle - |11\rangle}{\sqrt{2}}, \tag{6.52}$$

as the logical basis states. Within $\mathcal{V}$, the operator $\hat{X} \otimes \hat{X}$ has eigenstates $|\bar{0}\rangle$ and $|\bar{1}\rangle$ with eigenvalues ± 1, respectively, and hence behaves like the Pauli Z operator on a single qubit. In this sense, we put $\bar{Z} = \hat{X} \otimes \hat{X}$ and regard it as the *logical* Pauli Z operator. On the other hand, $\hat{Z} \otimes \hat{I}$ flips $|\bar{0}\rangle$ to $|\bar{1}\rangle$ and vice versa. Furthermore, it fixes the relative phase between the basis states. In these senses, it behaves like the Pauli X operator on a single qubit. Thus, we put $\bar{X} = \hat{Z} \otimes \hat{I}$ and regard it as the logical Pauli X operator. Note that $\bar{Z}$ and $\bar{X}$ commute with all elements of the stabilizer S and hence preserve the stabilized subspace $\mathcal{V}$. Nevertheless, unlike the operators in the stabilizer S, they act non-trivially on $\mathcal{V}$ and move around the quantum states inside $\mathcal{V}$.

More generally, the logical basis states $|\bar{x}\rangle$ associated with binary strings $\bar{x} := (\bar{x}_1 \bar{x}_2 \ldots \bar{x}_{n-k})_2$ can be designated by two sets of logical operators $\bar{Z}_1, \ldots, \bar{Z}_{n-k}$ and $\bar{X}_1, \ldots, \bar{X}_{n-k}$. While a measurement of the logical Pauli Z operators $\bar{Z}_j$ distinguishes the logical basis states, that is,

$$\bar{Z}_j |\bar{x}\rangle = \bar{x}_j |\bar{x}\rangle, \tag{6.53}$$

the logical Pauli X operators $\bar{X}_j$ fix the relative phases of the logical basis states by imposing

$$\bar{X}_j |\bar{x}_1, \ldots, \bar{x}_j, \ldots, \bar{x}_{n-k}\rangle = |\bar{x}_1, \ldots, \bar{x}_j \oplus 1, \ldots, \bar{x}_{n-k}\rangle. \tag{6.54}$$

Of course, given a stabilizer, one can construct logical operators without referring to the logical basis states. The necessary condition is that the logical operators should preserve the given subspace $\mathcal{V}$, transforming states within $\mathcal{V}$. This implies that any logical operator, $\bar{Z}_j$ or $\bar{X}_j$, must commute with all generators of stabilizer S. Nevertheless, the logical operators must not belong to the stabilizer and should be independent of the generators of the stabilizer. Otherwise, their operations on $\mathcal{V}$ are all trivial. Note that for a given logical basis, the choice of the logical operators is not unique. For example, for the same basis logical states in (6.52), one can choose $\bar{Z} = -\hat{Y} \otimes \hat{Y}$ and $\bar{X} = \hat{I} \otimes \bar{Z}$. Mathematically, equivalent logical operators belong to the same coset with respect to S. Given a set of independent generators of the stabilizer, there is a systematic method to construct logical operators utilizing Gottesman vectors. We thus refer interested readers to Cleve & Gottesman (1997).

6.3.3 Unitary Gates in Stabilizer Formalism

Consider a unitary operation $\hat{U}$ on a quantum system, and suppose that S stabilizes a given subspace $\mathcal{V}$. The unitary operation $\hat{U}$ transforms $\mathcal{V}$ onto $\hat{U}\mathcal{V}$, and S does not stabilizes the new subspace any longer. Then, what is the stabilizer of the new subspace $\hat{U}\mathcal{V}$?

Suppose that an operator $\hat{G}$ is an element of the stabilizer S. Then, $\hat{G} |\psi\rangle = |\psi\rangle$ for any quantum state $|\psi\rangle$ in the subspace $\mathcal{V}$. We observe that

$$\hat{U}\hat{G}\hat{U}^\dagger \hat{U} |\psi\rangle = \hat{U}\hat{G} |\psi\rangle = \hat{U} |\psi\rangle. \tag{6.55}$$

This means that $\hat{U}\hat{G}\hat{U}^\dagger$ stabilizes $\hat{U} |\psi\rangle$, and hence the whole subspace $\hat{U}\mathcal{V}$. In other words, the group $\hat{U}S\hat{U}^\dagger$ is the stabilizer of the new subspace $\hat{U}\mathcal{V}$. Now, suppose that

$\{\hat{G}_j\}$ is a generating set of stabilizer $\mathcal{S}$. Since any element of $\mathcal{S}$ is generated by $\hat{G}_j$, any operator of the form $\hat{U}\hat{G}\hat{U}^\dagger$ for $\hat{G} \in \mathcal{S}$ is generated by $\hat{U}\hat{G}_j\hat{U}^\dagger$. That is, the new stabilizer $\hat{U}\mathcal{S}\hat{U}^\dagger$ is generated by $\hat{U}\hat{G}_j\hat{U}^\dagger$. To sum up, under a unitary operation $\hat{U}$, the generators $\hat{G}_j$ of the stabilizer are transformed as follows

$$\hat{G}_j \mapsto \hat{U}\hat{G}_j\hat{U}^\dagger. \tag{6.56}$$

As the unitary operation $\hat{U}$ changes logical basis states, the logical operators must accordingly transform. Following the above line of arguments for the generators of the stabilizer, we see that the logical operators also transform in the same manner

$$\bar{X}_j \mapsto \hat{U}\bar{X}_j\hat{U}^\dagger, \quad \bar{Z}_j \mapsto \hat{U}\bar{Z}_j\hat{U}^\dagger. \tag{6.57}$$

So far, the discussion has been completely general. However, we are mainly interested in stabilizer subgroups of the Pauli group. For an element $\hat{G}$ of the Pauli group, in general, $\hat{U}\hat{G}\hat{U}^\dagger$ does not belong to the Pauli group and cannot be expressed with a tensor product of single-qubit Pauli operators. It would be useful to focus on unitary operators $\hat{U}$ the conjugation by which transforms the tensor products of single-qubit Pauli operators to such products. This is the subject of the next subsection.

6.3.4 Clifford Group

The *Clifford group* $\mathcal{C}(n)$ on n qubits is the group of unitary operators $\hat{U}$ that leave the Pauli group $\mathcal{P}(n)$ invariant under conjugation, that is,

$$\mathcal{C}(n) := \{\hat{U} : \hat{U}\mathcal{P}(n)\hat{U}^\dagger = \mathcal{P}(n)\}. \tag{6.58}$$

Mathematically, the Clifford group is the *normalizer* of the Pauli group in the unitary group $U(2^n)$. The elements of the Clifford group on n qubits are called n-qubit *Clifford operators*. As for the Pauli group, we will often omit the number n of qubits from the Clifford group specification $\mathcal{C}(n)$ and write $\mathcal{C}$ when there is no risk of confusion.

The Clifford group was first introduced in 1998 by Daniel Gottesman (Gottesman, 1998) when he reformulated the quantum error-correction codes in terms of stabilizers. The Clifford group is useful and interesting since stabilizers remain subgroups of the Pauli group under conjugation by its elements.

The Clifford group $\mathcal{C}(n)$ is generated by Hadamard gates and quadrant phase gates[3] on individual qubits, and CNOT (or equivalently CZ) gates on all pairs of qubits. To see this, first consider single-qubit transformations in the Clifford group $\mathcal{C}(1)$. By definition, the *conjugation* by a single-qubit operator in $\mathcal{C}(1)$ permutes the

---

[3] The quadrant phase gate is defined in Eq. (2.28).

Pauli operators $\hat{X}$, $\hat{Y}$ and $\hat{Z}$ up to a phase factor—note that the conjugation by a single-qubit unitary operator cannot alter the identity operator $\hat{I}$. Mathematically, it is equivalent to an one-to-one mapping $\mathbb{Z}_4 \rightarrow \mathbb{Z}_4$ with 0 fixed.[4] Geometrically, it can be regarded as an axes-permuting rotation in the Bloch space [see Fig. 6.2b]. Single-qubit transformations in the Clifford group thus correspond to the six symmetry transformations in the *point group*

$$D_3 := \left\{ \hat{I}, \hat{M}_x, \hat{M}_y, \hat{M}_z, \hat{R}, \hat{R}^2 \right\}, \tag{6.59}$$

which leaves an equilateral triangle invariant [see Fig. 6.2a]. Here $\hat{M}_x$, $\hat{M}_y$, and $\hat{M}_z$ denote the mirror reflections about the planes through the vertex X, Y, and Z, respectively, of the triangle. $\hat{R}$ represents the rotation by an angle of $2\pi/3$ around the axis through the center of the triangle. For example, consider the Hadamard gate. It is a rotation by an angle of π around the axis $(1, 0, 1)$ in the Bloch space. It follows from Eq. (2.25) (or a simple inspection) that the Hadamard gate exchanges $\hat{X}$ and $\hat{Z}$, $\hat{H}\hat{X}\hat{H}^\dagger = \hat{Z}$ and $\hat{H}\hat{Z}\hat{H}^\dagger = \hat{X}$, while it keeps $\hat{Y}$ intact, $\hat{H}\hat{Y}\hat{H}^\dagger = -\hat{Y}$. In short,

$$\hat{H} : (\hat{X}, \hat{Y}, \hat{Z}) \mapsto (\hat{Z}, -\hat{Y}, \hat{X}), \tag{6.60}$$

and the conjugation by $\hat{H}$ corresponds to the symmetry transformation $\hat{M}_y$ in D_3. Similarly, the conjugation by the quadrant phase gate $\hat{Q}$ exchanges $\hat{X}$ and $\hat{Y}$,

$$\hat{Q} : (\hat{X}, \hat{Y}, \hat{Z}) \mapsto (\hat{Y}, -\hat{X}, \hat{Z}), \tag{6.61}$$

and is equivalent to the symmetry transformation $\hat{M}_z$ in D_3. Changing the axes before and after $\hat{M}_y$, it follows from Eq. (2.25) that $\hat{M}_x$ is equivalent to $\hat{Q}\hat{H}\hat{Q}^\dagger$. Finally, the equivalent of $\hat{R}$ in D_3 is the rotation by $2\pi/3$ around the axis $(1, 1, 1)$ in the Bloch space, and it is identical to $\hat{H}\hat{Q}^\dagger$. Overall, the correspondence between the single-qubit Clifford group and the point group D_3 is summarized as

$$\hat{M}_z \leftrightarrow \hat{Q}, \quad \hat{M}_y \leftrightarrow \hat{H}, \quad \hat{M}_x \leftrightarrow \hat{Q}\hat{H}\hat{Q}^\dagger, \quad \hat{R} \leftrightarrow \hat{H}\hat{Q}^\dagger. \tag{6.62}$$

This asserts that the single-qubit Clifford group is generated by $\hat{H}$ and $\hat{Q}$.

The single-qubit Clifford group is rather simple. It is generated by the Hadamard gate and the quadrant phase gate.

```
In[•]:= gnr = GroupGenerators@CliffordGroup[S]
```
```
Out[•]= {S^H, S^S}
```

In fact, it is equivalent to the point group D_3. The point group consists of the six symmetry transformation {I, Mx, My, Mz, R, R**R}.

---

[4] Here $\mathbb{Z}_m := \{0, 1, \ldots, m - 1\}$ is merely a set of m elements without any further structure. The notation is more commonly used for a *cyclic group* of order m as in Eq. (C.10).

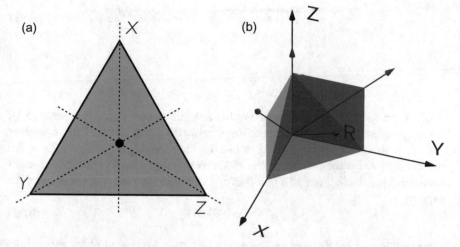

Fig. 6.2 **a** An equilateral triangle, which is invariant under the symmetry transformations in point group D_3. The dashed lines indicate the mirror planes in which the object is symmetric. The point in the middle indicates the rotation by angle $2\pi/3$. **b** Rotations in the Bloch space that permute the axes

```
My = Elaborate@S[6];
Mz = Elaborate@S[7];
Mx - Mz ** My ** Dagger[Mz];
R = My ** Dagger[Mz];
```

Note that the *conjugation* (not to be confused with conjugate) by an operator is a special case of supermap with a single unitary Kraus element.

```
Conjugation[op_] := Supermap[op]
```

Take a look how the Pauli operators transform under the conjugation by single-qubit operations in the Clifford group.

```
In[·]:= ops = {S[1], S[2], S[3]};
       Thread[ops → Conjugation[Mx]@ops] // PauliForm // TableForm
       Thread[ops → Conjugation[My]@ops] // PauliForm // TableForm
       Thread[ops → Conjugation[Mz]@ops] // PauliForm // TableForm
Out[·]//TableForm=
       X → -X
       Y → Z
       Z → Y
Out[·]//TableForm=
       X → Z
       Y → -Y
       Z → X
Out[·]//TableForm=
       X → Y
       Y → -X
       Z → Z
```

```
In[·]:=  Thread[ops → Conjugation[R]@ops] // PauliForm // TableForm
         Thread[ops → Conjugation[R ** R]@ops] // PauliForm // TableForm
```

Out[·]//TableForm=
     X → Y
     Y → Z
     Z → X

Out[·]//TableForm=
     X → Z
     Y → X
     Z → Y

Let us move on, and examine two-qubit operations in the Clifford group $\mathcal{C}(2)$. The conjugation by a two-qubit operation transforms a pair of four Pauli operators $\hat{I}, \hat{X}, \hat{Y}, \hat{Z}$ to another pair, just like a one-to-one mapping $\mathbb{Z}_4 \times \mathbb{Z}_4 \to \mathbb{Z}_4 \times \mathbb{Z}_4$. Apparently, local combinations of the Hadamard and quadrant phase gates cannot transform a pair including $\hat{I}$ to a pair that does not, nor vice versa. For example, the mapping

$$\hat{Y} \otimes \hat{I} \mapsto \hat{Z} \otimes \hat{Y} \tag{6.63}$$

is impossible with local combinations of Hadamard and quadrant phase gates alone since

$$(\hat{U} \otimes \hat{V})(\hat{Y} \otimes \hat{I})(\hat{U} \otimes \hat{V})^{\dagger} = (\hat{U}\hat{Y}\hat{U}^{\dagger}) \otimes \hat{I} \tag{6.64}$$

for all unitary operators $\hat{U}$ and $\hat{V}$. This is where the CNOT gate plays a critical role. The conjugation by the CNOT gate controlled by the first qubit on the second qubit maps

$$\begin{aligned}
\hat{X} \otimes \hat{I} &\mapsto \hat{X} \otimes \hat{X}, \\
\hat{Z} \otimes \hat{I} &\mapsto \hat{Z} \otimes \hat{I}, \\
\hat{I} \otimes \hat{X} &\mapsto \hat{I} \otimes \hat{X}, \\
\hat{I} \otimes \hat{Z} &\mapsto \hat{Z} \otimes \hat{Z}.
\end{aligned} \tag{6.65}$$

It is interesting to note that under the CNOT gate, the Pauli X operator "propagates" *forward* (from the control to target qubit) while the Pauli Z operator propagates *backward* (from the target to control qubit). The mappings of pairs involving $\hat{Y}$ can be deduced from the above correspondence using the identity $\hat{Y} = i\hat{X}\hat{Z}$. Since a product of Pauli operators can be mapped to any product of them as long as the product does not involve $\hat{I}$, a single CNOT gate is sufficient to complement the actions of the Hadamard and quadrant phase gates. For example, one way to achieve the transformation in Eq. (6.63) is applying successively the conjugations by $\hat{Q}^{\dagger} \otimes \hat{I}$, CNOT, $\hat{H} \otimes \hat{Q}$ leading to

$$\hat{Y} \otimes \hat{I} \to \hat{X} \otimes \hat{I} \to \hat{X} \otimes \hat{X} \to \hat{Z} \otimes \hat{Y}. \tag{6.66}$$

The same arguments apply for any number of qubits, and Hadamard, quadrant phase, and CNOT gates are sufficient to generate any operation in the Clifford group.

Here demonstrated is the correspondence between one-to-one mappings $\mathbb{Z}_4 \times \mathbb{Z}_4 \to \mathbb{Z}_4 \times \mathbb{Z}_4$ and the two-qubit operations in the Clifford group.

```
In[·]:= gnr = CNOT[S[1], S[2]]
Out[·]= CNOT[{S₁}, {S₂}]
```

Under the conjugation by the CNOT gate, the Pauli operators transform as the following.

```
In[·]:= ops = Multiply @@@ Tuples@{S[1, Full], S[2, Full]};
        Timing[new = Elaborate@Thread[ops → Conjugation[gnr]@ops];]
        new[[2 ;; ;; 4]] // PauliForm // TableForm
Out[·]= {0.124038, Null}
Out[·]//TableForm=
        I⊗X → I⊗X
        X⊗X → X⊗I
        Y⊗X → Y⊗I
        Z⊗X → Z⊗X
```

Combinations of the CNOT, Hadamard, and quadrant phase gates generate all transformations preserving the Pauli group. Here is one example.

```
gnr = CNOT[S[1], S[2]] ** S[1, 6] ** Dagger[S[1, 7]] ** S[2, 7] // Elaborate;
ops = {S[1, 1], S[2, 1], S[1, 3], S[2, 3]};
```

```
In[·]:= Thread[ops → Conjugation[gnr]@ops] // PauliForm // TableForm
Out[·]//TableForm=
        X⊗I → Y⊗X
        I⊗X → Z⊗Y
        Z⊗I → X⊗X
        I⊗Z → Z⊗Z
```

Let us now rigorously prove that a Clifford group is generated by Hadamard and quadrant phase gates on single qubits and CNOT gates on pairs of qubits: The inductive proof is constructive, and it turns out to be useful to find implementations of the operations in the Clifford group (Gottesman, 1998).

Let $\hat{U}$ be an element of the $(n + 1)$-qubit Clifford group $\mathcal{C}(n + 1)$. Without loss of generality, one can assume that

$$\hat{U}(\hat{Z} \otimes \hat{J})\hat{U}^\dagger = \hat{X} \otimes \hat{A}, \quad \hat{U}(\hat{X} \otimes \hat{J})\hat{U}^\dagger = \hat{Z} \otimes \hat{B}, \tag{6.67}$$

where $\hat{J} := \hat{I}^{\otimes n}$, and $\hat{A}$ and $\hat{B}$ are products of the Pauli operators, $\hat{A}, \hat{B} \in \mathcal{P}(n)$. If $\hat{U}$ does not satisfy the above properties, one can always rearrange qubits and apply additional single-qubit operations so that the modified operation assume the forms (see Problem 6.6). It is also possible to assume that

$$\hat{U}(\hat{Z} \otimes \hat{J})\hat{U}^\dagger = \hat{X} \otimes \hat{A}, \quad \hat{U}(\hat{X} \otimes \hat{J})\hat{U}^\dagger = \hat{I} \otimes \hat{B}, \tag{6.68}$$

but here we will exploit the forms in (6.67). Define an operator $\hat{V}$ acting on n qubits by

$$\hat{V}|y\rangle := \sum_{x=0}^{2^n-1} |x\rangle \left(\langle 0| \otimes \langle x|\right)\hat{U}(|0\rangle \otimes |y\rangle)) \tag{6.69}$$

for $y = 0, 1, 2, \ldots, 2^n - 1$. In effect, $\hat{V}$ projects out of $\hat{U}$ the component corresponding to $|1\rangle$ on the first qubit. One can show that derived from $\hat{U}$, the new operator $\hat{V}$ is an element of the n-qubit Clifford group $\mathcal{C}(n)$ (see Problem 6.7). In other words, the conjugation by $\hat{V}$ preserves the n-qubit Pauli group $\mathcal{P}(n)$. The goal is then to express $\hat{U}$ in terms of $\hat{A}$, $\hat{B}$, and $\hat{V}$. In short, $\hat{U}$ can be implemented in the following quantum circuit

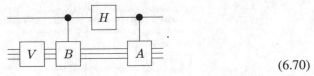

$$(6.70)$$

As $\hat{A}$ and $\hat{B}$ are products of the Pauli operators, the controlled-$\hat{A}$ and controlled-$\hat{B}$ gate can be implemented with the Hadamard, quadrant phase, and CNOT gates. Therefore, $\hat{U}$ can be achieved in a similar manner as long as $\hat{V}$ is.

The only remaining task is to show that $\hat{U}$ is equivalent to the quantum circuit in (6.70). Let us consider a general $(n + 1)$-qubit state

$$|\Psi\rangle = \frac{|0\rangle \otimes |\alpha\rangle + |1\rangle \otimes |\beta\rangle}{\sqrt{2}}, \tag{6.71}$$

where $|\alpha\rangle$ and $|\beta\rangle$ are n-qubit states. Upon the operation of $\hat{U}$ on it, the first term transforms to

$$\hat{U}(|0\rangle \otimes |\alpha\rangle) = |0\rangle \otimes |\alpha_0\rangle + |1\rangle \otimes |\alpha_1\rangle . \tag{6.72}$$

Since $\hat{Z} \otimes \hat{J}$ does not alter the first term, we also obtain

$$\begin{aligned}
\hat{U}(|0\rangle \otimes \hat{\alpha}) &= \hat{U}(\hat{Z} \otimes \hat{J})\hat{U}^\dagger \hat{U}(|0\rangle \otimes |\alpha\rangle) \\
&= (\hat{X} \otimes \hat{A})(|0\rangle \otimes |\alpha_0\rangle + |1\rangle \otimes |\alpha_1\rangle) \\
&= |0\rangle \otimes \hat{A}|\alpha_1\rangle + |1\rangle \otimes \hat{A}|\alpha_0\rangle .
\end{aligned} \tag{6.73}$$

The two results in (6.72) and (6.73) must be identical, so the two states $|\alpha_0\rangle$ and $|\alpha_1\rangle$ should be related to each other by $\hat{A}$

$$\hat{A}|\alpha_0\rangle = |\alpha_1\rangle , \quad \hat{A}|\alpha_1\rangle = |\alpha_0\rangle . \tag{6.74}$$

In passing, these relations are consistent with the fact that $\hat{A}$ is a product of the Pauli operators. Putting (6.74) back into (6.72),

$$\hat{U}(|0\rangle \otimes |\alpha\rangle) = |0\rangle \otimes |\alpha_0\rangle + |1\rangle \otimes \hat{A}|\alpha_0\rangle = (\hat{I} \otimes \hat{J} + \hat{X} \otimes \hat{A})(|0\rangle \otimes \hat{V}|\alpha\rangle) , \tag{6.75}$$

where for the second equality, we have used the fact that $|\alpha_0\rangle = \hat{V}|\alpha\rangle$ by definition of the operator $\hat{V}$ in (6.69).

Next, we examine how the second term transforms under the action of $\hat{U}$. We observe that

$$\hat{U}(|1\rangle \otimes |\beta\rangle) = \hat{U}(\hat{X} \otimes \hat{J})(|0\rangle \otimes |\beta\rangle) = (\hat{Z} \otimes \hat{B})\hat{U}(|0\rangle \otimes |\beta\rangle)$$
$$= (\hat{Z} \otimes \hat{B})(\hat{I} \otimes \hat{J} + \hat{X} \otimes \hat{A})(|0\rangle \otimes \hat{V} |\beta\rangle), \tag{6.76}$$

where we have used (6.75) for the last equality. We note that $\hat{X} \otimes \hat{A}$ and $\hat{Z} \otimes \hat{B}$ anti-commute since $\hat{U}(\hat{Z} \otimes \hat{J})\hat{U}^\dagger$ and $\hat{U}(\hat{X} \otimes \hat{J})\hat{U}^\dagger$ obviously do. This leads to

$$\hat{U}(|1\rangle \otimes |\beta\rangle) = (\hat{I} \otimes \hat{J} - \hat{X} \otimes \hat{A})(\hat{Z} \otimes \hat{B})(|0\rangle \otimes \hat{V} |\beta\rangle)$$
$$= (\hat{I} \otimes \hat{J} - \hat{X} \otimes \hat{A})(|0\rangle \otimes \hat{B}\hat{V} |\beta\rangle). \tag{6.77}$$

Putting (6.75) and (6.77) together, we have

$$\hat{U} |\Psi\rangle = \frac{|0\rangle \otimes \hat{V} |\alpha\rangle + |1\rangle \otimes \hat{A}\hat{V} |\alpha\rangle + |0\rangle \otimes \hat{B}\hat{V} |\beta\rangle + |1\rangle \otimes \hat{A}\hat{B}\hat{V} |\beta\rangle}{\sqrt{2}}, \tag{6.78}$$

which is exactly the result produced by the quantum circuit in (6.70).

The peculiarity of the Clifford group is most pronounced in the *Gottesman–Knill theorem* (Gottesman, 1999). It states that the so-called *Clifford circuits*,[5] quantum circuits consisting of unitary operators from the Clifford group, can be efficiently simulated on a classical computer. This sounds surprising because the Clifford operators are sufficient to generate a rich range of quantum effects including the Greenberger–Horne–Zeilinger (GHZ) experiment (Greenberger et al., 1989), quantum teleportation, and super dense coding. They are also sufficient to encode and decode quantum error-correction codes (Sect. 6.4). Nevertheless, the theorem indicates that the Clifford group "falls short of the full power of quantum computation." Indeed, we recall that in addition to Clifford gates, we need octant phase gates or Toffoli gates to implement an arbitrary unitary gate to an arbitrary accuracy (Sect. 2.4).

How is it possible to simulate Clifford circuits efficiently on a classical computer? The key point is that stabilizer formalism tracks the generators of the stabilizer rather than the quantum states as in the usual methods. To follow the evolution of the quantum states, one needs to keep track of an exponential number of complex amplitudes. However, for a system of n qubits, there are an order of n generators of the stabilizer because any finite group $\mathcal{G}$ has a generating set of at most $\log_2 |\mathcal{G}|$ elements (Theorem C.2). Each generator takes $2n$ bits for the Pauli X or Y operators on n qubits and an additional bit for the phase factor ± 1. Overall, one needs an order of n^2 bits to track the evolution of the stabilizer. We have seen that the Clifford group is generated by the Hadamard, quadrant phase, and CNOT gates and that the

---

[5] The Gottesman–Knill theorem is more general and also applies to so-called *stabilizer circuits*. In addition to Clifford gates, a stabilizer circuit includes Clifford gates conditioned on the outcomes of the measurement of Pauli operators.

generators under conjugation by these elementary gates can be updated in polynomial time, $\mathcal{O}(n)$ time, for the gate.

A specific algorithm to simulate Clifford circuits was proposed in Aaronson & Gottesman (2004). Recently, a simulation algorithm was also proposed to conduct a fast simulation of the stabilizer circuits using graph states.

6.3.5   Measurements in Stabilizer Formalism

Upon the measurement of an observable $\hat{M}$, a quantum state $|\psi\rangle$ "collapses" to one of the eigenstates of $\hat{M}$. When $\hat{M}$ is a tensor product of the Pauli operators, $\hat{M} \in \mathcal{P}(n)$, it has two eigenvalues ± 1. The measurement process is described by the projectors $\hat{P}_{\pm} = (\hat{I} \pm \hat{M})/2$ onto the eigenspaces belonging to eigenvalues ± 1. The description so far has been in the Schrödinger picture. How do the stabilizer and the logical operators change under the measurement following the stabilizer formalism?

Suppose that $\mathcal{S}$ is the stabilizer of a subspace $\mathcal{V}$ of interest. Recall from Sect. 6.3.1 that since the observable $\hat{M}$ is a member of the Pauli group $\mathcal{P}(n)$, it either commutes or anti-commutes with any specific operator in stabilizer $\mathcal{S}$. This leaves two possibilities, $\hat{M}$ either anti-commutes with some elements of $\mathcal{S}$ or commutes with every element of $\mathcal{S}$. Let us first examine the former case. We designate by $\hat{G}$ one of the elements that anti-commute with $\hat{M}$. The post-measurement state becomes $\hat{P}_{\pm} |\psi\rangle$ depending on the result ± 1 of the measurement. We note that

$$\hat{G}\hat{P}_{-} |\psi\rangle = \frac{1}{2}\hat{G}(\hat{I} - \hat{M}) |\psi\rangle = \frac{1}{2}(\hat{I} + \hat{M})\hat{G} |\psi\rangle = \hat{P}_{+} |\psi\rangle. \tag{6.79}$$

In other words, after the measurement of $\hat{M}$, we can make it sure that the post-measurement state is always $\hat{P}_{+} |\psi\rangle$, if necessary, by applying $\hat{G}$. With this practice assumed, the relevant subspace after the measurement is now $\hat{P}_{+}\mathcal{V}$. For any state $|\psi\rangle$ in $\mathcal{V}$, the post-measurement state $\hat{P}_{+} |\psi\rangle$ is an eigenstate of $\hat{M}$ belonging to the eigenvalue $+1$. This means that $\hat{M}$ is an element of the stabilizer $\mathcal{S}'$ of the new subspace $\hat{P}_{+}\mathcal{V}$. On the other hand, operator $\hat{G}$ cannot belong to $\mathcal{S}'$ any longer. Any element of $\mathcal{S}$ is still a member of $\mathcal{S}'$ as long as it commutes with $\hat{M}$. If some element $\hat{G}'$ in $\mathcal{S}$ does not commute with $\hat{M}$, then $\hat{G}\hat{G}'$ does and hence is a member of $\mathcal{S}'$. It is physically important to note that by construction, the observable $\hat{M}$ commutes with all elements of $\mathcal{S}'$.

Now, let us turn to the case where $\hat{M}$ commutes with every member of $\mathcal{S}$. We have to further distinguish two sub-cases. First, $\hat{M}$ itself belongs to $\mathcal{S}$. In this case, the measurement of $\hat{M}$ gives outcome $+1$ with unit probability, and nothing happens to the quantum states nor the stabilizer. The remaining case is when $\hat{M}$ commutes with all elements of $\mathcal{S}$ and yet does not belong to $\mathcal{S}$. In this case, measuring $\hat{M}$ reveals some information of the quantum state, and the subspace $\mathcal{V}$ cannot maintain its structure, so the stabilizer formalism breaks down in some sense. For example, suppose that we measure one of the logical operators, say, $\hat{M} = \bar{Z}$. If the state is

$|\psi\rangle = |\bar{0}\rangle c_0 + |\bar{1}\rangle c_1$, then measuring $\bar{Z}$ causes $|\psi\rangle$ to collapse to either $|\bar{0}\rangle$ or $|\bar{1}\rangle$, breaking the encoding. Such a measurement must be performed only at the end of whole computation. Such an observable $\hat{M}$ may occur as a Kraus element of the error process, rather than a pure measurement. Then it corresponds to an unrecoverable error.

The new logical operators can be obtained in a similar way. If they commute with $\hat{M}$, they remain unaffected. If there is any logical operator $\bar{L}$ that does not commute with $\hat{M}$, then it can be replaced with $\hat{G}\bar{L}$ since the latter certainly commutes with $\hat{M}$. It is straightforward to check that the new logical operators constructed this way satisfy the same commutation relations as the original logical operators do, and they commute with all elements in the new stabilizer S'.

In summary, one can track the evolution of the stabilizer and the logical operators of a given subspace under a measurement of an observable $\hat{M}$ according to the following rules:

1. First, identify a generator $\hat{G}$ of the stabilizer S that anti-commutes with $\hat{M}$.
2. Next, pick up new generators of the stabilizer. First, replace $\hat{G}$ with $\hat{M}$ in the list of generators of S. Second, for each $\hat{G}'$ of other generators of S, keep it if it commutes with $\hat{M}$ and replace it with $\hat{G}\hat{G}'$ if it anti-commutes with $\hat{M}$.
3. Finally, we choose a new set of logical operators. For each logical operator $\bar{L}$ ($\bar{X}_j$ or $\bar{Z}_j$), keep it if it commutes with $\hat{M}$, but replace it with $\hat{G}\bar{L}$ if it anti-commutes with $\hat{M}$.

Recall that the rules assume that after measurement of the observable $\hat{M}$, generator $\hat{G}$ is operated on the system conditioned on the measurement outcome. This way, the post-measurement state remains to be stabilized by the newly constructed stabilizer regardless of the measurement outcome. Since the operation of $\hat{G}$ is conditioned on classical data (the measurement outcome), it is analogous to classical feedback control. In this sense we call it a *semiclassical* feedback control.

6.3.6 Examples

The above analyses of unitary gates and measurements in stabilizer formalism are useful to analyze what happens upon combining specific measurements and operations. In particular, if a part of the quantum register starts in a known quantum state, for example, initialized in $|0\rangle \otimes |0\rangle \otimes \cdots \otimes |0\rangle$, it can often be described by a stabilizer. Let us take two heuristic examples, which were worked out in detail in Gottesman (1999).

The first example is the quantum teleportation discussed in Sect. 4.1 and is summarized in the following quantum circuit

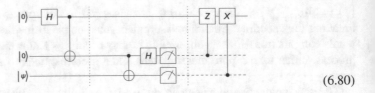

$$(6.80)$$

The protocol is implemented in three stages which are distinguished by the vertical dashed lines in the quantum circuit. The first stage is to share the entangled pair of qubits. The second in the middle corresponds to the Bell measurement. In the final stage, the measurement outcomes are transmitted to Alice, and Alice adjusts the state of her qubit according to the classical information. In order to reinterpret the quantum teleportation protocol by following the stabilizer formalism, we rearrange the quantum circuit in (6.80) into

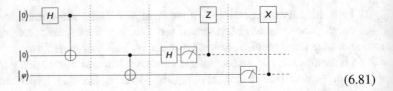

$$(6.81)$$

At the point marked by the first vertical line, the relevant stabilizer $\mathcal{S}$ is generated by

$$\hat{G}_1 = \hat{X} \otimes \hat{X} \otimes \hat{I}, \quad \hat{G}_2 = \hat{Z} \otimes \hat{Z} \otimes \hat{I}. \tag{6.82}$$

We pick up two logical operators

$$\bar{X} = \hat{I} \otimes \hat{I} \otimes \hat{X}, \quad \bar{Z} = \hat{I} \otimes \hat{I} \otimes \hat{Z}. \tag{6.83}$$

One can convince oneself that they commute with all the generators (and hence all elements) of the stabilizer. Under the CNOT gate from qubit B to C, the generators and the logical operators are transformed as

$$\begin{cases} \hat{G}_1 \mapsto \hat{X} \otimes \hat{X} \otimes \hat{X}, \\ \hat{G}_2 \mapsto \hat{Z} \otimes \hat{Z} \otimes \hat{I}; \end{cases} \quad \begin{cases} \bar{X} \mapsto \hat{I} \otimes \hat{I} \otimes \hat{X}, \\ \bar{Z} \mapsto \hat{I} \otimes \hat{Z} \otimes \hat{Z}. \end{cases} \tag{6.84}$$

Next, we measure the observable $\hat{X}$ on the qubit B, $\hat{M} \equiv \hat{I} \otimes \hat{X} \otimes \hat{I}$. Note that $\hat{M}$ anti-commutes with the generator $\hat{G}_2$ while it commutes with $\hat{G}_1$. We thus have to apply $\hat{G}_2$ conditioned on the measurement outcome. We update $\hat{G}_2$ by replacing it with $\hat{M}$. It also anti-commutes with the logical operator $\bar{Z}$ whereas it commutes with $\bar{X}$. We need to replace the logical operator $\bar{Z}$ by $\hat{G}_2 \bar{Z}$. Thus, the two replacements leads to new generators and logical operators as

$$\begin{cases} \hat{G}_1 \mapsto \hat{X} \otimes \hat{X} \otimes \hat{X}, \\ \hat{G}_2 \mapsto \hat{I} \otimes \hat{X} \otimes \hat{I}; \end{cases} \begin{cases} \bar{X} \mapsto \hat{I} \otimes \hat{I} \otimes \hat{X}, \\ \bar{Z} \mapsto \hat{Z} \otimes \hat{I} \otimes \hat{Z}. \end{cases} \tag{6.85}$$

At this stage, it is safe to drop the qubit B off since its state does not change nor is used after the measurement outcome has been recorded. Finally, we measure the observable $\hat{Z}$ on the qubit C, $\hat{M} \equiv \hat{I} \otimes \hat{I} \otimes \hat{Z}$. We note that $\hat{M}$ anti-commutes with generator $\hat{G}_1$ and the logical operator $\bar{X}$. In accordance with the same prescription as above, we apply $\hat{G}_1$ conditioned on the measurement outcome, and the generators and the logical operators are updated as follows

$$\begin{cases} \hat{G}_1 \mapsto \hat{I} \otimes \hat{I} \otimes \hat{Z}, \\ \hat{G}_2 \mapsto \hat{I} \otimes \hat{X} \otimes \hat{I}; \end{cases} \begin{cases} \bar{X} \mapsto \hat{X} \otimes \hat{X} \otimes \hat{I}, \\ \bar{Z} \mapsto \hat{Z} \otimes \hat{I} \otimes \hat{Z}. \end{cases} \tag{6.86}$$

A quantum circuit that follows the transformations in (6.85) and (6.86) more closely would look as the following

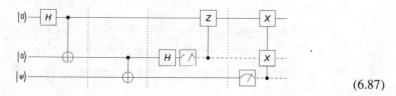

$$\tag{6.87}$$

However, the operation $\hat{X}$ on the second qubit is meaningless since the qubit was already measured. Therefore, the quantum circuit is equivalent to the standard model for quantum teleportation.

We discuss again quantum teleportation, this time, in the stabilizer formalism.

```
In[ ]:= Let[Complex, c]
       in = ProductState[S[3] → c@{0, 1}, "Label" → Ket[ψ]]
Out[ ]= (c₀ |0⟩ + c₁ |1⟩)_S₃
```

This is a quantum circuit model for quantum teleportation. It consists of three stages separated by the vertical dashed lines in the quantum circuit model. The first stage is to share the entangled pair of qubits. The second in the middle corresponds to the Bell measurement. The final stage transmits the measurement outcomes to Alice, and Alice adjusts the state of her qubit according to the classical information.

```
In[ ]:= qc = QuantumCircuit[in, LogicalForm[Ket[], S@{1, 2}], S[1, 6],
    CNOT[S[1], S[2]], "Separator", "Spacer", CNOT[S[2], S[3]],
    S[2, 6], {Measurement[S[2]], Measurement[S[3]]}, "Separator",
    ControlledU[S[2], S[1, 3]], ControlledU[S[3], S[1, 1]], "Invisible" → S[1.5]]
```

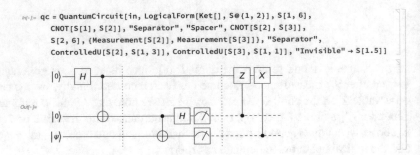

This rearranges the quantum circuit elements to re-analyze the quantum circuit model in stabilizer formalism.

```
In[ ]:= qc2 = QuantumCircuit[
    in, LogicalForm[Ket[], S@{1, 2}],
    S[1, 6], CNOT[S[1], S[2]], "Separator", "Spacer",
    CNOT[S[2], S[3]], "Separator",
    S[2, 6], Measurement[S[2]], ControlledU[S[2], S[1, 3]], "Separator",
    Measurement[S[3]], ControlledU[S[3], S[1, 1]]]
]
```

Now we test the result by performing the quantum circuit model many times.

```
In[ ]:= out = Table[ExpressionFor[qc2], {4}] /.
    c[0] * Conjugate[c[0]] + c[1] * Conjugate[c[1]] → 1;
    LogicalForm@KetFactor[#, S@{2, 3}] & /@ out // TableForm
```

$$Out[ ]//TableForm=$$
$$\left|0_{S_2}0_{S_3}\right\rangle \otimes \left(c_0 \left|0_{S_1}\right\rangle + c_1 \left|1_{S_1}\right\rangle\right)$$
$$\left|0_{S_2}1_{S_3}\right\rangle \otimes \left(c_0 \left|0_{S_1}\right\rangle + c_1 \left|1_{S_1}\right\rangle\right)$$
$$\left|1_{S_2}1_{S_3}\right\rangle \otimes \left(c_0 \left|0_{S_1}\right\rangle + c_1 \left|1_{S_1}\right\rangle\right)$$
$$\left|0_{S_2}1_{S_3}\right\rangle \otimes \left(c_0 \left|0_{S_1}\right\rangle + c_1 \left|1_{S_1}\right\rangle\right)$$

Each time the final state on the first qubit is the same regardless of the outcome of the measurements on the second and third qubit.

```
In[ ]:= out = Table[ExpressionFor[qc2], {4}] /.
    c[0] * Conjugate[c[0]] + c[1] * Conjugate[c[1]] → 1;
    LogicalForm@KetFactor[#, S@{2, 3}] & /@ out // TableForm
```

$$Out[ ]//TableForm=$$
$$\left|1_{S_2}0_{S_3}\right\rangle \otimes \left(c_0 \left|0_{S_1}\right\rangle + c_1 \left|1_{S_1}\right\rangle\right)$$
$$\left|1_{S_2}1_{S_3}\right\rangle \otimes \left(c_0 \left|0_{S_1}\right\rangle + c_1 \left|1_{S_1}\right\rangle\right)$$
$$\left|1_{S_2}1_{S_3}\right\rangle \otimes \left(c_0 \left|0_{S_1}\right\rangle + c_1 \left|1_{S_1}\right\rangle\right)$$
$$\left|1_{S_2}1_{S_3}\right\rangle \otimes \left(c_0 \left|0_{S_1}\right\rangle + c_1 \left|1_{S_1}\right\rangle\right)$$

```
In[ ]:= out = Table[ExpressionFor[qc2], {4}] /.
    c[0] * Conjugate[c[0]] + c[1] * Conjugate[c[1]] → 1;
    LogicalForm@KetFactor[#, S@{2, 3}] & /@ out // TableForm
```

$$Out[ ]//TableForm=$$
$$\left|0_{S_2}1_{S_3}\right\rangle \otimes \left(c_0 \left|0_{S_1}\right\rangle + c_1 \left|1_{S_1}\right\rangle\right)$$
$$\left|0_{S_2}0_{S_3}\right\rangle \otimes \left(c_0 \left|0_{S_1}\right\rangle + c_1 \left|1_{S_1}\right\rangle\right)$$
$$\left|0_{S_2}0_{S_3}\right\rangle \otimes \left(c_0 \left|0_{S_1}\right\rangle + c_1 \left|1_{S_1}\right\rangle\right)$$
$$\left|1_{S_2}1_{S_3}\right\rangle \otimes \left(c_0 \left|0_{S_1}\right\rangle + c_1 \left|1_{S_1}\right\rangle\right)$$

The second example is a remote implementation of a CNOT gate using the pre-shared entangled pair of qubits: Suppose that Alice and Bob are located far away from each other, and there is no quantum channel available between them. Fortunately, they had shared and are maintaining a maximally entangled pair of qubits, and there is a classical channel available for communication between them. Alice wants to implement a CNOT gate targeted on Bob's qubit, not the one involved in the shared entangled pair, controlled by her own qubit. The protocol is summarized in the following quantum circuit.

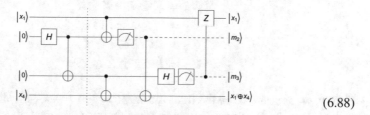

$$(6.88)$$

The initial stage marked by the vertical dashed line is just to share an entangled pair, and we start the analysis from this point. Then, the generators of the stabilizer are given by

$$\begin{cases} \hat{G}_1 = \hat{I} \otimes \hat{X} \otimes \hat{X} \otimes \hat{I}, \\ \hat{G}_2 = \hat{I} \otimes \hat{Z} \otimes \hat{Z} \otimes \hat{I}. \end{cases} \tag{6.89}$$

We also pick four logical operators to describe two *logical* qubits, one on Alice's side (A) and the other on Bob's side (B),

$$\begin{cases} \bar{X}_A = \hat{X} \otimes \hat{I} \otimes \hat{I} \otimes \hat{I}, \\ \bar{Z}_A = \hat{Z} \otimes \hat{I} \otimes \hat{I} \otimes \hat{I}, \\ \bar{X}_B = \hat{I} \otimes \hat{I} \otimes \hat{I} \otimes \hat{X}, \\ \bar{Z}_B = \hat{I} \otimes \hat{I} \otimes \hat{I} \otimes \hat{Z}. \end{cases} \tag{6.90}$$

The two subsequent CNOT gates transform the generators as

$$\begin{cases} \hat{G}_1 = \hat{I} \otimes \hat{X} \otimes \hat{X} \otimes \hat{X}, \\ \hat{G}_2 = \hat{Z} \otimes \hat{Z} \otimes \hat{Z} \otimes \hat{I}, \end{cases} \tag{6.91}$$

and the logical operators as

$$\begin{cases} \bar{X}_A = \hat{X} \otimes \hat{X} \otimes \hat{I} \otimes \hat{I}, \\ \bar{Z}_A = \hat{Z} \otimes \hat{I} \otimes \hat{I} \otimes \hat{I}, \\ \bar{X}_B = \hat{I} \otimes \hat{I} \otimes \hat{I} \otimes \hat{X}, \\ \bar{Z}_B = \hat{I} \otimes \hat{I} \otimes \hat{Z} \otimes \hat{Z}. \end{cases} \tag{6.92}$$

Next we measure $\hat{Z}$ on the second qubit, $\hat{M} = \hat{I} \otimes \hat{Z} \otimes \hat{I} \otimes \hat{I}$. $\hat{M}$ anti-commutes with $\hat{G}_1$ and $\bar{X}_A$. Therefore, we apply $\hat{G}_1$ conditioned on the measurement outcome, and update the generators and logical operators as

$$\begin{cases} \hat{G}_1 = \hat{I} \otimes \hat{Z} \otimes \hat{I} \otimes \hat{I}, \\ \hat{G}_2 = \hat{Z} \otimes \hat{Z} \otimes \hat{Z} \otimes \hat{I}, \end{cases} \tag{6.93}$$

and

$$\begin{cases} \bar{X}_A = \hat{X} \otimes \hat{I} \otimes \hat{X} \otimes \hat{X}, \\ \bar{Z}_A = \hat{Z} \otimes \hat{I} \otimes \hat{I} \otimes \hat{I}, \\ \bar{X}_B = \hat{I} \otimes \hat{I} \otimes \hat{I} \otimes \hat{X}, \\ \bar{Z}_B = \hat{I} \otimes \hat{I} \otimes \hat{Z} \otimes \hat{Z}. \end{cases} \tag{6.94}$$

Finally, we measure $\hat{X}$ on the third qubit, $\hat{M} = \hat{I} \otimes \hat{I} \otimes \hat{X} \otimes \hat{I}$. $\hat{M}$ anti-commutes with $\hat{G}_2$ and $\bar{Z}_B$. We apply $\hat{G}_2$ conditioned on the measurement outcome. The generators and logical operators are transformed as

$$\begin{cases} \hat{G}_1 = \hat{I} \otimes \hat{Z} \otimes \hat{I} \otimes \hat{I}, \\ \hat{G}_2 = \hat{I} \otimes \hat{I} \otimes \hat{X} \otimes \hat{I}, \end{cases} \tag{6.95}$$

and

$$\begin{cases} \bar{X}_A = \hat{X} \otimes \hat{I} \otimes \hat{X} \otimes \hat{X}, \\ \bar{Z}_A = \hat{Z} \otimes \hat{I} \otimes \hat{I} \otimes \hat{I}, \\ \bar{X}_B = \hat{I} \otimes \hat{I} \otimes \hat{I} \otimes \hat{X}, \\ \bar{Z}_B = \hat{Z} \otimes \hat{Z} \otimes \hat{I} \otimes \hat{Z}. \end{cases} \tag{6.96}$$

At this stage, the generators are meaningless since the second and third qubits are now classical after the two measurements. If one followed the above steps of analysis faithfully, she would get a quantum circuit of the form

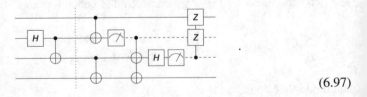

$$\tag{6.97}$$

However, the classically-conditioned operation of $\hat{X}$ on the third qubit right before the measurement of $\hat{X}$ does not affect the result. On the other hand, the classically-conditioned operation of $\hat{Z}$ on the second qubit is meaningless because the qubit has already been measured.

Here we investigate the procedure to remotely implement a CNOT gate.

This is the quantum circuit model we want to analyze.

```
qc = QuantumCircuit[S[2, 6], CNOT[S[2], S[3]], "Separator", "Spacer",
    {CNOT[S[1], S[2]], CNOT[S[3], S[4]]},
    Measurement[S[2]], CNOT[S[2], S@{3, 4}],
    S[3, 6], Measurement[S[3]], ControlledU[S[3], S[1, 3]]];
```

This is the general form of the input state.

```
Let[Integer, x, m]
in = LogicalForm[Ket[S[1] → x[1], S[4] → x[4]], S@{1, 2, 3, 4}];
```

This is the expected form of the output state.

```
out = Ket[S[1] → x[1], S@{2, 3} → m@{2, 3}, S[4] → Mod[x[1] + x[4], 2]];
```

This shows the quantum circuit model with the expected output states specified.

```
In[ ]:= qc1 = QuantumCircuit[in, qc, out, "PortSize" → {0.65, 2}, "Invisible" → S[2.5]]
```

Out[ ]=

Check whether the quantum circuit model works as expected.

```
In[ ]:= bs = Basis[S@{1, 4}];
new = qc ** bs;
Thread[LogicalForm[bs] → LogicalForm[new, S@{1, 2, 3, 4}]] // TableForm
```

Out[ ]//TableForm=

$$\begin{aligned}
|0_{S_1} 0_{S_4}\rangle &\rightarrow |0_{S_1} 1_{S_2} 1_{S_3} 0_{S_4}\rangle \\
|0_{S_1} 1_{S_4}\rangle &\rightarrow |0_{S_1} 1_{S_2} 0_{S_3} 1_{S_4}\rangle \\
|1_{S_1} 0_{S_4}\rangle &\rightarrow |1_{S_1} 1_{S_2} 1_{S_3} 1_{S_4}\rangle \\
|1_{S_1} 1_{S_4}\rangle &\rightarrow |1_{S_1} 0_{S_2} 0_{S_3} 0_{S_4}\rangle
\end{aligned}$$

6.4 Stabilizer Codes

Stabilizer codes are a class of quantum codes with construction based on the stabilizer formalism. The idea was put forward by Gottesman (Gottesman, 1996, 1998).

How does the stabilizer construction work? Let S be the stabilizer of a code space $\mathcal{V}$. Suppose that $\mathcal{V}$ is corrupted by an error operator $\hat{E}$ in the Pauli group $\mathcal{P}(n)$. What would happen to the code space? There are three distinctive cases[6]: First, $\hat{E}$ anti-commutes with one[7] of the generators, say $\hat{G}$, of the stabilizer S. In this

---

[6] The same classification of relationships between an error operator and a given stabilizer was used to analyze the effects of measurements in Sect. 6.3.5.

[7] Without loss of generality, we can assume that $\hat{E}$ anti-commutes with only one of the generator.

case, $\hat{E}$ takes the code space $\mathcal{V}$ to a subspace orthogonal to $\mathcal{V}$. To see this, suppose that $|\beta\rangle$ is a vector in the code space, and $|\beta'\rangle := \hat{E}|\beta\rangle$ is its corrupted state. Note that $\hat{G}\hat{E} + \hat{E}\hat{G} = 0$, and hence $\langle\alpha|\,\hat{G}\hat{E}\,|\beta\rangle + \langle\alpha|\,\hat{E}\hat{G}\,|\beta\rangle = 0$ for all $|\alpha\rangle$ in the code space. As both $|\alpha\rangle$ and $|\beta\rangle$ are stabilized by $\hat{G}$, we observe that $\langle\alpha|\beta'\rangle = 0$. Therefore, in this case, one can detect the error, and correct it. Second, $\hat{E}$ belongs to the stabilizer $\mathcal{S}$. In this case, $\hat{E}$ itself stabilizes the code space, and the code space remains intact without any corruption. Third, $\hat{E}$ commutes with all elements of the stabilizer, but does not belong to the stabilizer. This is the most dangerous case. As we have seen in Sect. 6.3.5, measuring $\hat{E}$ can reveal some information of the quantum states in the code space. This implies that $\hat{E}$ breaks the code space, and the corruption cannot be corrected.

The above observation already suggests that for an error to be correctable, it needs to either anti-commute with a generator of the stabilizer or belong to the stabilizer. However, there are usually a set of errors, and we want to protect quantum states against their combinations. Hence the error-correction conditions for stabilizer codes are a bit more involved. Nevertheless, the error-correction conditions in Sect. 6.2.1 translate more transparently to stabilizer formalism: Let $\mathcal{S}$ be the stabilizer of a code space $\mathcal{V}$. Suppose that the errors are described by a set $\{\hat{E}_\mu\}$ of operators in the Pauli group $\mathcal{P}(n)$. The necessary and sufficient condition for the code to correct the errors is that for all μ and ν, $\hat{E}_\mu^\dagger\hat{E}_\nu$ either belongs to $\mathcal{S}$ or anti-commutes with at least one generator of $\mathcal{S}$. Remember that the identity operator is always one of the error operators and thus the condition includes the above observation.

How do we actually perform a recovery operation to correct the error? Let $\{\hat{G}_1, \ldots, \hat{G}_k\}$ be a set of independent generators of the stabilizer $\mathcal{S}$. Suppose that $\{\hat{E}_\mu\}$ is the set of error operators that are *correctable* by the code. The first step is to detect error syndrome. This is achieved by measuring the generators $\hat{G}_1, \ldots, \hat{G}_k$. Each outcome ± 1 serves as a symptom of the errors. If an error $\hat{E}_\mu$ occurs, then it causes a syndrome $(\varepsilon_1^\mu, \ldots, \varepsilon_k^\mu)$. The error and the syndrome are related by

$$\hat{E}_\mu\hat{G}_j\hat{E}_\mu^\dagger = \varepsilon_j^\mu\hat{G}_j. \tag{6.98}$$

When a syndrome is associated with a single error $\hat{E}_\mu$, we can correct it simply by applying $\hat{E}_\mu^\dagger$. What if two different errors $\hat{E}_\mu$ and $\hat{E}_\nu$ cause the same syndrome? In other words, suppose that

$$(\varepsilon_1^\mu, \ldots, \varepsilon_k^\mu) = (\varepsilon_1^\nu, \ldots, \varepsilon_k^\nu). \tag{6.99}$$

In this case, we have

$$\hat{E}_\mu\hat{P}\hat{E}_\mu^\dagger = \hat{E}_\nu\hat{P}\hat{E}_\nu^\dagger, \tag{6.100}$$

hence

$$(\hat{E}_\mu^\dagger\hat{E}_\nu)\hat{P}(\hat{E}_\mu^\dagger\hat{E}_\nu)^\dagger = \hat{P}, \tag{6.101}$$

where $\hat{P}$ is the projection onto the code space. This means that $\hat{E}_\mu^\dagger \hat{E}_\nu$ is a member of the stabilizer. That is, even if it was the error $\hat{E}_\nu$ that actually occurred, it can still be corrected by applying $\hat{E}_\mu^\dagger$. Therefore, however many errors are associated with a given syndrome, one can choose any error $\hat{E}_\mu$ among them to apply $\hat{F}_\mu^\dagger$ to recover the original state.

One remaining task for a given stabilizer code is the encoding procedure. How can one prepare a system of n physical qubits, say, in the logical basis state $|\bar{0}\cdots\bar{0}\rangle$? This requires a procedure without referring to an explicit form of the logical basis states. One of the simplest approaches is to measure the generators $\hat{G}_1,\ldots,\hat{G}_k$ and the logical operators $\hat{Z}_1,\ldots,\hat{Z}_{n-k}$.[8] Recall that the combined set $\{\hat{G}_1,\ldots,\hat{G}_k,\hat{Z}_1,\ldots,\hat{Z}_{n-k}\}$ consists of all mutually commuting and independent operators. Thus, it generates a stabilizer subgroup stabilizing a one-dimensional subspace spanned by nothing but the desired state $|\bar{0}\cdots\bar{0}\rangle$. Therefore, one can regard a measurement of the generators and logical operators as the error syndrome detection. Then, an initialization to the desired state corresponds to the recovery procedure discussed above.

Let us take the encoding procedure in more explicit steps: Each measurement of $\hat{G}_1,\ldots,\hat{G}_k$ and $\hat{Z}_1,\ldots,\hat{Z}_{n-k}$ yields measurement outcome ± 1. Depending on the measurement results, the final state $|\psi\rangle$ after all measurements is stabilized by $\pm\hat{G}_1,\ldots,\pm\hat{G}_k,\pm\hat{Z}_1,\ldots,+\hat{Z}_{n-k}$. We then recall (see Sect. 6.3.1) that there exists a Pauli operator $\hat{G}$ that anti-commutes with only a particular one of the independent operators but commutes with all others. We find such a Pauli operator for each observable with measurement outcome -1, and apply those Pauli operators on $|\psi\rangle$. Once the state $|\bar{0}\cdots\bar{0}\rangle$ has been achieved, other logical basis states can be obtained by applying the logical operators $\bar{X}_1,\ldots,\bar{X}_{n-k}$ as necessary.

In the remainder of the section, we provide some examples of stabilizer codes. They were constructed in a different scheme, and are reconstructed here by following stabilizer formalism.

6.4.1 Bit-Flip Code

First consider the three-qubit bit-flip correction code. The code space is spanned by two logical basis states, $|\bar{0}\rangle := |000\rangle$ and $|\bar{1}\rangle := |111\rangle$. The stabilizer is generated by $\hat{Z}_1\hat{Z}_2$ and $\hat{Z}_2\hat{Z}_3$. Here $\hat{X}_j$, $\hat{Y}_j$, and $\hat{Z}_j$ denote the Pauli operators acting on qubit j. In this code, the logical operators are given by $\bar{Z} = \hat{Z}_1\hat{Z}_2\hat{Z}_3$ and $\bar{X} = \hat{X}_1\hat{X}_2\hat{X}_3$.

The code can correct the bit-flip errors in $\{\hat{I}, \hat{X}_1, \hat{X}_2, \hat{X}_3\}$. The error syndrome is thus detected by measuring the generators $\hat{Z}_1\hat{Z}_2$ and $\hat{Z}_2\hat{Z}_3$. Table 6.1 shows the error syndromes for bit-flip errors on different qubits.

---

[8] Encoding and decoding a general stabilizer code using unitary gates are also possible and explained in detail in Cleve & Gottesman (1997).

Table 6.1 The error syndrome of the bit-flip correction code

$\hat{Z}_1\hat{Z}_2$	$\hat{Z}_2\hat{Z}_3$	Error
1	1	$\hat{I}$
1	−1	$\hat{X}_3$
−1	1	$\hat{X}_1$
−1	−1	$\hat{X}_2$

It is heuristic to consider an error described by $\hat{Z}_1$. It commutes with both generators $\hat{Z}_1\hat{Z}_2$ and $\hat{Z}_2\hat{Z}_3$ but is not an element of the stabilizer. This means that the code is not protected against the error $\hat{Z}_1$. This is also the case with $\hat{Z}_2$ and $\hat{Z}_3$, which is not surprising because these operators correspond to phase-flip errors. The code is designed only for bit-flip errors and does not protect against phase-flip errors.

Here are the generators of the stabilizer of the code space for the bit-flip correction code.

```
In[ ]:= gnr = {S[1, 3] ** S[2, 3], S[2, 3] ** S[3, 3]};
        gnr // PauliForm
```

```
Out[ ]= {Z ⊗ Z ⊗ I, I ⊗ Z ⊗ Z}
```

These are error operators correctable by the code.

```
In[ ]:= err = {1, S[1, 1], S[2, 1], S[3, 1]};
        err // PauliForm
```

```
Out[ ]= {I ⊗ I ⊗ I, X ⊗ I ⊗ I, I ⊗ X ⊗ I, I ⊗ I ⊗ X}
```

```
In[ ]:= chk = Union[Multiply @@@ Choices[err, {2}]];
        chk // PauliForm
```

```
Out[ ]= {I ⊗ I ⊗ I, X ⊗ X ⊗ I, X ⊗ I ⊗ X, I ⊗ X ⊗ X, X ⊗ I ⊗ I, I ⊗ X ⊗ I, I ⊗ I ⊗ X}
```

Check and confirm that the errors are indeed correctable. To do it, we use `GottesmanTest`.

```
In[ ]:= ? GottesmanTest
```

```
Out[ ]=
```
> Symbol
>
> GottesmanTest [a, b] returns 1 if the two operators a and b
>   commute with each other, −1 if they anti−commute, and 0 otherwise.

As one can see, there is only one combination, `I⊗I⊗I`, of the error operators that commutes with all generators of the stabilizer. But it belongs to the stabilizer.

```
In[ ]:= mat = Outer[GottesmanTest, gnr, chk];
        TableForm[mat, TableHeadings → {PauliForm@gnr, PauliForm@chk},
          TableAlignments → Right]
```

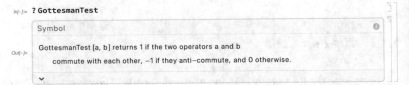

	I⊗I⊗I	X⊗X⊗I	X⊗I⊗X	I⊗X⊗X	X⊗I⊗I	I⊗X⊗I	I⊗I⊗X
Z⊗Z⊗I	1	1	−1	−1	−1	−1	1
I⊗Z⊗Z	1	−1	−1	1	1	−1	−1

This table displays the error syndrome of the bit-flip correction code.

```
In[ ]:= syndrome = Map[(# ** gnr ** #) / gnr &, err];
       TableForm[syndrome, TableAlignments → Right,
         TableHeadings → {PauliForm@err, PauliForm@gnr}]
```

Out[ ]//TableForm=

	Z⊗Z⊗I	I⊗Z⊗Z
I⊗I⊗I	1	1
X⊗I⊗I	-1	1
I⊗X⊗I	-1	-1
I⊗I⊗X	1	-1

6.4.2 Phase-Flip Code

Let us now turn to the three-qubit phase-flip correction code. The code space is spanned by the two logical basis states $|\bar{0}\rangle := |+++\rangle$ and $|\bar{1}\rangle := |---\rangle$, where $|\pm\rangle := (|0\rangle \pm |1\rangle)/\sqrt{2}$. The stabilizer is generated by $\hat{X}_1\hat{X}_2$ and $\hat{X}_2\hat{X}_3$. In this code, the logical operators are given by $\bar{Z} = \hat{X}_1\hat{X}_2\hat{X}_3$ and $\bar{X} = \hat{Z}_1\hat{Z}_2\hat{Z}_3$.

The code can correct the phase-flip errors in $\{\hat{I}, \hat{Z}_1, \hat{Z}_2, \hat{Z}_3\}$. The error syndrome is detected by measuring the generators $\hat{X}_1\hat{X}_2$ and $\hat{X}_2\hat{X}_3$. Table 6.2 shows the error syndromes for phse-flip errors on different qubits.

The above accounts are a direct translation of the phase-flip correction code presented in Sect. 6.1.2. However, the bit-flip and phase-flip correction codes are closely related to each other as suggested by a comparison of the error syndromes in Tables 6.1 and 6.2 of the two codes. One can make the relation clearer and more rigorous in stabilizer formalism: The code space of the phase-flip correction code is an image of a unitary transformation, more specifically, a Clifford operator $\hat{U} := \hat{H}_1\hat{H}_2\hat{H}_3$,

$$|+++\rangle = \hat{U}\,|000\rangle \,, \quad |---\rangle = \hat{U}\,|111\rangle \,. \tag{6.102}$$

In accordance with the transformation rules for stabilizers under unitary transformations discussed in Sect. 6.3.3, the generators of the phase-flip correction code are related to those of the bit-flip correction code by the same Clifford operator,

$$\hat{X}_1\hat{X}_2 = \hat{U}\,\hat{Z}_1\hat{Z}_2\,\hat{U}^\dagger\,, \quad \hat{X}_2\hat{X}_3 = \hat{U}\,\hat{Z}_2\hat{Z}_3\,\hat{U}^\dagger\,. \tag{6.103}$$

Table 6.2 The error syndrome of the phase-flip correction code

$\hat{X}_1\hat{X}_2$	$\hat{X}_2\hat{X}_3$	Error
1	1	$\hat{I}$
1	-1	$\hat{Z}_3$
-1	1	$\hat{Z}_1$
-1	-1	$\hat{Z}_2$

The error syndromes in Tables 6.1 and 6.2 are certainly related to each other by the same Clifford operator $\hat{U}$.

Here are the generators of the stabilizer of the code space for the bit-flip correction code.

```
In[·]:= gnr = {S[1, 1] ** S[2, 1], S[2, 1] ** S[3, 1]};
       gnr // PauliForm
```
```
Out[·]= {X ⊗ X ⊗ I, I ⊗ X ⊗ X}
```

These are error operators correctable by the code.

```
In[·]:= err = {1, S[1, 3], S[2, 3], S[3, 3]};
       err // PauliForm
```
```
Out[·]= {I ⊗ I ⊗ I, Z ⊗ I ⊗ I, I ⊗ Z ⊗ I, I ⊗ I ⊗ Z}
```

```
In[·]:= chk = Union[Multiply @@@ Choices[err, {2}]];
       chk // PauliForm
```
```
Out[·]= {I ⊗ I ⊗ I, Z ⊗ Z ⊗ I, Z ⊗ I ⊗ Z, I ⊗ Z ⊗ Z, Z ⊗ I ⊗ I, I ⊗ Z ⊗ I, I ⊗ I ⊗ Z}
```

Check and confirm that the errors are indeed correctable. As one can see, there is only one combination, I⊗I⊗I, of the error operators that commutes with all generators of the stabilizer. But it belongs to the stabilizer.

```
In[·]:= mat = Outer[GottesmanTest, gnr, chk];
       TableForm[mat, TableHeadings → {PauliForm@gnr, PauliForm@chk},
         TableAlignments → Right]
```

Out[·]//TableForm=	I⊗I⊗I	Z⊗Z⊗I	Z⊗I⊗Z	I⊗Z⊗Z	Z⊗I⊗I	I⊗Z⊗I	I⊗I⊗Z
X⊗X⊗I	1	1	-1	-1	-1	-1	1
I⊗X⊗X	1	-1	-1	1	1	-1	-1

This table displays the error syndrome of the phase-flip correction code.

```
In[·]:= syndrome = Map[(# ** gnr ** #) / gnr &, err];
       TableForm[syndrome, TableAlignments → Right,
         TableHeadings → {PauliForm@err, PauliForm@gnr}]
```

Out[·]//TableForm=	X⊗X⊗I	I⊗X⊗X
I⊗I⊗I	1	1
Z⊗I⊗I	-1	1
I⊗Z⊗I	-1	-1
I⊗I⊗Z	1	-1

6.4.3 Nine-Qubit Code

Let us now combine the bit-flip and phase-flip correction codes to get Shor's nine-qubit code to correct arbitrary single-qubit errors. A set of independent generators of the stabilizer of the code is listed in Table 6.3. The logical operators are given by

$$\bar{Z} = \hat{X} \otimes \hat{X} \otimes \hat{X} \otimes \hat{X} \otimes \hat{X} \otimes \hat{X} \otimes \hat{X} \otimes \hat{X} \otimes \hat{X},$$
$$\bar{X} = \hat{Z} \otimes \hat{Z} \otimes \hat{Z} \otimes \hat{Z} \otimes \hat{Z} \otimes \hat{Z} \otimes \hat{Z} \otimes \hat{Z} \otimes \hat{Z}.$$
$$(6.104)$$

The errors are described by the set of single-qubit Pauli operators

Table 6.3 A set of independent generators of the stabilizer for Shor's nine-qubit code

Symbol	Operator
$\hat{G}_1$	$\hat{Z} \otimes \hat{Z} \otimes \hat{I} \otimes \hat{I} \otimes \hat{I} \otimes \hat{I} \otimes \hat{I} \otimes \hat{I} \otimes \hat{I}$
$\hat{G}_2$	$\hat{I} \otimes \hat{Z} \otimes \hat{Z} \otimes \hat{I} \otimes \hat{I} \otimes \hat{I} \otimes \hat{I} \otimes \hat{I} \otimes \hat{I}$
$\hat{G}_3$	$\hat{I} \otimes \hat{I} \otimes \hat{I} \otimes \hat{Z} \otimes \hat{Z} \otimes \hat{I} \otimes \hat{I} \otimes \hat{I} \otimes \hat{I}$
$\hat{G}_4$	$\hat{I} \otimes \hat{I} \otimes \hat{I} \otimes \hat{I} \otimes \hat{Z} \otimes \hat{Z} \otimes \hat{I} \otimes \hat{I} \otimes \hat{I}$
$\hat{G}_5$	$\hat{I} \otimes \hat{I} \otimes \hat{I} \otimes \hat{I} \otimes \hat{I} \otimes \hat{I} \otimes \hat{Z} \otimes \hat{Z} \otimes \hat{I}$
$\hat{G}_6$	$\hat{I} \otimes \hat{I} \otimes \hat{I} \otimes \hat{I} \otimes \hat{I} \otimes \hat{I} \otimes \hat{I} \otimes \hat{Z} \otimes \hat{Z}$
$\hat{G}_7$	$\hat{X} \otimes \hat{X} \otimes \hat{X} \otimes \hat{X} \otimes \hat{X} \otimes \hat{X} \otimes \hat{I} \otimes \hat{I} \otimes \hat{I}$
$\hat{G}_8$	$\hat{I} \otimes \hat{I} \otimes \hat{I} \otimes \hat{X} \otimes \hat{X} \otimes \hat{X} \otimes \hat{X} \otimes \hat{X} \otimes \hat{X}$

$$\left\{ \hat{I}_j, \hat{X}_j, \hat{Y}_j, \hat{Z}_j : j = 1, 2, \ldots, 9 \right\}. \tag{6.105}$$

It is rather tedious and yet straightforward to check by inspection that the above set of errors does satisfy the error-correction conditions discussed at the beginning of the section.

Shor's code needs 9 qubits to encode physical single-qubit quantum states.

```
jj = Range[9];
```

This is a generating set of the stabilizer of the code.

```
In[·]:= gnr = {
    S[1, 3] ** S[2, 3], S[2, 3] ** S[3, 3],
    S[4, 3] ** S[5, 3], S[5, 3] ** S[6, 3],
    S[7, 3] ** S[8, 3], S[8, 3] ** S[9, 3],
    Multiply @@ S[{1, 2, 3, 4, 5, 6}, 1],
    Multiply @@ S[{4, 5, 6, 7, 8, 9}, 1]}
```

$Out[·]= \left\{ S_1^z S_2^z, \; S_2^z S_3^z, \; S_4^z S_5^z, \; S_5^z S_6^z, \; S_7^z S_8^z, \; S_8^z S_9^z, \; S_1^x S_2^x S_3^x S_4^x S_5^x S_6^x, \; S_4^x S_5^x S_6^x S_7^x S_8^x S_9^x \right\}$

These are logical operators of the code.

```
In[·]:= opX = Multiply @@ S[jj, 1]
       opZ = Multiply @@ S[jj, 3]
```

$Out[·]= S_1^x S_2^x S_3^x S_4^x S_5^x S_6^x S_7^x S_8^x S_9^x$

$Out[·]= S_1^z S_2^z S_3^z S_4^z S_5^z S_6^z S_7^z S_8^z S_9^z$

These are the error operators correctable by the code.

```
In[·]:= err = Prepend[S[jj, All], 1]
```

$Out[·]= \{ 1, \; S_1^x, \; S_1^y, \; S_1^z, \; S_2^x, \; S_2^y, \; S_2^z, \; S_3^x, \; S_3^y, \; S_3^z, \; S_4^x, \; S_4^y, \; S_4^z,$
$\quad S_5^x, \; S_5^y, \; S_5^z, \; S_6^x, \; S_6^y, \; S_6^z, \; S_7^x, \; S_7^y, \; S_7^z, \; S_8^x, \; S_8^y, \; S_8^z, \; S_9^x, \; S_9^y, \; S_9^z \}$

In order to check that the above errors are indeed correctable, we consider combinations of two error operators.

```
In[·]:= chk = Union[Multiply @@@ Choices[err, {2}]];
       Length[chk]
Out[·]:= 379
```

This calculates the commutation of the above combinations with the generators of the stabilizer.

```
In[·]:= Timing[mat = Outer[GottesmanTest, chk, gnr];]
Out[·]:= {5.48404, Null}
```

This displays a *part* of the commutation relation. One can see that each of them either commutes or anti-commutes with the generators.

```
In[·]:= TableForm[mat[[ ;; 5, ;; 6]],
         TableAlignments → Right, TableHeadings → {chk[[ ;; 5]], gnr}]
Out[·]//TableForm=
```

	$S_2^z S_2^z$	$S_2^z S_3^z$	$S_4^z S_5^z$	$S_5^z S_6^z$	$S_7^z S_8^z$	$S_8^z S_9^z$
1	1	1	1	1	1	1
$S_1^x S_2^x$	1	-1	1	1	1	1
$S_1^x S_2^y$	1	-1	1	1	1	1
$S_1^x S_2^z$	-1	1	1	1	1	1
$S_1^x S_3^x$	-1	-1	1	1	1	1

The most dangerous case is that the check operator commutes with all of the generators but does not belong to the stabilizer. To check if there is such case, we examine the check operators that commute with all of the generators.

```
In[·]:= kk = Catenate@MapIndexed[If[ContainsOnly[#1, {1}], #2, Nothing] &, mat]
Out[·]:= {1, 52, 55, 118, 223, 226, 262, 313, 316, 325}
```

This shows that such check operators actually belong to the stabilizer. The considered errors are thus correctable by the code.

```
In[·]:= chk[[kk]]
Out[·]:= {1, S_1^z S_2^z, S_1^z S_3^z, S_2^z S_3^z, S_4^z S_5^z, S_4^z S_6^z, S_5^z S_6^z, S_7^z S_8^z, S_7^z S_9^z, S_8^z S_9^z}
```

This table displays a *part* of the error syndrome of the code.

```
In[·]:= syndrome = Map[(# ** gnr ** #) / gnr &, err];
       Style[TableForm[syndrome[[ ;; 8]],
         TableAlignments → Right, TableHeadings → {err, gnr}], Small]
```

	$S_2^z S_2^z$	$S_2^z S_3^z$	$S_4^z S_5^z$	$S_5^z S_6^z$	$S_7^z S_8^z$	$S_8^z S_9^z$	$S_1^z S_1^z S_3^z S_4^z S_5^z S_6^z$	$S_4^z S_5^z S_6^z S_7^z S_8^z S_9^z$
1	1	1	1	1	1	1	1	1
S_1^x	-1	1	1	1	1	1	1	1
S_1^y	-1	1	1	1	1	1	-1	1
S_1^z	1	1	1	1	1	1	-1	1
S_2^x	-1	-1	1	1	1	1	1	1
S_2^y	-1	-1	1	1	1	1	-1	1
S_2^z	1	1	1	1	1	1	-1	1
S_3^x	1	-1	1	1	1	1	1	1

6.4.4 Five-Qubit Code

The smallest possible code to protect the single-qubit states against generic errors on single qubits is the five-qubit code discovered independently by Bennett et al. (1996) and by Laflamme et al. (1996). It belongs to a wider class of codes called *Calderbank–Shor–Steane (CSS) codes* (Calderbank & Shor, 1996; Steane, 1996). CSS codes were initially developed in analogy with the classical linear codes. However, one can also regard these as a special type of stabilizer codes.

The stabilizer of the five-qubit code is generated by the following four generators

$$
\begin{aligned}
\hat{G}_1 &= \hat{X} \otimes \hat{Z} \otimes \hat{Z} \otimes \hat{X} \otimes \hat{I}, \\
\hat{G}_2 &= \hat{I} \otimes \hat{X} \otimes \hat{Z} \otimes \hat{Z} \otimes \hat{X}, \\
\hat{G}_3 &= \hat{X} \otimes \hat{I} \otimes \hat{X} \otimes \hat{Z} \otimes \hat{Z}, \\
\hat{G}_4 &= \hat{Z} \otimes \hat{X} \otimes \hat{I} \otimes \hat{X} \otimes \hat{Z}.
\end{aligned}
\tag{6.106}
$$

It can correct arbitrary errors on single qubits,

$$
\{\hat{I}_j, \hat{X}_j, \hat{Y}_j, \hat{Z}_j : j = 1, 2, 3, 4, 5\},
\tag{6.107}
$$

which can be confirmed by checking the commutation relations of the combinations of errors in (6.107) with the generators in (6.106). One can choose

$$
\bar{Z} = \hat{Z} \otimes \hat{Z} \otimes \hat{Z} \otimes \hat{Z} \otimes \hat{Z}, \quad \bar{X} = \hat{X} \otimes \hat{X} \otimes \hat{X} \otimes \hat{X} \otimes \hat{X}
\tag{6.108}
$$

as the logical operators of the code.

We consider a code encoded on 5 qubits.

```
jj = Range[5];
```

Here are the generators of the stabilizer of the five-qubit code.

```
In[ ]:= gnr = {
    S[1, 1] ** S[2, 3] ** S[3, 3] ** S[4, 1],
    S[2, 1] ** S[3, 3] ** S[4, 3] ** S[5, 1],
    S[1, 1] ** S[3, 1] ** S[4, 3] ** S[5, 3],
    S[1, 3] ** S[2, 1] ** S[4, 1] ** S[5, 3]};
PauliForm[gnr]
```

```
Out[ ]= {X⊗Z⊗Z⊗X⊗I, I⊗X⊗Z⊗Z⊗X, X⊗I⊗X⊗Z⊗Z, Z⊗X⊗I⊗X⊗Z}
```

These are logical operators of the code.

```
In[ ]:= opX = Multiply @@ S[jj, 1];
       PauliForm[opX]
       opZ = Multiply @@ S[jj, 3];
       PauliForm[opZ]
```

Out[ ]= X ⊗ X ⊗ X ⊗ X ⊗ X

Out[ ]= Z ⊗ Z ⊗ Z ⊗ Z ⊗ Z

Here are the error operators correctable by the code.

```
In[ ]:= err = Prepend[S[jj, All], 1];
       PauliForm[err]
```

Out[ ]= {I⊗I⊗I⊗I⊗I, X⊗I⊗I⊗I⊗I, Y⊗I⊗I⊗I⊗I, Z⊗I⊗I⊗I⊗I,
        I⊗X⊗I⊗I⊗I, I⊗Y⊗I⊗I⊗I, I⊗Z⊗I⊗I⊗I, I⊗I⊗X⊗I⊗I,
        I⊗I⊗Y⊗I⊗I, I⊗I⊗Z⊗I⊗I, I⊗I⊗I⊗X⊗I, I⊗I⊗I⊗Y⊗I,
        I⊗I⊗I⊗Z⊗I, I⊗I⊗I⊗I⊗X, I⊗I⊗I⊗I⊗Y, I⊗I⊗I⊗I⊗Z}

```
In[ ]:= chk = Union[Multiply @@@ Choices[err, {2}]];
       Length[chk]
```

Out[ ]= 121

```
In[ ]:= Timing[mat = Outer[GottesmanTest, chk, gnr];]
```

Out[ ]= {1.12877, Null}

This displays a *part* of the commutation relations of some of the check operators. One can see that each of them either commutes or anti-commutes with the generators.

```
In[ ]:= TableForm[mat[[ ;; 5]], TableAlignments → Right,
         TableHeadings → PauliForm@{chk[[ ;; 5]], gnr}]
```

Out[ ]//TableForm=

	X⊗Z⊗Z⊗X⊗I	I⊗X⊗Z⊗Z⊗X	X⊗I⊗X⊗Z⊗Z	Z⊗X⊗I⊗X⊗Z
I⊗I⊗I⊗I⊗I	1	1	1	1
X⊗X⊗I⊗I⊗I	-1	1	1	-1
X⊗Y⊗I⊗I⊗I	-1	-1	1	1
X⊗Z⊗I⊗I⊗I	1	-1	1	1
X⊗I⊗X⊗I⊗I	-1	-1	1	-1

The most dangerous case is that the check operator commutes with all of the generators but does not belong to the stabilizer. To check if there is such case, we examine the check operators that commute with all of the generators.

```
In[ ]:= kk = Catenate@MapIndexed[If[ContainsOnly[#1, {1}], #2, Nothing] &, mat]
```

Out[ ]= {1}

This shows that such check operators actually belong to the stabilizer. The considered errors are thus correctable by the code.

```
In[ ]:= chk[[kk]]
```

Out[ ]= {1}

This shows the error syndrome of the code upon the measurement of the generators of the stabilizer.

```
In[·]:= syndrome = Map[(# ** gnr ** #) / gnr &, err];
    TableForm[syndrome, TableAlignments → Right,
        TableHeadings → {PauliForm@err, PauliForm@gnr}]
```

Out[·]//TableForm=

	X⊗Z⊗Z⊗X⊗I	I⊗X⊗Z⊗Z⊗X	X⊗I⊗X⊗Z⊗Z	Z⊗X⊗I⊗X⊗Z
I⊗I⊗I⊗I⊗I	1	1	1	1
X⊗I⊗I⊗I⊗I	1	1	1	-1
Y⊗I⊗I⊗I⊗I	-1	1	-1	-1
Z⊗I⊗I⊗I⊗I	-1	1	-1	1
I⊗X⊗I⊗I⊗I	-1	1	1	1
I⊗Y⊗I⊗I⊗I	-1	-1	1	-1
I⊗Z⊗I⊗I⊗I	1	-1	1	-1
I⊗I⊗X⊗I⊗I	-1	-1	1	1
I⊗I⊗Y⊗I⊗I	-1	-1	-1	1
I⊗I⊗Z⊗I⊗I	1	1	-1	1
I⊗I⊗I⊗X⊗I	1	-1	-1	1
I⊗I⊗I⊗Y⊗I	-1	-1	-1	-1
I⊗I⊗I⊗Z⊗I	-1	1	1	-1
I⊗I⊗I⊗I⊗X	1	1	-1	-1
I⊗I⊗I⊗I⊗Y	1	-1	-1	-1
I⊗I⊗I⊗I⊗Z	1	-1	1	1

As mentioned before, in stabilizer formalism, any meaningful procedure can be performed without referring to the logical basis states. Nevertheless, it is heuristically interesting to evaluate the logical basis states explicitly. These can be found by simultaneously diagonalizing the generators of the stabilizer and the logical operator $\bar{Z}$. For the five-qubit code, they are given (unnormalized) by

$$
\begin{aligned}
|\bar{0}\rangle = &|00000\rangle - |00011\rangle + |00101\rangle - |00110\rangle + |01001\rangle + |01010\rangle \\
& - |01100\rangle - |01111\rangle - |10001\rangle + |10010\rangle + |10100\rangle - |10111\rangle \\
& - |11000\rangle - |11011\rangle - |11101\rangle - |11110\rangle
\end{aligned}
$$

(6.109)

and

$$
\begin{aligned}
|\bar{1}\rangle = &- |00001\rangle - |00010\rangle - |00100\rangle - |00111\rangle - |01000\rangle + |01011\rangle \\
& + |01101\rangle - |01110\rangle - |10000\rangle - |10011\rangle + |10101\rangle + |10110\rangle \\
& - |11001\rangle + |11010\rangle - |11100\rangle + |11111\rangle .
\end{aligned}
$$

(6.110)

One can see that the description of the code space in terms of the so-called 'codewords'—the logical basis states of the code space—is rather complicated. This example demonstrates how compact the description in stabilizer formalism is.

Let us evaluate explicitly the logical basis states of the code. This is a brute-force approach. A more sophisticated method is to follow the encoding procedure discussed in the main text.

From the generating set of the stabilizer given above, we can construct the elements of the stabilizer explicitly.

```
In[·]:= new=Union@Join[gnr,Multiply@@@Tuples[gnr,2]];
       grp=Union@Join[new,Multiply@@@Tuples[new,2]]
       Length[grp]
```

$$Out[\cdot]= \left\{1,\ S_1^x\,S_2^x\,S_3^y\,S_5^y,\ S_1^x\,S_2^y\,S_4^y\,S_5^x,\ S_1^x\,S_2^z\,S_3^z\,S_4^x,\ S_1^x\,S_3^z\,S_4^z\,S_5^x,\right.$$
$$S_1^y\,S_2^x\,S_3^x\,S_4^y,\ S_1^y\,S_2^y\,S_3^z\,S_5^z,\ S_1^y\,S_2^z\,S_4^z\,S_5^y,\ S_1^y\,S_3^y\,S_4^x\,S_5^x,\ S_1^z\,S_2^x\,S_4^x\,S_5^z,$$
$$\left.S_1^z\,S_2^y\,S_3^y\,S_4^z,\ S_1^z\,S_2^z\,S_3^x\,S_5^x,\ S_1^z\,S_3^z\,S_4^y\,S_5^y,\ S_2^z\,S_3^z\,S_4^x\,S_5^x,\ S_2^y\,S_3^x\,S_4^x\,S_5^y,\ S_2^z\,S_3^y\,S_4^y\,S_5^z\right\}$$

$$Out[\cdot]= 16$$

Recall that the logical Z operator is given as fallows.

```
In[·]:= opZ = Multiply @@ S[jj, 3]
```

$$Out[\cdot]= S_1^z\,S_2^z\,S_3^z\,S_4^z\,S_5^z$$

We now want to get the simultaneous eigenstates of the stabilizer and the logical Z operator.

```
ops = Append[grp, opZ];
mat = Matrix@Rest@ops;
```

```
In[·]:= {val, vec} = CommonEigensystem[mat];
       Style[TableForm[val[[-5 ;;]], TableAlignments → Right], Small]
```

$$Out[\cdot]=\ \begin{array}{cccccccccccccccc} 1 & 1 & -1 & -1 & 1 & 1 & -1 & -1 & 1 & 1 & -1 & -1 & 1 & 1 & -1 & -1 \\ 1 & 1 & 1 & 1 & -1 & -1 & -1 & -1 & -1 & -1 & -1 & -1 & 1 & 1 & 1 & -1 \\ 1 & 1 & 1 & 1 & -1 & -1 & -1 & -1 & -1 & -1 & -1 & -1 & 1 & 1 & 1 & 1 \\ 1 & 1 & 1 & 1 & 1 & 1 & 1 & 1 & 1 & 1 & 1 & 1 & 1 & 1 & 1 & -1 \\ 1 & 1 & 1 & 1 & 1 & 1 & 1 & 1 & 1 & 1 & 1 & 1 & 1 & 1 & 1 & 1 \end{array}$$

We see that the code space is spanned by the last two eigenvectors. The last one belongs to the eigenvalue 1 of the logical Z operator.

```
In[·]:= bs = Basis[S@jj];
       ket0 = -vec[[-1]].bs;
       ket0 // SimpleForm
```

$$Out[\cdot]=\ \frac{|00000\rangle}{4} - \frac{|00011\rangle}{4} + \frac{|00101\rangle}{4} - \frac{|00110\rangle}{4} + \frac{|01001\rangle}{4} + \frac{|01010\rangle}{4} - \frac{|01100\rangle}{4} - \frac{|01111\rangle}{4} -$$
$$\frac{|10001\rangle}{4} + \frac{|10010\rangle}{4} + \frac{|10100\rangle}{4} - \frac{|10111\rangle}{4} - \frac{|11000\rangle}{4} + \frac{|11011\rangle}{4} + \frac{|11101\rangle}{4} - \frac{|11110\rangle}{4}$$

```
In[·]:= ket1 = vec[[-2]].bs;
       ket1 // SimpleForm
```

$$Out[\cdot]=\ -\frac{1}{4}\,|00001\rangle - \frac{|00010\rangle}{4} - \frac{|00100\rangle}{4} - \frac{|00111\rangle}{4} - \frac{|01000\rangle}{4} + \frac{|01011\rangle}{4} + \frac{|01101\rangle}{4} - \frac{|01110\rangle}{4} -$$
$$\frac{|10000\rangle}{4} - \frac{|10011\rangle}{4} + \frac{|10101\rangle}{4} + \frac{|10110\rangle}{4} - \frac{|11001\rangle}{4} + \frac{|11010\rangle}{4} - \frac{|11100\rangle}{4} + \frac{|11111\rangle}{4}$$

This confirms that they are indeed eigenstates of the logical Z operator with proper eigenvalues.

```
In[·]:= (opZ ** ket0) / ket0
       (opZ ** ket1) / ket1 // Simplify
```

$$Out[\cdot]= 1$$

$$Out[\cdot]= -1$$

This confirms that the logical X operator flips the logical basis states as expected.

```
In[·]:= opX ** ket0 / ket1 // Simplify
       opX ** ket1 / ket0 // Simplify
```

$$Out[\cdot]= 1$$

$$Out[\cdot]= 1$$

Fig. 6.3 An arrange of qubits (black dots) on a square lattice on the surface of a torus to construct a toric code

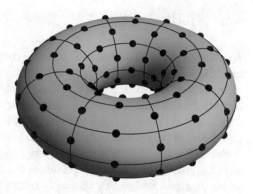

6.5   Surface Codes

All discussions of quantum error-correction codes so far implicitly assumed that one can apply a quantum gate on any pair of qubits with uniform fidelity regardless of the spatial separation of the qubits. Clearly, doing so is unrealistic. In a realistic quantum computer, a qubit is coupled directly to only a few qubits in close proximity. A quantum gate on two qubits that are not directly coupled to each other is performed through virtual gate operations through other qubits between them. The fidelity of such a quantum gate is naturally much worse. The *surface codes* introduced by Kitaev in 1997 are an interesting subclass of stabilizer codes that exhibit "locality" properties (Kitaev, 1997, 2003; Dennis et al., 2002; Bravyi and Kitaev, 1998; Freedman, 2001). Surface codes rely only on local operations and measurements, and one only needs to operate a quantum gate or a measurement on the neighboring qubits at a time. More recently, a fault-tolerant quantum computation with only local operations has been shown to be possible based on the surface codes (Fowler et al., 2012).

Another notable thing is that the surface codes feature topological properties. For a surface code, the operators in the stabilizer and logical operators are governed by the topology of the underlying array of qubits. This is not surprising as surface codes have been derived from the models arranged on the surface of a torus (Kitaev, 2003; Dennis et al., 2002) that exhibit topological order. However, the crucial point is that surface codes are tolerant to local errors (Dennis et al., 2002; Wang et al., 2003). The operational locality mentioned above and the tolerance to local errors are two of the most significant advantages of surface codes.

Here we introduce two families of surface codes. These are called *toric codes* and *planar codes* after the underlying geometry of the arrangement of their qubits to construct the codes. The toric codes are difficult to implement because of the required periodic boundary conditions, yet toric codes are conceptually simple and provide the key ideas of surface codes. For this reason, we first start with toric codes and then move on to planar codes, where the requirement of the periodic boundary conditions has been removed.

6.5.1   Toric Codes

To construct a toric code, it is convenient to suppose that the qubits have been arranged on the edges of a square lattice on the surface of a torus as in Fig. 6.3. The geometry of the underlying lattice is equivalent to a square lattice on the flat plane with opposite boundaries regarded as identical. In other words, periodic boundary conditions are imposed in the horizontal and vertical directions as in Fig. 6.4. We want to construct generators of the stabilizer and the logical operators.

We define two types of operators, *plaquette* and *vertex operators*. Plaquette operators are associated with the plaquettes of the underlying lattice: Consider a plaquette p, for example, denoted by an empty circle in blue in Fig. 6.4a. It has four edges and hence four qubits 1, 2, 3, 4 on them surrounding it. The plaquette operator $\hat{P}_p$ associated with plaquette p is defined by the product of the Pauli Z operators on the qubits on those edges,

$$\hat{P}_p = \hat{Z}_{p,1}\hat{Z}_{p,2}\hat{Z}_{p,3}\hat{Z}_{p,4}. \tag{6.111}$$

Vertex operators are attributed to the vertices of the underlying lattice: Consider a vertex v on the lattice, say, the one marked by a red open circle in Fig. 6.4a. It has four edges connected to the vertex and on each of the edges there resides a qubit. The vertex generator $\hat{V}_v$ associated with the vertex v is defined by the product of the Pauli X operators on the four qubits,

$$\hat{V}_v = \hat{X}_{v,1}\hat{X}_{v,2}\hat{X}_{v,3}\hat{X}_{v,4}. \tag{6.112}$$

Obviously, all plaquette operators commute with each other, and likewise, all vertex operators commute with each other. Interesting to note is that any plaquette operator commutes with every vertex operator because a plaquette and a vertex share either two edges or none. When they share two edges, the two Pauli Z operators from the plaquette operator commute with the two Pauli X operators from the vertex operator.

The stabilizer of the codes is defined to be generated by the set of all plaquette and vertex operators. But how many of the generators are independent? This question is important since it determines how many qubits the code can encode. Note that a product of plaquette operators associated with neighboring plaquettes has the Pauli Z operators only on the edges of the closed boundary of the included plaquettes, as illustrated by the blue closed loop in Fig. 6.4b. There is no non-trivial contribution from the shared edges because $\hat{Z}^2 = \hat{I}$. Similarly, a product of the vertex operators associated with neighboring vertices has the Pauli X operators only on the edges crossed by the closed boundary, as illustrated by the orange dashed line in Fig. 6.4b, of the involved vertices because $\hat{X}^2 = \hat{I}$. Now, consider the product of all plaquette operators. The boundary embracing all the plaquettes on the lattice is the boundary of the lattice itself. However, due to the periodic boundary conditions, the boundary is trivial, meaning that the product of all plaquette operators is the identity, $\prod_p \hat{P}_p = \hat{I}$. In the same manner, it follows that the product of all vertex operators is the identity as well, $\prod_p \hat{V}_v = \hat{I}$. Therefore, one plaquette or vertex operator can be expressed

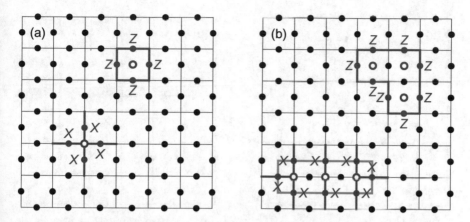

Fig. 6.4 A toric code on a square lattice with periodic boundary conditions, with opposite boundaries regarded as identical. **a** A plaquette operator (blue) and a vertex operator (red). The labels "X" and "Z" refer to the Pauli X and Z operators, respectively, on the qubit located on the edge. The plaquette and vertex operators on the lattice generate the stabilizer of the toric code. **b** Some elements of the stabilizer of a toric code. A product of the neighboring plaquette operators results in a closed loop of edges (in blue) on which the Pauli Z operators act. Similarly, a product of the neighboring vertex operators results in a closed loop of dual edges (dashed in orange) on which the Pauli X operators act

as the product of the rest of the same type. In other words, all but one plaquette operators are independent of each other. This is also the case for vertex operators.

On an $L \times L$ lattice, there are $2L^2$ edges. On the other hand, there are L^2 plaquettes and L^2 vertices, and overall there are $2L^2$ plaquette or vertex operators. However, as pointed out above, among L^2 plaquette operators, only $L^2 - 1$ of them are independent. Likewise, out of L^2 vertex operators, only $L^2 - 1$ of them are independent of each other. In short, the stabilizer of the code is generated by $2(L^2 - 1)$ independent generators. This implies that the code space is 2^2-dimensional and can encode two logical qubits.

Since the code encodes two logical qubits, we need to construct two logical Z operators and two logical X operators. Recall (see Sect. 6.3.2) that logical operators must commute with all elements of the stabilizer since they must preserve the code space, but they do not belong to the stabilizer. Let us consider a line of edges running all the way from the bottom boundary to the top boundary as depicted by a thick vertical in blue in Fig. 6.5a. Let $\bar{Z}_1$ be the product of the Pauli Z operators on all edges along the line. It follows that $\bar{Z}_1$ commute with all elements of the stabilizer because the line shares two edges (if at all) with any vertex operator—it commutes trivially with all plaquette operators. However, it cannot be a member of the stabilizer because it cannot be expressed as a product of plaquette operators since the line cannot be a boundary of a region on the lattice. This can be seen more clearly on the surface of the underlying torus shown in Fig. 6.5c. The line is a closed loop through the hole of the torus, and it is homologically different from the boundary loop of any region of the surface. Therefore, we can take $\bar{Z}_1$ as one of the logical Z operators.

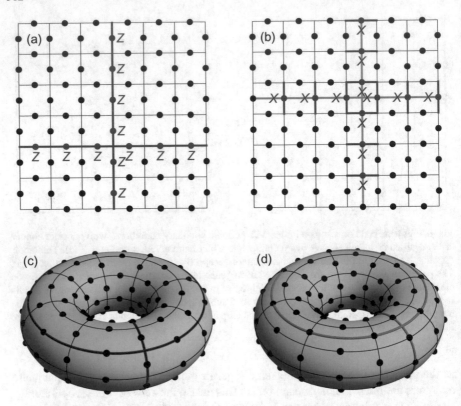

Fig. 6.5 Logical operators in a toric code. **a** Two inequivalent logical Z operators that correspond to two homologically different loops encircling the hole of torus (**c**). **b** Two inequivalent logical X operators corresponding to the two homologically different loops on the surface of torus (**d**)

The other logical Z operator can be constructed from a line running from the left to right boundary, as depicted by the thick horizontal line in blue in Fig. 6.5a. That is, we define $\bar{Z}_2$ to be the product of the Pauli Z operators on all edges along the line. The line encloses the hole of the torus as illustrated in Fig. 6.5c. Following the same arguments as shown above, we see that $\bar{Z}_2$ commutes with all elements of the stabilizer but does not belong to the stabilizer. We take $\bar{Z}_2$ as the second logical Z operator.

Logical X operators can also be constructed in a similar way. Consider a line running vertically from the bottom to top boundary through the plaquettes as depicted by the vertical orange dashed line in Fig. 6.5b. We define $\bar{X}_1$ to be the product of the Pauli X operators on all the edges crossing the line. $\bar{X}_2$ is defined in the same manner with a line through plaquettes from the left to right boundary of the lattice. For the same reason given above, the two operators $\bar{X}_1$ and $\bar{X}_2$ commute with all elements of the stabilizer but are not members of the stabilizer. Therefore, we take them as logical X operators of the code.

Recall that the choice of logical operators is not unique. In this case, difference in the choice of logical operators correspond to different lines connecting opposite boundaries. The lines do not need to be straight as long as they connect opposite boundaries. For example, instead of ax straight line for $\bar{Z}_1$, one can choose any line connecting the bottom and top boundaries. The resulting operator $\bar{Z}_1'$ is obtained from $\bar{Z}_1$ by multiplying the plaquette operators. In this sense, all logical operators associated with lines that can be smoothly deformed to each other on the surface of the underlying torus can be regarded as equivalent.

Consider a toric code on a 3x3 square lattice on the surface of a torus.

```
$L = 3;
jj = Range[0, $L - 1];
```

The qubits on the horizontal and vertical edges are labelled by S and T, respectively.

```
Let[Qubit, S, T]
```

This defines plaquette operators.

```
A[{i_, j_}] :=
 S[i, j, 3] ** S[i, Mod[j + 1, $L], 3] ** T[i, j, 3] ** T[Mod[i + 1, $L], j, 3]
```

Here are the *independent* plaquette operators.

In[ ]:= `AA = Most@Flatten@Table[A@{i, j}, {i, jj}, {j, jj}]`

Out[ ]= $\{S_{0,0}^z\, S_{0,1}^z\, T_{0,0}^z\, T_{1,0}^z,\ S_{0,1}^z\, S_{0,2}^z\, T_{0,1}^z\, T_{1,1}^z,\ S_{0,0}^z\, S_{0,2}^z\, T_{0,2}^z\, T_{1,2}^z,\ S_{1,0}^z\, S_{1,1}^z\, T_{1,0}^z\, T_{2,0}^z,$
$S_{1,1}^z\, S_{1,2}^z\, T_{1,1}^z\, T_{2,1}^z,\ S_{1,0}^z\, S_{1,2}^z\, T_{1,2}^z\, T_{2,2}^z,\ S_{2,0}^z\, S_{2,1}^z\, T_{0,0}^z\, T_{2,0}^z,\ S_{2,1}^z\, S_{2,2}^z\, T_{0,1}^z\, T_{2,1}^z\}$

In[ ]:= `Length[AA]`

Out[ ]= 8

This defines vertex operators.

```
B[{i_, j_}] :=
 S[i, j, 1] ** S[Mod[i - 1, $L], j, 1] ** T[i, j, 1] ** T[i, Mod[j - 1, $L], 1]
```

Here are the *independent* vertex operators.

In[ ]:= `BB = Most@Flatten@Table[B@{i, j}, {i, jj}, {j, jj}]`

Out[ ]= $\{S_{0,0}^x\, S_{2,0}^x\, T_{0,0}^x\, T_{0,2}^x,\ S_{0,1}^x\, S_{2,1}^x\, T_{0,0}^x\, T_{0,1}^x,\ S_{0,2}^x\, S_{2,2}^x\, T_{0,1}^x\, T_{0,2}^x,\ S_{0,0}^x\, S_{1,0}^x\, T_{1,0}^x\, T_{1,2}^x,$
$S_{0,1}^x\, S_{1,1}^x\, T_{1,0}^x\, T_{1,1}^x,\ S_{0,2}^x\, S_{1,2}^x\, T_{1,1}^x\, T_{1,2}^x,\ S_{1,0}^x\, S_{2,0}^x\, T_{2,0}^x\, T_{2,2}^x,\ S_{1,1}^x\, S_{2,1}^x\, T_{2,0}^x\, T_{2,1}^x\}$

In[ ]:= `Length[BB]`

Out[ ]= 8

The plaquette operators commute with vertex operators.

In[·]:= Outer[GottesmanTest, AA, BB] // MatrixForm

Out[·]//MatrixForm=
$$\begin{pmatrix} 1 & 1 & 1 & 1 & 1 & 1 & 1 & 1 \\ 1 & 1 & 1 & 1 & 1 & 1 & 1 & 1 \\ 1 & 1 & 1 & 1 & 1 & 1 & 1 & 1 \\ 1 & 1 & 1 & 1 & 1 & 1 & 1 & 1 \\ 1 & 1 & 1 & 1 & 1 & 1 & 1 & 1 \\ 1 & 1 & 1 & 1 & 1 & 1 & 1 & 1 \\ 1 & 1 & 1 & 1 & 1 & 1 & 1 & 1 \\ 1 & 1 & 1 & 1 & 1 & 1 & 1 & 1 \end{pmatrix}$$

Here are two logical Z operators of the code.

In[·]:= opZ = {
    Multiply @@ Table[S[i, 1, 3], {i, jj}],
    Multiply @@ Table[T[0, j, 3], {j, jj}]
    }

Out[·]= $\left\{ S_{0,1}^{z} S_{1,1}^{z} S_{2,1}^{z}, T_{0,0}^{z} T_{0,1}^{z} T_{0,2}^{z} \right\}$

In[·]:= Outer[GottesmanTest, opZ, AA] // MatrixForm
    Outer[GottesmanTest, opZ, BB] // MatrixForm

Out[·]//MatrixForm=
$$\begin{pmatrix} 1 & 1 & 1 & 1 & 1 & 1 & 1 & 1 \\ 1 & 1 & 1 & 1 & 1 & 1 & 1 & 1 \end{pmatrix}$$

Out[·]//MatrixForm=
$$\begin{pmatrix} 1 & 1 & 1 & 1 & 1 & 1 & 1 & 1 \\ 1 & 1 & 1 & 1 & 1 & 1 & 1 & 1 \end{pmatrix}$$

Here are two logical X operators of the code.

In[·]:= opX = {
    Multiply @@ Table[T[i, 2, 1], {i, jj}],
    Multiply @@ Table[S[1, j, 1], {j, jj}]
    }

Out[·]= $\left\{ T_{0,2}^{x} T_{1,2}^{x} T_{2,2}^{x}, S_{1,0}^{x} S_{1,1}^{x} S_{1,2}^{x} \right\}$

In[·]:= Outer[GottesmanTest, opX, AA] // MatrixForm
    Outer[GottesmanTest, opX, BB] // MatrixForm

Out[·]//MatrixForm=
$$\begin{pmatrix} 1 & 1 & 1 & 1 & 1 & 1 & 1 & 1 \\ 1 & 1 & 1 & 1 & 1 & 1 & 1 & 1 \end{pmatrix}$$

Out[·]//MatrixForm=
$$\begin{pmatrix} 1 & 1 & 1 & 1 & 1 & 1 & 1 & 1 \\ 1 & 1 & 1 & 1 & 1 & 1 & 1 & 1 \end{pmatrix}$$

6.5.2   *Planar Codes*

Now, let us turn to planar codes. For a planar code, we consider a square lattice without imposing a periodic boundary condition. Instead, we need two different types of boundaries. In Fig. 6.6, the left and right boundaries end with edges while the top and bottom boundaries are terminated with vertices. The former are called *smooth boundaries* and the latter *rough boundaries*.

The plaquette and vertex operators associated with plaquettes and vertices interior of the lattice are defined the same as for toric codes. However, plaquettes and vertices at the boundaries have different surrounding edges, and the corresponding operators require slightly modified definitions. A plaquette on a rough edge has only three

Fig. 6.6 (A planar surface code on a square lattice with rough and smooth boundaries. Shown are a plaquette operator associated with a plaquette (blue empty circle) at the rough boundary at the top and a vertex operator associated with a vertex (red empty circle) at the smooth boundary on the left. The labels "X" and "Z" refer to the Pauli X and Z operators, respectively, on the qubit located on the edge

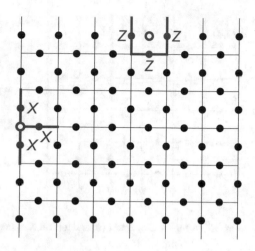

edges around it, and the plaquette operator associated with such a vertex is given by a product of three Pauli Z operators,

$$\hat{P}_p = \hat{Z}_{p,1}\hat{Z}_{p,2}\hat{Z}_{p,3}. \tag{6.113}$$

Likewise, a vertex at smooth boundaries has only three edges connected to it, and the vertex operator associated with such a vertex is given by a product of three Pauli X operators

$$\hat{V}_v = \hat{X}_{v,1}\hat{X}_{v,2}\hat{X}_{v,3}. \tag{6.114}$$

All plaquette and vertex operators in the interior to and at the boundary of the lattice generate the stabilizer. Without a periodic boundary condition imposed in the toric codes, all plaquette and vertex operators are independent of each other. How many of them are there? For a $L \times L$ square lattice with two rough and two smooth boundaries, there are L^2 plaquettes and $(L+1)(L-1)$ vertices. On the other hand, there are $L(L+1) + L(L-1)$ edges and hence the same number of physical qubits on the lattice. This implies that the planar code can encode one logical qubit.

Let us construct logical operators $\bar{Z}$ and $\bar{X}$ on the code space. Consider a line running from the bottom to top boundaries with an example shown in Fig. 6.7a. Recall that both are rough boundaries. The line share two edges, if at all, with a vertex operator regardless whether it is in the interior or at the smooth boundaries. Therefore, if we define $\bar{Z}$ to be the product of the Pauli Z operators on the edges along the line, then $\bar{Z}$ commutes with all generators of the stabilizer. Ending at opposite boundaries, however, $\bar{Z}$ cannot be generated by any combination of plaquette operators. This concludes that $\bar{Z}$ can be chosen as the logical Z operator. Similarly, we can choose a line running through plaquettes from the left to right edges as depicted by the orange dashed line in Fig. 6.7b. We define the logical operator $\bar{X}$ to be the product of the Pauli X operators on the edges crossing the line. Since it shares two edges, if at

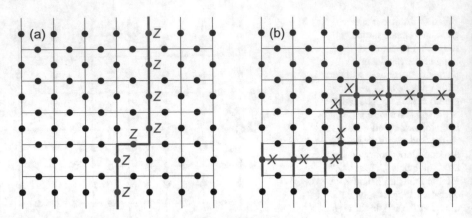

Fig. 6.7   A logical Z operator (**a**) and logical X operator (**b**) in a planar surface code

all, with any plaquette (either in the interior or at the rough boundaries), $\bar{X}$ certainly commutes with all plaquette operators, but it is independent of any plaquette or vertex operators.

So far we have constructed surface codes associated with square lattices, but doing so is not restricted to square lattices (Dennis et al., 2002). One can construct a surface code for any tessellation of a surface. One can also consider surfaces of a higher genus. For a closed orientable surface of genus g, $2g$ qubits can be encoded. For planar codes, we may consider a surface with e distinct rough edges separated by e distinct smooth edges. Then $e - 1$ qubits can be encoded, associated with the lines that connect one rough edge with any of the others. Punching holes in the lattice is another way to encode more logical qubits as shown in Fig. 6.8. Suppose that the outer boundary of the lattice is smooth and that there are h holes with smooth boundaries. Then the code can encode h qubits. For each hole, a closed loop on the lattice that encloses the hole is associated with a logical Z operator, and a line through the plaquettes—a line on the dual lattice—from the boundary of the hole to the outer boundary is associated with a logical X operator.

6.5.3   Recovery Procedure

As a special type of stabilizer code, the recovery procedure of surface codes also follows the general prescription: First measure the generators to detect the error syndrome and apply necessary Pauli operators to correct the affected qubits.

For surface codes, the generators of the stabilizer are either plaquette or vertex operators. A local measurement of those operators can be achieved by using ancillary qubits on the plaquettes or vertices of the underlying lattice (Fowler et al., 2012). Let us first consider a particular plaquette. We assume an ancillary qubit at the center of the plaquette and prepare it initially in the state $|0\rangle$. We then apply a unitary gate

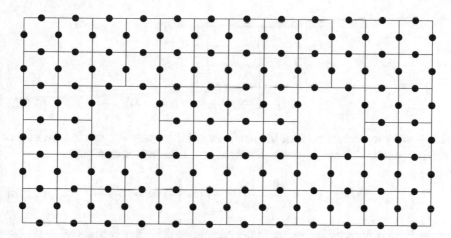

Fig. 6.8 A surface code on a lattice with a smooth outer boundary and two holes with smooth boundaries. It can encode two qubits

described by the following quantum circuit

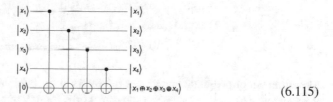

$$(6.115)$$

The output state of the ancillary qubit is given by

$$|x_1 \oplus x_2 \oplus x_3 \oplus x_4\rangle \tag{6.116}$$

if the physical qubits $1, 2, 3, 4$ on the surrounding edges is initially in the state $|x_1 x_2 x_3 x_4\rangle$. Therefore, measuring the ancillary qubit in the standard basis gives one of the eigenvalues

$$x_1 \oplus x_2 \oplus x_3 \oplus x_4 = \pm 1 \tag{6.117}$$

of the plaquette operator $\hat{P} = \hat{Z}_1 \hat{Z}_2 \hat{Z}_3 \hat{Z}_4$.

Measurement of the vertex operator $\hat{V} = \hat{X}_1 \hat{X}_2 \hat{X}_3 \hat{X}_4$ associated with a given vertex with four edges $1, 2, 3, 4$ connected to it can be achieved through a slightly different unitary gate. We assume an ancillary qubit located at the vertex. Recalling that $\hat{V} = \hat{H}^{\otimes 4} \hat{P} \hat{H}^{\otimes 4}$, one can see that the unitary gate described by the quantum circuit of the form

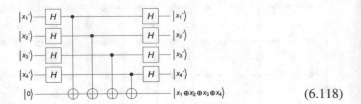

$$(6.118)$$

followed by a measurement on the ancillary qubit leads to a measurement of the vertex operator $\hat{V}$. Here the states $|x'\rangle$ for $x = 0, 1$ denote the eigenstates

$$|0'\rangle \equiv |+\rangle := \frac{|0\rangle + |1\rangle}{\sqrt{2}}, \quad |1'\rangle \equiv |-\rangle := \frac{|0\rangle - |1\rangle}{\sqrt{2}} \qquad (6.119)$$

of the Pauli X operator belonging to the eigenvalue ± 1. They correspond to $|x\rangle$ but are defined in the rotated frame, $|x'\rangle = \hat{H} |x\rangle$. Using the identities in Eq. (2.44), one can simplify the above quantum circuit to the form (see Problem 6.10)

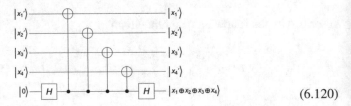

$$(6.120)$$

Measurement of plaquette and vertex operators at the boundaries can be achieved in the same way as well because the scheme works for any number of physical qubits.

The error syndrome of the surface code features an interesting topological nature. A phase-flip error $\hat{Z}$ on a qubit, say, qubit 1 in Fig. 6.9a can be diagnosed by measuring two vertex operators at the two ends of the edge where the corrupted qubit is located. The error is corrected simply by applying $\hat{Z}$ on qubit 1. Note that a series of phase-flip errors on qubits 2, 3, and 4 give the same error syndrome, of course, diagnosed by the same vertex operators as the phase-flip error on qubit 1. As we discussed for the general error recovery procedure for stabilizer codes in Sect. 6.4, even if the actual errors occurred on qubits 2, 3, and 4, the errors can still be fixed by applying $\hat{Z}$ on qubit 1. In this case, it is because the application of $\hat{Z}$ on qubit 1, 2, 3, and 4 corresponds to applying the plaquette operator $\hat{P} = \hat{Z}_1 \hat{Z}_2 \hat{Z}_3 \hat{Z}_4$ associated with the plaquette [shaded in blue in Fig. 6.9a] surrounded by the four edges. Recall that plaquette operators belong to the stabilizer of the surface code (in this particular example, we assume a toric code), and they operate trivially on the code space.

Another more general example is shown in the left lower corner of Fig. 6.9a. It depicts two cases of correlated phase-flip errors, one on the qubits on the blue solid line and the other on the qubits on the blue dashed line. Both cases give the same error syndrome, which is detected by the vertex operators at the two ends shared by the two lines. In fact, any line connecting the same ends gives the same error syndrome as long as the line can be smoothly deformed on the surface of the torus,

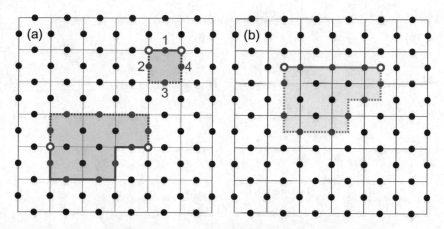

Fig. 6.9 a Phase-flip errors (blue dots on thick blue edges) and vertex defects (blue empty circles) associated with them. **b** Bit-flip errors (red dots on thick orange lines) and plaquette defects (red empty circles) associated with them

recalling that the square lattice with periodic boundary conditions on a flat plane is equivalent to the one the surface of a torus.[9] In this sense, it is convenient to regard that phase-flip errors on neighboring edges create *vertex defects* at the ends of the line of the affected edges. Vertex defects can be detected by measuring vertex operators. The phase flip errors associated with a line connecting a pair of vertex defects[10] can be corrected by applying $\hat{Z}$ on the qubits on any line that can be smoothly deformed from the line.

Similarly, bit-flip errors $\hat{X}$ on a line segment through plaquettes can be regarded to create *plaquette defects* at the end of the line segment. An example is illustrated in Fig. 6.9b. The bit-flip errors on the qubits along the orange solid line and those along the orange dashed line give the same error syndrome because the two lines share the same ends and can be smoothly deformed to each other without cutting the lines.

The error syndrome in the surface codes has even deeper topological aspects, and further, the vertex and plaquette defects have a physical interpretation as "particles" (Kitaev, 2003). Consider the examples illustrated in Fig. 6.10. By applying $\hat{Z}$ on a qubit next to one of the vertex defects in Fig. 6.10a, we can "move" the vertex defect to a neighboring vertex as in Fig. 6.10b. Similarly, one can also move a plaquette defect by applying $\hat{X}$ on a qubit next to the plaquette defect. At the bottom of Fig. 6.10a, there are two plaquette defects associated with a line of bit-flip errors (orange line). The plaquette defect on the right has moved up by one lattice unit and to the right by one lattice unit as in Fig. 6.10b. It has been achieved by applying two $\hat{X}$ operators consecutively on two qubits next to the plaquette defect.

---

[9] In planar codes, a line of phase-flip errors should not hit a rough boundary.

[10] In toric codes, vertex defects can only be created in pairs. In planar codes, single vertex defects can be created when the line of the phase-flip errors ends at a rough boundary.

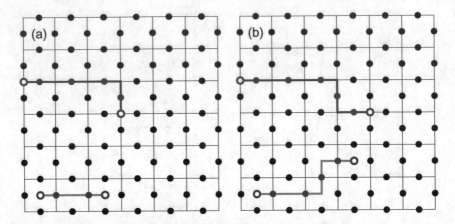

Fig. 6.10 Vertex (blue empty circles) and plaquette (red empty circles) defects before (**a**) and after (**b**) defects have moved. One can move a vertex defect by applying $\hat{Z}$ on a qubit next to it and a plaquette defect by applying $\hat{X}$ next to it

Fig. 6.11 Exchange of the vertex and plaquette defects. In the lower example, a vertex defect moves and returns to its original vertex without enclosing any other defect. It has no effect. In the upper example, the path along which the right vertex defect has traveled encloses a plaquette vertex. It causes a phase shift of -1

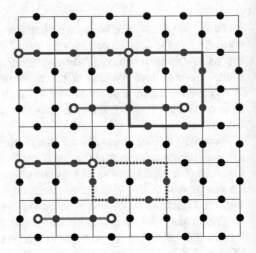

The motion of the vertex and plaquette defects (or particles) features fascinating topological properties as illustrated in Fig. 6.11. Consider first the lower example in Fig. 6.11. The vertex defect at the right end of a line of the phase-flip errors is moved around two plaquettes. The overall effect is thus equivalent to applying two vertex operators enclosed by the path and is trivial on the code space. Now let us turn to the upper example in Fig. 6.11. In this case, the vertex defect on the right travels around four plaquettes and returns to its original vertex. However, a crucial difference here is that the path encloses a plaquette defect. What is the overall effect of the motion of the vertex defect? One can clearly see in this particular example that the path of the vertex defect shares one edge (at the point where the orange and blue line cross) with the line (in orange) of the bit-flip errors. However, this is completely general and the path must share an odd number of edges with the line of the bit-flip errors as long as it closes an odd number of plaquette defects. Then, due to the anti-commutation of $\hat{X}$ and $\hat{Z}$, such a path causes a phase shift of -1. Topologically, it is said that the path cannot shrink smoothly to a point because of the enclosed plaquette defects. In many other respects, the vertex and plaquette defects behave like particles of different species with exotic statistical properties.

The manipulation of the vertex and plaquette defects described above is an example of topological quantum computation, the account of which has long been deferred since Sect. 3.3. As a matter of fact, it was the first proposal for topological quantum computation, and it has inspired a huge number of works on the subject. More recently, the idea was further developed for topological quantum computation with Majorana fermions—fermions that are anti-particles of their own—on more realistic physical systems (Kitaev, 2001; Alicea et al., 2011). The latter works lead to a fairly dramatic quest for Majorana fermions, and a growing amount of experimental evidence has accumulated on the existence of Majorana fermions as elementary excitations in condensed matter systems (Mourik et al., 2012; Deng et al., 2012; Das et al., 2012; Nadj-Perge et al., 2014).

Problems

6.1. Let $\mathcal{V}$ be a code space, and suppose that $\mathcal{S}$ is the stabilizer subgroup of the Pauli group with respect to $\mathcal{V}$. Show that any two elements of $\mathcal{S}$ commute.

6.2. Let $\hat{G}_1, \hat{G}_2, \ldots, \hat{G}_k$ be elements of the Pauli group $\mathcal{P}(n)$ on n qubits. Show that $\hat{G}_1, \hat{G}_2, \ldots, \hat{G}_k$ are *independent* if and only if the corresponding Gottesman vectors [see (6.41)] $\left|\hat{G}_1\right\rangle, \left|\hat{G}_2\right\rangle, \ldots, \left|\hat{G}_k\right\rangle$ are *linearly independent* of each other in $(\mathbb{Z}_2)^{2n}$ regarded as a vector space over $\mathbb{Z}_2$.

6.3. Let $\hat{G}_1, \hat{G}_2, \ldots, \hat{G}_k$ be mutually *independent* elements of the Pauli group $\mathcal{P}(n)$ on n qubits. Show that there exists an element $\hat{G}$ in $\mathcal{P}(n)$ such that $\hat{G}$ anti-commutes with one and only one of $\hat{G}_1, \hat{G}_2, \ldots, \hat{G}_k$ and commutes with the rest.

6.4. Identify the elements of the single-qubit Clifford group $\mathcal{C}(1)$.
Hint: In total, there are 192 elements in $\mathcal{C}(1)$, generated by the Hadamard gate $\hat{H}$ and the quadrant phase gate $\hat{Q}$. Also note that the elements appear with physically irrelevant phase factors $\pm 1, \pm i, \pm e^{i\pi/4}, \pm e^{-i\pi/4}$. This example illustrates how efficient the description of a group is in terms of its generators.

6.5. Let $\hat{A}$ and $\hat{B}$ be the CNOT gates on the same pair of qubits but with the control and target qubit exchanged. Show that $\hat{A}$ and $\hat{B}$ are not independent generators of the Clifford group.

6.6. Let $\hat{U}$ be an element of the $(n+1)$-qubit Clifford group $\mathcal{C}(n+1)$. Show that by applying single-qubit operations and/or rearranging the qubits if necessary, one can always put both $\hat{U}(\hat{Z} \otimes \hat{J})\hat{U}^\dagger$ and $\hat{U}(\hat{X} \otimes \hat{J})\hat{U}^\dagger$ simultaneously in the form

$$\hat{U}(\hat{Z} \otimes \hat{J})\hat{U}^{\dagger} = \hat{X} \otimes \hat{A}, \quad \hat{U}(\hat{X} \otimes \hat{J})\hat{U}^{\dagger} = \hat{Z} \otimes \hat{B}, \qquad (6.121)$$

where $\hat{J} := \hat{I}^{\otimes n}$ and $\hat{A}, \hat{B} \in \mathcal{P}(n)$.

Note: It is also possible to put them simultaneously in the form

$$\hat{U}(\hat{Z} \otimes \hat{J})\hat{U}^{\dagger} = \hat{X} \otimes \hat{A}, \quad \hat{U}(\hat{X} \otimes \hat{J})\hat{U}^{\dagger} = \hat{I} \otimes \hat{B}. \qquad (6.122)$$

6.7. Let $\hat{U}$ be an element of the $(n+1)$-qubit Clifford group $\mathcal{C}(n+1)$. Define an operator $\hat{V}$ acting on n qubits by

$$\hat{V}|y\rangle := \sum_{x=0}^{2^{n}-1} |x\rangle \left(\langle 0| \otimes \langle x| \right)\hat{U}(|0\rangle \otimes |y\rangle)). \qquad (6.123)$$

Show that $\hat{V}$ is an element of the n-qubit Clifford group $\mathcal{C}(n)$.

6.8. Suppose that you are given a unitary operator $\hat{U}$ operating on n qubits. Explain a procedure to test whether $\hat{U}$ is a Clifford operator or not.

6.9. Consider a toric code associated with a 3×3 square lattice on the surface of a torus. Find the four logical basis states of the code space.
Remark: There is no need to list the logical basis states explicitly to implement a surface code. This is a heuristic exercise.

6.10. Show that the two quantum circuits in Eqs. (6.118) and (6.120) are equivalent. Hint: Recall the identities in Eq. (2.44).

Chapter 7
Quantum Information Theory

How many (classical or quantum) bits would we need in order to store information from a given source? How much information could we reliably transmit via a noisy communication channel? What are the minimal resources necessary to transform one form of information into another? These are the key questions that information theory, classical or quantum, addresses. In essence, it is concerned with identifying and quantifying fundamental resources—but not specific methods, such as algorithms, or equipment—for generating, storing, manipulating, and transmitting the information. Information refers to the state of a physical system, and as Landauer (1991) uttered, it is physical. As such, the answers to the above questions must be different depending on the underlying physical principles. Quantum information theory is concerned with quantum mechanics. This new principle enriches classical information theory that is based on classical mechanics, bringing about fresh possibilities.

This chapter starts by introducing the notion of *entropy*, first classical and then quantum entropies. The entropy quantifies information and is the most fundamental concept in information theory. The chapter next discusses *quantum entanglement* as a physical resource. As mentioned above, quantum information theory is far richer than its classical counterpart. Among many fundamental and technical differences, quantum entanglement makes the most striking one, as an intriguing resource that is not available in classical information. Not surprisingly, quantum entanglement forms the vast majority of topics studied in quantum information theory.

This chapter is sketchier than the others and aims to be a quick introduction to quantum information theory. It surveys only the most basic notions and ideas of quantum information theory. It is partly because unlike more traditional disciplines of physics, quantum information theory could look rather disordered or disoriented,

---

Supplementary Information The online version contains supplementary material available at https://doi.org/10.1007/978-3-030-91214-7_7.

M.-S. Choi, *A Quantum Computation Workbook*, https://doi.org/10.1007/978-3-030-91214-7_7

especially when encountered at first, full of seemingly unrelated subjects. Once accustomed to the basic ideas and concepts introduced in this chapter, one can delve deeper into various advanced subjects by resorting to other more specialized texts including Chapters 11 and 12 of Nielsen & Chuang (2011).

7.1   Shannon Entropy

Entropy is a measure of uncertainty. Entropy is an essential concept ubiquitously used in every science that has statistical elements. Examples include classical and quantum statistical mechanics, classical and quantum information theory, data analysis, machine learning, and many more.

7.1.1   Definition

In classical information, entropy is concerned with random variables. Suppose that X is a random variable taking a value from a discrete set $\{x\}$. The value of X is not fixed but determined randomly by a probability distribution, $p(x)$. Intuitively, the uncertainty in X is lower if the probability distribution has a narrow peak at a particular value x. For example, consider a binary case $\{0, 1\}$. When $p(0) = 1$ and $p(1) = 0$, we know the value of X with certainty. In this case, there is no uncertainty at all. On the contrary, the uncertainty is high when the probability is broadly distributed. For example, in the binary case, if $p(0) = p(1) = 1/2$, it is hard to guess the value of X. The uncertainty is maximum in this case.

To quantify such a tendency of the uncertainty in a random variable, we define the entropy by

$$H(X) \equiv H(\{p(x)\}) := -\sum_x p(x) \log p(x). \tag{7.1}$$

It is called the *Shannon entropy* associated with the probability distribution $\{p(x)\}$. The logarithm concerning entropies are taken to base 2. When $p(x) = 0$ for some values x, then the logarithm alone is ill-defined. However, the function $x \log x$ converges to 0, $x \log x \to 0^-$, as $x \to 0^+$. We thus define that $0 \log 0 = 0$. The entropy is associated with the probability distribution rather than a particular random variable or the indices of the values. In this respect, the notation $H(\{p(x)\})$ is more appropriate than $H(X)$. Nevertheless, for notational convenience, we will often use both notations interchangeably. The Shannon entropy in (7.1) puts the foundation of all variants of entropy. Every variant of entropy is conceptually equivalent to (7.1).

To check whether the definition in (7.1) reflects the expected tendency of uncertainty, consider again the binary random variable with the value set $\{0, 1\}$. Let $p(0) = p$ and $p(1) = 1 - p$. Figure 7.1 shows the Shannon entropy $H(\{p, 1 - p\})$ as a function of p. We observe that $H(\{0, 1\}) = 0$. It is consistent because $X = 1$

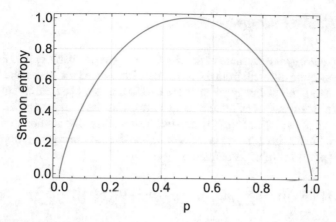

Fig. 7.1 The Shannon entropy $H(\{p, 1-p\})$ associated with a binary probability distribution $\{p(0) = p, p(1) = 1 - p\}$ of a random variable with values in $\{0, 1\}$

with certainty in this case. Similarly, $H(\{1, 0\}) = 0$ because in this case $X = 0$ without ambiguity. On the other hand, $H(\{p, 1 - p\})$ has the maximum $\log 2$ at $p = 1/2$. This is also consistent with our expectation because $p = 1/2$ is the most uncertain situation.

So far, entropy has been regarded as a measure of uncertainty. It may sound contradictory at first glance, but a valuable alternative interpretation is that entropy quantifies the information content in the state of a physical system. Suppose that we have learned the value x of a random variable X. Compared to the case we do not learn the value, our knowledge of the random variable has increased. We then ask the question, "How much of information, on average, have we acquired by learning the value of X?" Since the Shannon entropy is the amount of uncertainty before we learn the value and it is the amount of uncertainty that is cleared by learning the value, the information gain is also given by the Shannon entropy. The two interpretations thus provide two alternative views of entropy: One can regard the entropy as a measure of the uncertainty *before* we learn the value of X. But one can also regard it as a measure of the information gained *after* we learn the value of X.

A basic yet significant property of the Shannon entropy is the inequality

$$0 \leq H(X) \leq \log d, \tag{7.2}$$

where d is the number of values X can take. The entropy becomes maximum, $H(X) = \log d$, if and only if the probability is uniformly distributed. We defer proving the property until we introduce relative entropy since the proof is simplest by using the quantity.

7.1.2 Relative Entropy

The relative entropy is a measure of the closeness of two probability distributions. In character, it is quite different from the previously introduced measures, the Euclidean distance (5.189), the Kolmogorov distance (5.195), or the classical fidelity (5.227). It behaves, in many respects, more like the entropy as the name suggests.

Suppose that $\{p(x)\}$ and $\{q(x)\}$ are two probability distributions of the same random variable X with values in $\{x\}$. The *relative Shannon entropy* or simply *relative entropy* of $\{p(x)\}$ to $\{q(x)\}$ is defined by

$$H(\{p(x)\} \| \{q(x)\}) := \sum_x p(x) \log \frac{p(x)}{q(x)} = -H(\{p(x)\}) - \sum_x p(x) \log q(x).$$

(7.3)

As before, it is assumed that $0 \log 0 = 0$. In addition, we define $p(x) \log 0 = -\infty$ if $p(x) > 0$.

Why is the relative entropy a good measure of closeness of two probability distributions? Or, is it at all? For example, consider the two probability distributions

$$\{p(x)\} = \left\{ \frac{2}{3}, \frac{1-\epsilon}{3}, \frac{\epsilon}{3} \right\}, \quad \{q(x)\} = \left\{ \frac{2}{3}, \frac{1}{3}, 0 \right\}$$

(7.4)

with $0 < \epsilon \ll 1$. The two distributions look reasonably similar to each other. Yet, we have $H(\{p(x)\} \| \{q(x)\}) = \infty$ for any arbitrarily small yet finite ϵ. For comparison, consider another probability distribution

$$\{r(x)\} = \left\{ \frac{1}{6}, \frac{5}{6}, 0 \right\}.$$

(7.5)

It looks considerably different from both $\{p(x)\}$ and $\{q(x)\}$. However, we have

$$H(\{r(x)\} \| \{q(x)\}) \approx 1.4183.$$

(7.6)

Why should we regard that $\{p(x)\}$ is more different from $\{q(x)\}$ than $\{r(x)\}$ is?

Another weird property of the relative entropy, as a distance measure, is that it is asymmetric about $\{p(x)\}$ and $\{q(x)\}$. For the above example, we already observed

$$H(\{p(x)\} \| \{q(x)\}) = \infty,$$

(7.7)

whereas we find

$$H(\{q(x)\} \| \{p(x)\}) \approx \frac{\epsilon}{3 \log 2} \approx 0.$$

(7.8)

Because of the above drawbacks, the relative entropy is not particularly useful or interesting in its own right. There is a separate reason why it is used so broadly in investigating entropic quantities and relations. At the core of the reason is *Gibbs' inequality*

$$H(\{p(x)\} \parallel \{q(x)\}) \geq 0 \tag{7.9}$$

with equality attained if and only if the two probability distributions $\{p(x)\}$ and $\{q(x)\}$ are identical. A widely popular technique is to express entropic quantities in terms of the relative entropy or regard them as special cases of the relative entropy, and then use the inequality (7.9) as necessary.

As an example, let us now prove (7.2); the maximum entropy is $\log d$ for a random variable X taking d distinct values. Let $\{p(x)\}$ be the probability distribution associated with X. In addition, we consider a uniform distribution $q(x) = 1/d$. According to Gibbs' inequality (7.9), we have

$$H(\{p(x)\} \parallel \{q(x)\}) = -H(\{p(x)\}) + \log d \geq 0. \tag{7.10}$$

The equality is attained if and only if $p(x)$ is uniform just like $q(x)$, $p(x) = 1/d$.

Let us close this subsection with a proof of Gibbs' inequality. It is heuristic because some of its elements are frequently used to study other entropic quantities and relations. We first recall the properties of a *convex* function. A function $f(x)$ is said to be convex if it satisfies

$$f((1-t)x + ty) < (1-t)f(x) + tf(y) \tag{7.11}$$

for all $0 \leq t \leq 1$ and all x and y in the domain of the function. It is often expressed in a slightly more general but equivalent form

$$f(\textstyle\sum_j p(j)x_j) \leq \sum_j p(j)f(x_j) \tag{7.12}$$

for all x_j and all $p(j)$ such that $0 \leq p(j) \leq 1$ and $\sum_j p(j) = 1$. If $f(x)$ is differentiable, then one can express the convexity condition in the form

$$f(x) \geq f(y) + f'(y)(x - y) \tag{7.13}$$

for all x and y. Geometrically, it means that the graph of $f(x)$ lies above all its tangents. Now note that $f(x) = -\log x$ is a convex function. Then we apply the condition (7.13) to this negative logarithm function putting $y = 1$ to get

$$-\log x \geq \frac{1-x}{\log_e 2}. \tag{7.14}$$

Finally, we examine the relative entropy and obtain

$$H(\{p(x)\} \parallel \{q(x)\}) = -\sum_x p(x) \log \frac{q(x)}{p(x)} \geq \sum_x p(x)\left(1 - \frac{q(x)}{p(x)}\right), \tag{7.15}$$

where we have used (7.14). Then it is clear that

$$H(\{p(x)\} \| \{q(x)\}) \geq \sum_x (p(x) - q(x)) = 1 - 1 = 0. \qquad (7.16)$$

7.1.3 Mutual Information

So far we have considered a single random variable. Let us now consider two random variables X and Y. Suppose that they randomly take values in the sets $\{x\}$ and $\{y\}$, respectively, governed by the *joint probability distribution* $p(x, y)$. As long as one is concerned with the total uncertainty in X and Y or the total information gain after one learns the both values of X and Y, there is nothing new. The quantity in question is nothing but the Shannon entropy associated with the joint probability distribution

$$H(X, Y) \equiv H(\{p(x, y)\}) = -\sum_{x,y} p(x, y) \log p(x, y). \qquad (7.17)$$

To distinguish it from other entropies to be introduced below, we call it *joint Shannon entropy* or simply *joint entropy*.

With multiple random variables, the more interesting question is, how much do we get to know about X (or how much our uncertainty in $\hat{X}$ still remains) when we have learned the value of Y? By learning the value of Y, the amount of information we acquire is the Shannon entropy,

$$H(Y) = H(\{p(y)\}) \qquad (7.18)$$

associated with the probability distribution of Y, $p(y) = \sum_x p(x, y)$. In other words, the total uncertainty $H(X, Y)$ about X and Y has decreased by the same amount of uncertainty. The *Shannon entropy of X conditional on knowing Y* or simply *conditional entropy* of X on Y,

$$H(X|Y) := H(X, Y) - H(Y), \qquad (7.19)$$

is the amount of uncertainty still remaining about X given that we know the value of Y. One can further justify (7.19) as a measure of conditional uncertainty. Initially, that is, before we learn the value of Y, the probability distribution of X is given by $p(x) = \sum_y p(x, y)$. After we learn the value of Y, the random variable X is now governed by the conditional probability distribution $p(x|y)$. Therefore, one can estimate the average amount of uncertainty about X still remaining after we know Y, $H(X|Y)$, by the Shannon entropy associated with the conditional probability distribution $p(x|y)$,

$$H(X|Y) = \langle H(\{p(x|y)\}) \rangle_Y, \qquad (7.20)$$

where the average is over all possible values of Y with respect to the probability distribution $p(y)$. By Bayes' rule, the conditional probability is given by

$$p(x|y) = \frac{p(y|x)p(x)}{p(y)} = \frac{p(x, y)}{p(y)}. \qquad (7.21)$$

Putting it into (7.20), we have

$$
\begin{aligned}
H(X|Y) &= -\sum_y p(y) \sum_x \frac{p(x, y)}{p(y)} \log \frac{p(x, y)}{p(y)} \\
&= -\sum_{x,y} p(x, y) \log p(x, y) + \sum_y p(y) \log p(y) \qquad (7.22) \\
&= H(X, Y) - H(Y),
\end{aligned}
$$

reproducing the defining expression in (7.19).

If X and Y are statistically independent, then learning the value of Y does not help know about X at all. In this case, one thus expects that the remaining uncertainty $H(X|Y)$ after knowing Y is the same as the Shannon entropy $H(\{p(x)\})$ associated with the initial probability distribution $p(x)$ of X. This is indeed the case since with $p(x, y) = p(x)p(y)$, we have $H(X, Y) = H(X) + H(Y)$ and hence $H(X|Y) = H(X)$. On the other hand, if the two variables are correlated with each other, then learning the value of Y reduces the uncertainty about X, on average, below the initial level. For example, consider X and Y taking values from a common value set $\{1, 2, \ldots, 6\}$. Suppose that $p(x, y) = \delta_{x,y}/6$. Once the value of Y is known, we know the value of X with certainty and there remains no uncertainty about X.

Therefore, the difference between the initial amount of uncertainty about X and the amount of uncertainty about X remaining after learning the value of Y measures the correlation between the two variables. We call it the *mutual information* of X and Y, and denote it by $H(X{:}Y)$. In other words, the mutual information

$$H(X{:}Y) := H(X) - H(X|Y), \qquad (7.23)$$

quantifies the common information shared by X and Y. From the definition (7.19) of conditional entropy, we have identities

$$
\begin{aligned}
H(X{:}Y) &= H(X) + H(Y) - H(X, Y) &\qquad (7.24a) \\
&= H(X) - H(X|Y) &\qquad (7.24b) \\
&= H(Y) - H(Y|X). &\qquad (7.24c)
\end{aligned}
$$

From the above arguments leading to the definition of mutual information, it is physically clear that the conditional entropy $H(X|Y)$ cannot exceed $H(x)$ (Problem 7.1 for a mathematical proof). This implies the *subadditivity*

$$H(X, Y) \leq H(X) + H(Y) \tag{7.25}$$

with equality if and only if X and Y are statistically independent. For the same reason, the mutual information $H(X{:}Y)$ cannot exceed $H(X)$ nor $H(Y)$ (Problem 7.1 for a mathematical proof),

$$H(X{:}Y) \leq H(X), \tag{7.26}$$

which is equivalent to

$$H(X|Y) \geq 0. \tag{7.27}$$

Equality in each of the above two inequalities is attained if and only if X is a function of Y, $X = f(Y)$, that is, if X and Y are perfectly correlated.

There are many more interesting and useful properties of entropies and mutual information. Here we have kept the discussion minimal. Interested readers may refer to Wehrl (1978).

7.1.4 Data Compression

As an example of application of the Shannon entropy, let us consider a problem of data compression. Suppose that we have a binary string of of length n,

$$x_1 x_2 \cdots x_n \tag{7.28}$$

with each 'character' $x_k \in \{0, 1\}$. Is it possible to *compress* the string into a shorter string that still carries the same information? If it is the case, one can save the physical resource, say, to store the message.

To mathematically model the problem, we regard each character x_k for $k = 1, 2, \ldots, n$ as a realization of a random variable X_k. We assume that the random variables X_k are statistically independent and identically distributed with the probabilities $p(0) = p$ and $p(1) = 1 - p$.

The basic idea to compress the message is to classify strings into *typical* and *atypical* strings. For large n, a string such as $00 \cdots 0$ or $11 \cdots 1$ is highly unlikely, and we regard it atypical. Typical strings contain, on average, np 0s and $n(1 - p)$ 1s. We do not need to store every string of characters. We only store typical strings. The probability for an actual string to be atypical vanishes for large n.

How many distinct typical strings are there then? Before we estimate the number N_{typical} of such strings,[1] we first calculate the probability $p(x_1, x_2, \ldots, x_n)$ for a particular typical string $x_1 x_2 \cdots x_n$. Since the characters are independent and identically distributed,

$$p(x_1, \ldots, x_n) = p(x_1)p(x_2) \cdots p(x_n) \approx p^{np}(1 - p)^{n(1-p)} \tag{7.29}$$

---

[1] One can directly estimate N_{typical} for the binary case as in Problem 7.2.

as long as $x_1 x_2 \cdots x_n$ is a typical string. We observe that the probabilities to get different typical strings are uniform, independent of the string. This implies that $N_{\text{typical}} = 1/p(x_1, \ldots, x_n)$. From (7.29), we then have

$$\log N_{\text{typical}} \approx - \log p(x_1, \ldots, x_n)$$
$$= n \left[-p \log p - (1-p) \log(1-p) \right] = nH(\{p, 1-p\}).$$
(7.30)

In short, there are $2^{nH(\{p,1-p\})}$ typical strings. This implies that we only need $nH(\{p, 1-p\})$ bits to store n-bit strings. We save $n[1 - H(\{p, 1-p\})]$ bits. This is what Shannon's *noiseless coding theorem* essentially states.

In the above argument, we have considered binary strings. It can be easily extended for an alphabet of d characters, $\{c_1, c_2, \ldots, c_d\}$. The random variables modeling different characters are statistically independent and governed by the same probability distribution $\{p(x) : x = c_1, c_2, \ldots, c_d\}$. Typical strings contain $np(x)$ x's for $x = c_1, c_2, \ldots, c_d$. The probability for a particular string is then given by

$$p(x_1, x_2, \ldots, x_n) = \prod_{k=1}^{n} p(x_k) \approx \prod_{x=c_1}^{c_d} p(x)^{np(x)}$$
(7.31)

as long as the string $x_1 x_2 \cdots x_n$ is typical. Therefore, we have

$$\log N_{\text{typical}} = -\log p(x_1, x_2, \ldots, x_n) = -n \sum_x p(x) \log p(x) = nH(\{p(x)\}).$$
(7.32)

This implies that we need a physical resource to store only $2^{nH(\{p(x)\})}$ instead of all d^n strings.

7.2 Von Neumann Entropy

The Shannon entropy is a measure of uncertainty associated with a classical probability distribution. We now generalize the notion to quantum states described by density operators.

7.2.1 Definition

A classical probability distribution is associated with a random variable choosing values from a set of outcomes. In quantum mechanics, a density operator is associated with an ensemble of quantum states. One can imagine a quantum information source that generates messages of "letters" each of which is chosen from the ensemble.

The *von Neumann entropy* of a quantum state $\hat{\rho}$ is defined by

$$S(\hat{\rho}) := -\mathrm{Tr}\hat{\rho}\log\hat{\rho}, \tag{7.33}$$

where the logarithm is with respect to base 2. A basic approach to evaluate the expression in (7.33) is to take the spectral decomposition

$$\hat{\rho} = \sum_j |\lambda_j\rangle \lambda_j \langle\lambda_j|, \tag{7.34}$$

where λ_j are the eigenvalues of $\hat{\rho}$ and $|\lambda_j\rangle$ are corresponding eigenvectors. Given that $\hat{\rho}$ is a positive operator, all the eigenvalues are non-negative, $\lambda_j \geq 0$. Then, the von Neumann entropy is given by

$$S(\hat{\rho}) = -\sum_j \lambda_j \log \lambda_j. \tag{7.35}$$

In comparison with the Shannon entropy in (7.1), the eigenvalues play the role of probabilities. For zero eigenvalues ($\lambda_j = 0$), we put $0\log 0 = 0$.

For example, consider the density operator

$$\hat{\rho} = \frac{1}{2}\hat{I} + \frac{1}{4}\hat{Y} \doteq \frac{1}{4}\begin{bmatrix} 2 & -i \\ i & 2 \end{bmatrix}. \tag{7.36}$$

It has two eigenvalues $1/4$ and $3/4$, and the von Neumann entropy is given by

$$S(\hat{\rho}) = \frac{1}{2} - \frac{3}{4}\log\frac{3}{4}. \tag{7.37}$$

The von Neumann entropy of any pure state $|\psi\rangle$ vanishes,

$$S(|\psi\rangle\langle\psi|) = 0 \tag{7.38}$$

because the only non-vanishing eigenvalue of $|\psi\rangle\langle\psi|$ is 1. Physically, this is because there is no ambiguity in the quantum state at hand. On the other hand, the von Neumann entropy of the maximally mixed state $\hat{I}/d$, where d is the dimension of the Hilbert space, is given by

$$S(\hat{I}/d) = \log d. \tag{7.39}$$

Recalling from (7.2) that the Shannon entropy is the biggest for the uniform distribution, one can expect that $\log d$ is the maximum value that the von Neumann entropy can take. Indeed, we will later see that

$$0 \leq S(\hat{\rho}) \leq \log d \tag{7.40}$$

for any quantum state.

We consider a simple example. The density operator describes an ensemble of two pure states $|0\rangle$ and $|+\rangle := (|0\rangle + |1\rangle)/\sqrt{2}$.

```
In[ ]:= vec = (Ket[] + Ket[S → 1]) / Sqrt[2];
        rho = p * Dyad[Ket[], Ket[]] + (1 - p) * Dyad[vec, vec]
        Matrix[rho] // MatrixForm
```

$$Out[ ]= (1-p)\left(\frac{|0_S\rangle\langle 0_S|}{2} + \frac{|0_S\rangle\langle 1_S|}{2} + \frac{|1_S\rangle\langle 0_S|}{2} + \frac{|1_S\rangle\langle 1_S|}{2}\right) + p|\_\rangle\langle\_|$$

```
Out[ ]//MatrixForm=
```

$$\begin{pmatrix} \frac{1-p}{2} + p & \frac{1-p}{2} \\ \frac{1-p}{2} & \frac{1-p}{2} \end{pmatrix}$$

Here are the eigenvalues of the density operator.

```
In[ ]:= Plot[ProperValues[rho], {p, 0, 1}, FrameLabel → {"p", "Eigenvalues"}]
```

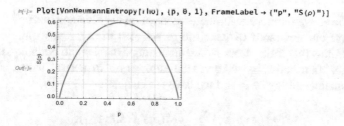

Here is the von Neumann entropy of the density operator as a function of p.

```
In[ ]:= Plot[VonNeumannEntropy[rho], {p, 0, 1}, FrameLabel → {"p", "S(ρ)"}]
```

As another example, we consider a interpolation between the completely random state I/2 and a pure state (I+ **a·S**)/2 with ‖**a**‖=1.

```
In[ ]:= rho = 1 / 2 + (1 - p) {Sin[θ] × Cos[ϕ], Sin[θ] × Sin[ϕ], Cos[θ]}.S[All] / 2
        Tr[Matrix[rho]] // Simplify
```

$$Out[ ]= \frac{1}{2} + \frac{1}{2} \times (1-p)\left(Cos[\theta] S^z + Cos[\phi] S^x Sin[\theta] + S^y Sin[\theta] \times Sin[\phi]\right)$$

$$Out[ ]= 1$$

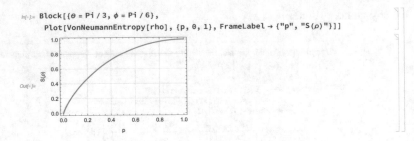

```
In[-]:= Block[{θ = Pi / 3, φ = Pi / 6},
         Plot[VonNeumannEntropy[rho], {p, 0, 1}, FrameLabel → {"p", "S(ρ)"}]]
```

A convex linear superposition gives rise to another quantum states,

$$\hat{\rho} = \sum_j p_j \hat{\rho}_j \tag{7.41}$$

where p_j are probabilities, that is, $0 \leq p_j \leq 1$ and $\sum_j p_j = 1$. Intuitively, we expect that the uncertainty about $\hat{\rho}$ is larger than the average uncertainty of individual $\hat{\rho}_j$ because we are not only ignorant of each state $\hat{\rho}_j$ but also of which of those states to be selected upon a random choice. This implies that the von Neumann entropy is a concave function of density operator,

$$\sum_j p_j S(\hat{\rho}_j) \leq S(\textstyle\sum_j p_j \hat{\rho}_j). \tag{7.42}$$

Following the above intuitive argument, we may add the Shannon entropy $H(\{p_j\})$ to account for our ignorance of which to be selected from the ensemble of states $\hat{\rho}_j$. One can show that if the states $\hat{\rho}_j$ are all distinguishable with certainty, then the addition of the contribution indeed gives the entropy of $\hat{\rho}$. In general, it overestimates the von Neumann entropy of $\hat{\rho}$, and we have the inequality

$$S(\textstyle\sum_j p_j \hat{\rho}_j) \leq \sum_j p_j S(\hat{\rho}_j) + H(\{p_j\}). \tag{7.43}$$

To prove the inequalities (7.42) and (7.43) rigorously, we would need more mathematical tools and entropic relations to be discussed below. Interested readers are encouraged to try Problem 7.6 after we introduce required materials.

7.2.2 Relative Entropy

We introduce the *relative von Neumann entropy* as the quantum counterpart of the relative Shannon entropy. One can regard it as a measure of the closeness of two quantum states just as the relative Shannon entropy is a measure of closeness of two classical probability distribution. However, like the relative Shannon entropy, von Neumann entropy is of little use in its own right. Rather, it provides powerful mathematical techniques that can be employed to investigate various entropic quantities and relations.

Let $\hat{\rho}$ and $\hat{\sigma}$ be two density operators on a vector space. We define the *relative von Neumann entropy* or simply *relative entropy* of $\hat{\rho}$ to $\hat{\sigma}$ by

$$S(\hat{\rho}\|\hat{\sigma}) := \text{Tr}\hat{\rho}\log\hat{\rho} - \text{Tr}\hat{\rho}\log\hat{\sigma} = -S(\hat{\rho}) - \text{Tr}\hat{\rho}\log\hat{\sigma}. \tag{7.44}$$

We have already explained how to calculate the first term. The second term in (7.44) is more complicated to calculate. One way is to use the spectral decomposition of $\hat{\sigma}$

$$\hat{\sigma} = \sum_j |s_j\rangle s_j \langle s_j| \tag{7.45}$$

to have

$$S(\hat{\rho}\|\hat{\sigma}) = -S(\hat{\rho}) - \sum_j \langle s_j|\hat{\rho}|s_j\rangle \log s_j. \tag{7.46}$$

We note that when an eigenvalue s_j of $\hat{\sigma}$ vanishes with the finite expectation value $\langle s_j|\hat{\rho}|s_j\rangle$, the second term diverges to ∞. In general, the relative entropy diverges when the support of $\hat{\rho}$—the subspace belonging to non-zero eigenvalues—has a finite overlap with the null space of $\hat{\sigma}$—the subspace belonging to eigenvalue zero.

For example, consider the two quantum states

$$\hat{\rho} = \frac{1}{2}\hat{I} + \frac{1}{4}\hat{Z} \doteq \begin{bmatrix} 3/4 & 0 \\ 0 & 1/4 \end{bmatrix}, \quad \hat{\sigma} = \frac{1}{2}\hat{I} + \frac{1}{4}\hat{X} \doteq \begin{bmatrix} 1/2 & 1/4 \\ 1/4 & 1/2 \end{bmatrix}. \tag{7.47}$$

Both have the same eigenvalues 3/4 and 1/4 and hence the same von Neumann entropy

$$S(\hat{\rho}) = S(\hat{\sigma}) = \frac{1}{2} - \frac{3}{4}\log\frac{3}{4}. \tag{7.48}$$

On the other hand, the eigenvectors of $\hat{\sigma}$ are $|\pm\rangle := (|0\rangle \pm |1\rangle)/\sqrt{2}$. Therefore, the relative entropy is given by

$$S(\hat{\rho}\|\hat{\sigma}) = -S(\hat{\rho}) - \langle+|\hat{\rho}|+\rangle\log\frac{3}{4} - \langle-|\hat{\rho}|-\rangle\log\frac{1}{4} = \frac{1}{2} - \frac{1}{4}\log\frac{3}{4}. \tag{7.49}$$

Consider a quantum state described by the following density operator.

```
In[·]:= rho = 1 / 2 + S[3] / 4;
        rho // Matrix // MatrixForm
Out[·]//MatrixForm=
        ( 3/4   0  )
        ( 0    1/4 )
```

Here is the von Neumann entropy of the above state.

In[ ]:= **VonNeumannEntropy[rho]**

Out[ ]= $\dfrac{1}{2} + \dfrac{3\,\mathrm{Log}\left[\frac{4}{3}\right]}{4\,\mathrm{Log}[2]}$

Consider another density operator.

In[ ]:= **sig = S[6] ** rho ** S[6];**
sig // Matrix // MatrixForm

Out[ ]//MatrixForm=
$\begin{pmatrix} \frac{1}{2} & \frac{1}{4} \\ \frac{1}{4} & \frac{1}{2} \end{pmatrix}$

It has the same eigenvalues as rho and hence the same von Neumann entropy.

In[ ]:= **VonNeumannEntropy[sig]**

Out[ ]= $\dfrac{1}{2} + \dfrac{3\,\mathrm{Log}\left[\frac{4}{3}\right]}{4\,\mathrm{Log}[2]}$

This gives the relative entropy of rho to sig.

In[ ]:= **VonNeumannEntropy[rho, sig]**

Out[ ]= $\dfrac{1}{2} - \dfrac{\mathrm{Log}\left[\frac{4}{3}\right]}{4\,\mathrm{Log}[2]}$

When $\hat{\rho}$ and $\hat{\sigma}$ are diagonal simultaneously,

$$\hat{\rho} = \sum_j |\lambda_j\rangle r_j \langle \lambda_j|, \quad \hat{\sigma} = \sum_j |\lambda_j\rangle s_j \langle \lambda_j| \tag{7.50}$$

the relative von Neumann entropy is simple to evaluate and reduces to the relative Shannon entropy in (7.3),

$$S(\hat{\rho}\|\hat{\sigma}) = H(\{r_j\} \| \{s_j\}). \tag{7.51}$$

The power of the relative entropy as a mathematical tool comes from *Klein's inequality*

$$S(\hat{\rho}\|\hat{\sigma}) \geq 0 \tag{7.52}$$

with equality if and only if $\hat{\rho} = \hat{\sigma}$, to be compared to Gibbs' inequality (7.9) for the relative Shannon entropy. Like Gibbs' inequality, one can prove Klein's inequality based on the convexity of the logarithm function. But it is slightly more complicated since $\hat{\rho}$ and $\hat{\sigma}$ do not commute with each other in general. Interested readers are referred to Problem 7.3.

For a basic application of Klein's inequality (7.52), we put $\hat{\sigma} = \hat{I}/d$, the maximally mixed state, into the inequality. It follows that

$$S(\hat{\rho}\|\hat{I}/d) = -S(\hat{\rho}) - Tr\,\hat{\rho}\log(\hat{I}/d) = -S(\hat{\rho}) + \log d \geq 0 \tag{7.53}$$

with equality if and only if $\hat{\rho} = \hat{I}/d$. It proves the deferred claim in (7.40).

7.2.3 Quantum Mutual Information

For the Shannon entropy (Sect. 7.1.3), we introduced the joint and conditional entropies by considering multiple random variables. To generalize those concepts to the quantum entropy, we consider a composite system of two subsystems A and B.

Suppose that $\hat{\rho}_{AB}$ is a density operator of the total system. The *quantum joint entropy*, or simply *joint entropy* if there is no risk of confusion, of A and B is nothing but the von Neumann entropy associated with ρAB,

$$S(A, B) := S(\hat{\rho}_{AB}) = -\text{Tr}\hat{\rho}_{AB} \log \hat{\rho}_{AB}. \tag{7.54}$$

For a product state $\hat{\rho}_{AB} = \hat{\rho}_A \otimes \hat{\rho}_B$, the joint entropy is the sum of the entropies of the individual entropies

$$S(\hat{\rho}_A \otimes \hat{\rho}_B) = S(\hat{\rho}_A) + S(\hat{\rho}_B). \tag{7.55}$$

In analogy with the classical case, we also define the *quantum conditional entropy* of A on B by

$$S(A|B) = S(A, B) - S(B), \tag{7.56}$$

to be compared to (7.19), and the *quantum mutual information* of A and B by

$$S(A{:}B) := S(A) + S(B) - S(A, B) \tag{7.57a}$$
$$= S(A) - S(A|B) \tag{7.57b}$$
$$= S(B) - S(B|A), \tag{7.57c}$$

to be compared to (7.24). Here $S(A) := S(\hat{\rho}_A)$ and $S(B) := S(\hat{\rho}_B)$ are associated with the reduced density operators

$$\hat{\rho}_A := \underset{B}{\text{Tr}}\, \hat{\rho}_{AB}, \quad \hat{\rho}_B := \underset{A}{\text{Tr}}\, \hat{\rho}_{AB}, \tag{7.58}$$

respectively. In comparison with the classical case, $\hat{\rho}_{AB}$ corresponds to a joint probability distribution $p(x, y)$, and $\hat{\rho}_A$ and $\hat{\rho}_B$ to $p(x)$ and $p(y)$, respectively. Note that for a pure state, we have (Problem 7.4)

$$S(A) = S(B) \tag{7.59}$$

and hence

$$S(A{:}B) = 2S(A) = 2S(B). \tag{7.60}$$

It is worth noticing that the physical interpretations of classical joint and conditional entropies defy our expectation when applied to the quantum joint and conditional entropies. For example, unlike the solid inequality $H(X) \leq H(X, Y)$

for classical random variables X and Y, the analogous inequality $S(A) \leq S(A, B)$ does not hold. Equivalently, the conditional entropy is not necessarily positive, $S(A|B) \not\geq 0$. For a counter example, consider the quantum state $(|00\rangle + |11\rangle)/\sqrt{2}$. It is a pure state and certainly $S(A, B) = 0$. However, $\hat{\rho}_A = \hat{\rho}_B = \hat{I}/2$ and hence $S(A) = S(B) = 1 > S(A, B)$.

Nevertheless, the quantum mutual information is always non-negative. It leads to the *subadditivity*

$$S(\hat{\rho}_{AB}) \leq S(\hat{\rho}_A) + S(\hat{\rho}_B). \tag{7.61}$$

This can be easily seen by employing Klein's inequality (7.52) (Problem 7.5).

The subadditivity (7.61) further implies an interesting property of the von Neumann entropy. Introduce a third system C, and consider a purification $|\Psi\rangle_{ABC}$ of $\hat{\rho}_{AB}$ into $\mathcal{V}_A \otimes \mathcal{V}_B \otimes \mathcal{V}_C$. We employ the subadditivity inequality (7.61) to get

$$S(A, C) \leq S(A) + S(C). \tag{7.62}$$

Since $|\Psi\rangle_{ABC}$ is a pure state,

$$S(A, B) = S(C), \quad S(A, C) = S(B). \tag{7.63}$$

Putting it into (7.62), we have

$$S(B) - S(A) \leq S(A, B). \tag{7.64}$$

Similarly, from $S(B, C) \leq S(B) + S(C)$, we have

$$S(A) - S(B) \leq S(A, B). \tag{7.65}$$

Combining the above two inequalities into one, we have the so-called *triangle inequality* for the von Neumann entropy

$$|S(\hat{\rho}_A) - S(\hat{\rho}_B)| \leq S(\hat{\rho}_{AB}). \tag{7.66}$$

There are many more interesting and useful properties of the quantum joint and conditional entropies and mutual information. Here we have discussed a minimal set of them. Interested readers are referred to Wehrl (1978).

7.3   Entanglement and Entropy

Interacting systems, classical or quantum, tend to build correlation among them. *Entanglement* is a type of correlation. But it exhibits many interesting correlation effects that are specifically quantum and cannot be simulated classically. Many protocols for quantum information processing, quantum communication, and quantum cryptography utilize entanglement as the key resource.

In this section, we discuss how to characterize, manipulate, and quantify entanglement. The subjects pose many open questions and are currently under active research. Therefore, our discussion will remain at the elementary and introductory level. Motivated readers are referred to Horodecki et al. (2009).

7.3.1 What is Entanglement?

We have already introduced the mathematical definition of entanglement in Sect. 1.1. However, here we state the definition again for convenience.

We consider a composite system AB consisting of two subsystems A and B. The two systems are associated with the Hilbert space V and W, respectively. We suppose that $\{|v_i\rangle\}$ and $\{|w_j\rangle\}$ are orthonormal bases of V and W, respectively. We often use the shorthand notation $|v_i w_j\rangle \equiv |v_i\rangle \otimes |w_j\rangle$ for the tensor-product basis states. Entanglement across more than two parties is also an important subject, but here we will only discuss bipartite case.

Suppose that $|\Psi\rangle_{AB}$ is a pure state of the composite system. $|\Psi\rangle_{AB}$ is a separable state if it can be written in a product form

$$|\Psi\rangle_{AB} = |\psi\rangle_A \otimes |\phi\rangle_B . \tag{7.67}$$

Otherwise, it is an entangled state.

As for mixed states, the separability allows not only product states but also any convex linear combination of product states. In other words, a mixed state $\hat{\rho}_{AB}$ of the composite system is said to be separable if it can be written in the form

$$\hat{\rho}_{AB} = \sum_{ij} p_j \hat{\rho}_j \otimes \hat{\sigma}_j, \tag{7.68}$$

where $\hat{\rho}_j$ and $\hat{\sigma}_j$ are density operators on A and B, respectively, and $0 \leq p_j \leq 1$ and $\sum_j p_j = 1$. If not, $\hat{\rho}_{AB}$ is said to be entangled.

7.3.2 Separability Tests

Given a quantum state, how can we tell if it is separable or entangled? For pure states, the Schmidt decomposition provides a straightforward method. We have already discussed it briefly in Sect. 1.1.1 and explain it in detail in Appendix A.6.1.

For a mixed state, there is no simple method available to check if it is separable or entangled. Here we introduce a few well-known methods.

A powerful method is to use the partial transposition (Appendix B.4). It was discovered by Peres (1996) and has inspired many other methods for separability tests of mixed states. Let $\hat{L}$ be a linear operator on the composition system. In the standard tensor-product basis, the partial transposition of $\hat{L}$ with respect to B is defined by the relation

$$\langle v_i w_j | \hat{L}^{T_B} | v_k w_l \rangle = \langle v_i w_l | \hat{L} | v_k w_j \rangle. \tag{7.69}$$

The partial transposition $\hat{L}^{T_A}$ of $\hat{L}$ with respect to A is defined analogously. The so-called *positive partial transposition criterion* states that if $\hat{\rho}_{AB}$ is separable, then the partial transposition $\hat{\rho}_{AB}^{T_B}$ of $\hat{\rho}$ is positive. As an example, consider a *Werner state*,

$$\hat{\rho}_{AB} = \frac{1}{4} \hat{I} \otimes \hat{I} - \frac{p}{4} \left(\hat{X} \otimes \hat{X} + \hat{Y} \otimes \hat{Y} + \hat{Z} \otimes \hat{Z} \right). \tag{7.70}$$

The partial transposition of it with respect to B is given by

$$\begin{aligned}
\hat{\rho}_{AB}^{T_B} &= \frac{1}{4} \hat{I} \otimes \hat{I}^T - \frac{p}{4} \left(\hat{X} \otimes \hat{X}^T + \hat{Y} \otimes \hat{Y}^T + \hat{Z} \otimes \hat{Z}^T \right) \\
&= \frac{1}{4} \hat{I} \otimes \hat{I} - \frac{p}{4} \left(\hat{X} \otimes \hat{X} - \hat{Y} \otimes \hat{Y} + \hat{Z} \otimes \hat{Z} \right).
\end{aligned} \tag{7.71}$$

A direct inspection of the eigenvalues of $\hat{\rho}_{AB}^{T_B}$ reveals that for $p > 1/3$, two of the eigenvalues become negative. The positive partial transposition criterion is violated and the state must be an entangled state.

We consider a Werner state, and apply the positive partial transposition criterion.

We first prepare a single state.

```
In[ ]:= singlet = (Ket[S[2] → 1] - Ket[S[1] → 1]) / Sqrt[2];
        singlet // LogicalForm
```

$$Out[ ]= \frac{\left| 0_{S_1} 1_{S_2} \right\rangle - \left| 1_{S_1} 0_{S_2} \right\rangle}{\sqrt{2}}$$

A Werner state is a convex linear superposition of the single state and the completely random mixture.

```
In[ ]:= rho = p * Dyad[singlet, singlet] + (1 - p) / 4 // Elaborate
        rho // Matrix // MatrixForm
```

$$Out[ ]= \frac{1}{4} - \frac{1}{4} p\, S_1^x S_2^x - \frac{1}{4} p\, S_1^y S_2^y - \frac{1}{4} p\, S_1^z S_2^z$$

$$Out[ ]//MatrixForm=
\begin{pmatrix}
\frac{1}{4} - \frac{p}{4} & 0 & 0 & 0 \\
0 & \frac{1}{4} + \frac{p}{4} & -\frac{p}{2} & 0 \\
0 & -\frac{p}{2} & \frac{1}{4} + \frac{p}{4} & 0 \\
0 & 0 & 0 & \frac{1}{4} - \frac{p}{4}
\end{pmatrix}$$

This is the partial transposition of rho with respect to the second qubit.

```
In[·]:= new = PartialTranspose[rho, S[?]] // Elaborate
       new // Matrix // MatrixForm
```

$$Out[·]= \frac{1}{4} - \frac{1}{4} p\, S_1^x S_2^x + \frac{1}{4} p\, S_1^y S_2^y - \frac{1}{4} p\, S_1^z S_2^z$$

Out[·]//MatrixForm=

$$\begin{pmatrix} \frac{1}{4} - \frac{p}{4} & 0 & 0 & -\frac{p}{2} \\ 0 & \frac{1}{4} + \frac{p}{4} & 0 & 0 \\ 0 & 0 & \frac{1}{4} + \frac{p}{4} & 0 \\ -\frac{p}{2} & 0 & 0 & \frac{1}{4} - \frac{p}{4} \end{pmatrix}$$

This shows the eigenvalues of the partial transposition of **rho**. For p<1/3, all eigenvalues are positive and the positive partial transposition is satisfied. However, two of the eigenvalues get negative for p>1/3, and hence the state must be entangled.

```
In[·]:= Plot[Evaluate@ProperValues[new], {p, 0, 1},
       FrameLabel → {"p", "Eigenvalues"}, PlotLegends → Automatic]
```

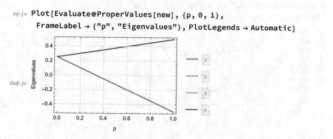

We apply the positive partial transposition criterion to another example.

We first prepare a single state.

```
In[·]:= singlet = (Ket[S[2] → 1] - Ket[S[1] → 1]) / Sqrt[2];
       singlet // LogicalForm
```

$$Out[·]= \frac{\left|0_{S_1} 1_{S_2}\right\rangle - \left|1_{S_1} 0_{S_2}\right\rangle}{\sqrt{2}}$$

The state we are concerned with is a mixture of the singlet state and a fully polarized state, $p\,|S\rangle\langle S| + (1-p)\,|00\rangle\langle 00|$.

```
In[·]:= rho =
       p * Dyad[singlet, singlet] + (1 - p) Dyad[Ket[], Ket[], S@{1, 2}] // Elaborate
       rho // Matrix // Simplify // MatrixForm
```

$$Out[·]= \frac{1}{4} - \frac{1}{4} p\, S_1^x S_2^x - \frac{1}{4} p\, S_1^y S_2^y + \frac{1}{4}(1-2p)\, S_1^z S_2^z + \frac{1}{4}(1-p)\, S_1^z + \frac{1}{4}(1-p)\, S_2^z$$

Out[·]//MatrixForm=

$$\begin{pmatrix} 1-p & 0 & 0 & 0 \\ 0 & \frac{p}{2} & -\frac{p}{2} & 0 \\ 0 & -\frac{p}{2} & \frac{p}{2} & 0 \\ 0 & 0 & 0 & 0 \end{pmatrix}$$

This is the partial transposition of **rho** with respect to the second qubit.

```
In[ ]:= new = PartialTranspose[rho, S[2]] // Elaborate
       new // Matrix // Simplify // MatrixForm
```

$$Out[ ]= \frac{1}{4} - \frac{1}{4} \, p \, S_1^x \, S_2^x + \frac{1}{4} \, p \, S_1^y \, S_2^y + \frac{1}{4} \, (1-2\,p) \, S_1^z \, S_2^z + \frac{1}{4} \, (1-p) \, S_1^z + \frac{1}{4} \, (1-p) \, S_2^z$$

```
Out[ ]//MatrixForm=
```

$$\begin{pmatrix} 1-p & 0 & 0 & -\frac{p}{2} \\ 0 & \frac{p}{2} & 0 & 0 \\ 0 & 0 & \frac{p}{2} & 0 \\ -\frac{p}{2} & 0 & 0 & 0 \end{pmatrix}$$

This shows that for all 0<p≤1, the partial transposition has a negative eigenvalue. Therefore, the state must be entangled.

```
In[ ]:= Plot[Evaluate@ProperValues[new], {p, 0, 1},
       FrameLabel → {"p", "Eigenvalues"}, PlotLegends → Automatic]
```

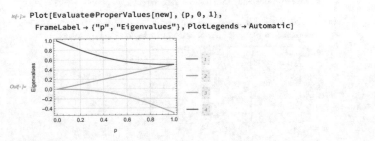

Unfortunately, the positive partial transposition is a necessary but not sufficient condition for separability. Nevertheless, it is a very strong condition and gives a good guide for separability of the quantum state in question. Moreover, it is a necessary and sufficient condition in the $2 \otimes 2$ and $2 \otimes 3$ cases, that is, when dim $V = 2$ and dim $W = 2, 3$.

Later, people realized that it was not by chance that the positive partial transposition criterion gives a strong necessary condition for separability. The crucial point is that the transposition, regarded as a supermap, is positive but not completely positive (Appendix B.4). Let $\mathscr{F}$ be the transposition supermap, $\mathscr{F} : \hat{A} \mapsto \hat{A}^T$. Then, the partial transposition corresponds to $\mathscr{I} \otimes \mathscr{F}$ (or $\mathscr{F} \otimes \mathscr{I}$). It turns out that any positive but not completely positive supermap $\mathscr{F}$ gives rise to a necessary condition for separability: If $\hat{\rho}_{AB}$ is separable, then $(\mathscr{I} \otimes \mathscr{F})(\hat{\rho})$ is positive.

We now introduce the method based on *entanglement witness*. An observable $\hat{W}$ is called an entanglement witness if it has at least one negative eigenvalue and satisfies

$$(\langle\alpha| \otimes \langle\beta|) \hat{W} (|\alpha\rangle \otimes |\beta\rangle) \geq 0 \tag{7.72}$$

for all pure states $|\alpha\rangle$ and $|\beta\rangle$ of A and B, respectively. A quantum state $\hat{\rho}_{AB}$ is a separable state if

$$\mathrm{Tr}\,\hat{W}\hat{\rho}_{AB} \geq 0 \tag{7.73}$$

for all entanglement witnesses. On the other hand, the entanglement of $\hat{\rho}_{AB}$ is detected by an entanglement witness $\hat{W}$ if and only if $\mathrm{Tr}\,\hat{W}\hat{\rho}_{AB} < 0$. This method works for a geometrical reason. The set of separable states is a convex subset in the entire Hilbert space $V \otimes W$ and can be described by hyperplanes. An entanglement witness represents a hyperplane.

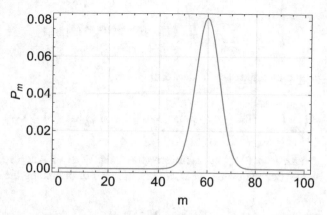

Fig. 7.2 The probability P_m for outcome m from the measurement of the total spin along the z-axis $\hat{S}^z_{\text{total}} = \sum_j \hat{S}^z_j$ on Alice's qubits. The plot is for $n = 100$ and $p = 3/5$

7.3.3 Entanglement Distillation

Suppose that Alice and Bob share a pair of entangled but not maximally entangled state. By now they are separated far way and no quantum channel is available. Luckily, they can still communicate through a classical channel. Is it possible for them to transform their partially entangled state to a maximally entangled state? The answer is no. Local operations and classical communications (often called LOCC) cannot generate or increase entanglement.

Now suppose that they have a large number n of copies of a partially entangled state $|\Psi\rangle$. Although they cannot transform each pair to a maximally entangled state, they can extract a smaller number k ($k < n$) of maximally entangled pairs. This is called the *entanglement distillation*. The entanglement distillation is not only useful for practical applications but also valuable for a fundamental reason. It can quantify entanglement. We will come back to this point later.

So, how can Alice and Bob extract maximally entangled pairs from partially entanglemed pairs using only local operations and classical communications? For an illustration,[2] we suppose that the state of each partially entangled pair is

$$|\Psi\rangle = |0\rangle \otimes |0\rangle \sqrt{1-p} + |1\rangle \otimes |1\rangle \sqrt{p}. \tag{7.74}$$

The overall state of the n pairs is given by

$$|\Psi\rangle^{\otimes n} = \left(|0\rangle \otimes |0\rangle \sqrt{1-p} + |1\rangle \otimes |1\rangle \sqrt{p}\right)^{\otimes n}. \tag{7.75}$$

The binomial expansion of the right-hand side of the above expression leads to

---

[2] Here we follow the method in Preskill (1998).

$$|\Psi\rangle^{\otimes n} = \sum_{m=0}^{n} \binom{n}{m}^{1/2} (1-p)^{(n-m)/2} p^{m/2} |\Psi_m\rangle, \tag{7.76}$$

where $|\Psi_m\rangle$ is an entangled state of the form

$$|\Psi_m\rangle = \binom{n}{m}^{-1/2} (|00\cdots\rangle \otimes |00\cdots\rangle + |01\cdots\rangle \otimes |01\cdots\rangle + \cdots) \tag{7.77}$$

with Schmidt rank $R = \binom{n}{m}$, and each term has exactly m of Alice's (and Bob's) qubits in 1. Now Alice measures the *total* spin of her n qubits along the z-axis

$$\hat{S}^z_{\text{total}} := \sum_{j=1}^{n} \hat{S}^z_j. \tag{7.78}$$

The possible outcomes are $m = 0, 1, \ldots, n$. For each m, the post-measurement state is nothing but $|\Psi_m\rangle$. The probability for a particular outcome m is given by

$$P_m = \binom{n}{m}(1-p)^{n-m} p^m. \tag{7.79}$$

As demonstrated in Fig. 7.2, P_m has a sharp peak at $m = np$ for large n. Therefore, from now on, we will assume that $m = np$.

Next, Alice and Bob are to convert $|\Psi_m\rangle$ to independent copies of maximally entangled pairs. Suppose that $R \approx \binom{n}{np} = 2^k$ for some integer $k \leq n$. Then, recalling (2.38) and Fig. 2.4, it is possible to transform the state to k copies of maximally entangled pairs. To do that, both Alice and Bob perform a transformation bringing the 2^k-dimensional support of her and his respective density operators to the Hilbert space of k qubits. In general, R is not close to a power of 2. However, one can increase the chance by dividing the entire pairs into ℓ blocks of n pairs. Alice perform the previously prescribed measurement on each block. Then, the total Schmidt rank becomes

$$R = \binom{n}{m_1}\binom{n}{m_2}\cdots\binom{n}{m_\ell} \tag{7.80}$$

with every $m_j \approx np$. For some ℓ, R now approaches 2^k.

How many maximally entangled pairs can one extract from $n\ell$ partially entangled pairs? Recall (Problem 7.2) that for large n,

$$\binom{n}{m} \approx 2^{nH(p)}, \tag{7.81}$$

where

$$H(p) := H(\{p, 1-p\}) = -p \log p - (1-p) \log(1-p) \tag{7.82}$$

is the binary entropy function. Then we have

$$R \approx 2^{n\ell H(p)} \approx 2^k, \tag{7.83}$$

which implies that the ratio is governed by the von Neumann entropy

$$\frac{k}{n\ell} = H(p) = S(\hat{\rho}_A). \tag{7.84}$$

In the above demonstration, we have used a particular form of partially entangled state. However, the result holds true in general: Given n copies of partially entangled pure state $|\Psi\rangle_{AB}$, one can extract at most $k_{max} = nS(\hat{\rho}_A)$ maximally entangled pairs. The ratio k_{max}/n in the large n limit thus characterizes to what extent the state $|\Psi\rangle_{AB}$ is entangled. For pure state, it happens to be given by the von Neumann entropy. In the subsection, we will see that this is a good measure of entanglement for mixed states as well.

7.3.4 Entanglement Measures

For both operational purposes and fundamental reasons, it is desired to quantify entanglement in a quantum state. For a pure state $|\Psi\rangle_{AB}$, the Schmidt decomposition

$$|\Psi\rangle_{AB} = \sum_j |\alpha_j\rangle \otimes |\beta_j\rangle s_j \tag{7.85}$$

is a good starting point. We see that the more uniformly the Schmidt values $\{s_j\}$ are distributed, the more extensive the entanglement is. Therefore, it is reasonable to quantify the entanglement of $|\Psi\rangle_{AB}$ by the Shannon entropy $H(\{s_j\})$ associated with the distribution $\{s_j\}$. Since it is identical to the von Neumann entropy of subsystems (Problem 7.4),

$$S(\hat{\rho}_A) = S(\hat{\rho}_B) = H(\{s_j\}), \tag{7.86}$$

we define the entanglement measure by

$$E := S(\hat{\rho}_A) = S(\hat{\rho}_B). \tag{7.87}$$

It is called the *entropy of entanglement*. In the previous subsection, we have seen that $S(\hat{\rho}_A)$ also determines the entanglement distillation ratio for pure states.

The initial idea to quantify entanglement was elevated by the view of entanglement as a resource for quantum communication. Since a maximally entangled state can quantum teleport one qubit, it is natural to compare the given state with the maximally entangled state. Many entanglement measures are based on this idea. The most common examples are the *distillable entanglement* and *entanglement cost*.

The distillable entanglement arises in the situation that we have already encountered in the previous subsection. There we have considered only pure states. Now consider a partially entangled mixed state $\hat{\rho}_{AB}$. Suppose that Alice and Bob share n copies of pairs with each in $\hat{\rho}_{AB}$. We ask what is the maximum number k_{max} of maximally entangled pairs that can be extracted from the partially entangled pairs. The distillable entanglement is defined by the ratio,

$$E_D := k_{max}/n \tag{7.88}$$

in the large n limit.

The entanglement cost arises in the opposite situation. Suppose that Alice and Bob share k maximally entangled pairs. Alice and Bob want to generate n pairs with each in a given state $\hat{\rho}_{AB}$. We ask what is the minimum value k_{min} of k? We define the entanglement cost by

$$E_C := k_{min}/n \tag{7.89}$$

in the large n limit.

For pure states, both are identical to the entropy of entanglement,

$$E_D = E_C = E = S(\hat{\rho}_A) = S(\hat{\rho}_B). \tag{7.90}$$

Unfortunately, for mixed state, both E_D and E_C are usually very difficult to calculate.

Another noteworthy entanglement measure is the *logarithmic negativity*. It is inspired by the positive partial transposition criterion in Sect. 7.3.2, and defined by

$$E_N := \log \|\hat{\rho}_{AB}^{T_B}\|_{tr}, \tag{7.91}$$

where $\| \cdot \|_{tr}$ denotes the trace norm defined in (5.167). Unlike other entanglement measures for mixed states, the logarithmic negativity is easy to evaluate. However, it does not reduce to the entropy of entanglement for all pure states.

We have briefly introduced a couple of widely known entanglement measures. Horodecki et al. (2009) and Plenio & Virmani (2007) give thorough discussions of various entanglement measures and their properties.

Problems

7.1. Let X and Y be two random variables. Show that

$$0 \le H(X|Y) \le H(X), \quad 0 \le H(Y|X) \le H(Y). \tag{7.92}$$

7.2. Consider binary string of length n. Within the model described in Sect. 7.1.4, the number of distinct *typical* strings is of an order of the binomial coefficient,

$$N_{\text{typical}} \approx \binom{n}{np} = \frac{n!}{(np)!(n-np)!}. \tag{7.93}$$

Show that

$$\log N_{\text{typical}} \approx n H(\{p, 1-p\}). \tag{7.94}$$

Hint: Use *Stirling's approximation*, $\log x! \approx x \log x - x + \mathcal{O}(\log x)$.

7.3. **(Klein's trace inequality)** Suppose that $f(x)$ is a differentiable *convex* function of a real variable x. Recall that the graph of $f(x)$ lies above all its tangents, that is,

$$f(x) \geq f(y) + f'(y)(x-y) \tag{7.95}$$

for all x and y. Show that

$$\text{Tr} f(\hat{A}) \geq \text{Tr} f(\hat{B}) + \text{Tr} f'(\hat{B})(\hat{A} - \hat{B}) \tag{7.96}$$

for all observables $\hat{A}$ and $\hat{B}$ on a finite-dimensional vector space. The inequality in (7.96) is often called *Klein's trace inequality*.
Similarly, if $f(x)$ is *concave*, then

$$\text{Tr} f(\hat{A}) \leq \text{Tr} f(\hat{B}) + \text{Tr} f'(\hat{B})(\hat{A} - \hat{B}) \tag{7.97}$$

for all observables $\hat{A}$ and $\hat{B}$.

7.4. Consider a pure state $|\Psi\rangle_{AB}$ of a composite system consisting of A and B. Show that

$$S(\hat{\rho}_A) = S(\hat{\rho}_B). \tag{7.98}$$

7.5. Using Klein's inequality (7.52), prove the subadditivity

$$S(\hat{\rho}_{AB}) \leq S(\hat{\rho}_A) + S(\hat{\rho}_B). \tag{7.99}$$

7.6. Here we are to prove the concavity of entropy (7.42). Denote the given system by A and introduce an ancillary system B. Suppose the composite system AB is in the state

$$\hat{\rho}_{AB} = \sum_{ij} p_j \hat{\rho}_j \otimes |b_j\rangle\langle b_j| \tag{7.100}$$

where $\{|b_j\rangle\}$ is an orthonormal basis of the Hilbert space associated with B. Use the subadditivity (7.61) to prove the concavity of entropy.

Correction to: A Quantum Computation Workbook

Correction to:
M.-S. Choi, *A Quantum Computation Workbook*,
https://doi.org/10.1007/978-3-030-91214-7

In the original version of the book, the author's belated corrections have been incorporated. The book has been updated with the changes.

The updated original version of the book can be found at
https://doi.org/10.1007/978-3-030-91214-7

Appendix A
Linear Algebra

Linear algebra is an elementary language to mathematically describe quantum mechanics. This appendix summarizes the concepts, definitions, theorems, and properties of linear algebra that are frequently used in quantum information physics. In most cases, we omit rigorous proofs to focus on main concepts.

Many textbooks on linear algebra focus on arithmetic techniques related to the properties of matrices, such as the Gauss elimination and the formula of determinant. In most areas of physics, notably in quantum mechanics, more relevant are the fundamental concepts and algebraic structures of vector spaces and linear operators. Lang (1987)—an abridged edition Lang (1986) is also available—is one of the textbooks that introduce and discuss the latter subjects.

A.1 Vectors

A.1.1 Vector Space

One of the most distinguished features of quantum states compared to classical states is superposition inherited from the wave-particle duality. It is thus natural to describe quantum states mathematically by vectors. Vectors can be multiplied by numbers (called scalars) and added with each other, exactly the way the superposition principle dictates. We first need a field, a set of scalars with addition and multiplication.

Definition A.1 (*field*) A set $\mathbb{F}$ of elements is called a *field* if it satisfies the following conditions:

(a) (addition) If $x, y \in \mathbb{F}$, then $x + y \in \mathbb{F}$.
(b) (multiplication) If $x, y \in \mathbb{F}$, then $xy \in \mathbb{F}$.
(c) (zero) $0 \in \mathbb{F}$ and $1 \in \mathbb{F}$.
(d) (inverse) If $x \in \mathbb{F}$, then $-x \in \mathbb{F}$.

© The Editor(s) (if applicable) and The Author(s), under exclusive license to Springer Nature Switzerland AG 2022, corrected publication 2023
M.-S. Choi, *A Quantum Computation Workbook*,
https://doi.org/10.1007/978-3-030-91214-7

(e) (inverse) If $x \in \mathbb{F}$ and $x \neq 0$, then $x^{-1} \in \mathbb{F}$.

The elements of the given field are called the *scalars*.

In the simplest terms, vectors represent physical quantities with both magnitude and direction. In quantum mechanics, more important feature of vectors is superposition. Here is the formal definition with mathematical rigor.

Definition A.2 (*vector space*) A set $\mathcal{V}$ of elements is called a *vector space* over a field $\mathbb{F}$, if it satisfies the following conditions:

(a) If $v_1, v_2 \in \mathcal{V}$, then $v_1 + v_2 \in \mathcal{V}$.
(b) If $v_1 \in \mathcal{V}$ and $x \in \mathbb{F}$, then $x v_2 \in \mathcal{V}$.
(c) If $v_1, v_2, v_3 \in \mathcal{V}$, then $(v_1 + v_2) + v_3 = v_1 + (v_2 + v_3)$.
(d) There is an element $0 \in \mathcal{V}$ (a *null vector*) such that for all $v_2 \in \mathcal{V}$ $0 + v_2 = v_2 + 0 = v_2$. Note that both "zero" in $\mathbb{F}$ or the null vector in $\mathcal{V}$ are denoted by '0'.
(e) For a given $v_2 \in \mathcal{V}$, there exists $-v_2 \in \mathcal{V}$ such that $v_2 + (-v_2) = 0$. The subtraction between vectors are defined by $v_1 - v_2 := v_1 + (-v_2)$.
(f) For $x, y \in \mathbb{F}$ and $v_1, v_2 \in \mathcal{V}$,

$$x(v_1 + v_2) = x v_1 + x v_2, \tag{A.1a}$$
$$(x + y)v_2 = x v_2 + y v_2, \tag{A.1b}$$
$$(xy)v_2 = x(y v_2). \tag{A.1c}$$

The elements of a vector space are called *vectors*.

Common examples include vector spaces over the fields of rational numbers ($\mathbb{Q}$), real numbers ($\mathbb{R}$), complex numbers ($\mathbb{C}$), and quarternions ($\mathbb{H}$). Note that the set of integer numbers ($\mathbb{Z}$) with the standard arithmetic rules is not a field. As is the case mostly in quantum mechanics, *vector spaces will be assumed to be over the field of complex numbers $\mathbb{C}$ throughout this book unless mentioned otherwise.*

Definition A.3 (*linear independence*) Let $\mathcal{V}$ be a vector space. Vectors $v_1, \ldots, v_n$ in $\mathcal{V}$ are said to be *linearly dependent* of each other if there exits a non-trivial solution $z_1, \ldots, z_n \in \mathbb{C}$ to the equation

$$z_1 v_1 + \cdots + z_n v_n = 0. \tag{A.2}$$

If not, they are said to be *linearly independent*.

A.1.2 Hermitian Product

In a vector space, the multiplication of vectors has not be defined. The *inner product* gives a special kind of multiplication between vectors and provides the vector space with a geometric structure, i.e., the *orthogonality* of vectors.

The inner product is usually a bilinear mapping. In many fields of physics (e.g., quantum mechanics), however, we will be dealing with vector spaces over $\mathbb{C}$ (the field of complex numbers). To preserve the notion of positive definiteness, we need to adopt a slightly different definition of the inner product. This modified inner product is called a Hermitian product to distinguish it from a usual inner product.

Definition A.4 (*Hermitian product*) Let V be a vector space. A *Hermitian product* on V is a function $\langle \cdot, \cdot \rangle$ from $V \times V$ to $\mathbb{C}$ satisfying the following conditions:

(a) For all $u, v, w \in V$, $\langle u, v + w \rangle = \langle u, v \rangle + \langle u, w \rangle$.
(b) For all $z \in \mathbb{C}$ and $v, w \in V$, $\langle zv, w \rangle = z^* \langle v, w \rangle$ and $\langle v, zw \rangle = z \langle v, w \rangle$.
(c) For all $v, w \in V$, $\langle v, w \rangle = \langle w, v \rangle^*$.
(d) $\langle v, v \rangle \geq 0$ for all $v \in V$, and $\langle v, v \rangle > 0$ if $v \neq 0$.[1]

Two vectors v and w are said to be *orthogonal* if $\langle v, w \rangle = 0$. On the other hand, they are said to be *parallel* if they are linearly dependent.

One of the most fundamental properties of Hermitian product is the following inequality.

Theorem A.5 (Cauchy-Schwarz inequality) *Let $\langle \cdot, \cdot \rangle$ be a Hermitian product on a vector space V. The so-called* Cauchy-Schwarz inequality

$$|\langle v, w \rangle|^2 \leq \langle v, v \rangle \langle w, w \rangle \tag{A.3}$$

holds for any vectors v and w in V. Equality holds if and only if v and w are linearly dependent (i.e., parallel).

To prove the Cauchy-Schwarz inequality, consider the vector defined by

$$u := v - w \frac{\langle w, v \rangle}{\langle w, w \rangle}. \tag{A.4}$$

Geometrically, it corresponds to the difference between v and the component of v projected onto $|w\rangle$. Hence u is orthogonal to w, $\langle u, w \rangle = 0$. We have

$$\langle u, u \rangle = \langle v, v \rangle - \frac{\langle v, w \rangle \langle w, v \rangle}{\langle w, w \rangle}. \tag{A.5}$$

Since $\langle u, u \rangle \geq 0$, the above equation leads to the inequality. Furthermore, we recall that $\langle u, u \rangle = 0$ if and only if $u = 0$. It immediately implies that the equality in (A.5) holds if and only if

$$v = w \frac{\langle w, v \rangle}{\langle w, w \rangle}. \tag{A.6}$$

Therefore, v and w must be linearly dependent.

---

[1] A Hermitian product satisfying this condition is said to be *positive definite*. We include this condition here because it is required for most applications in quantum mechanics.

The geometric structure due to the Hermitian product allows defining the *magnitude* or *norm* of the vectors. For a vector v, we let $\|v\|$ denote the norm of v and define it by

$$\|v\| := \sqrt{\langle v, v \rangle}. \tag{A.7}$$

This norm is called the *canonical norm* associated with the Hermitian product $\langle \cdot, \cdot \rangle$. It satisfies, as it should as a norm, the *triangle inequality*

$$\|v + w\| \leq \|v\| + \|w\| \tag{A.8}$$

for all vectors v and w. In quantum mechanics, a norm is essential for the probabilistic interpretation (Sect. 1.3).

A norm, in turn, provides a *distance* measure. A natural way to measure the distance $D(v, w)$ between two vectors v and w is to use the norm,

$$D(v, w) := \|v - w\| \equiv \sqrt{\langle v - w, v - w \rangle}. \tag{A.9}$$

The above distance measure is called the *canonical distance* associated with the Hermitian product $\langle \cdot, \cdot \rangle$. The triangle inequality (A.8) for the norm directly implies the analogous inequality for the distance,

$$\|u - w\| \leq \|u - v\| + \|v - w\| \tag{A.10}$$

for all vectors u, v, and w. A distance measure is useful in quantifying the closeness of two quantum states (Sects. 5.5.1 and 5.5.2).

A Hermitian product provides another way of measuring how close two state vectors are through the notion of *fidelity* (Sect. 5.5.3). The fidelity between two vectors v and w is defined to be $|\langle v, w \rangle|$.

A.1.3   Basis

Definition A.6 (*basis*) Let $\mathcal{V}$ be a vector space. If every element of $\mathcal{V}$ is a linear combination of $v_1, \ldots, v_n$, then $v_1, \ldots, v_n$ are said to *span* (or *generate*) the vector space $\mathcal{V}$. The set $\{v_1, \ldots, v_n\} \subset \mathcal{V}$ is called a *basis* of $\mathcal{V}$ if $v_1, \ldots, v_n$ span $\mathcal{V}$ and are linearly independent. The number of elements in a basis of $\mathcal{V}$ is called the *dimension* of $\mathcal{V}$ and denoted by $\dim \mathcal{V}$.

Quantum states are described by a state vector in a *Hilbert space*. A Hilbert space is a vector space, usually infinite dimensional, with additional analytic properties provided by the notion of completeness. However, as long as the dimension is finite, there is no distinction between the Hilbert space and the vector space. Unless mentioned otherwise explicitly, we assume that vector spaces are finite dimensional.

When a basis (recall Definition A.6) $\{v_1, v_2, \ldots, v_n\}$ spanning $\mathcal{V}$ satisfies

$$\langle v_i, v_j \rangle = \delta_{ij}, \tag{A.11}$$

it is called an *orthonormal basis*. For a finite dimensional vector space $\mathcal{V}$, one can always find an orthogonal basis as long as $\mathcal{V} \neq \{0\}$.

A choice of basis is arbitrary and one can change the basis to another: Suppose that $\mathcal{A} = \{v_1, v_2, \ldots, v_n\}$ and $\mathcal{B} = \{w_1, w_2, \ldots, w_n\}$ are two bases of the same vector space $\mathcal{V}$. As both $\mathcal{A}$ and $\mathcal{B}$ are bases, one can expand each vector w_j in $\mathcal{B}$ in the vectors v_i in $\mathcal{A}$ (and vice versa):

$$w_j = \sum_i v_i U_{ij}, \tag{A.12}$$

where U_{ij} are complex coefficients. The matrix $U := [U_{ij}]$ composed of these coefficients characterizes the relation between the two bases, and it must be invertible since the elements in each basis are linearly independent of each other. The relation is particularly simple when the bases are *orthonormal*: As $\mathcal{A}$ is orthonormal, the coefficients U_{ij} can be obtained by

$$U_{ij} = \langle v_i, w_j \rangle. \tag{A.13}$$

More importantly, one can show that U is a *unitary* matrix (see also Theorem A.15).

A.1.4 Representations

Given a fixed basis $\{v_1, v_2, \ldots, v_n\}$ of a vector space $\mathcal{V}$, any vector $\alpha \in \mathcal{V}$ is *uniquely* specified by the coefficients $\alpha_j \in \mathbb{C}$ in the expansion

$$\alpha = v_1\alpha_1 + v_2\alpha_2 + \cdots v_n\alpha_n. \tag{A.14}$$

The column vector consisting of $\alpha_1, \ldots, \alpha_n$ is said to be the *representation* of the vector α in the basis, and denoted by

$$\alpha \doteq \begin{bmatrix} \alpha_1 \\ \alpha_2 \\ \vdots \\ \alpha_n \end{bmatrix}. \tag{A.15}$$

When the basis $\{v_1, v_2, \ldots, v_n\}$ is *orthonormal*, the expansion coefficients α_j are obtained directly by means of the Hermitian product, $\alpha_j = \langle v_j, \alpha \rangle$. Hence, the vector is represented by the expansion

$$\alpha = \sum_j v_j \langle v_j, \alpha \rangle \tag{A.16}$$

or, equivalently, by the column vector

$$\alpha \doteq \begin{bmatrix} \langle v_1, \alpha \rangle \\ \langle v_2, \alpha \rangle \\ \vdots \\ \langle v_n, \alpha \rangle \end{bmatrix}. \tag{A.17}$$

Consider another vector $\beta \in V$ and suppose that $\beta_j := \langle v_j, \beta \rangle$ is its representation in the same orthonormal basis. Then, the Hermitian product $\langle \alpha, \beta \rangle$ can be evaluated using their column-vector representations

$$\langle \alpha, \beta \rangle = \sum_j \alpha_j^* \beta_j = \begin{bmatrix} \alpha_1^* & \alpha_2^* & \cdots & \alpha_n^* \end{bmatrix} \begin{bmatrix} \beta_1 \\ \beta_2 \\ \vdots \\ \beta_n \end{bmatrix}, \tag{A.18}$$

where we have used the identity $\langle \alpha, v_j \rangle = \langle v_j, \alpha \rangle^* = \alpha_j^*$.

Upon the change of basis, the representations of vectors also change. Suppose that a vector $\alpha \in V$ is represented by

$$\alpha \doteq \begin{bmatrix} \alpha_1 \\ \alpha_2 \\ \vdots \\ \alpha_n \end{bmatrix} \tag{A.19}$$

in the basis $\{v_i\}$, and by

$$\alpha \doteq \begin{bmatrix} \alpha_1' \\ \alpha_2' \\ \vdots \\ \alpha_n' \end{bmatrix} \tag{A.20}$$

in the basis $\{w_j\}$. The relation between the two representations is fixed by the relation between the two bases. Let us take a closer look at the relation when the two bases are orthonormal. We note from (A.16) that

$$\alpha_k' = \langle w_k, \alpha \rangle = \sum_{j=1}^n \langle w_k, v_j \rangle \langle v_j, \alpha \rangle = \sum_k U_{kj} \alpha_j, \tag{A.21}$$

where we have put $U_{kj} := \langle w_k, v_j \rangle$. We have seen in Eq. (A.13) that the matrix U is unitary. Therefore, we see that the representations in two different orthonormal bases are related by a unitary matrix as

$$
\begin{bmatrix} \alpha_1' \\ \alpha_2' \\ \vdots \\ \alpha_n' \end{bmatrix} = \begin{bmatrix} U_{11} & U_{12} & \cdots & U_{1n} \\ U_{21} & U_{22} & \cdots & U_{2n} \\ \vdots & \vdots & \ddots & \vdots \\ U_{n1} & U_{n2} & \cdots & U_{nn} \end{bmatrix} \begin{bmatrix} \alpha_1 \\ \alpha_2 \\ \vdots \\ \alpha_n \end{bmatrix}. \tag{A.22}
$$

A.2 Linear Operators

In quantum mechanics, the evolution of quantum states and the properties of physical quantities are described by linear operators. Linear operators are special kind of linear mappings.

A.2.1 Linear Maps

As already mentioned, the most important algebraic property of a vector space is the superposition. Therefore, the mapping preserving this property from one vector space to another plays an important role in the theory of linear algebra.

Definition A.7 (*linear map*) Let V and W be vector spaces. A mapping (or simply map)

$$
\hat{L} : V \to W, \quad v \mapsto \hat{L}v \tag{A.23}
$$

is said to be *linear* if it satisfies the following two properties:

(a) For any $v, w \in V$, we have $\hat{L}(v + w) = \hat{L}v + \hat{L}w$.
(b) For all $z \in \mathbb{C}$ and $v \in V$ we have $\hat{L}(zv) = z(\hat{L}v)$.

When $V = W$, the map is called a *linear operator* on V.

Since a linear map preserves superposition, it is completely determined by specifying how it maps just the basis vectors. The following theorem summarizes the property.

Theorem A.8 *Let V and W be vector spaces. Let $\{v_1, \ldots, v_n\}$ be a basis of V, and $w_1, \ldots, w_n \in W$ be arbitrary vectors—not to be necessarily distinctive nor to form a basis. Then there exists a unique linear map $\hat{L} : V \to W$ such that $w_j = \hat{L}v_j$ for all $j = 1, \ldots, n$.*

Proof of the theorem highlights the implications of linearity. Let us take a look at a proof. Define a map $\hat{A} : V \to W$ by the associations

$$\hat{A}v_j \mapsto w_j \tag{A.24}$$

and

$$\hat{A}(v_1 z_1 + \cdots v_n z_n) = w_1 z_1 + \cdots w_n z_n \tag{A.25}$$

for all $z_1, \ldots, z_n \in \mathbb{C}$. $\hat{A}$ is linear by construction, and we have shown that there exists a linear map satisfying the required condition. Now suppose that two linear maps $\hat{A}$ and $\hat{B}$ satisfy the condition. Let $v = v_1 z_1 + \cdots v_n z_n \in \mathcal{V}$ with $z_j \in \mathbb{C}$. Note that

$$\hat{B}v = (\hat{B}v_1)z_1 + \cdots + (\hat{B}v_n)z_n = w_1 z_1 + \cdots + w_n z_n = \hat{A}v. \tag{A.26}$$

Since v is arbitrary, we conclude that $\hat{A} = \hat{B}$.

A.2.2 Representations

As asserted by Theorem A.8, a linear map $\hat{L} : \mathcal{V} \to \mathcal{W}$ is completely determined by specifying how it maps each element of a basis $\{v_j : j = 1, \ldots, n\}$ of $\mathcal{V}$. Expanding the result $\hat{L}v_j$ in a basis $\{w_i : i = 1, \ldots, m\}$ of $\mathcal{W}$ as

$$\hat{L}v_j = \sum_i w_i L_{ij}, \tag{A.27}$$

we can equivalently say that $\hat{L}$ is *uniquely* specified by the coefficients $L_{ij} \in \mathbb{C}$. The $m \times n$ matrix composed of the coefficients is called the matrix representation of $\hat{L}$ in the bases $\{v_j\}$ and $\{w_i\}$. $\hat{L}$ being represented by the matrix L is denoted by

$$\hat{L} \doteq \begin{bmatrix} L_{11} & L_{12} & \cdots \\ L_{21} & L_{22} & \cdots \\ \vdots & \vdots & \ddots \end{bmatrix}. \tag{A.28}$$

When the bases $\{v_j\}$ and $\{w_i\}$ are *orthonormal*, the matrix elements L_{ij} can be obtained by means of the Hermitian product, $L_{ij} = \langle w_i, \hat{L}v_j \rangle$, and hence

$$\hat{L}v_j = \sum_i w_i \langle w_i, \hat{L}v_j \rangle. \tag{A.29}$$

In quantum mechanics, one has to frequently calculate the matrix representations of linear maps in orthonormal bases, and the procedure is summarized in the following table:

$$\hat{L} \doteq \begin{array}{c|cc} & v_1 & v_2 & \cdots \\ \hline w_1 & \langle w_1, \hat{L}v_1 \rangle & \langle w_1, \hat{L}v_2 \rangle & \cdots \\ w_2 & \langle w_2, \hat{L}v_1 \rangle & \langle w_2, \hat{L}v_2 \rangle & \cdots \\ \vdots & \vdots & \vdots & \ddots \end{array} \tag{A.30}$$

When $\mathcal{V} = \mathcal{W}$, a linear operator is represented by a square matrix.

It is important to note that the matrix representation of a linear map depends on the choice of bases of $\mathcal{V}$ and $\mathcal{W}$. How does the matrix representation change when we choose different bases? Here let us focus on linear operators ($\mathcal{V} = \mathcal{W}$). Suppose that $\{v_i\}$ and $\{w_j\}$ are two different orthonormal bases of $\mathcal{V}$. Since they are both orthonormal, there must be a unitary operator (Definition A.13) $\hat{U}$ such that

$$w_j = \hat{U}v_j = \sum_i v_i U_{ij}, \tag{A.31}$$

where $U_{ij} := \langle v_i, w_j \rangle$ is a unitary matrix (see Eq. (A.12)). Suppose that L_{ij} be the matrix representation of a linear operator $\hat{L}$ in the basis $\{v_i\}$. What is the matrix representation L'_{ij} of $\hat{L}$ in the new basis $\{w_j\}$? Using the relation (A.31), we obtain the new matrix representation

$$L'_{ij} = \langle w_i, \hat{L}w_j \rangle = \sum_{kl} U^*_{ik} \langle v_i, \hat{L}v_l \rangle U_{lj} = \sum_{kl} U^*_{ik} L_{il} U_{lj}. \tag{A.32}$$

In other words, the matrix representations in two different bases are related with each other by

$$L' = U^\dagger L U. \tag{A.33}$$

A.2.3 Hermitian Conjugate of Operators

On a vector space equipped with a Hermitian product, given a linear operator $\hat{L}$ one can define another linear operator $\hat{L}^\dagger$ naturally related to $\hat{L}$. Hermitian conjugates of operators greatly simplify the evaluations of operator-related expressions and the spectral analysis of them.

Theorem A.9 (Hermitian conjugate) *Let $\mathcal{V}$ and $\mathcal{W}$ be vector spaces over $\mathbb{C}$ equipped with Hermitian products. Let $\hat{L} : \mathcal{V} \to \mathcal{W}$ be a linear map. Then, the following statements hold true:*

(a) There exists a unique linear map $\hat{L}^\dagger : \mathcal{W} \to \mathcal{V}$ such that

$$\langle w, \hat{L}v \rangle_\mathcal{W} = \langle \hat{L}^\dagger w, v \rangle_\mathcal{V} \tag{A.34}$$

for all $v \in \mathcal{V}$ and $w \in \mathcal{W}$.

(b) $(\hat{L}^\dagger)^\dagger$ *exists as well, and is identical to* $\hat{L}$, $(\hat{L}^\dagger)^\dagger = \hat{L}$.

The linear operator $\hat{L}^\dagger$ *is called the* Hermitian conjugate *of* $\hat{L}$.

The matrix representation of a linear map is unique, and one can also define the Hermitian conjugate in terms of the matrix representation. Let $\hat{L} : \mathcal{V} \to \mathcal{W}$ be represented by

$$\hat{L}v_j = \sum_i w_i L_{ij}\,. \tag{A.35}$$

Then, $\hat{L}^\dagger : \mathcal{W} \to \mathcal{V}$ is defined by

$$\hat{L}^\dagger w_j = \sum_i v_i L_{ji}^*\,. \tag{A.36}$$

That is, the matrix representation of $\hat{L}^\dagger$ is the conjugate-transposition of the matrix representation of $\hat{L}$. For finite-dimensional vector spaces, the two definitions are equivalent.

For an operator on a vector space, its Hermitian conjugate also acts on the same vector space and enables to characterize the operator itself.

Definition A.10 *(normal operator)* A linear operator $\hat{L}$ on a vector space $\mathcal{V}$ is said to be *normal* if $[\hat{L}^\dagger, \hat{L}] = 0$.

The two most important examples of normal operators are Hermitian operators and unitary operators.

In quantum mechanics, the linear operator representing a physical quantity should be Hermitian:

Definition A.11 *(Hermitian operator)* A linear operator $\hat{H}$ on a vector space is called *Hermitian* if

$$\langle \hat{H}v, w \rangle = \langle v, \hat{H}w \rangle\,, \quad \forall v, w \in \mathcal{V}. \tag{A.37}$$

There is a simple test for a Hermitian operator on a finite dimensional vector space:

Theorem A.12 *Let* $\mathcal{V}$ *be a vector space. An operator* $\hat{H}$ *on* $\mathcal{V}$ *is Hermitian if and only if* $\langle v, \hat{H}v \rangle \in \mathbb{R}$ *for all* $v \in \mathcal{V}$.

Another type of operators one encounters very frequently in quantum mechanics are unitary operators, *norm-preserving and invertible* linear maps.

Definition A.13 *(unitary operator)* Let $\mathcal{V}$ be a vector space $\mathcal{V}$ equipped with a Hermitian product. A linear operator $\hat{U}$ is said to be *unitary* when it maps $\mathcal{V}$ onto the whole of $\mathcal{V}$ and preserve the norm. That is, $\hat{U}\mathcal{V} = \mathcal{V}$ and $\langle \hat{U}v, \hat{U}v \rangle = \langle v, v \rangle$ for all $v \in \mathcal{V}$.

A unitary operator is characterized by the fact that its Hermitian conjugate is identical to its inverse.

Theorem A.14 *If $\hat{U}$ is a unitary operator on V, then*

$$\hat{U}^\dagger \hat{U} = \hat{U}\hat{U}^\dagger = 1. \tag{A.38}$$

In fact, Eq. (A.38) can be used as an alternative definition of a unitary operator.

The unique linear map in Theorem A.8 becomes a unitary operator when the image vectors form another orthonormal basis of the same vector space.

Theorem A.15 *Let V be a vector space. Let $\{v_1, \ldots, v_n\}$ and $\{w_1, \ldots, w_n\}$ be orthonormal bases of V. Then there exists a unique unitary operator $\hat{U}$ on V such that $w_j = \hat{U}v_j$ for all $j = 1, \ldots, n$.*

We have already seen in Eq. (A.12) that two orthonormal bases are related by a unitary matrix. Theorem A.15 just asserts it again. Indeed, if U is the matrix representation of $\hat{U}$ in the basis $\{v_i\}$, then

$$w_j = \hat{U}v_j = \sum_i v_i U_{ij} . \tag{A.39}$$

Theorem A.15 is even more general and hold for any orthonormal subsets:

Theorem A.16 *Let V is a Hilbert space equipped with a Hermitian product $\langle \cdot, \cdot \rangle$, and $U \subset V$ a subspace. Suppose $\hat{U} : U \to V$ is a linear operator which preserves the Hermitian product. That is, for any $u, u' \in U$,*

$$\langle \hat{U}u, \hat{U}u' \rangle = \langle u, u' \rangle . \tag{A.40}$$

There exists a unitary operator $\hat{V} : V \to V$ which extents $\hat{U}$. That is, $\hat{V}u = \hat{U}u$ for all $u \in U$ and $\hat{V}$ is defined on the entire space V.

An immediate consequence of Theorem A.16 is that for any pair of vectors v and w one can always find a unitary operator $\hat{U}$ such that $w = \hat{U}v$.

A.3   Dirac's Bra-Ket Notation

For a given vector space V, one can construct another special vector space V^* associated with V, consisting of all linear mappings from V to $\mathbb{C}$, $V^* := \{\phi : V \to \mathbb{C}\}$. It is called the *dual space* of V. With a fixed vector $v \in V$ and the Hermitian product, one can define a linear mapping $\phi_v : V \to \mathbb{C}$ by the relation $\phi_v(w) := \langle v, w \rangle$. Certainly, ϕ_v is an element of V^*. This way, by choosing different vectors from V, one can define a particular kind of linear mappings belonging to V^*. Now the key observation is that in fact, any linear mapping in V^* is of this kind. That is, there is a one-to-one correspondence $v \leftrightarrow \phi_v$ between V and V^*. ϕ_v is called the *dual vector* of v.

In Dirac's bra-ket notation, the dual ϕ_v is denoted by $\langle v|$ whereas the native vector v is denoted by $|v\rangle$; hence the name. It is just a simple notational change. However, it simplifies most evaluations in quantum mechanics so greatly that it is widely used. *Throughout the book, we will almost always be using the bra-ket notation.*

When $\{|\alpha_1\rangle, \ldots, |\alpha_n\rangle\}$ is an orthonormal basis of $\mathcal{V}$, $\{\langle\alpha_1|, \ldots, \langle\alpha_n|\}$ is also an orthonormal basis of $\mathcal{V}^*$ and is called the *dual basis* of the former. More importantly, the two bases satisfy the relation

$$\langle\alpha_i|\alpha_j\rangle = \delta_{ij}. \tag{A.41}$$

Suppose that a vector $|v\rangle \in \mathcal{V}$ is expanded as

$$|v\rangle = \sum_j |\alpha_j\rangle v_j, \quad v_j \in \mathbb{C}. \tag{A.42}$$

Then, its dual vector $\langle v|$ is given by

$$\langle v| = \sum_j v_j^* \langle\alpha_j|. \tag{A.43}$$

Armed with the basic properties of the bra-ket notation, now consider a combination of the form $|v\rangle\langle v'|$, where $|v\rangle, |v'\rangle \in \mathcal{V}$: We regard it as an operator on $\mathcal{V}$ defined by the association

$$|v\rangle\langle v'| : |u\rangle \mapsto |v\rangle\langle v'|u\rangle \tag{A.44}$$

for all $|u\rangle \in \mathcal{V}$. A simple inspection (see Definition A.9) shows that its Hermitian conjugate $(|v\rangle\langle v'|)^\dagger$ is just given by

$$(|v\rangle\langle v'|)^\dagger = |v'\rangle\langle v|. \tag{A.45}$$

If one constructs $|\alpha_i\rangle\langle\alpha_j|$ out of a basis $\{|\alpha_j\rangle\}$ of $\mathcal{V}$, then any linear map on $\mathcal{V}$ can be expressed in terms of them, and they form a basis of the vector space $\mathcal{L}(\mathcal{V})$ of linear operators (Appendix B.1). In particular, if the basis $\{|\alpha_j\rangle\}$ is *orthonormal*, then a linear operator $\hat{L}$ on $\mathcal{V}$ with the matrix representation L in the same basis is equivalent to

$$\hat{L} = \sum_{ij} |\alpha_i\rangle L_{ij} \langle\alpha_j|, \tag{A.46}$$

and, accordingly, its Hermitian conjugate $\hat{L}^\dagger$ to

$$\hat{L}^\dagger = \sum_{ij} |\alpha_i\rangle L_{ji}^* \langle\alpha_j|. \tag{A.47}$$

Both expansions can be verified by evaluating the matrix elements in the basis and applying the orthogonality relation (A.41). An interesting result is that for *any* orthonormal basis $\{|\alpha_j\rangle\}$ of $\mathcal{V}$, the linear combination

$$\hat{I} = \sum_j |\alpha_j\rangle\langle\alpha_j| \tag{A.48}$$

is equal to the identity operator on $\mathcal{V}$. It is called the *completeness relation*. The orthogonality relation (A.41) and the completeness relation (A.48), which are mutually complementary, together empower the bra-ket notation.

Consider a system of two qubits. The associated Hilbert space is spanned by the standard basis.

```
In[·]:= bs = Basis[2]
```
$$\text{Out[·]= } \{|0, 0\rangle, |0, 1\rangle, |1, 0\rangle, |1, 1\rangle\}$$

Consider a Hermitian operator, physically, corresponding to the Heisenberg exchange interaction between two S=1/2 spins.

```
In[·]:= op = Pauli[1, 1] + Pauli[2, 2] + Pauli[3, 3]
```
$$\text{Out[·]= } \sigma^x \otimes \sigma^x + \sigma^y \otimes \sigma^y + \sigma^z \otimes \sigma^z$$

This shows the matrix representation of the operator.

```
In[·]:= mat = Matrix[op];
        mat // MatrixForm
```
$$\text{Out[·]//MatrixForm=} \begin{pmatrix} 1 & 0 & 0 & 0 \\ 0 & -1 & 2 & 0 \\ 0 & 2 & -1 & 0 \\ 0 & 0 & 0 & 1 \end{pmatrix}$$

This is an expansion of the operator in the bra-ket notation.

```
In[·]:= op2 = MultiplyDot[bs, mat, Dagger[bs]]
```
$$\text{Out[·]= } |0, 0\rangle\langle 0, 0| - |0, 1\rangle\langle 0, 1| + 2|0, 1\rangle\langle 1, 0| + $$
$$2|1, 0\rangle\langle 0, 1| - |1, 0\rangle\langle 1, 0| + |1, 1\rangle\langle 1, 1|$$

Verify the above expansion by evaluating the matrix representation.

```
In[·]:= mat2 = Outer[Multiply, Dagger[bs], op2 ** bs];
        mat2 // MatrixForm
```
$$\text{Out[·]//MatrixForm=} \begin{pmatrix} 1 & 0 & 0 & 0 \\ 0 & -1 & 2 & 0 \\ 0 & 2 & -1 & 0 \\ 0 & 0 & 0 & 1 \end{pmatrix}$$

More generally, given two vector spaces $\mathcal{V}$ and $\mathcal{W}$, the combination $|w\rangle\langle v|$ with for $|v\rangle \in \mathcal{V}$ and $|w\rangle \in \mathcal{W}$ is a linear map $\mathcal{V} \to \mathcal{W}$ with an association similar to that in Eq. (A.44). Its Hermitian conjugate, $(|w\rangle\langle v|)^\dagger : \mathcal{W} \to \mathcal{V}$, is given by $(|w\rangle\langle v|)^\dagger = |v\rangle\langle w|$.

To illustrate the power of the bra-ket notation, let us consider a linear map $\hat{L} : \mathcal{V} \to \mathcal{W}$ and examine $\hat{L}|\alpha_j\rangle$: Let $\{|\alpha_j\rangle : j = 1, \ldots, m\}$ and $\{|\beta_k\rangle : k = 1, \ldots, n\}$ be orthonormal bases of $\mathcal{V}$ and $\mathcal{W}$, respectively. As $\hat{L}|\alpha_j\rangle$ is an element of $\mathcal{W}$, it should not be affected by the identity operator $\hat{I}_\mathcal{W}$ on $\mathcal{W}$, $\hat{L}\,|\alpha_j\rangle = \hat{I}_\mathcal{W}\hat{L}\,|\alpha_j\rangle$. Using the completeness relation (A.48) (replacing $|\alpha_j\rangle$ with $|\beta_k\rangle$)), one can get

$$\hat{L}\,|\alpha_j\rangle = \hat{I}_\mathcal{W}\hat{L}\,|\alpha_j\rangle = \sum_k |\beta_k\rangle\,\langle\beta_k|\,\hat{L}\,|\alpha_j\rangle. \tag{A.49}$$

Noting that $L_{kj} = \langle\beta_k|\,\hat{L}\,|\alpha_j\rangle$ is the matrix elements of the representation of $\hat{L}$ in the given bases [see Eq. (A.29)], one recovers the defining relation [see Eq. (A.27)]

$$\hat{L}\,|\alpha_j\rangle = \sum_k |\beta_k\rangle\,L_{kj} \tag{A.50}$$

for the matrix representation of $\hat{L}$. Given the matrix representation L_{kj}, one can further expand $\hat{L}$ in terms of the bra-ket notation as

$$\hat{L} = \sum_{kj} |\beta_k\rangle\,L_{kj}\,\langle\alpha_j|\,. \tag{A.51}$$

Once expanded in the bra-ket notation, its Hermitian conjugate $\hat{L}^\dagger : \mathcal{W} \to \mathcal{V}$ reads as

$$\hat{L}^\dagger = \sum_{kj} |\alpha_j\rangle\,L_{kj}^*\,\langle\beta_k|\,. \tag{A.52}$$

A.4 Spectral Theorems

The eigenvalues and eigenvectors of normal operators (Definition A.10) exhibit particularly useful properties. They are frequently used in quantum mechanics and simplify many calculations and analyses. Here we summarize the properties of the eigenvalues and eigenvectors of normal operators, especially, Hermitian and unitary operators. Although we will mainly focus on the spectral properties, we will also discuss some other related properties.

A.4.1 Spectral Decomposition

The following theorems summarize the properties of the eigenvectors and eigenvalues of normal operators, especially, Hermitian, positive, and unitary operators:

Theorem A.17 *Let $\hat{A}$ be a normal operator (Definition A.10) on a vector space $\mathcal{V}$.*

(a) Eigenstates of $\hat{A}$ belonging to distinct eigenvalues are orthogonal to each other.
(b) The set of all eigenvectors of $\hat{A}$ spans $\mathcal{V}$.

Theorem A.17 enables to expand a normal operator in terms of its eigenvectors and eigenvalues using Dirac's bra-ket notation. Suppose that a normal operator $\hat{A}$ on a vector space $\mathcal{V}$ has eigenvectors $|a\rangle$ and the corresponding eigenvalues a. If some eigenvalues are degenerate, that is, there are more than one linearly independent eigenvectors belonging to the same eigenvalue, we choose mutually orthogonal eigenvectors, which is always possible. Then, the normalized eigenvectors form an orthonormal basis, which is called the *eigenbasis* from $\hat{A}$ of the vector space $\mathcal{V}$. Then the matrix representation of $\hat{A}$ in the eigenbasis from itself must be diagonal with the diagonal elements given by the eigenvalues. Hence, in Dirac's bra-ket notation, $\hat{A}$ can be expanded as

$$\hat{A} = \sum_a |a\rangle\, a\, \langle a| , \qquad (A.53)$$

where the sum is over all eigenvalues of $\hat{A}$. The expansion is called the *spectral decomposition* of $\hat{A}$.

Theorem A.18 *(a) A Hermitian operator $\hat{H}$ is normal.*
(b) Every eigenvalue of a Hermitian operator is real.

In quantum mechanics, the operators usually describe the changes of state of a system in terms of the transformation of vectors. However, there is also a class of operators, called density operators, that describe the mixed state of the system (Sect. 1.1.2). For a proper statistical interpretation of mixed states, the density operators are required to satisfy certain properties. They are Hermitian and positive among other properties. It motivates the following definition.

Definition A.19 *(positive operator)* Let $\hat{H}$ be a *Hermitian* operator on a vector space $\mathcal{V}$.

(a) $\hat{H}$ is said to be *positive* (or more specifically, *positive definite*) if $\langle v|\hat{H}|v\rangle > 0$ for all non-vanishing vectors $|v\rangle \in \mathcal{V}$. A positive operator is denoted as $\hat{H} > 0$.
(b) $\hat{H}$ is said to be *positive semidefinite* or *non-negative* if $\langle v|\hat{H}|v\rangle \geq 0$ for all non-vanishing vectors $|v\rangle \in \mathcal{V}$. It is denoted as $\hat{H} \geq 0$.

Very often in physics, positive definite and positive semidefinite operators are not distinguished, and both types are simply called positive operators.

Theorem A.20 Let $\hat{H}$ be a Hermitian operator on a vector space. Then,

(a) $\hat{H}$ is positive if and only if every eigenvalue of it is positive;
(b) $\hat{H}$ is positive-semidefinite if and only if the eigenvalues are non-negative.

A common example of positive operator is an operator of the form

$$\hat{A} = |\psi\rangle\, \langle\psi| \qquad (A.54)$$

for some vector $|\psi\rangle$. It has a single non-zero eigenvalue 1, and the corresponding eigenvector is $|\psi\rangle$. Any vector $|\varphi\rangle$ orthogonal to $|\psi\rangle$ is an eigenvector of $\hat{A}$ belonging to the (degenerate) eigenvalue 0, $\hat{A}|\varphi\rangle = 0$. Hence, $\hat{A} \geq 0$. More generally, consider an operator of the form

$$\hat{A} := \sum_j |\psi_j\rangle\langle\psi_j| \tag{A.55}$$

for an arbitrary set $\{|\psi_j\rangle\}$ of vectors (not necessarily orthogonal nor normalized). One can show that $\hat{A} \geq 0$.

For any linear operator $\hat{A}$, the compositions $\hat{A}^\dagger \hat{A}$ and $\hat{A}\hat{A}^\dagger$ give positive operators as well. To see it, note that

$$\langle v| \hat{A}^\dagger \hat{A} |v\rangle = \|\hat{A}|v\rangle\|^2 \geq 0 \tag{A.56}$$

for any $|v\rangle$. Therefore, $\hat{A}^\dagger \hat{A} \geq 0$. One can use a similar argument to convince oneself that $\hat{A}\hat{A}^\dagger \geq 0$.

Sometimes, it is useful to normalize the eigenvectors $|a\rangle$ of a *positive operator* (Definition A.19 and Theorem A.20) with their own (positive) eigenvalues a so that $\langle a|a\rangle = a$. In this case, the spectral decomposition of a positive operator $\hat{A}$ is given by

$$\hat{A} = \sum_a |a\rangle\langle a|, \tag{A.57}$$

This form of the spectral decomposition for a positive operator should not be confused with the completeness relation (A.48), where an orthonormal basis is used. For a *positive semidefinite operator*, there only appear eigenvectors with positive eigenvalues in the summation in (A.57), and the eigenvectors with zero eigenvalue are automatically dropped.

Theorem A.21 *Let $\hat{U}$ be a unitary operator on a vector space $\mathcal{V}$. Then, every eigenvalue of $\hat{U}$ is of the form $e^{i\phi}$ with $\phi \in \mathbb{R}$.*

For example, consider an operator $\hat{U}$ with the matrix representation

$$\hat{U} \doteq \begin{bmatrix} \cos\phi & -\sin\phi \\ \sin\phi & \cos\phi \end{bmatrix} \tag{A.58}$$

in the logical basis $|0\rangle$ and $|1\rangle$. It has two eigenvectors

$$|L/R\rangle := \frac{|0\rangle \pm i|1\rangle}{\sqrt{2}} \tag{A.59}$$

with eigenvalues $e^{\pm i\phi}$. $\hat{U}$ has the spectral decomposition

$$\hat{U} = |L\rangle e^{-i\phi}\langle L| + |R\rangle e^{+i\phi}\langle R|. \tag{A.60}$$

A.4.2   Functions of Operators

The spectral decomposition provides a convenient way to define *functions of a normal operator*. Let $f : \mathcal{D} \to \mathbb{C}$ be a function of complex variable defined in a domain $\mathcal{D} \subset \mathbb{C}$. Suppose that $\hat{A}$ be a normal operator with all eigenvalues $a \in \mathcal{D}$. Then the function $f(\hat{A})$ of the operator $\hat{A}$—another operator on the same vector space derived from $\hat{A}$—is defined by [to be compared with (A.53)]

$$f(\hat{A}) := \sum_a |a\rangle f(a) \langle a| . \tag{A.61}$$

Surprisingly, most students try to define a function of an operator by means of the Taylor series expansion involving multiple powers of the operator. In most cases, however, it is very difficult to convince oneself whether the series is actually converging or not. Even if it does, it is tremendously difficult to figure out the behavior of the resulting operator, not to speak of the evaluation of the series itself. In contrast, the definition in (A.61) is well-defined as long as $f(z)$ of complex variable z is well defined, and often provides clear physical meaning of the resulting operator $f(\hat{A})$. Above all, the evaluation is straightforward and, in many cases, simple. The definition applies only to normal operators. However, it is not a serious restriction since in most physics applications, it is normal operators that we want to transform by means of already existing functions.

Consider again a Hermitian operator describing the Heisenberg exchange interaction between two S=1/2 spins.

```
In[·]:= opH = -J * (Pauli[1, 1] + Pauli[2, 2] + Pauli[3, 3])
```
$$\text{Out[·]= } -J \left(\sigma^x \otimes \sigma^x + \sigma^y \otimes \sigma^y + \sigma^z \otimes \sigma^z \right)$$

We want to consider functions of operators, for example, opH in particular. To do it, it is most efficient to proceed with the spectral decomposition of the operator.

```
In[·]:= {val, vec} = ProperSystem[opH]
```
$$\text{Out[·]= } \left\{ \{3\,J, -J, -J, -J\}, \left\{ -|0, 1\rangle + |1, 0\rangle, |1, 1\rangle, |0, 1\rangle + |1, 0\rangle, |0, 0\rangle \right\} \right\}$$

The eigenvectors are orthogonal, but not properly normalized.

```
In[·]:= Outer[Multiply, Dagger[vec], vec] // MatrixForm
```
Out[·]//MatrixForm=
$$\begin{pmatrix} 2 & 0 & 0 & 0 \\ 0 & 1 & 0 & 0 \\ 0 & 0 & 2 & 0 \\ 0 & 0 & 0 & 1 \end{pmatrix}$$

Normalize them, and check again.

In[ ]:= `nvec = vec / Sqrt[{2, 1, 2, 1}]`
         `Outer[Multiply, Dagger[nvec], nvec] // MatrixForm`

Out[ ]= $\left\{ \dfrac{-|0, 1\rangle + |1, 0\rangle}{\sqrt{2}},\ |1, 1\rangle,\ \dfrac{|0, 1\rangle + |1, 0\rangle}{\sqrt{2}},\ |0, 0\rangle \right\}$

Out[ ]//MatrixForm=

$$\begin{pmatrix} 1 & 0 & 0 & 0 \\ 0 & 1 & 0 & 0 \\ 0 & 0 & 1 & 0 \\ 0 & 0 & 0 & 1 \end{pmatrix}$$

Suppose we want to get the exponential function of opH. For example, the time-evolution operator is given by

In[ ]:= `opU = MultiplyExp[-I t opH]`

Out[ ]= $e^{i J t \left(\sigma^x \otimes \sigma^x + \sigma^y \otimes \sigma^y + \sigma^z \otimes \sigma^z \right)}$

This uses the spectral decomposition.

In[ ]:= `newU = Total@Multiply[nvec, Exp[-I t val], Dagger[nvec]]`

Out[ ]= $e^{i J t} |0, 0\rangle \langle 0, 0| + \dfrac{1}{2} e^{i J t} |0, 1\rangle \langle 0, 1| +$

$\dfrac{1}{2} e^{-3 i J t} |0, 1\rangle \langle 0, 1| + \dfrac{1}{2} e^{i J t} |0, 1\rangle \langle 1, 0| -$

$\dfrac{1}{2} e^{-3 i J t} |0, 1\rangle \langle 1, 0| + \dfrac{1}{2} e^{i J t} |1, 0\rangle \langle 0, 1| - \dfrac{1}{2} e^{-3 i J t} |1, 0\rangle \langle 0, 1| +$

$\dfrac{1}{2} e^{i J t} |1, 0\rangle \langle 1, 0| + \dfrac{1}{2} e^{-3 i J t} |1, 0\rangle \langle 1, 0| + e^{i J t} |1, 1\rangle \langle 1, 1|$

This converts the bra-ket expression into a form in terms of the Pauli operators.

In[ ]:= `newU2 = Elaborate@ExpressionFor@Matrix[newU]`

Out[ ]= $\dfrac{1}{4} e^{-3 i J t} \left(1 + 3 e^{4 i J t} \right) \sigma^0 \otimes \sigma^0 + \dfrac{1}{4} e^{-3 i J t} \left(-1 + e^{4 i J t} \right) \sigma^x \otimes \sigma^x +$

$\dfrac{1}{4} e^{-3 i J t} \left(-1 + e^{4 i J t} \right) \sigma^y \otimes \sigma^y + \dfrac{1}{4} e^{-3 i J t} \left(-1 + e^{4 i J t} \right) \sigma^z \otimes \sigma^z$

In many cases, MultiplyExp can be further evaluated by means of Elaborate.

In[ ]:= `opU2 = Elaborate@Elaborate[opU]`

Out[ ]= $\dfrac{1}{4} e^{-3 i J t} \left(1 + 3 e^{4 i J t} \right) \sigma^0 \otimes \sigma^0 + \dfrac{1}{4} e^{-3 i J t} \left(-1 + e^{4 i J t} \right) \sigma^x \otimes \sigma^x +$

$\dfrac{1}{4} e^{-3 i J t} \left(-1 + e^{4 i J t} \right) \sigma^y \otimes \sigma^y + \dfrac{1}{4} e^{-3 i J t} \left(-1 + e^{4 i J t} \right) \sigma^z \otimes \sigma^z$

A.5 Factorization of Operators

Matrices can be decomposed into several factors. As linear operators are represented by matrices, the methods directly also applies to linear operators. We have already seen one form, the spectral decomposition, in Sect. A.4.1. Here we introduce a few more forms.

Theorem A.22 (singular-value decomposition) *Let A be an $m \times n$ matrix of complex numbers. It can be written in the form*

$$A = USV^\dagger, \tag{A.62}$$

where U and V are $m \times m$ and $n \times n$ unitary matrices, respectively, and S is an $m \times n$ diagonal matrix of the form

$$S = \begin{bmatrix} s_1 & & \\ & s_2 & \\ & & \ddots \end{bmatrix} \tag{A.63}$$

with $s_j \geq 0$. Positive values of s_j are called the singular values *of A.*

For a linear operator $\hat{A}$ on a vector space $\mathcal{V}$, one can apply the above theorem to its matrix representation A,

$$\hat{A} = \sum_{ij} |i\rangle A_{ij} \langle j| = \sum_{ijk} |i\rangle U_{ij} s_j V^\dagger_{jk} \langle k|. \tag{A.64}$$

It results in the singular-value decomposition of the operator $\hat{A}$

$$\hat{A} = \hat{U} \hat{S} \hat{V}^\dagger, \tag{A.65}$$

where

$$\hat{U} := \sum_{ij} |i\rangle U_{ij} \langle j|, \quad \hat{S} := \sum_j |j\rangle s_j \langle j|, \quad \hat{V} := \sum_{ij} |i\rangle V_{ij} \langle j|. \tag{A.66}$$

Since the both matrices U and V are unitary, the operators $\hat{U}$ and $\hat{V}$ are also unitary. Another useful form of the singular-value decomposition is

$$\hat{A} = \sum_j |u_j\rangle s_j \langle v_j|, \tag{A.67}$$

where

$$|u_j\rangle := \sum_i |i\rangle U_{ij}, \quad \langle v_j| := \sum_i |i\rangle V_{ij}. \tag{A.68}$$

The unitarity of the matrices U and V implies that

$$\langle u_i|u_j\rangle = \delta_{ij}, \quad \langle v_i|v_j\rangle = \delta_{ij}. \tag{A.69}$$

Theorem A.23 (polar decomposition) *Any $n \times n$ matrix A of complex numbers can be written in the form*

$$A = UP = QV,\tag{A.70}$$

where U and V are unitary matrices, and $P := \sqrt{A^\dagger A}$ *and* $Q := \sqrt{AA^\dagger}$ *are positive-definite matrices.*

The above theorem has been stated for matrices. However, it is directly translated to linear operators. In other words, any linear operator $\hat{A}$ on a vector space $\mathcal{V}$ can be written in the form

$$\hat{A} = \hat{U}\hat{P} = \hat{Q}\hat{V},\tag{A.71}$$

where $\hat{P} := \sqrt{\hat{A}^\dagger \hat{A}}$ and $\hat{Q} := \sqrt{\hat{A}\hat{A}^\dagger}$ and positive operators, and $\hat{U}$ and $\hat{V}$ are unitary operators.

The polar decomposition is easily conceivable in terms of the singular-value decomposition. Suppose that

$$\hat{A} = \hat{U}'\hat{S}\hat{V}'^\dagger\tag{A.72}$$

is a singular-value decomposition of a linear operator $\hat{A}$. Then we observe that

$$\hat{A} = \hat{U}'\hat{V}'^\dagger\,\hat{V}'\hat{S}\hat{V}'^\dagger = (\hat{U}'\hat{V}'^\dagger)\sqrt{\hat{A}^\dagger\hat{A}},\tag{A.73}$$

which is of the first form in (A.71). Here recall from Sect. A.4.2 that for a positive operator $\hat{C}$ (such as $\hat{A}^\dagger\hat{A}$ and $\hat{A}\hat{A}^\dagger$) with spectral decomposition

$$\hat{C} = \sum_c |c\rangle\, c\, \langle c| \quad (c \geq 0),\tag{A.74}$$

the new operator $\sqrt{\hat{C}}$ is given by

$$\sqrt{\hat{C}} = \sum_c |c\rangle\, \sqrt{c}\, \langle c|.\tag{A.75}$$

Similarly, we rearrange the factors as

$$\hat{A} = \hat{U}'\hat{S}\hat{U}'^\dagger\,\hat{U}'\hat{V}'^\dagger = \sqrt{\hat{A}\hat{A}^\dagger}\,(\hat{U}'\hat{V}'^\dagger).\tag{A.76}$$

It is of the second form in (A.71).

A.6 Tensor-Product Spaces

When there are more than one systems, each system is associated with a different vector space. Even for a single system, independent degrees of freedom (such as external and internal degrees of freedom) are associated with different vector spaces.

Then the vector space of the total system should be constructed by the vector spaces associated with the individual systems or degrees of freedom. Mathematically, such a construction is called the tensor product of the constituent vector spaces.

A.6.1   Vectors in a Product Space

Let $\mathcal{V}$ and $\mathcal{W}$ be vector spaces, and $\{|v_i\rangle\}$ and $\{|w_j\rangle\}$ their respective bases. The tensor product $\mathcal{V} \otimes \mathcal{W}$ is a vector space spanned by the basis

$$\{|v_i\rangle \otimes |w_j\rangle : i = 1, \ldots, \dim \mathcal{V}; \; j = 1, \ldots, \dim \mathcal{W}\} \qquad (A.77)$$

We call it the *standard tensor-product basis* or simply *standard basis* of $\mathcal{V} \otimes \mathcal{W}$. Obviously, the dimension of $\mathcal{V} \otimes \mathcal{W}$ is given by $\dim(\mathcal{V} \otimes \mathcal{W}) = (\dim \mathcal{V})(\dim \mathcal{W})$. The symbol '$\otimes$' in the basis states separates $|v_i\rangle$ and $|w_j\rangle$ from each other clearly indicating which vector space they are from. In many cases, when there is no risk of confusion, it is dropped and the product states are simply written as $|v_i\rangle |w_j\rangle$ or even $|v_i w_j\rangle$.

Here is the standard basis (product basis) for a (unlabelled) two-qubit system.

```
In[·]:=  bs = Basis[2];
         bs // LogicalForm
Out[·]=  {|0, 0⟩, |0, 1⟩, |1, 0⟩, |1, 1⟩}
```

The Hermitian products $\langle \cdot, \cdot \rangle_\mathcal{V}$ and $\langle \cdot, \cdot \rangle_\mathcal{W}$ in the corresponding spaces are inherited to the tensor-product space to give the standard Hermitian product

$$((\langle v_i| \otimes \langle w_j|)(|v_k\rangle \otimes |w_l\rangle)) = \langle v_i|v_k\rangle_\mathcal{V} \langle w_j|w_l\rangle_\mathcal{W}. \qquad (A.78)$$

Expanded in the standard basis, a vector

$$|\Psi\rangle = \sum_{ij} |v_i\rangle \otimes |w_j\rangle \, \Psi_{ij} \in \mathcal{V} \times \mathcal{W} \qquad (A.79)$$

involves $M \times N$ terms in general, where $M := \dim \mathcal{V}$ and $N := dim\mathcal{W}$. It can be reduced to a form with much less terms. To see this, rewrite the matrix Ψ consisting of the expansion coefficients Ψ_{ij} by means of the singular-value decomposition (Theorem A.22)

$$\Psi = U \Sigma V^\dagger, \qquad (A.80)$$

where U is a $M \times M$ unitary matrix, V a $N \times N$ matrix, and Σ a $M \times N$ diagonal matrix. The diagonal elements s_j of Σ are all non-negative, and the number R of non-zero elements cannot be greater than $\min(M, N)$. Putting (A.80) back into (A.79) leads to the so-called *Schmidt decomposition*

$$|\Psi\rangle = \sum_{j=1}^{R} |\alpha_j\rangle \otimes |\beta_j\rangle s_j, \quad |\alpha_j\rangle := \sum_i |v_i\rangle U_{ij}, \quad |\beta_j\rangle := \sum_i |w_i\rangle V_{ij}^*. \quad (A.81)$$

Note that $\langle\alpha_i|\alpha_j\rangle = \delta_{ij}$ and $\langle\beta_i|\beta_j\rangle = \delta_{ij}$ because U and V are unitary, and that $\sum_{j=1}^{R} s_j^2 = 1$ ($0 < s_j < 1$) if $|\Psi\rangle$ is normalized.

The number R is called the *Schmidt rank* or *Schmidt number* of the vector $|\Psi\rangle$. When $R = 1$, $|\Psi\rangle$ is factorized as $|\Psi\rangle = |\alpha_1\rangle \otimes |\beta_1\rangle$, and is said to be *separable*. Otherwise, it cannot be factorized and it is called an *entangled vector*. The Schmidt decomposition is a convenient method to test whether a vector is separable or entangled.

Consider an arbitrary vector in the tensor-product space.

In[ ]:= `cc = Re@RandomVector[4];`
`vec = bs.cc;`
`vec // LogicalForm`

Out[ ]:= $-0.392437 \,|0, 0\rangle + 0.309225 \,|0, 1\rangle - 0.352869 \,|1, 0\rangle + 0.733405 \,|1, 1\rangle$

This is its Schmidt decomposition and shows that the state vector is entangled.

In[ ]:= `{ww, uu, vv} = SchmidtDecomposition[vec, {1}, {2}];`
`ww // Normal`
`uu`
`vv`

Out[ ]:= $\{0.935711, 0.190978\}$

Out[ ]:= $\{0.504017 \,|0\rangle + 0.863694 \,|1\rangle, -0.863694 \,|0\rangle + 0.504017 \,|1\rangle\}$

Out[ ]:= $\{-0.537096 \,|0\rangle + 0.843521 \,|1\rangle, 0.843521 \,|0\rangle + 0.537096 \,|1\rangle\}$

`SchmidtForm` presents the Schmidt decomposition in a more intuitively-appealing form. For a thorough analysis of the result, use `SchmidtDecomposition`.

In[ ]:= `new = SchmidtForm[vec, {1}, {2}]`

Out[ ]:= $0.190978 \,(-0.863694 \,|0\rangle + 0.504017 \,|1\rangle) \otimes (0.843521 \,|0\rangle + 0.537096 \,|1\rangle) +$
$0.935711 \,(0.504017 \,|0\rangle + 0.863694 \,|1\rangle) \otimes (-0.537096 \,|0\rangle + 0.843521 \,|1\rangle)$

Check whether the two vectors are the same or not.

In[ ]:= `vec - ReleaseTimes[new] // Garner // Chop`

Out[ ]:= `0`

A.6.2 Operators on a Product Space

Let $\hat{A}$ and $\hat{B}$ be linear operators on $\mathcal{V}$ and $\mathcal{W}$, respectively. Then the *tensor-product operator* $\hat{A} \otimes \hat{B}$ is a linear operator on the tensor-product space $\mathcal{V} \otimes \mathcal{W}$ defined by the association

$$(\hat{A} \otimes \hat{B})(|v_i\rangle \otimes |w_j\rangle) = (\hat{A}|v_i\rangle) \otimes (\hat{B}|w_j\rangle). \tag{A.82}$$

Suppose that $\hat{A}$ and $\hat{B}$ are represented by matrices A and B, respectively, i.e.,

$$\hat{A}|v_j\rangle = \sum_i |v_i\rangle A_{ij}, \quad \hat{B}|v_j\rangle = \sum_i |w_i\rangle B_{ij}. \tag{A.83}$$

Then it follows from (A.82) that

$$(\hat{A} \otimes \hat{B})(|v_i\rangle \otimes |w_j\rangle) = \left(\sum_k |v_k\rangle A_{ki}\right) \otimes \left(\sum_l |w_l\rangle B_{lj}\right) = \sum_{kl} |v_k\rangle \otimes |w_l\rangle A_{ki} B_{lj}. \tag{A.84}$$

This implies that the matrix representation of $\hat{A} \otimes \hat{B}$ in the standard product basis is given by the *direct product* $A \otimes B$ of the two matrices.

In general, a linear operator $\hat{C}$ on $\mathcal{V} \otimes \mathcal{W}$ is not a single product but a sum of such products. How many terms are there? Suppose that the matrix $C_{ij;kl}$ is the matrix representation of $\hat{C}$ in the standard product basis,

$$\hat{C}|v_k w_l\rangle = \sum_{ij} |v_i v_j\rangle C_{ij;kl}, \tag{A.85}$$

or equivalently,

$$\hat{C} = \sum_{ij;kl} |v_i w_j\rangle \langle v_k w_l| \, C_{ij;kl} = \sum_{ij;kl} |v_i\rangle \langle v_k| \otimes |w_j\rangle \langle w_l| \, C_{ij;kl}. \tag{A.86}$$

The $M^2 \times N^2$ matrix $G_{ik;jl} := C_{ij;kl}$ with collective indices (ik) and (jl) has a singular value decomposition

$$G_{ik;jl} = \sum_{\mu} U_{ik;\mu} \gamma_\mu V_{\mu;jl}^\dagger, \quad \gamma_\mu \ge 0. \tag{A.87}$$

Defining

$$\hat{A}_\mu := \sum_{ik} |v_i\rangle \langle v_k| \, U_{ik;\mu}, \quad \hat{B}_\mu := \sum_{jl} |w_j\rangle \langle w_l| \, V_{jl;\mu}^*, \tag{A.88}$$

leads to the expression

$$\hat{C} = \sum_{\mu} \hat{A}_\mu \otimes \hat{B}_\mu \gamma_\mu, \tag{A.89}$$

which is in direct analogy with the Schmidt decomposition (A.81) of a vector in the tensor-product space. Here the number of non-vanishing singular values γ_μ is less than or equal to $\min(M^2, N^2)$.

Appendix B
Superoperators

A superoperator is a linear operator acting on a vector space of linear operators. As the concept of vectors is completely general, at a first glance there seems to be no reason why one should reserve a distinctive name for such operators and devote additional discussions. A considerable amount of interest in superoperators came with the booming of quantum information theory in the 1990s when it became clear that superoperators are important in the study of entanglement. Since then, mathematical theories on superoperators have been developed at a notably fast pace and have been applied to a wide range of subjects in quantum computation and quantum information. In this appendix, we briefly survey the properties of superoperators and provide some mathematical tools for the studies of entanglement and decoherence (see Chap. 5).

B.1 Operators as Vectors

The addition of two operators acting on a vector space as well as the multiplication of an operator by a scalar are defined in a natural and straightforward way. That is, operators on a vector space form a vector space themselves. The vector space of operators is not merely a mathematical generalization, but has an important physical relevance. In quantum physics, a mixed state, a statistical mixture of pure quantum states, is described by a so-called density operator.

There is another rather mathematically motivated and yet physically important fact that makes regarding operators as vectors very useful: Any unitary operator $\hat{U}$ can be written in the form $\hat{U} = \exp(i\hat{H})$, where $\hat{H}$ is an Hermitian operator. To describe any physical process, one has to deal with unitary operators. Because of the defining constraint, $\hat{U}^\dagger\hat{U} = \hat{I}$, it is often difficult to directly handle unitary operators. In most case, it is much more convenient and easier to handle Hermitian operators and to consider the exponential function of them. As there is no constraint—apart from the rather trivial Hermiticity condition—for Hermitian operators, it is natural to

© The Editor(s) (if applicable) and The Author(s), under exclusive license to Springer Nature Switzerland AG 2022, corrected publication 2023
M.-S. Choi, *A Quantum Computation Workbook*,
https://doi.org/10.1007/978-3-030-91214-7

express them as linear combinations of some basis elements. It makes the handling of physically relevant operators much more tractable.

In this appendix, we will discuss the general structure of the vector space of all linear operators on a vector space. Before we discuss more general vector spaces of linear operators, let us first consider vector spaces of matrices.

Exercise B.1 (*matrices as vectors*) Consider the set $\mathcal{M}_n$ of all $n \times n$ complex matrices.

(a) Show that $\mathcal{M}_n$ is a vector space.
(b) What is the dimension of $\mathcal{M}_n$.
(c) Define a *Hermitian product* in $\mathcal{M}_n$.
(d) Construct an *orthogonal basis* of $\mathcal{M}_n$.

Here is an even more specific example.

Example B.2 (*Pauli decomposition*) Consider $\mathcal{M}_2$.

(a) Show that the four Pauli matrices $\hat{\sigma}^\mu$ ($\mu = 0, 1, 2, 3$) form an orthogonal basis.
(b) Given an arbitrary matrix $L \in \mathcal{M}_2$, expand it in terms of the Pauli matrices. That is, find the most general form of L in terms of the Pauli matrices.
(c) Find the most general form of a Hermitian matrix $H \in \mathcal{M}_2$.
(d) Find the most general form of a unitary matrix $U \in \mathcal{M}_2$.

Let us demonstrate that any 2×2 matrix can be written as a linear superposition of the Pauli matrices. Consider an arbitrary 2×2 matrix.

```
In[ ]:= L = {{1, 2 I},
            {-I, 3}};
        L // MatrixForm
Out[ ]//MatrixForm=
        ( 1   2 i )
        ( -i   3  )
```

ExpressionFor converts a matrix into an operator expression in terms of the Pauli operators -- the Pauli operators are the operator forms of the Pauli matrices.

```
In[ ]:= op = ExpressionFor[L]
        Elaborate[op]
Out[ ]= 2 σ⁰ - σᶻ + 2 i σ⁺ - i σ⁻
Out[ ]= 2 σ⁰ + (i σˣ)/2 - (3 σʸ)/2 - σᶻ
```

The symbols σ^μ in the above are the displayed form of Pauli.

```
In[ ]:= InputForm[op]
Out[ ]//InputForm=
        2*Pauli[0] - Pauli[3] + (2*I)*Pauli[4] - I*Pauli[5]
```

ThePauli is the matrix form of Pauli. The following statement reconstructs the original matrix.

```
In[ ]:= new = 2 ThePauli[0] + (I / 2) ThePauli[1] - (3 / 2) ThePauli[2] - ThePauli[3];
        new // MatrixForm
Out[ ]//MatrixForm=
        ( 1   2 i )
        ( -i   3  )
```

The conversion of Pauli to the corresponding matrix -- ThePauli -- can be achieved by simply using Matrix.

```
In[·]:= new2 = Matrix[op];
       new2 // MatrixForm
Out[·]//MatrixForm=
```
$$\begin{pmatrix} 1 & 2i \\ -i & 3 \end{pmatrix}$$

Let us analyse the above demonstration in more detail. Here are the Pauli matrices. They form a basis of $\mathcal{M}_2$.

```
In[·]:= bs = ThePauli /@ {0, 1, 2, 3};
       MatrixForm /@ bs
```
$$Out[·]= \left\{ \begin{pmatrix} 1 & 0 \\ 0 & 1 \end{pmatrix}, \begin{pmatrix} 0 & 1 \\ 1 & 0 \end{pmatrix}, \begin{pmatrix} 0 & -i \\ i & 0 \end{pmatrix}, \begin{pmatrix} 1 & 0 \\ 0 & -1 \end{pmatrix} \right\}$$

The function PauliDecompose returns the expansion coefficients in the Pauli basis.

```
In[·]:= cc = PauliDecompose[L]
```
$$Out[·]= \left\{ 2, \frac{i}{2}, -\frac{3}{2}, -1 \right\}$$

Indeed, the coefficients reconstructs the original matrix.

```
In[·]:= new = cc.bs;
       new // MatrixForm
Out[·]//MatrixForm=
```
$$\begin{pmatrix} 1 & 2i \\ -i & 3 \end{pmatrix}$$

```
In[·]:= L - new // Chop
Out[·]= {{0, 0}, {0, 0}}
```

Let us further consider a 2×2 Hermitian matrix.

```
In[·]:= H = RandomHermitian[];
       H // MatrixForm
Out[·]//MatrixForm=
```
$$\begin{pmatrix} 0.558658 + 0. i & -0.89873 - 0.66087 i \\ -0.89873 + 0.66087 i & 0.120512 + 0. i \end{pmatrix}$$

```
In[·]:= H - Topple[H] // Chop // MatrixForm
Out[·]//MatrixForm=
```
$$\begin{pmatrix} 0 & 0 \\ 0 & 0 \end{pmatrix}$$

It is noted that the expansion coefficients are all real.

```
In[·]:= cc = PauliDecompose[H] // Chop
Out[·]= {0.339585, -0.89873, 0.66087, 0.219073}
```

Finally, consider a 2×2 unitary matrix.

```
In[·]:= U = RandomUnitary[];
       U // MatrixForm
Out[·]//MatrixForm=
```
$$\begin{pmatrix} -0.35467 + 0.843542 i & -0.196382 - 0.352251 i \\ -0.116783 + 0.386016 i & 0.526328 + 0.748553 i \end{pmatrix}$$

```
In[·]:= Topple[U].U // Chop // MatrixForm
Out[·]//MatrixForm=
```
$$\begin{pmatrix} 1. & 0 \\ 0 & 1. \end{pmatrix}$$

One can see that the column vector of the expansion coefficients is normalized.

```
In[·]:= cc = PauliDecompose[U]
Out[·]= {0.085829 + 0.796047 i, -0.156582 + 0.0168825 i,
         0.369134 - 0.0397996 i, -0.440499 + 0.0474942 i}
```

```
In[·]:= Conjugate[cc].cc // Chop
Out[·]= 1.
```

In the above demonstration, we have used the Pauli operators for *unlabelled* qubits. One could use the Pauli operators for qubits with labels. Let us consider a system of two qubits, which are denoted by the symbol S.

```
Let[Qubit, S]
```

Consider an arbitrary 2×2 matrix.

```
In[·]:= mat = RandomInteger[{-3, 3}, {2, 2}];
        mat // MatrixForm
Out[·]//MatrixForm=
```
$$\begin{pmatrix} 1 & 0 \\ 1 & -1 \end{pmatrix}$$

This converts the matrix into an operator expression in terms of the Pauli operators on the labelled qubits. Here S[μ] corresponds to Pauli[μ] acting on the the qubit S[None].

```
In[·]:= op = Elaborate@ExpressionFor[mat, S[None]]
Out[·]=
```
$$\frac{S^x}{2} - \frac{i\, S^y}{2} + S^z$$

The operator expression can be converted back to a matrix by using Matrix.

```
In[·]:= new = Matrix[op];
        new // MatrixForm
Out[·]//MatrixForm=
```
$$\begin{pmatrix} 1 & 0 \\ 1 & -1 \end{pmatrix}$$

Definition B.1 (*vector space of linear maps*) Let $\mathcal{V}$ and $\mathcal{W}$ be vector spaces. Let $\mathcal{L}(\mathcal{V}, \mathcal{W})$ be the set of all linear maps $\hat{L} : \mathcal{V} \to \mathcal{W}$. Equip it with a natural multiplication of linear map $\hat{L}$ by scalars $x \in \mathbb{C}$ as

$$(x\hat{L})\,|v\rangle := x(\hat{L}\,|v\rangle) \tag{B.1}$$

for all $|v\rangle \in \mathcal{V}$. Also define the sum of two linear maps by

$$(\hat{L} + \hat{M})\,|v\rangle := \hat{L}\,|v\rangle + \hat{M}\,|v\rangle \tag{B.2}$$

for all $|v\rangle \in \mathcal{V}$. Then the set $\mathcal{L}(\mathcal{V}, \mathcal{W})$ forms a vector space. When $\mathcal{V} = \mathcal{W}$, $\mathcal{L}(\mathcal{V}) := \mathcal{L}(\mathcal{V}, \mathcal{V})$ is the vector space of all linear *operators* on $\mathcal{V}$.

Let $\{|v_1\rangle, \ldots, |v_m\rangle\}$ and $\{|w_1\rangle, \ldots, |w_n\rangle\}$ be orthonormal bases of $\mathcal{V}$ and $\mathcal{W}$, respectively. A natural choice for basis of $\mathcal{L}(\mathcal{V}, \mathcal{W})$ is

$$\{|w_j\rangle\langle v_i| : i = 1, \ldots, m; \ j = 1, \ldots, n\}. \tag{B.3}$$

$\mathcal{L}(\mathcal{V}, \mathcal{W})$ also needs a Hermitian product. In the same spirit as the Hilbert-Schmidt inner product[2] of matrices, a natural choice of the Hermitian product in $\mathcal{L}(\mathcal{V}, \mathcal{W})$ inherited from the Hermitian products in $\mathcal{V}$ and $\mathcal{W}$ is that

$$\langle \hat{A}, \hat{B} \rangle := \mathrm{Tr}\hat{A}^\dagger\hat{B} = \sum_j \langle v_j, \hat{A}^\dagger\hat{B}v_j\rangle_\mathcal{V} = \sum_j \langle \hat{A}v_j, \hat{B}v_j\rangle_\mathcal{W}. \tag{B.4}$$

It is called the *Hilbert-Schmidt inner product* or *Frobenius inner product*. With respect to this Hermitian product, the basis in (B.3) is orthonormal.

For the vector space $\mathcal{L}(\mathcal{V})$ of all *operators* on $\mathcal{V}$, equipped with the Hilbert-Schmidt inner product in (B.4), another choice of basis other than the standard basis (B.3) is widely used. It is to pick the identity operator $\hat{I}$ as an element of the basis. Then every other element in the basis must be *traceless*. For example, let $\mathcal{S}$ be a two-dimensional Hilbert space, and in $\mathcal{L}(\mathcal{S})$ the four Pauli operators form such a basis, $\left\{\hat{I} \equiv \hat{S}^0, \hat{S}^x, \hat{S}^y, \hat{S}^z\right\}$. Obviously, the non-identity three Pauli operators are traceless, $\mathrm{Tr}\hat{S}^\mu = 0$ for $\mu = x, y, z$. Any operator $\hat{A} \in \mathcal{L}(\mathcal{S})$ is expanded in the four Pauli operators

$$\hat{A} = \hat{S}^0\alpha_0 + \hat{S}^x\alpha_x + \hat{S}^y\alpha_y + \hat{S}^z\alpha_z \quad (\alpha_\mu \in \mathbb{C}). \tag{B.5}$$

The expansion coefficients can be obtained using the orthogonality of the basis and the Hilbert-Schmidt inner product,

$$\alpha_\mu = \frac{1}{2}\mathrm{Tr}\hat{S}^\mu\hat{A} \tag{B.6}$$

(recall that the Pauli operators are all Hermitian). We have already observed this fact in Exercise B.2 using the matrix form of the Pauli operators.

Example B.3 Consider the Hilbert space $\mathcal{S} \otimes \mathcal{S}$ associated with a system of two qubits. Show that the products of the Pauli operators

$$\left\{\hat{S}_1^\mu \otimes \hat{S}_2^\nu\right\} \tag{B.7}$$

form an orthogonal basis of $\mathcal{L}(\mathcal{S} \otimes \mathcal{S})$.

Solution: Consider an arbitrary 4×4 matrix.

---

[2] In mathematics, it is often called the *Frobenius inner product*.

```
In[ ]:= mat = RandomInteger[{-3, 3}, {4, 4}];
        mat // MatrixForm
```

$$\begin{pmatrix} -2 & -1 & -1 & 1 \\ 1 & 0 & 1 & -1 \\ 2 & 2 & 0 & 2 \\ -2 & -3 & 0 & -3 \end{pmatrix}$$

This converts the matrix into an operator expression in terms of the products of the Pauli operators, which are represented by `Pauli[μ,ν,...]`.

```
In[ ]:= op = Elaborate@ExpressionFor[mat]
```

$$Out[ ]= -\frac{5}{4}\sigma^0\otimes\sigma^0 + \frac{\sigma^0\otimes\sigma^x}{2} + \frac{\sigma^0\otimes\sigma^z}{4} - \frac{3\sigma^x\otimes\sigma^0}{4} + \frac{\sigma^x\otimes\sigma^x}{2} + i\,\sigma^x\otimes\sigma^y + \frac{5\sigma^x\otimes\sigma^z}{4} - \frac{1}{4}i\,\sigma^y\otimes\sigma^0 +$$
$$\frac{1}{2}i\,\sigma^y\otimes\sigma^x + \sigma^y\otimes\sigma^y - \frac{5}{4}i\,\sigma^y\otimes\sigma^z + \frac{\sigma^z\otimes\sigma^0}{4} - \frac{\sigma^z\otimes\sigma^x}{2} - i\,\sigma^z\otimes\sigma^y - \frac{5\sigma^z\otimes\sigma^z}{4}$$

This converts the operator expression back into the original matrix.

```
In[ ]:= new = Matrix[op];
        new // MatrixForm
```

$$\begin{pmatrix} -2 & -1 & -1 & 1 \\ 1 & 0 & 1 & -1 \\ 2 & 2 & 0 & 2 \\ -2 & -3 & 0 & -3 \end{pmatrix}$$

In the above demonstration, we have used the Pauli operators for *unlabelled* qubits. One could use the Pauli operators for qubits with labels. Let us consider a system of two qubits, which are denoted by the symbol S and the flavor indices.

`Let[Qubit, S]`

This converts the matrix into an operator expression in terms of the Pauli operators on the labelled qubits. Here $S[1,\mu]$ corresponds to $Pauli[\mu,0]$ acting on the first qubit and $S[2,\mu]$ to $Pauli[0,\mu]$ on the second qubit.

```
In[ ]:= op = ExpressionFor[mat, S[{1, 2}, None]]
        Elaborate[op]
```

$$Out[ ]= -\frac{5}{4} - \frac{5}{4}S_1^z S_2^z - \frac{3}{2}S_1^z S_2^+ + \frac{1}{2}S_1^z S_2^- + S_1^+ S_2^+ + S_1^+ S_2^- +$$
$$\frac{5}{2}S_1^- S_2^z + 2 S_1^- S_2^+ - 2 S_1^- S_2^- + \frac{S_2^z}{4} - S_1^- - \frac{S_1^-}{2} + \frac{S_2^z}{4} + \frac{S_2^+}{2} + \frac{S_2^-}{2}$$

$$Out[ ]= -\frac{5}{4} + \frac{1}{2}S_1^x S_2^x + i\,S_1^x S_2^y + \frac{5}{4}S_1^x S_2^z + \frac{1}{2}i\,S_1^y S_2^x + S_1^y S_2^y -$$
$$\frac{5}{4}i\,S_1^y S_2^z - \frac{1}{2}S_1^z S_2^x - i\,S_1^z S_2^y - \frac{5}{4}S_1^z S_2^z - \frac{3S_1^x}{4} - \frac{i\,S_1^y}{4} + \frac{S_1^z}{4} + \frac{S_2^x}{2} + \frac{S_2^z}{4}$$

The operator expression can be converted back to a matrix by using Matrix.

```
In[ ]:= new = Matrix[op];
        new // MatrixForm
```

$$\begin{pmatrix} -2 & -1 & -1 & 1 \\ 1 & 0 & 1 & -1 \\ 2 & 2 & 0 & 2 \\ -2 & -3 & 0 & -3 \end{pmatrix}$$

B.2   Superoperators

As the operators on vector spaces are vectors themselves, one can consider a linear map $\mathscr{F} : \mathcal{L}(V) \to \mathcal{L}(W)$ from operators on V to those on W. We call it a *super mapping* or *supermap* to distinguish it from one between simple vectors. Physically, supermaps are most relevant when input operators represent mixed states, that is, when they are density operators.

In many cases, the input and output spaces are identical, $V = W$. In such a case, $\mathscr{F}$ itself is an operator—an operator on operators—and is called a *superoperator* on V. Superoperators are useful to mathematically describe the evolution of open quantum systems, i.e., systems interacting with other surrounding systems.

B.2.1   *Matrix Representation*

How can a supermap be characterized? Recall that a linear map of simple vectors is characterized by its matrix representation. Upon a choice of bases $\{|v_j\rangle\}$ and $\{|w_i\rangle\}$ of V and W, respectively, $\hat{L} : V \to W$ is completely specified by

$$\hat{L}\,|v_j\rangle = \sum_i |w_i\rangle\, L_{ij} \,. \tag{B.8}$$

For a supermap $\mathscr{F} : \mathcal{L}(V) \to \mathcal{L}(W)$, the involved spaces $\mathcal{L}(V)$ and $\mathcal{L}(W)$ have additional algebraic structures, and there are several ways to characterize it at different levels. One straightforward way to characterize a supermap is to take a plain analogy of the above matrix representation. Recall that $|v_k\rangle\langle v_l|$ and $|w_i\rangle\langle w_j|$ form the standard bases of $\mathcal{L}(V)$ and $\mathcal{L}(W)$, respectively. For each $|v_k\rangle\langle v_l|$, $\mathscr{F}(|v_k\rangle\langle v_l|)$ belongs to $\mathcal{L}(W)$ and is expanded in the standard basis $\{|w_i\rangle\langle w_j|\}$ [see Eq. (B.3)] as (notice the order of indices in $C_{ik;jl}$)

$$\mathscr{F}(|v_k\rangle\langle v_l|) = \sum_{ij} |w_i\rangle\langle w_j|\, C_{ik;jl} \,, \quad C_{ik;jl} \in \mathbb{C} \,. \tag{B.9}$$

Here the matrix C—regarding (ik) and (jl) as collective indices—is called the *Choi matrix* associated with the supermap $\mathscr{F}$, and it completely characterizes the supermap $\mathscr{F}$. For an arbitrary linear operator $\hat{\rho} := \sum_{kl} |v_k\rangle\langle v_l| \rho_{kl} \in \mathcal{L}(V)$, its image through $\mathscr{F}$ is given by

$$\hat{\sigma} := \mathscr{F}(\hat{\rho}) = \sum_{kl} \mathscr{F}(|v_k\rangle\langle v_l|)\rho_{kl} = \sum_{ij} \sum_{kl} |w_i\rangle\langle w_j|\, C_{ik;jl}\, \rho_{kl} \,. \tag{B.10}$$

This implies that the matrix elements (in the standard tensor-product basis) of $\hat{\sigma}$ and $\hat{\rho}$ are related to each other by the Choi matrix as

$$\sigma_{ij} = \sum_{kl} C_{ik;jl}\,\rho_{kl}. \tag{B.11}$$

We consider supermaps transforming operators of the form

```
In[ ]:= Let[Complex, ρ]
        rho = ρ@{0, 1, 2, 3}.S[Full]
Out[ ]= S⁰ ρ₀ + Sˣ ρ₁ + Sʸ ρ₂ + Sᶻ ρ₃
```

Here is a supermap, specified by a set of three Kraus elements.

```
In[ ]:= ops = {2 S[4], S[5], S[6]};
        spr = Supermap[ops]
Out[ ]= Supermap[{2 S⁺, S⁻, Sᴴ}]
```

Under the supermap, the operator rho transforms as follows.

```
In[ ]:= new = spr[rho] // Elaborate
Out[ ]= -Sʸ ρ₂ + 1/2 × (7 ρ₀ - 3 ρ₃) + Sᶻ ( 3 ρ₀/2 + ρ₁ - 5 ρ₃/2 ) + Sˣ ρ₃
```

This gives the Choi matrix corresponding to the supermap.

```
In[ ]:= tsr = ChoiMatrix[spr];
        Dimensions@tsr
Out[ ]= {2, 2, 2, 2}
```

The Choi matrix provides a matrix representation of the supermap. This is illustrated by the transformation of the elements of the matrix representation of the operator rho.

```
In[ ]:= new2 = TensorContract[TensorProduct[tsr, Matrix@rho], {{2, 5}, {4, 6}}];
        new2 // Simplify // MatrixForm
Out[ ]//MatrixForm=
        ( 5 ρ₀ + ρ₁ - 4 ρ₃      i ρ₂ + ρ₃    )
        ( -i ρ₂ + ρ₃          2 ρ₀ - ρ₁ + ρ₃ )
```

```
In[ ]:= new - ExpressionFor[new2, S] // Elaborate
Out[ ]= 0
```

Let us take a few examples: First, consider a supermap of the simplest form

$$\mathcal{F}(\hat{\rho}) = \hat{A}\hat{\rho}\hat{B}^{\dagger}, \tag{B.12}$$

where $\hat{A}, \hat{B} \in \mathcal{L}(\mathcal{V}, \mathcal{W})$. Then the Choi matrix of $\mathcal{F}$ is given by

$$C_{ij;kl} = A_{ij}B_{kl}^{*}. \tag{B.13}$$

Next, consider a supermap of the form

$$\mathcal{F}(\hat{\rho}) = -i[\hat{H}, \hat{\rho}]. \tag{B.14}$$

It is the coherent part of the Lindblad equation (see Sect. 5.4). One can use the result in (B.13) putting either $\hat{A} = \hat{I}$ or $\hat{B} = \hat{I}$. It immediately follows that the Choi matrix for the supermap specified in (B.14)

$$C_{ij;kl} = -i\left(H_{ij}\delta_{kl} - \delta_{ij}H_{kl}^*\right). \tag{B.15}$$

B.2.2 Operator-Sum Representation

Another method to characterize a supermap is the so-called operator-sum representation, and it turns out to be extremely useful in many areas of physics, including quantum information theory and quantum statistical mechanics. Putting the identity $\rho_{kl} = \langle v_k|\hat{\rho}|v_l\rangle$ back into (B.10), one gets

$$\mathscr{F}(\hat{\rho}) = \sum_{ij}\sum_{kl} |w_i\rangle \langle v_k|\hat{\rho}|v_l\rangle \langle w_j| \, C_{ik;jl}. \tag{B.16}$$

Now identify $\hat{E}_{ik} := |w_i\rangle \langle v_k|$ as a linear map from $\mathcal{V}$ to $\mathcal{W}$, $\hat{E}_{ik} \in \mathcal{L}(\mathcal{V}, \mathcal{W})$. Similarly, $|v_l\rangle \langle w_j| \in \mathcal{L}(\mathcal{W}, \mathcal{V})$ and it is identical to $\hat{E}_{jl}^\dagger$. Hence

$$\mathscr{F}(\hat{\rho}) = \sum_{ij}\sum_{kl} \hat{E}_{ik}\hat{\rho}\hat{E}_{jl}^\dagger C_{ik;jl}. \tag{B.17'}$$

With the notation of collective indices $\mu \equiv (ik)$ and $\nu \equiv (jl)$, the supermap $\mathscr{F}$ takes the operator-sum representation

$$\mathscr{F}(\hat{\rho}) = \sum_{\mu=1}^{MN}\sum_{\nu=1}^{MN} \hat{E}_\mu \hat{\rho} \hat{E}_\nu^\dagger C_{\mu\nu}, \tag{B.17}$$

where $M := \dim \mathcal{V}$ and $N := \dim \mathcal{W}$. Diagrammatically, it is depicted as

$$
\begin{array}{ccc}
\mathcal{V} & \xleftarrow{\hat{E}_\nu^\dagger} & \mathcal{W} \\
\downarrow{\scriptstyle\hat{\rho}} & & \downarrow{\scriptstyle\mathscr{F}(\hat{\rho})} \\
\mathcal{V} & \xrightarrow{\hat{E}_\mu} & \mathcal{W}
\end{array}
\tag{B.18}
$$

In Eqs. (B.17) and (B.17'), a standard basis $\left\{\hat{E}_\mu\right\}$ has been chosen in $\mathcal{L}(\mathcal{V}, \mathcal{W})$. But one can choose any basis, which leads to the following theorem.

Theorem B.2 *Let $\mathcal{V}$ and $\mathcal{W}$ be vector spaces. If $\mathscr{F} : \mathcal{L}(\mathcal{V}) \to \mathcal{L}(\mathcal{W})$ is a supermap, then there exist $\hat{F}_\mu \in \mathcal{L}(\mathcal{V}, \mathcal{W})$ such that*

$$\mathscr{F}(\hat{\rho}) = \sum_{\mu=1}^{MN} \sum_{\nu=1}^{MN} \hat{F}_\mu \hat{\rho} \hat{F}_\nu^\dagger \, C_{\mu\nu}, \quad C_{\mu\nu} \in \mathbb{C}. \tag{B.19}$$

Consider a supermap specified by a set of operator and a matrix of coefficients.

```
In[·]:= ops = {I, S[1], S[2], S[3]}
Out[·]= {i, Sˣ, Sʸ, Sᶻ}
```

```
In[·]:= Let[Complex, c]
        cc = Array[c, {4, 4}];
        cc // MatrixForm
Out[·]//MatrixForm=
```
$$\begin{pmatrix} c_{1,1} & c_{1,2} & c_{1,3} & c_{1,4} \\ c_{2,1} & c_{2,2} & c_{2,3} & c_{2,4} \\ c_{3,1} & c_{3,2} & c_{3,3} & c_{3,4} \\ c_{4,1} & c_{4,2} & c_{4,3} & c_{4,4} \end{pmatrix}$$

Here is the supermap. It represents a map from operators to other operators.

```
In[·]:= spr = Supermap[ops, cc]
Out[·]= Supermap[{i, Sˣ, Sʸ, Sᶻ}, {{c_{1,1}, c_{1,2}, c_{1,3}, c_{1,4}},
        {c_{2,1}, c_{2,2}, c_{2,3}, c_{2,4}}, {c_{3,1}, c_{3,2}, c_{3,3}, c_{3,4}}, {c_{4,1}, c_{4,2}, c_{4,3}, c_{4,4}}}]
```

```
In[·]:= Let[Complex, ρ]
        rho = ρ[[{0, 1, 2, 3}]].S[Full]
Out[·]= S⁰ ρ₀ + Sˣ ρ₁ + Sʸ ρ₂ + Sᶻ ρ₃
```

This is the result acting the supermap on the above operator.

```
In[·]:= new = spr[rho]
```
$$\begin{aligned}
Out[\cdot]= \; & c_{2,2}\,\rho_0 + c_{3,3}\,\rho_0 + c_{4,4}\,\rho_0 + c_{1,1}\,S^0\,\rho_0 + i\,c_{1,2}\,\rho_1 - i\,c_{2,1}\,\rho_1 - i\,c_{3,4}\,\rho_1 + i\,c_{4,3}\,\rho_1 + \\
& i\,c_{1,3}\,\rho_2 + i\,c_{2,4}\,\rho_2 - i\,c_{3,1}\,\rho_2 - i\,c_{4,2}\,\rho_2 + i\,c_{1,4}\,\rho_3 - i\,c_{2,3}\,\rho_3 + i\,c_{3,2}\,\rho_3 - i\,c_{4,1}\,\rho_3 + \\
& S^x\,(i\,c_{1,2}\,\rho_0 - i\,c_{2,1}\,\rho_0 + i\,c_{3,4}\,\rho_0 - i\,c_{4,3}\,\rho_0 + c_{1,1}\,\rho_1 + c_{2,2}\,\rho_1 - c_{3,3}\,\rho_1 - c_{4,4}\,\rho_1 - \\
& \quad c_{1,4}\,\rho_2 + c_{2,3}\,\rho_2 + c_{3,2}\,\rho_2 - c_{4,1}\,\rho_2 + c_{1,3}\,\rho_3 + c_{2,4}\,\rho_3 + c_{3,1}\,\rho_3 + c_{4,2}\,\rho_3) + \\
& S^y\,(i\,c_{1,3}\,\rho_0 - i\,c_{2,4}\,\rho_0 - i\,c_{3,1}\,\rho_0 + i\,c_{4,2}\,\rho_0 + c_{1,4}\,\rho_1 + c_{2,3}\,\rho_1 + c_{3,2}\,\rho_1 + c_{4,1}\,\rho_1 + \\
& \quad c_{1,1}\,\rho_2 - c_{2,2}\,\rho_2 + c_{3,3}\,\rho_2 - c_{4,4}\,\rho_2 - c_{1,2}\,\rho_3 - c_{2,1}\,\rho_3 + c_{3,4}\,\rho_3 + c_{4,3}\,\rho_3) + \\
& S^z\,(i\,c_{1,4}\,\rho_0 + i\,c_{2,3}\,\rho_0 - i\,c_{3,2}\,\rho_0 - i\,c_{4,1}\,\rho_0 - c_{1,3}\,\rho_1 + c_{2,4}\,\rho_1 - c_{3,1}\,\rho_1 + c_{4,2}\,\rho_1 + \\
& \quad c_{1,2}\,\rho_2 + c_{2,1}\,\rho_2 + c_{3,4}\,\rho_2 + c_{4,3}\,\rho_2 + c_{1,1}\,\rho_3 - c_{2,2}\,\rho_3 - c_{3,3}\,\rho_3 + c_{4,4}\,\rho_3)
\end{aligned}$$

We are often interested in mapping density operators—not just any operators. In this case, the relevant supermaps are required to preserve the properties of density operators—density operators are *Hermitian* and in particular *positive* (Definition A.19). The condition to preserve Hermiticity simplifies the representation (B.17) further.

Theorem B.3 *Let $\mathcal{V}$ and $\mathcal{W}$ be vector spaces, equipped with Hermitian products. Let $\mathscr{F} : \mathcal{L}(\mathcal{V}) \to \mathcal{L}(\mathcal{W})$ be a supermap. If $\mathscr{F}(\hat{\rho})$ is Hermitian for every Hermitian $\hat{\rho} \in \mathcal{L}(\mathcal{V})$, then there exist $\hat{F}_\mu \in \mathcal{L}(\mathcal{V}, \mathcal{W})$ such that*

$$\mathscr{F}(\hat{\rho}) = \sum_{\mu=1}^{MN} \epsilon_\mu \hat{F}_\mu \hat{\rho} \hat{F}_\mu^\dagger, \tag{B.20}$$

where $\epsilon_\mu = \pm 1$ and the linear maps $\hat{F}_\mu$ are orthogonal to each other, $Tr\hat{F}_\mu^\dagger \hat{F}_\nu = 0$ for $\mu \neq \nu$.

Consider a supermap specified by a set of operator and a vector of coefficients.

```
In[·]:= ops = {I, S[1], S[2], S[3]}
Out[·]= {i, S^x, S^y, S^z}
```

```
In[·]:= Let[Real, c]
        cc = Array[c, 4];
        cc // MatrixForm
```
Out[·]//MatrixForm=
$$\begin{pmatrix} c_1 \\ c_2 \\ c_3 \\ c_4 \end{pmatrix}$$

Here is the supermap. It represents a map from operators to other operators.

```
In[·]:= spr = Supermap[ops, cc]
Out[·]= Supermap[{i, S^x, S^y, S^z}, {c_1, c_2, c_3, c_4}]
```

```
In[·]:= Let[Complex, ρ]
        rho = ρ[{0, 1, 2, 3}].S[Full]
Out[·]= S^0 ρ_0 + S^x ρ_1 + S^y ρ_2 + S^z ρ_3
```

This is the result acting the supermap on the above operator.

```
In[·]:= new = spr[rho]
Out[·]= (c_2 + c_3 + c_4) ρ_0 + c_1 S^0 ρ_0 + (c_1 + c_2 - c_3 - c_4) S^x ρ_1 +
        (c_1 - c_2 + c_3 - c_4) S^y ρ_2 + (c_1 - c_2 - c_3 + c_4) S^z ρ_3
```

In the representation (B.20), all numerical factors have been absorbed into the operators $\hat{F}_\mu$ leaving only possibly negative signs in ϵ_μ. An immediate question is, what condition should a supermap satisfy to have $\epsilon_\mu = 1$ for all μ? Would the condition to preserve positivity be sufficient to guarantee it? Unfortunately, the positivity-preserving condition does not bring any meaningful simplification, and a much stronger condition is required:

Definition B.4 (*completely positive supermap*) Let V and W be vector spaces and $\mathscr{F} : \mathcal{L}(V) \to \mathcal{L}(W)$ a supermap. $\mathscr{F}$ is said to be *completely positive* if $\mathscr{F} \otimes \mathscr{I} : \mathcal{L}(V \otimes \mathcal{E}) \to \mathcal{L}(W \otimes \mathcal{E})$ is positive[3] for any vector space $\mathcal{E}$, where $\mathscr{I}$ denotes the identity superoperator on $\mathcal{L}(\mathcal{E})$.

Physically, the vector space $\mathcal{E}$ is associated with an environment (Sect. 5.2). $\mathscr{F} \otimes \mathscr{I}$ acts non-trivially only on V associated with the system but trivially on $\mathcal{E}$. To be physically meaningful, $\mathscr{F} \otimes \mathscr{I}$ is expected preserve the properties, especially positivity, of density operators $\hat{\rho}$ on $V \otimes \mathcal{E}$. Note that $\hat{\rho}$ may contain a considerable amount of entanglement due to prior interactions between the system and the environment.

---

[3] Here "positive" actually means "positivity-preserving".

An important example of a supermap that is *not* completely positive is *transposition*. For an operator $\hat{A} = \sum_{ij} |v_i\rangle A_{ij} \langle v_j|$ on $\mathcal{V}$, the transposition $\hat{A}^T$ of $\hat{A}$ is defined by the matrix transposition of A,

$$\hat{A}^T := \sum_{ij} |v_i\rangle A_{ji} \langle v_j| = \sum_{ij} |i\rangle A_{ij}^T \langle v_j|. \tag{B.21}$$

The corresponding superoperator $\mathscr{F}$ is thus given by

$$\mathscr{F}(|v_i\rangle \langle v_j|) := |v_j\rangle \langle v_i|. \tag{B.22}$$

Consider an entangled state $|\Phi\rangle = |0\rangle \otimes |0\rangle + |1\rangle \otimes |1\rangle$ in $\mathcal{V} \otimes \mathcal{E}$. Clearly,

$$|\Phi\rangle \langle \Phi| = |00\rangle \langle 00| + |00\rangle \langle 11| + |11\rangle \langle 00| + |11\rangle \langle 11| \tag{B.23}$$

is positive (Definition A.19). Now let us inspect

$$(\mathscr{F} \otimes \mathscr{I})(|\Phi\rangle \langle \Phi|) = |00\rangle \langle 00| + |10\rangle \langle 01| + |01\rangle \langle 10| + |11\rangle \langle 11|. \tag{B.24}$$

The superoperator $\mathscr{F} \otimes \mathscr{I}$ corresponds to partial transposition (further details to be discussed in Appendix B.4). The matrix representation in the tensor-product basis

$$(\mathscr{F} \otimes \mathscr{I})(|\Phi\rangle \langle \Phi|) \doteq \begin{bmatrix} 1 & & & \\ & 0 & 1 & \\ & 1 & 0 & \\ & & & 1 \end{bmatrix} \tag{B.25}$$

reveals that the resulting operator has an eigenvalue -1 and cannot be positive by Theorem A.20. This concludes that transposition is not a completely positive superoperator.

The operator-sum representation in (B.19) or (B.20) is further simplified for completely positive supermaps. The following example exhibits the motivation.

Exercise B.4 For any linear maps $\hat{F}_\mu \in \mathcal{L}(\mathcal{V}, \mathcal{W})$, the supermap $\mathscr{F} : \mathcal{L}(\mathcal{V}) \to \mathcal{L}(\mathcal{W})$ defined by

$$\mathscr{F}(\hat{\rho}) := \sum_\mu \hat{F}_\mu \hat{\rho} \hat{F}_\mu^\dagger \tag{B.26}$$

is completely positive.

Note that the linear maps $\hat{F}_\mu$ in (B.26) are completely arbitrary. They do not have to be orthogonal to each other, $\mathrm{Tr}\hat{F}_\mu^\dagger \hat{F}_\nu \neq 0$, nor to span the space $\mathcal{L}(\mathcal{V}, \mathcal{W})$. The following theorem confirms that any completely positive supermap takes the above form in Eq. (B.26). In fact, for a given supermap, one can find more compact and refined linear maps to represent it with.

Theorem B.5 (Kraus representation theorem) *Let V and W be vector spaces, and* $\mathscr{F} : L(V) \to L(W)$ *be a supermap. Then the following statement are equivalent:*

(a) $\mathscr{F}$ *is completely positive.*
(b) For any $\hat{\rho} \in L(V)$, the effect $\mathscr{F}(\hat{\rho})$ can be written as

$$\mathscr{F}(\hat{\rho}) = \sum_{\mu=0}^{m-1} \hat{F}_\mu \hat{\rho} \hat{F}_\mu^\dagger , \tag{B.27}$$

where $m \leq (\dim V)(\dim W)$ and $\hat{F}_\mu : V \to W$ are mutually orthogonal linear maps—$\mathrm{Tr}\hat{F}_\mu^\dagger \hat{F}_\nu = 0$ for all $\mu \neq \nu$.
(c) For any $\hat{\rho} \in L(V)$, the effect $\mathscr{F}(\hat{\rho})$ can be written as a finite sum of the form

$$\mathscr{F}(\hat{\rho}) = \sum_\mu \hat{F}_\mu \hat{\rho} \hat{F}_\mu^\dagger , \tag{B.28}$$

where $\hat{F}_\mu : V \to W$ are (arbitrary) linear maps.

The expressions (B.27) and (B.28) are called the *Kraus operator-sum representation* or simply the *Kraus representation* of the completely positive supermap $\mathscr{F}$. The linear maps $\hat{F}_\mu$ are called the *Kraus elements* or the *Kraus maps* (the *Kraus operators* when $V = W$). The *orthogonal* Kraus elements in Eq. (B.27) are optimal in the sense that the sum has the least possible number of terms.

A completely positive supermap can be specified by a set of Kraus elements.

In[ ]:= `ops = { S[4], S[5], S[3]}`
Out[ ]:= $\{S^+, S^-, S^z\}$

Here is the supermap. It represents a map from operators to other operators.

In[ ]:= `spr = Supermap[ops]`
Out[ ]:= $\mathrm{Supermap}\left[\{S^+, S^-, S^z\}\right]$

In[ ]:= `Let[Complex, ρ]`
   `rho = ρ[{0, 1, 2, 3}].S[Full]`
Out[ ]:= $S^0 \, \rho_0 + S^x \, \rho_1 + S^y \, \rho_2 + S^z \, \rho_3$

This is the result acting the supermap on the above operator.

In[ ]:= `new = spr[rho]`
Out[ ]:= $2 \, \rho_0 - S^x \, \rho_1 - S^y \, \rho_2$

It is fairly obvious that a supermap expressed in the form (B.27) or (B.28) is completely positive. One can prove the converse starting from (B.17′).

Exercise B.5 Using the representation in (B.17') and requiring the positivity of $(\mathscr{F} \otimes \mathscr{I})(|\Phi\rangle \langle\Phi|)$ with

$$|\Phi\rangle := \sum_j |v_j\rangle \otimes |v_j\rangle \in \mathcal{V} \otimes \mathcal{V}, \tag{B.29}$$

prove that a completely positive map has the Kraus representation of the form (B.27).

B.2.3   Choi Isomorphism

A less widely known yet intriguing method to characterize a supermap is provided by the *Choi isomorphism* (also known as Jamiolkowski, Choi-Jamiolkowski or Jamiolkowski-Choi isomorphism).[4]

Before we discuss the Choi isomorphism of supermaps, let us first examine the same isomorphism of linear maps. Let $\hat{A} : \mathcal{V} \to \mathcal{W}$ be a linear map. Recall that it is completely characterized by the $n \times m$ matrix A such that

$$\hat{A} = \sum_{kj} |w_k\rangle A_{kj} \langle v_j|, \tag{B.30}$$

where $\{v_j\}$ and $|w_k\rangle$ are bases of $\mathcal{V}$ and $\mathcal{W}$, respectively. Now note that the same matrix defines a vector

$$|A\rangle := \sum_{kj} \langle w_k| \otimes |v_j\rangle A_{kj} \tag{B.31}$$

in the tensor-product space $\mathcal{W} \otimes \mathcal{V}$. The correspondence $\hat{A} \leftrightarrow |A\rangle$ turns out to be an isomorphism between $\mathcal{L}(\mathcal{V}, \mathcal{W})$ and $\mathcal{W} \otimes \mathcal{V}$. The isomorphism is called the Choi isomorphism and $|A\rangle$ is called the *Choi vector* associated with the linear map $\hat{A}$. Looking almost trivial at a first glance, the isomorphism brings about several interesting things. To see it, consider a maximally entangled state

$$|\Phi\rangle := \sum_k |v_k\rangle \otimes |v_k\rangle \in \mathcal{V} \otimes \mathcal{V}. \tag{B.32}$$

in the tensor-product space $\mathcal{V} \otimes \mathcal{V}$. First, observe that the Choi vector $|A\rangle$ of a linear map $\hat{A}$ is given by

$$|A\rangle = (\hat{A} \otimes \hat{I}) |\Phi\rangle . \tag{B.33}$$

This is depicted in the following quantum circuit

---

[4] There are subtle but important differences between the Choi and Jamiolkowski isomorphism; see Jiang et al. (2013).

$$|A\rangle = |\Phi\rangle \left\{ \boxed{\hat{A}} \right. \tag{B.34}$$

The isomorphism preserves the Hermitian products in $\mathcal{L}(V, W)$ and $W \otimes V$, that is, $\langle \hat{A}, \hat{B} \rangle = \mathrm{Tr}\hat{A}^\dagger \hat{B} = \langle A|B\rangle$ for all linear maps $\hat{A}$ and $\hat{B}$. Furthermore, for an arbitrary state $|\psi\rangle = \sum_j |v_j\rangle \psi_j \in V$, define its conjugate state by

$$|\psi^*\rangle := \sum_j |v_j\rangle \psi_j^*. \tag{B.35}$$

Then, it follows that

$$\hat{A} |\psi\rangle = \langle \psi^*| (\hat{A} \otimes \hat{I}) |\Phi\rangle = \langle \psi^*|A\rangle, \tag{B.36}$$

where the Hermitian product on the right-hand side is applied partially and only on V, and the remaining part is a vector belonging to W. In quantum circuit model, it is depicted as

$$\tag{B.37}$$

where the quantum circuit element —▷◁— represents the projection onto the state specified at the output port. Interestingly, the result is not affected whether the projection onto $|\psi^*\rangle$ is made before or after the operation $\hat{A}$. This does not violate any physical principle as the two parts in $V \otimes V$ are separated spacelike.

Now let us turn to the Choi isomorphism between supermaps and operators: Consider again the maximally entangled state in Eq. (B.32) and operate an extended supermap $\mathscr{F} \otimes \mathcal{I} : \mathcal{L}(V \otimes V) \to \mathcal{L}(W \otimes V)$ on $|\Phi\rangle \langle \Phi|$ to get

$$\hat{C}_{\mathscr{F}} := (\mathscr{F} \otimes \mathcal{I})(|\Phi\rangle \langle \Phi|) = \sum_{kl} \mathscr{F}(|v_k\rangle \langle v_l|) \otimes |v_k\rangle \langle v_l|$$

$$= \sum_{ij} \sum_{kl} |w_i v_k\rangle \langle w_j v_l| \, C_{ik;jl}, \tag{B.38}$$

where C is the Choi matrix of $\mathscr{F}$; see Eq. (B.9). In quantum circuit model, it reads as

$$\hat{C}_{\mathscr{F}} = |\Phi\rangle \left\{ \boxed{\mathscr{F}} \right. \tag{B.39}$$

Clearly $\hat{C}_{\mathscr{F}}$ is an operator (not a superoperator) on $\mathcal{W} \otimes \mathcal{V}$ and the Choi matrix C is nothing but its matrix representation in the standard tensor-product basis. Hence $\hat{C}_{\mathscr{F}}$ is called the *Choi operator* associated with $\mathscr{F}$. It turns out that the correspondence $\mathscr{F} \leftrightarrow \hat{C}_{\mathscr{F}}$ by means of (B.38) is one-to-one and an isomorphism.[5] For any state $|\psi\rangle \in \mathcal{V}$, the effect $\mathscr{F}(|\psi\rangle \langle \psi|)$ of supermap $\mathscr{F}$ on the pure state can be obtained by means of the conjugate state $|\psi^*\rangle$ [see Eq. (B.35)] as

$$\mathscr{F}(|\psi\rangle \langle \psi|) = \langle \psi^* | \hat{C}_{\mathscr{F}} |\psi^*\rangle \,, \tag{B.40}$$

or, more generally, for any $\hat{\rho} = \sum_{ij} |v_i\rangle \langle v_j| \rho_{ij}$

$$\mathscr{F}(\hat{\rho}) = \mathrm{Tr}_{\mathcal{V}} \hat{\rho}^* \hat{C}_{\mathscr{F}} \,, \tag{B.41}$$

where $\hat{\rho}^* := \sum_{ij} |v_i\rangle \langle v_j| \rho_{ij}^*$. This has been depicted in the quantum circuit

$$\tag{B.42}$$

Furthermore, taking the matrix representation of each entity in the relation (B.41) confirms the linear transformation rule (B.11) between the matrix elements of $\hat{\rho}$ and $\hat{\sigma} := \mathscr{F}(\hat{\rho})$ The transformation rule in (B.11) is another illustration of the Choi isomorphism.

The Choi operator plays a key role in the *gate teleportation* protocol, and it provides an interesting proof of the Kraus-representation theorem (see Sect. 5.2).

B.3 Partial Trace

The tensor product (Appendix A.6) extends vectors and operators. A partial trace is effectively an inverse procedure and reduces operators on a tensor-product space to one of the component spaces.

Consider an operator $\hat{T}$ on a tensor product space $\mathcal{V} \otimes \mathcal{W}$. The *partial trace* over the space $\mathcal{W}$ is defined by

$$\mathop{\mathrm{Tr}}_{\mathcal{W}} \hat{T} := \sum_{ijk} |v_i\rangle \langle v_i w_j| \hat{T} |v_k w_j\rangle \langle v_k| \,, \tag{B.43}$$

---

[5] Here we have just defined an association $\mathscr{F} \mapsto \hat{C}_{\mathscr{F}}$. Given an operator $\hat{C} \in \mathcal{L}(\mathcal{W} \otimes \mathcal{V})$, one can also find the corresponding supermap $\mathscr{F}_{\hat{C}}$, that is, the association $\hat{C} \to \mathscr{F}_{\hat{C}}$ in the reverse direction; see Størmer (2013).

where $\{|v_i\rangle\}$ and $\{|w_j\rangle\}$ are given bases of V and W, respectively. The partial trace over V is defined analogously. The procedure is said to *trace out* the vector space W, and the resulting operator is called a *reduced operator* of $\hat{T}$. The reduced operator acts on V. Often Eq. (B.43) is casually written as

$$\operatorname*{Tr}_{W} \hat{T} = \sum_{j} \langle w_j | \hat{T} | w_j \rangle. \tag{B.44}$$

It should be understood that on the right-hand side, the Hermitian product is applied partially and only on W. The expression remains to be an operator on V, rather than a complex number.

Given the matrix representation $T_{ij;kl}$ of $\hat{T}$ in the tensor-product basis

$$\hat{T} |v_k w_l\rangle = \sum_{ij} |v_i w_j\rangle T_{ij;kl}, \tag{B.45}$$

one can obtain the matrix representation of the reduced operator $\hat{A} := \operatorname{Tr}_W \hat{T}$ in the basis $\{|v_i\rangle\}$ by

$$A_{ik} = \sum_{j} T_{ij;kj} . \tag{B.46}$$

Although we have defined partial trace in a chosen basis, it does not depend on the basis.

Consider an operator A on a three-qubit Hilbert space. Suppose that an 8×8 matrix A is its matrix representation.

```
In[·]:= A = ThePauli[1, 0, 0] + ThePauli[0, 2, 0] + ThePauli[2, 0, 2] + ThePauli[3, 0, 3];
        A // MatrixForm
Out[·]//MatrixForm=
        ⎛  1   0  -i   0   1  -1   0   0 ⎞
        ⎜  0  -1   0  -i   1   1   0   0 ⎟
        ⎜  i   0   1   0   0   0   1  -1 ⎟
        ⎜  0   i   0  -1   0   0   1   1 ⎟
        ⎜  1   1   0   0  -1   0  -i   0 ⎟
        ⎜ -1   1   0   0   0   1   0  -i ⎟
        ⎜  0   0   1   1   i   0  -1   0 ⎟
        ⎝  0   0  -1   1   0   i   0   1 ⎠
```

This is the reduced matrix after tracing out the second and third qubits.

```
In[·]:= A1 = PartialTrace[A, {2, 3}];
        A1 // MatrixForm
Out[·]//MatrixForm=
        ⎛ 0  4 ⎞
        ⎝ 4  0 ⎠
```

This traces out the second qubit.

```
In[·]:= A13 = PartialTrace[A, {2}];
        A13 // MatrixForm
Out[·]//MatrixForm=
        ⎛  2   0   2  -2 ⎞
        ⎜  0  -2   2   2 ⎟
        ⎜  2   2  -2   0 ⎟
        ⎝ -2   2   0   2 ⎠
```

Consider again an operator on a system of three *labelled* qubits.

```
In[ ]:= op = S[1, 1] + S[2, 2] + S[1, 2] ** S[3, 2] + S[1, 3] ** S[3, 3]
```
$$Out[ ]= S_1^y S_3^y + S_1^z S_3^z + S_1^x + S_2^y$$

In a tensor-product form of the Pauli operators, it reads as follows.

```
In[ ]:= PauliForm[op]
```
$$Out[ ]= I \otimes Y \otimes I + X \otimes I \otimes I + Y \otimes I \otimes Y + Z \otimes I \otimes Z$$

This is the reduced operator after tracing out the second and third qubits. It acts on the first qubit.

```
In[ ]:= A1 = PartialTrace[op, S@{2, 3}] // Elaborate
```
$$Out[ ]= 4 S_1^x$$

This is the reduced operator after tracing out the second qubit. It acts on the first and third qubit.

```
In[ ]:= A13 = PartialTrace[op, S[2]] // Elaborate
```
$$Out[ ]= 2 S_1^y S_3^y + 2 S_1^z S_3^z + 2 S_1^x$$

Partial trace is a *completely positive* supermap (Definition B.4). By Theorem B.5, one can prove it simply by constructing an operator-sum representation. In accordance with the definition of the partial trace in (B.43), we define Kraus elements

$$\hat{F}_j := \sum_i |v_i\rangle \langle v_i w_j| . \tag{B.47}$$

Then it follows that

$$\operatorname*{Tr}_{\mathcal{W}} \hat{T} = \sum_j \hat{F}_j \hat{T} \hat{F}_j^\dagger . \tag{B.48}$$

B.4   Partial Transposition

We conclude this appendix with a rather unusual mathematical tool—the *partial transposition*. As the name suggests, it applies the matrix transposition to the part of the matrix representation of an operator that corresponds to a certain subsystem. The resulting matrix gives a new operator associated with it. Roughly speaking,[6] the partial transformation would correspond to a time-reversal transformation only on the subsystem.

Partial transposition has attracted considerable attention thanks to the seminal work on the separability test of mixed states of a composite quantum system by Peres (1996) and Horodecki et al. (1996). Ever since then, it has been widely used

---

[6] Rigorously speaking, this statement is wrong because no anti-unitary transformation such as time reversal can be applied on a subpart of the system.

to study the structure of the tensor-product space of composite systems concerning various entanglement properties.

Consider an operator $\hat{A}$ on a tensor product space $\mathcal{V} \otimes \mathcal{W}$ with the matrix representation

$$\hat{A} = \sum_{ij;kl} |v_i w_j\rangle \langle v_k w_l| A_{ij;kl} \tag{B.49}$$

in the standard tensor-product basis, $\{|v_i w_j\rangle \equiv |v_i\rangle \otimes |w_j\rangle\}$. The *partial transposition* $\hat{A}^{T_\mathcal{V}}$ of $\hat{A}$ with respect to the subspace $\mathcal{V}$ is defined by

$$\hat{A}^{T_\mathcal{V}} := \sum_{ij;kl} |v_i w_j\rangle \langle v_k w_l| A_{kj;il} . \tag{B.50}$$

Equivalently, we define

$$\langle v_i w_j| \hat{A}^{T_\mathcal{V}} |v_k w_l\rangle = \langle v_k w_j| \hat{A} |v_i w_l\rangle . \tag{B.51}$$

The partial transposition with respect to $\mathcal{W}$ is defined analogously. In a fixed basis, partial transposition is defined entirely by the matrix representations. For the above case,

$$A_{ij;kl}^{T_\mathcal{V}} = A_{kj;il} . \tag{B.52}$$

For example, consider a linear operator on two qubits with the matrix representation

$$\hat{A} \doteq \begin{bmatrix} A_{0,0;0,0} & A_{0,0;0,1} & A_{0,0;1,0} & A_{0,0;1,1} \\ A_{0,1;0,0} & A_{0,1;0,1} & A_{0,1;1,0} & A_{0,1;1,1} \\ \hline A_{1,0;0,0} & A_{1,0;0,1} & A_{1,0;1,0} & A_{1,0;1,1} \\ A_{1,1;0,0} & A_{1,1;0,1} & A_{1,1;1,0} & A_{1,1;1,1} \end{bmatrix} . \tag{B.53}$$

We have

$$\hat{A}^{T_\mathcal{V}} \doteq \begin{bmatrix} A_{0,0;0,0} & A_{0,0;0,1} & A_{1,0;0,0} & A_{1,0;0,1} \\ A_{0,1;0,0} & A_{0,1;0,1} & A_{1,1;0,0} & A_{1,1;0,1} \\ \hline A_{0,0;1,0} & A_{0,0;1,1} & A_{1,0;1,0} & A_{1,0;1,1} \\ A_{0,1;1,0} & A_{0,1;1,1} & A_{1,1;1,0} & A_{1,1;1,1} \end{bmatrix} \tag{B.54}$$

$$\hat{A}^{T_\mathcal{W}} \doteq \begin{bmatrix} A_{0,0;0,0} & A_{0,1;0,0} & A_{0,0;1,0} & A_{0,1;1,0} \\ A_{0,0;0,1} & A_{0,1;0,1} & A_{0,0;1,1} & A_{0,1;1,1} \\ \hline A_{1,0;0,0} & A_{1,1;0,0} & A_{1,0;1,0} & A_{1,1;1,0} \\ A_{1,0;0,1} & A_{1,1;0,1} & A_{1,0;1,1} & A_{1,1;1,1} \end{bmatrix} . \tag{B.55}$$

It is important to remember that the partial transposition is basis-dependent.

Consider an operator on two qubits.

```
In[ ]:= op = S[1, 1] ** S[2, 2] + 3 I S[1, 2] ** S[2, 3]
```
$$Out[ ]= S_1^x S_2^y + 3 i S_1^y S_2^z$$

Here is the partial transpose with respect to the second qubit.

```
In[ ]:= new = PartialTranspose[op, S[2]] // Elaborate
```
$$Out[ ]= -S_1^x S_2^y + 3 i S_1^y S_2^z$$

Here is the partial transposition with respect to the both qubits.

```
In[ ]:= new = PartialTranspose[op, S@{1, 2}] // Elaborate
```
$$Out[ ]= -S_1^x S_2^y - 3 i S_1^y S_2^z$$

It must coincide with the (overall) transposition. It is indeed the case as one can see below.

```
In[ ]:= ExpressionFor[Transpose@Matrix[op], S@{1, 2}] // Elaborate
```
$$Out[ ]= -S_1^x S_2^y - 3 i S_1^y S_2^z$$

Consider again the above demonstration, now in terms of matrix representations.

```
In[ ]:= mat = ThePauli[1, 2] + 3 * I * ThePauli[2, 3];
         mat // MatrixForm
```
Out[ ]//MatrixForm=
$$\begin{pmatrix} 0 & 0 & 3 & -i \\ 0 & 0 & i & -3 \\ -3 & -i & 0 & 0 \\ i & 3 & 0 & 0 \end{pmatrix}$$

Here is the partial transpose with respect to the second qubit.

```
In[ ]:= new = PartialTranspose[mat, {2}];
         new // MatrixForm
```
Out[ ]//MatrixForm=
$$\begin{pmatrix} 0 & 0 & 3 & i \\ 0 & 0 & -i & -3 \\ -3 & i & 0 & 0 \\ -i & 3 & 0 & 0 \end{pmatrix}$$

Here is the partial transposition with respect to the both qubits. It must coincide with the (overall) transposition. It is indeed the case as one can see below.

```
In[ ]:= new = PartialTranspose[mat, {1, 2}];
         new // MatrixForm
```
Out[ ]//MatrixForm=
$$\begin{pmatrix} 0 & 0 & -3 & i \\ 0 & 0 & -i & 3 \\ 3 & i & 0 & 0 \\ -i & -3 & 0 & 0 \end{pmatrix}$$

A noteworthy property of partial transposition is that the partial transposition of a positive operator is not always positive (Definition A.19). For example, consider an entangled state $|\Phi\rangle = |0\rangle \otimes |0\rangle + |1\rangle \otimes |1\rangle$ in $\mathcal{V} \otimes \mathcal{W}$, and an associate operator

$$\hat{A} := |\Phi\rangle \langle \Phi| = |00\rangle \langle 00| + |00\rangle \langle 11| + |11\rangle \langle 00| + |11\rangle \langle 11|. \tag{B.56}$$

Clearly, $\hat{A}$ is positive [see Eqs. (A.54) and (A.55)]. However, the partial transposition $\hat{A}^{T_{\mathcal{V}}}$

$$\hat{A}^{T_{\mathcal{V}}} = |00\rangle \langle 00| + |10\rangle \langle 01| + |01\rangle \langle 10| + |11\rangle \langle 11| \tag{B.57}$$

has a negative eigenvalue -1 as one can see from the matrix representation

$$\hat{A}^{T_V} \doteq \left[\begin{array}{c|c|c} 1 & & \\ \hline & \begin{array}{cc} 0 & 1 \\ 1 & 0 \end{array} & \\ \hline & & 1 \end{array}\right]. \tag{B.58}$$

By Theorem A.20, $\hat{A}^{T_V}$ cannot be positive. On the other hand, the partial transposition preserves the positivity of any operator of the form

$$\hat{A} = \sum_j p_j \hat{\rho}_j \otimes \hat{\sigma}_j, \tag{B.59}$$

where where $\hat{\rho}_j$ and $\hat{\sigma}_j$ are positive operators on $\mathcal{V}$ and $\mathcal{W}$, respectively, and p_j are non-negative. The operator in (B.59) is a convex linear superposition of products of positive operators. Such operators represent *separable* mixed states (Sect. 1.1.2). Consequently, partial transposition provides a convenient necessary condition for separability of mixed states. Unfortunately, the condition is not a sufficient condition for the separability. In other words, the partial transpositions of some entangled states are positive.

Partial transposition naturally arises when one regards transposition as a supermap. For an operator $\hat{A} = \sum_{ij} |v_i\rangle A_{ij} \langle v_j|$ on $\mathcal{V}$, the transposition $\hat{A}^T$ of $\hat{A}$ is defined by the matrix transposition of A,

$$\hat{A}^T := \sum_{ij} |v_i\rangle A_{ji} \langle v_j| = \sum_{ij} |i\rangle A_{ij}^T \langle v_j|. \tag{B.60}$$

The corresponding superoperator $\mathcal{F}$ is thus given by

$$\mathcal{F}(|v_i\rangle \langle v_j|) := |v_j\rangle \langle v_i|. \tag{B.61}$$

The extended superoperator $\mathcal{F} \otimes \mathcal{I}$ corresponds to partial transposition. As shown above, partial transposition does not preserve the positivity of an operator. Hence, transposition is not a completely positive supermap (Definition B.4).

Appendix C
Group Theory

Group theory is the study of algebraic structures called groups. It has emerged as an abstraction unifying ideas of number theory, geometry, and the theory of algebraic equations. It forms the core part of abstract algebra.

The notion of group suits well the physical conception of symmetry, and group theory has provided valuable theoretical tools to exploit the symmetry of physical problems. Crystallography was the first in physics to use group theory extensively. With the advent of quantum mechanics, group theory has occupied a key position at the center stage of physical theories in various areas ranging from condensed matter physics to high-energy physics.

In quantum information and related areas, the use of group theory is not restricted to symmetry considerations. It facilitates and boosts investigations into the algebraic structures of multi-partite quantum states and quantum operations. For example, group theory lays out elegant and efficient algebraic tools to develop quantum algorithms, construct quantum error-correction codes, and analyze quantum cryptosystems.

In this appendix, we introduce briefly elementary concepts and theorems concerning groups. A great number of textbooks are available for more complete accounts, including Cornwell (1984)—or Cornwell (1997) if an abridged version is preferred—as well as Wigner (1959) for physicists.

C.1   The Concept

Mathematically, a group is a set of algebraic objects that must obey certain *axioms*. The development of the theory does not depend on the specific nature of the elements themselves, but in most physical applications these elements are transformations of one kind or another. The *multiplication* between objects is key to determine the structure of a group. Physically, it corresponds to the composition—successive application—of transformations.

© The Editor(s) (if applicable) and The Author(s), under exclusive license to Springer 395
Nature Switzerland AG 2022, corrected publication 2023
M.-S. Choi, *A Quantum Computation Workbook*,
https://doi.org/10.1007/978-3-030-91214-7

Definition C.1 (*group*) A set $\mathcal{G}$ of elements is called *group* if the following four *group axioms* are satisfied:

(a) There exists an operation which associates with every pair of elements $G_1 \in \mathcal{G}$ and $G_2 \in \mathcal{G}$ another element $G_3 \in \mathcal{G}$. This operation is called *multiplication* and is written as $G_3 = G_1 G_2$, G_3 being described as the "product of G_1 and G_2."

(b) For any three elements $G_1, G_2, G_3 \in \mathcal{G}$,

$$(G_1 G_2)G_3 = G_1(G_2 G_3) \,, \tag{C.1}$$

i.e., the multiplication is *associative*.

(c) There exists an *identity element* $E \in \mathcal{G}$ such that

$$GE = EG = G \tag{C.2}$$

for every element $G \in \mathcal{G}$.

(d) For each element $G \in \mathcal{G}$ there exists an *inverse element* $G^{-1} \in \mathcal{G}$ such that

$$GG^{-1} = G^{-1}G = E \,. \tag{C.3}$$

The *order* of a group $\mathcal{G}$ is defined to be the number of elements in $\mathcal{G}$, which may be finite, countably infinite, or even non-countably infinite. It is denoted by $|\mathcal{G}|$. A group with finite order is called a *finite group*.

One of the most frequently appearing examples of group in quantum information is *Pauli group*. The Pauli group on a single qubit consists of the Pauli operators I, X, Y, and Z. Explicitly, it is given by

$$\mathcal{P}(1) = \{\pm I, \pm iI, \pm X, \pm iX, \pm Y, \pm iY, \pm Z, \pm iZ\} \,. \tag{C.4}$$

The additional phase factors ± 1 and $\pm i$ are included because a multiplication of two Pauli operators may result in another Pauli operator with one of such phase factors.

As an example, consider the Pauli group on a single qubit.

```
In[-]:= grp = PauliGroup[S[1]]
Out[-]= PauliGroup[{S₁}]
```

```
In[-]:= elm = GroupElements@grp;
        elm // PauliForm
Out[-]= {I, X, Y, Z, -I, -X, -Y, -Z, i I, i X, i Y, i Z, -i I, -i X, -i Y, -i Z}
```

This shows a part of the group multiplication table in terms of the index of the elements in the group.

```
In[..]:= Lbl = GroupMultiplicationTable@grp;
TableForm[tbl[[ ;; 8, ;; 8]], TableHeadings → Automatic]
Out[..]//TableForm=
```

	1	2	3	4	5	6	7	8
1	1	2	3	4	5	6	7	8
2	2	1	12	15	6	5	16	11
3	3	16	1	10	7	12	5	14
4	4	11	14	1	8	15	10	5
5	5	6	7	8	1	2	3	4
6	6	5	16	11	2	1	12	15
7	7	12	5	14	3	16	1	10
8	8	15	10	5	4	11	14	1

The group multiplication table can be displayed explicitly in terms of the group elements themselves.

```
In[..]:= mat = Map[Part[elm, #] &, tbl, {2}];
TableForm[PauliForm@mat[[ ;; 8, ;; 8]],
TableHeadings → PauliForm@{elm, elm}, TableAlignments → Right]
Out[..]//TableForm=
```

	I	X	Y	Z	-I	-X	-Y	-Z
I	I	X	Y	Z	-I	-X	-Y	-Z
X	X	I	iZ	-iY	-X	-I	-iZ	iY
Y	Y	-iZ	I	iX	-Y	iZ	-I	-iX
Z	Z	iY	-iX	I	-Z	-iY	iX	-I
-I	-I	-X	-Y	-Z	I	X	Y	Z
-X	-X	-I	-iZ	iY	X	I	iZ	-iY
-Y	-Y	iZ	-I	-iX	Y	-iZ	I	iX
-Z	-Z	-iY	iX	-I	Z	iY	-iX	I

Definition C.2 (*generators*) Let $\mathcal{G}$ be a group. The elements $G_1, G_2, \ldots, G_k$ of $\mathcal{G}$ are said to *generate* the group $\mathcal{G}$ and are denoted by

$$\mathcal{G} = \langle\{G_1, G_2, \ldots, G_k\}\rangle \tag{C.5}$$

if every element of $\mathcal{G}$ can be expressed as a multiplication of them. The elements are called *generators* of $\mathcal{G}$.

The description of a group can be drastically simplified by the use of generators. For example, consider the Pauli group in (C.4). It contains 16 elements but is generated by three elements,

$$\mathcal{P}(1) = \langle\{X, Y, Z\}\rangle . \tag{C.6}$$

It is more pronounced when we consider the Pauli group $\mathcal{P}(2)$ on two qubits,

$$\mathcal{P}(2) = \{\pm I \otimes I, \pm iI \otimes I, \pm X \otimes I, \pm iX \otimes I, \pm X \otimes X, \pm iX \otimes X, \ldots\}. \tag{C.7}$$

There are 64 elements in $\mathcal{P}(2)$ whereas it is generated by just 6 elements

$$\mathcal{P}(2) = \langle\{X \otimes I, Y \otimes I, Z \otimes I, I \otimes X, I \otimes Y, I \otimes Z\}\rangle . \tag{C.8}$$

For a given group $\mathcal{G}$, the set of generators is not unique. For example, the Pauli group on a single qubit can also be generated by iI, X, and Z; $\mathcal{P}(1) = \langle\{iI, X, Z\}\rangle$.

Theorem C.3 *Let $\mathcal{G}$ be a finite group. There exists a set of $\log_2 |\mathcal{G}|$ elements that generates $\mathcal{G}$.*

The description of a group can be drastically simplified by using generators.

```
In[·]:= gnr = GroupGenerators@grp;
       PauliForm[gnr]
Out[·]= {X, Y, Z}
```

For another example, consider the Pauli group on two qubits.

```
In[·]:= elm = GroupElements@PauliGroup[S@{1, 2}];
       PauliForm[elm]
Out[·]= {I⊗I, I⊗X, I⊗Y, I⊗Z, X⊗I, X⊗X, X⊗Y, X⊗Z, Y⊗I, Y⊗X, Y⊗Y, Y⊗Z,
        Z⊗I, Z⊗X, Z⊗Y, Z⊗Z, -(I⊗I), -(I⊗X), -(I⊗Y), -(I⊗Z), -(X⊗I),
        -(X⊗X), -(X⊗Y), -(X⊗Z), -(Y⊗I), -(Y⊗X), -(Y⊗Y), -(Y⊗Z), -(Z⊗I),
        -(Z⊗X), -(Z⊗Y), -(Z⊗Z), i I⊗I, i I⊗X, i I⊗Y, i I⊗Z, i X⊗I, i X⊗X,
        i X⊗Y, i X⊗Z, i Y⊗I, i Y⊗X, i Y⊗Y, i Y⊗Z, i Z⊗I, i Z⊗X, i Z⊗Y, i Z⊗Z,
        -i I⊗I, -i I⊗X, -i I⊗Y, -i I⊗Z, -i X⊗I, -i X⊗X, -i X⊗Y, -i X⊗Z,
        -i Y⊗I, -i Y⊗X, -i Y⊗Y, -i Y⊗Z, -i Z⊗I, -i Z⊗X, -i Z⊗Y, -i Z⊗Z}
```

It has 64 elements. That is, the order of the Pauli group on two qubits is 64.

```
In[·]:= Length[elm]
Out[·]= 64
```

```
In[·]:= GroupOrder@PauliGroup[S@{1, 2}]
Out[·]= 64
```

The Pauli group on two qubits can be generated by just 6 elements.

```
In[·]:= gnr = GroupGenerators@PauliGroup[S@{1, 2}];
       PauliForm[gnr]
Out[·]= {X⊗I, Y⊗I, Z⊗I, I⊗X, I⊗Y, I⊗Z}
```

Definition C.4 (*Abelian group*) If all the elements of a group commute, the group is said to be *Abelian*.

A special example of Abelian group are *cyclic groups*. A cyclic group $\mathcal{G}$ consists of elements which can be obtained by raising one of them to successive powers, i.e.,

$$\mathcal{G} = \left\{ G, G^2, G^3, \ldots, G^n = E \right\},\qquad(C.9)$$

where n is some integer. It is thus generated by a single element, $\mathcal{G} = \langle\{G\}\rangle$; Any element of a cyclic group generates the group. A common example of cyclic group is

$$\mathbb{Z}_n := \{0, 1, 2, \ldots, n - 1\}\qquad(C.10)$$

with addition modulo n as the group multiplication. In particular, $\mathbb{Z}_2$ appears very often in wide areas of science.

Theorem C.5 (rearrangement theorem) *Let $\mathcal{G}$ be a group. For any fixed element $G \in \mathcal{G}$, the set $G\mathcal{G} := \left\{GG'|G' \in \mathcal{G}\right\}$ contains every element of $\mathcal{G}$ once and only once. The same holds for the set $\mathcal{G}G := \left\{G'G|G' \in \mathcal{G}\right\}$.*

Definition C.6 (*group homomorphism*) If $\phi : \mathcal{G} \to \mathcal{G}'$ is a mapping of a group $\mathcal{G}$ onto another group $\mathcal{G}'$ such that

$$\phi(G_1)\phi(G_2) = \phi(G_1 G_2) \tag{C.11}$$

for all $G_1, G_2 \in \mathcal{G}$, then ϕ is said to be a *homomorphic* mapping or a *homomorphism*. If ϕ is an one-to-one mapping, then it is called an *isomorphism*. When there exists an isomorphism from $\mathcal{G}$ onto $\mathcal{G}'$, $\mathcal{G}$ and $\mathcal{G}'$ are said to be *isomorphic* to each other and denoted by $\mathcal{G} \simeq \mathcal{G}'$.

C.2 Classes

A *class* of a group $\mathcal{G}$ is a subset of $\mathcal{G}$ having a certain property. This particular property makes classes play an important role in representation theory.

Definition C.7 (*conjugate elements and classes*) Let $\mathcal{G}$ be a group. An element $G' \in \mathcal{G}$ is said to be *conjugate* to another element G if there exists an element $H \in \mathcal{G}$ such that

$$G' = HGH^{-1}. \tag{C.12}$$

Obviously, if G' is conjugate to G, then G is also conjugate to G'. It is therefore permissible to talk of a set of *mutually* conjugate elements. A *class* of a group $\mathcal{G}$ is a set of mutually conjugate elements of $\mathcal{G}$.

Each class is completely determined by any member G of it. That is, given G, we obtain the whole class by forming the products

$$\mathcal{G}G\mathcal{G}^{-1} := \left\{ HGH^{-1} | H \in \mathcal{G} \right\}. \tag{C.13}$$

Theorem C.8 *Classes have the following properties:*

(a) Every element of a group $\mathcal{G}$ is a member of some class of $\mathcal{G}$.
(b) No element of $\mathcal{G}$ can be a member of two different classes of $\mathcal{G}$.
(c) The identity E of $\mathcal{G}$ always forms a class on its own.

The conjugation relation between group elements as defined in (C.12) has physical implications when applied to quantum mechanics. For example, if $\mathcal{G}$ is a group consisting of rotations, no class of $\mathcal{G}$ contains both proper and improper rotations. Moreover, in each class of proper rotations all rotations are through the same angle. Similarly, in each class of improper rotations the proper parts are all through the same angle.

C.3 Invariant Subgroups

The notion of invariant subgroups is usually introduced together with the notion of *cosets* to construct *quotient groups*. It is also closely related to the concept of *classes* as will be seen shortly. Therefore, invariant subgroups also play an important role in representation theory.

Definition C.9 (*subgroup*) Let $\mathcal{G}$ be a group and $\mathcal{H} \subset \mathcal{G}$. If $\mathcal{H}$ itself is a group (with the multiplication among its elements given by that of $\mathcal{G}$), then $\mathcal{H}$ is called a subgroup of the group $\mathcal{G}$.

Definition C.10 (*invariant subgroups*) A subgroup $\mathcal{H}$ of a group $\mathcal{G}$ is said to be *invariant* if

$$GHG^{-1} \in \mathcal{H} \tag{C.14}$$

for every $H \in \mathcal{H}$ and $G \in \mathcal{G}$. In many cases the property in Eq. (C.14) is written in a more compact form as

$$\mathcal{G}\mathcal{H}\mathcal{G}^{-1} = \mathcal{H} \tag{C.15}$$

with the abbreviation $\mathcal{G}\mathcal{H}\mathcal{G}^{-1} := \{GHG^{-1} | H \in \mathcal{H}, G \in \mathcal{G}\}$.

Invariant subgroups are sometimes called *normal subgroups* or *normal divisors* because the defining property in Eq. (C.14) is of the same form as Eq. (C.12). The following theorem ensures the connection between the *classes* and the invariant subgroups.

Theorem C.11 *A subgroup $\mathcal{H}$ of a group $\mathcal{G}$ is an invariant subgroup if and only if $\mathcal{H}$ consists entirely of* complete *classes of $\mathcal{G}$.*

A special kind of invariant subgroup is the *centre* of the group. This notion is closely related to the notion of *quotient groups*.

Definition C.12 (*centre of a group*) The *centre* $\mathcal{Z}$ of a group $\mathcal{G}$ is defined to be the subgroup consisting of *all* elements of $\mathcal{G}$ that commute with every element of $\mathcal{G}$.

$\mathcal{Z}$ is an Abelian invariant subgroup of $\mathcal{G}$. Moreover, any subgroup of $\mathcal{Z}$ is an Abelian invariant subgroup of $\mathcal{G}$ and called a *central invariant subgroup* of $\mathcal{G}$.

C.4 Cosets and Quotient Groups

A subgroup of a group can be used to decompose the group (as a set) into disjoint subsets called *cosets*. We have already seen in Sect. C.2 that conjugacy classes is another way to do it. However, unlike conjugacy classes, all cosets of a group are of equal size.

Definition C.13 (*coset*) Let $\mathcal{H}$ be a subgroup (not necessarily an invariant subgroup) of a group $\mathcal{G}$. Then for any fixed $G \in \mathcal{G}$ (which may or may not be member of $\mathcal{H}$) the set $\mathcal{H}G := \{HG : H \in \mathcal{H}\}$ is called the *right coset* of $\mathcal{H}$ with respect to G. Similarly, the set $G\mathcal{H} := \{GH : H \in \mathcal{H}\}$ is called the *left coset* of $\mathcal{H}$ with respect to G.

The properties of cosets are summarized in the following two theorems:

Theorem C.14 *Let $\mathcal{H}$ be a subgroup of a group $\mathcal{G}$.*

(a) *If $G \in \mathcal{H}$, then $\mathcal{H}G = \mathcal{H}$.*
(b) *If $G' \in \mathcal{H}G$, then $\mathcal{H}G' = \mathcal{H}G$.*
(c) *If $G \notin \mathcal{H}$, then $\mathcal{H}G$ is not a subgroup of $\mathcal{G}$.*
(d) *Every element of $\mathcal{G}$ is a member of some right coset.*
(e) *Two right cosets of $\mathcal{H}$ are either identical or have no elements in common.*
(f) *Any two elements HG and $H'G$ of $\mathcal{H}G$ are different provided that $H \neq H'$. In particular, if $\mathcal{H}$ is a finite subgroup of order $|\mathcal{H}|$, then $\mathcal{H}G$ contains $|\mathcal{H}|$ different elements.*
(g) *If $\mathcal{G}$ is a finite group of order $|\mathcal{G}|$ and $\mathcal{H}$ has order $|\mathcal{H}|$, then the number of distinct right cosets is $|\mathcal{G}|/|\mathcal{H}|$.*

Above the statements are for the right cosets, but every statement applies equally to left cosets.

Theorem C.15 (cosets and invariant subgroups) *The right and left cosets of a subgroup $\mathcal{H}$ of a group $\mathcal{G}$ are identical, that is,*

$$\mathcal{H}G = G\mathcal{H} \tag{C.16}$$

for all $G \in \mathcal{G}$, if and only if $\mathcal{H}$ is an invariant *subgroup.*

The property (b) in Theorem C.14 is particularly important. It shows that the same coset is formed starting from *any* member of the coset. All members of a coset therefore appear on an equal footing, so that *any* member of the coset can be taken as the *coset representative* that labels the coset and from which the coset can be constructed. Accordingly, it may be useful to ignore the internal structure of each right coset of a subgroup $\mathcal{H}$ and consider each coset as an *element* of the *set of distinct right cosets*. Indeed, such a set of cosets forms a group with the suitable definition of multiplication as stated in the following theorem:

Theorem C.16 (quotient group $\mathcal{G}/\mathcal{H}$) *The set of right cosets of an invariant subgroup $\mathcal{H}$ of a group $\mathcal{G}$ forms a group with the group multiplication defined by*

$$G_1\mathcal{H} \cdot G_2\mathcal{H} = (G_1G_2)\mathcal{H} \tag{C.17}$$

for any $G_1, G_2 \in \mathcal{G}$. This group is called a quotient group *(or* factor group*) and is denoted by $\mathcal{G}/\mathcal{H}$. Furthermore, if $\mathcal{G}$ is finite, then*[7]

---

[7] It follows from Theorem C.14 (g).

$$|\mathcal{G}/\mathcal{H}| = |\mathcal{G}|/|\mathcal{H}| \, . \tag{C.18}$$

The key idea behind the notion of quotient group is that two elements G, G' of $\mathcal{G}$ are regarded equivalent as long as there exists $H \in \mathcal{H}$ such that $G = G'H$ (or $H' \in \mathcal{H}$ such that $G = H'G'$).

Note that unless the subgroup $\mathcal{H}$ is an invariant subgroup, it is not guaranteed that Eq. (C.17) provides a meaningful and consistent definition for the group multiplication. One has first to show that Eq. (C.17) provides a meaningful definition in that if alternative coset representatives are chosen for the cosets on the left-hand side of the equation, the coset on the right-hand side remains unchanged. Suppose that G'_1 and G'_2 are alternative coset representatives for $\mathcal{H}G_1$ and $\mathcal{H}G_2$, respectively, so that $G'_1 \in \mathcal{H}G_1$ and $G'_2 \in \mathcal{H}G_2$. It has to be proved that $\mathcal{H}(G'_1 G'_2) = \mathcal{H}(G_1 G_2)$. As $G'_1 \in \mathcal{H}G_1$ and $G'_2 \in \mathcal{H}G_2$, there exist $H_1, H_2 \in \mathcal{H}$ such that $G'_1 = H_1 G_1$ and $G'_2 = H_2 G_2$ Furthermore, as $\mathcal{H}$ is invariant, $G_1\mathcal{H} = \mathcal{H}G_1$ and hence there exists $H'_2 \in \mathcal{H}$ such that $G_1 H_2 = H'_2 G_1$. Consequently, one has

$$\mathcal{H}G'_1 \cdot \mathcal{H}G'_2 = \mathcal{H}(H_1 G_1 H_2 G_2) = \mathcal{H}(H_1 H'_2 G_1 G_2) = \mathcal{H}(G_1 G_2) \, . \tag{C.19}$$

As an example, consider again the Pauli group on a single qubit

$$\mathcal{G} = \{\pm I, \pm iI, \pm X, \pm iX, \pm Y, \pm iY, \pm Z, \pm iZ\} \, , \tag{C.20}$$

which we have already discussed in Eq. (C.4). $\mathcal{Z} = \{I, -I, iI, -iI\}$ is an invariant subgroup of $\mathcal{G}$. In fact, $\mathcal{Z}$ is the center of $\mathcal{G}$ (Definition C.12). The quotient group $\mathcal{G}/\mathcal{Z}$ is given by

$$\mathcal{G}/\mathcal{Z} = \{\mathcal{Z}, X\mathcal{Z}, Y\mathcal{Z}, Z\mathcal{Z}\}. \tag{C.21}$$

In the quotient group $\mathcal{G}/\mathcal{Z}$, the elements $X, -X, iX$, and $-iX$ are not distinguished and regarded as equivalent. Likewise, the elements $Y, -Y, iY$, and $-iY$ are regarded equivalent. This is convenient, for example, when one is only interested in the commutation relation but not the explicit multiplications between elements.

C.5 Product Groups

Given two or more sets, it is common to construct a new set consisting of tuples of the elements from the given sets. Given two or more groups, one can construct a new group in an analogous manner with the group multiplication inherited from the given groups.

Definition C.17 Let $\mathcal{G}$ and $\mathcal{G}'$ be groups. The *direct product group* or simply *product group*, $\mathcal{G} \times \mathcal{G}'$, is defined as follows:

(a) The underlying set of $\mathcal{G} \times \mathcal{G}'$ consists of the ordered pair of elements from $\mathcal{G}$ and $\mathcal{G}'$, that is,

$$\mathcal{G} \times \mathcal{G}' := \{(G, G') | G \in \mathcal{G}, G' \in \mathcal{G}'\} \tag{C.22}$$

(b) For any $G_1, G_2 \in \mathcal{G}$ and $G'_1, G'_2 \in \mathcal{G}'$, the group multiplication is defined by

$$(G_1, G'_1)(G_2, G'_2) = (G_1 G_2, G'_1 G'_2). \tag{C.23}$$

It is straightforward to extend the definition to direct products of more than two groups by repeating the above operation.

The simplest example is the direct product of cyclic groups [see Eq. (C.10)], $\mathbb{Z}_{n_1} \times \mathbb{Z}_{n_2} \times \cdots \times \mathbb{Z}_{n_k}$. The resulting group is Abelian (Definition C.4). In fact, one can show that any Abelian group is isomorphic to $\mathbb{Z}_{n_1} \times \mathbb{Z}_{n_2} \times \cdots \times \mathbb{Z}_{n_k}$ for some $n_1, n_2, \ldots, n_k$. Interestingly, $\mathbb{Z}_{n_1} \times \mathbb{Z}_{n_2} \times \cdots \times \mathbb{Z}_{n_k}$ can also be regarded as a vector space over the field $\mathbb{Z}_2$. This fact provides convenient tools in studying the structure of the Pauli group (Sect. 6.3.1).

Appendix D
Mathematica Application Q3

Q3 is a Mathematica application to help study quantum information processing, quantum many-body systems, and quantum spin systems. It provides various tools and utilities for symbolic calculations and numerical simulations in these areas of quantum physics.

Q3 consists of several packages at different levels. Quisso, Fock, and Wigner are the three main packages, and they are devoted to the simulation of quantum information processing, quantum many-body systems, and quantum spin systems, respectively. They are based on another lower-level package, Pauli. Pauli itself provides useful tools to handle the Pauli matrices and operators for unlabelled qubits directly. However, it also lays out the programming structures and defines objects for the aforementioned three and other higher-level packages.

Q3 is distributed through the GitHub repository:

> https://github.com/quantum-mob/Q3App

D.1 Installation

Q3 provides two installation methods: The first is based on the paclet system that has recently been introduced by Wolfram Research. It is not only fully automatic but also convenient to get updates later on. But it is supported only for Mathematica 12.1 or later. If your copy of Mathematica is compatible, then just copy the following code and run it on your Mathemtica Notebook:

```
Module[
   { ps },
   ps = PacletSiteRegister[
      "https://github.com/quantum-mob/PacletServer/raw/main",
      "Quantum Mob Paclet Server"
      ];
```

M.-S. Choi, *A Quantum Computation Workbook*, https://doi.org/10.1007/978-3-030-91214-7

```
    PacletSiteUpdate[ps];
    PacletInstall["Q3"]
]
```

The other method is to download and copy the files to a proper folder—the traditional method. For details, take a look at the *installation guide* at:

https://github.com/quantum-mob/Q3App/blob/main/INSTALL.md

D.2   Quick Start

Once the application is installed, put

```
    "Q3" or "Q3/guide/Q3"
```

in the search field of the Wolfram Language Documentation Center (the help window of Mathematica) to get detailed technical information about the application.

Note that after installing the application, the first time you search for a keyword in Wolfram Language Documentation Center, Mathematica builds the search index of the new documentation files. It can take a few seconds to minutes depending on your computer. It happens only once (everytime you update the application).

Appendix E
Integrated Compilation of Demonstrations

QuantumWorkbook is a compilation of Mathematica Notebook files containing the Wolfram Language code files that have been used to generate the demonstrations in the book. Readers can try and modify the code themselves to build their own examples on the demonstrations and to experiment their fresh ideas.

QuantumWorkbook is distributed through the GitHub repository:

https://github.com/quantum-mob/QuantumWorkbook

E.1 Installation

Copy the following code, and just evaluate it in your Mathematica(R) Notebook:

```
Module[
  { ps },
  ps = PacletSiteRegister[
    "https://github.com/quantum-mob/PacletServer/raw/main",
    "Quantum Mob Paclet Server"
  ];
  PacletSiteUpdate[ps];
  PacletInstall["Q3"];
  PacletInstall["QuantumWorkbook"]
]
```

Note that along with QuantumWorkbook, it also installs the main application Q3 for your convenience.

This package may be modified for bug fixes and improvements. You may want to check for updates from time to time:

```
QuantumWorkbookCheckUpdate[]
```

© The Editor(s) (if applicable) and The Author(s), under exclusive license to Springer Nature Switzerland AG 2022, corrected publication 2023
M.-S. Choi, *A Quantum Computation Workbook*,
https://doi.org/10.1007/978-3-030-91214-7

In case there is an update, you can install it by using the following function:

```
QuantumWorkbookUpdate[]
```

E.2 Quick Start

Once **QuantumWorkbook** is installed, put

```
"QuantumWorkbook"
```

in the search field of the Wolfram Documentation Center (the Help window of Mathematica). You will see the table of contents of the workbook.

Appendix F
Solutions to Select Problems

F.1   The Postulates of Quantum Mechanics

F.2   Quantum Computation: Overview

Problem 2.4

First, note that

where note that the controlled-Z operation is "symmetric" in the roles of control and target; hence the circuit representation by two dots. Then it follows that

$$\tag{F.1}$$

Problem 2.7

The eigenstates of U are the same as those of the Pauli X operator.

```
In[-]:= U = Rotation[φ, S[1, 1]] // Elaborate
```

$$Out[-]= \; Cos\left[\frac{\phi}{2}\right] - i \; S_1^x \; Sin\left[\frac{\phi}{2}\right]$$

To see it, take an eigenstate of the Pauli X operator on qubit S[1,None] belonging to the eigenvalue 1.

In[·]:= `vec = S[1, 6] ** Ket[];`
`vec // LogicalForm`

Out[·]= $\dfrac{\left|0_{S_1}\right\rangle}{\sqrt{2}} + \dfrac{\left|1_{S_1}\right\rangle}{\sqrt{2}}$

This confirms that the vector is indeed the intended eigenstate.

In[·]:= `U ** vec // LogicalForm // TrigToExp // Simplify`

Out[·]= $\dfrac{e^{-\frac{i\phi}{2}}\left(\left|0_{S_1}\right\rangle + \left|1_{S_1}\right\rangle\right)}{\sqrt{2}}$

This checks for the other eigenstate belonging to the eigenvalue -1.

In[·]:= `vec = S[1, 6] ** Ket[S[1] → 1];`
`vec // LogicalForm`

Out[·]= $\dfrac{\left|0_{S_1}\right\rangle}{\sqrt{2}} - \dfrac{\left|1_{S_1}\right\rangle}{\sqrt{2}}$

In[·]:= `U ** vec // LogicalForm // TrigToExp // Simplify`

Out[·]= $\dfrac{e^{\frac{i\phi}{2}}\left(\left|0_{S_1}\right\rangle - \left|1_{S_1}\right\rangle\right)}{\sqrt{2}}$

The ancillary qubit takes a relative phase shift depending on in which eigenstate the native qubit is.

In[·]:= `Let[Real, ϕ];`
`qc = QuantumCircuit[LogicalForm[Ket[], S@{1, 2}], S[1, 6], "Separator",`
`  S[2, 6], ControlledU[S[2], Rotation[ϕ, S[1, 1]]], "Label" → "U"]]`

Out[·]=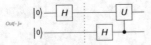

In[·]:= `out = ExpressionFor[qc] // TrigToExp;`
`KetFactor@out`

Out[·]= $\dfrac{1}{2}\, e^{-\frac{i\phi}{2}}\left(\left|0_{S_1}\right\rangle + \left|1_{S_1}\right\rangle\right) \otimes \left(e^{\frac{i\phi}{2}}\left|0_{S_2}\right\rangle + \left|1_{S_2}\right\rangle\right)$

Make a basis change to detect it.

In[·]:= `qc = QuantumCircuit[LogicalForm[Ket[], S@{1, 2}], S[1, 6], "Separator",`
`  S[2, 6], ControlledU[S[2], Rotation[ϕ, S[1, 1]]], "Label" → "U",`
`  "Separator", Rotation[ϕ / 2, S[2, 3]], S[2, 6], Measurement[S[2]]]`

Out[·]=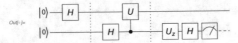

Finally, check the result.

In[·]:= `out = ExpressionFor[qc] // TrigToExp;`
`LogicalForm[KetFactor@out, S@{1, 2}]`

Out[·]= $\dfrac{e^{-\frac{i\phi}{4}}\left(\left|0_{S_1}0_{S_2}\right\rangle + \left|1_{S_1}0_{S_2}\right\rangle\right)}{\sqrt{2}}$

Problem 2.8

This constructs the desired quantum circuit model.

```
In[*]:= qc = QuantumCircuit[LogicalForm[Ket[], S@{1, 2, 3}],
        S[{1, 2, 3}, 6], ControlledU[S[2], Rotation[ϕ, S[1, 1]], "Label" → "U"],
        ControlledU[S[3], Rotation[2 ϕ, S[1, 1]], "Label" → "U²"],
        "Separator", {Rotation[θ, S[2, 3]], Rotation[Pi / 2, S[3, 3]]}, S[3, 6]]
```

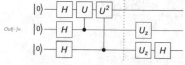

```
In[*]:= out = ExpressionFor[qc] // TrigToExp // Simplify;
        KetFactor@out /. ϕ → Pi / 2
```

$$\text{Out[*]=} \quad \frac{\left(\frac{1}{4} + \frac{i}{4}\right) e^{-\frac{3 i \pi}{4}} \left(\left|0_{S_1}\right\rangle + \left|1_{S_1}\right\rangle\right) \otimes \left(e^{\frac{i \pi}{4}} \left|0_{S_2}\right\rangle + \left|1_{S_2}\right\rangle\right) \otimes \left(2 \left|0_{S_3}\right\rangle\right)}{\sqrt{2}}$$

Problem 2.10 Suppose that the statement (2.10a) holds. Then

$$\hat{A}\hat{B} = \hat{I}, \quad \hat{A}\hat{X}\hat{B} = \hat{U}. \tag{F.2}$$

The first condition in the above implies that $\hat{B} = \hat{A}^\dagger$. Then put $\hat{W} = \hat{A}\hat{H}$, where $\hat{H}$ is the Hadamard gate, so that

$$\hat{W}\hat{Z}\hat{W}^\dagger = \hat{A}\hat{X}\hat{A}^\dagger = \hat{U}. \tag{F.3}$$

That is, the statement (2.10b) holds. Furthermore, observe that

$$\text{Tr}\hat{U} = \text{Tr}\hat{W}\hat{Z}\hat{W}^\dagger = \text{Tr}\hat{Z} = 0, \tag{F.4}$$

$$\det \hat{U} = \det \hat{W}\hat{Z}\hat{W}^\dagger = \det \hat{Z} = -1, \tag{F.5}$$

and hence the statement (2.10c) holds as well.

Next, suppose that $\hat{U} = \hat{W}\hat{Z}\hat{W}^\dagger$ for some unitary operator $\hat{W}$. Put $\hat{A} = \hat{W}\hat{H}$ and $\hat{B} = \hat{A}^\dagger$, which satisfy (F.2); that is, the statement (2.10a) holds. Furthermore, statement (2.10c) also holds as already shown above.

Finally, suppose that the statement (2.10c) holds: The condition $\text{Tr}\hat{U} = 0$ implies that

$$\hat{U} = u_x\hat{X} + u_y\hat{Y} + u_z\hat{Z} \quad (u_x, u_y, u_z \in \mathbb{C}). \tag{F.6}$$

The unitarity condition of $\hat{U}$ implies that all the phases of u_μ must be the same and that $\sum_\mu |u_\mu|^2 = 1$. This implies in turn that

$$\hat{U} = e^{i\phi}\hat{W}\hat{Z}\hat{W}^\dagger \tag{F.7}$$

for some $\phi \in \mathbb{R}$ and some unitary operator $\hat{W}$. The condition $\det \hat{U} = -1$ leads to $\phi = n\pi$ with $n \in \mathbb{Z}$. For $e^{i\phi} = -1$ (n odd), simply replace $\hat{W}$ with $\hat{W}' = \hat{W}\hat{X}$.

Therefore, regardless of $e^{i\phi} = \pm 1$, the statement (2.10b) holds. As we have already shown above that the statement (2.10b) implies (2.10a), this completes the proof.

F.3   Realizations of Quantum Comptuers

Problem 3.3 We define

$$|\Omega\rangle := \frac{|1\rangle\,\Omega_1 + |2\rangle\,\Omega_2}{\Omega}. \tag{F.8}$$

where $\Omega := \sqrt{\Omega_1^2 + \Omega_2^2}$. Then the Hamiltonian reads as

$$\hat{H} = \epsilon\,|\epsilon\rangle\,\langle\epsilon| - \frac{1}{2}\Omega\left(|\Omega\rangle\,\langle\epsilon| + |\epsilon\rangle\,\langle\Omega|\right). \tag{F.9}$$

(a) The expressions for the two eigenstates of $\hat{H}$ in (F.9) are formally the same as those in the main text (Sect. 3.3.1). The Hamiltonian involves only $|\epsilon\rangle$ and $|\Omega\rangle$. There must be one more state. It is the dark state of the system. We denote it by $|D\rangle$. The dark state must be orthogonal to the bright state $|\Omega\rangle$ as well as $|\epsilon\rangle$. Therefore, we can construct it as follows

$$|D\rangle = \frac{|1\rangle\,\Omega_2^* - |2\rangle\,\Omega_1^*}{\Omega}. \tag{F.10}$$

By inspection, one can confirm that it is orthogonal both to $|\Omega\rangle$ and trivially to $|\epsilon\rangle$. Using the parametrization suggested in the statement of the problem, we can rewrite $|D\rangle$ into the form

$$|D\rangle = \frac{|1\rangle\,\sin(\theta/2)e^{-i\phi/2} - |2\rangle\,\cos(\theta/2)e^{i\phi/2}}{\Omega}. \tag{F.11}$$

(b) The Abelian gauge potential is given by

$$A^\phi := \langle D|\frac{\partial}{\partial\phi}|D\rangle = \frac{i\cos\theta}{2}. \tag{F.12}$$

As it is Abelian, it is customary to define the Berry phase as $\gamma := -i\int\limits_0^{2\pi} d\phi A^\phi = \frac{1}{2}\cos\theta$.

(c) For the path where ϕ varies from 0 to 2π with θ fixed, the Abelian geometric phase is given by

$$U(\mathcal{C}) = e^{-i\gamma} = e^{-i\pi(\cos\theta)}. \tag{F.13}$$

Problem 3.4 A direct inspection would be sufficient. Here we slightly modify the quantum circuit into the form

$$(F.14)$$

We then use the already known result in (3.70) for the quantum circuit in Eq. (3.69). In this case, $\hat{U}_z$ is trivial, $\hat{U}_z = \hat{I}$, and the input state is $\hat{H}\,|\psi\rangle$. It leads to the output state on the second qubit

$$\frac{|0\rangle \otimes (\hat{H}\hat{H}\,|\psi\rangle) + |1\rangle \otimes (\hat{H}\hat{Z}\hat{H}\,|\psi\rangle)}{\sqrt{2}}. \qquad (F.15)$$

The identities, $\hat{H}^2 = \hat{I}$ and $\hat{H}\hat{Z}\hat{H} = \hat{X}$, lead to

$$\frac{|0\rangle \otimes |\psi\rangle + |1\rangle \otimes (\hat{X}\,|\psi\rangle)}{\sqrt{2}}. \qquad (F.16)$$

as expected.

Here is a quantum circuit model to implement the Pauli X gate based on measurement.

In[ ]:= `qc = QuantumCircuit[ProductState[S[1] → {c[0], c[1]}, "Label" → Ket["ψ"]],`
`LogicalForm[Ket[], S@{2}], S[{1, 2}, 6], CZ[S[1], S[2]], S[1, 6]]`

Out[ ]:=

In[ ]:= `out = Elaborate[qc];`
`KetFactor[out, S[1]]`

Out[ ]:= $|0_{S_1}\rangle \otimes \left(\dfrac{c_0\,|{-}\rangle}{\sqrt{2}} + \dfrac{c_1\,|1_{S_2}\rangle}{\sqrt{2}} \right) + |1_{S_1}\rangle \otimes \left(\dfrac{c_1\,|{-}\rangle}{\sqrt{2}} + \dfrac{c_0\,|1_{S_2}\rangle}{\sqrt{2}} \right)$

We go further to post-process the output state on the second qubit.

In[ ]:= `qc1 = QuantumCircuit[qc, Measurement[S[1]], ControlledU[S[1], S[2, 1]]]`

Out[ ]:=

In[ ]:= `new = Elaborate[qc1] /. {Conjugate[c[0]] × c[0] + Conjugate[c[1]] × c[1] → 1};`
`LogicalForm[KetFactor[new, S[1]], S@{1, 2}]`
Out[ ]:= $|0_{S_1}\rangle \otimes (c_0\,|0_{S_2}\rangle + c_1\,|1_{S_2}\rangle)$

Problem 3.6 Let $\hat{U}$ be a unitary operator on a finite-dimensional vector space $\mathcal{V}$. Then we have

$$\hat{U}\exp(\hat{A})\hat{U}^\dagger = \exp(\hat{U}\hat{A}\hat{U}^\dagger) \qquad (F.17)$$

for any linear operator $\hat{A}$ on $\mathcal{V}$.

(a) We note that $\hat{X}$ is both unitary and Hermitian. Furthermore, $\hat{X}\hat{Z}\hat{X} = -\hat{Z}$. These observations lead to

$$\hat{X}\exp(-i\hat{Z}\phi/2)\hat{X} = \exp(-i\hat{X}\hat{Z}\hat{X}\phi/2) = \exp(+i\hat{Z}\phi/2) = \hat{U}_z(-\phi). \quad \text{(F.18)}$$

(b) Similarly, $\hat{H}$ is both unitary and Hermitian. It also satisfies $\hat{H}\hat{Z}\hat{H} = \hat{X}$. It follows that

$$\hat{H}\exp(-i\hat{Z}\phi/2)\hat{H} = \exp(-i\hat{H}\hat{Z}\hat{H}\phi/2) = \exp(-i\hat{X}\phi/2) = \hat{U}_x(\phi). \quad \text{(F.19)}$$

F.4   Quantum Algorithms

Problem 4.1

Here is a particular classical oracle.

```
In[ ]:= f[0, 0] = 0
       f[0, 1] = 0
       f[1, 0] = 1
       f[1, 1] = 0
Out[ ]= 0

Out[ ]= 0

Out[ ]= 1

Out[ ]= 0

In[ ]:= cc = {1, 2};
       tt = {3};
       ct = Join[cc, tt];
       qc = QuantumCircuit[
         LogicalForm[Ket[S[tt] → 1], S[ct]],
         S[ct, 6],
         Oracle[f, S@cc, S@tt],
         S[tt, 6]
       ]
       out = ExpressionFor[qc];
       KetFactor[out, S[tt]] // LogicalForm
```

$$Out[ ]= |1_{s_3}\rangle \otimes \left(\frac{1}{2}\left(|0_{s_1}0_{s_2}\rangle + |0_{s_1}1_{s_2}\rangle - |1_{s_1}0_{s_2}\rangle + |1_{s_1}1_{s_2}\rangle\right)\right)$$

```
In[ ]:= bb = f @@@ IntegerDigits[Range[0, 2^2], 2, 2]
Out[ ]= {0, 0, 1, 0, 0}
```

Here is a particular classical oracle.

```
In[·]:= f[0, 0] = {0, 1}
        f[0, 1] = {1, 0}
        f[1, 0] = {0, 1}
        f[1, 1] = {1, 1}

Out[·]= {0, 1}

Out[·]= {1, 0}

Out[·]= {0, 1}

Out[·]= {1, 1}

In[·]:= cc = {1, 2};
        tt = {3, 4};
        ct = Join[cc, tt];
        qc = QuantumCircuit[
          LogicalForm[Ket[S[tt] → 1], S[ct]],
          S[ct, 6],
          Oracle[f, S@cc, S@tt],
          S[tt, 6]
          ]
        out = ExpressionFor[qc];
        KetFactor[out, S[tt]] // LogicalForm
```

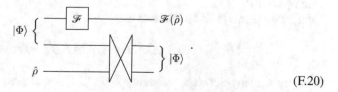

$$Out[·]= \left|1_{S_3}1_{S_4}\right\rangle \otimes \left(\frac{1}{2}\left(-\left|0_{S_1}0_{S_2}\right\rangle - \left|0_{S_1}1_{S_2}\right\rangle - \left|1_{S_1}0_{S_2}\right\rangle + \left|1_{S_1}1_{S_2}\right\rangle\right)\right)$$

```
In[·]:= bb = f@@@ IntegerDigits[Range[0, 2^2], 2, 2]
Out[·]= {{0, 1}, {1, 0}, {0, 1}, {1, 1}, {0, 1}}
```

F.5 Quantum Decoherence

Problem 5.2 We analyse the quantum circuit

$$|\Phi\rangle \left\{ \begin{array}{c} \boxed{\mathscr{F}} \qquad \mathscr{F}(\hat{\rho}) \\ \\ \end{array} \right.$$

$$\hat{\rho} \qquad\qquad\qquad \left. \begin{array}{c} \\ \\ \end{array} \right\} |\Phi\rangle$$

$$(F.20)$$

Before the projection onto $|\Phi\rangle := \frac{1}{\sqrt{d}}\sum_j |v_j\rangle \otimes |v_j\rangle$, where d is the dimension of each quantum register, the state reads as

$$\hat{R} = \frac{1}{d} \sum_{ij} \mathscr{F}(|v_i\rangle \langle v_j|) \otimes |v_i\rangle \langle v_j| \otimes \hat{\rho}. \tag{F.21}$$

The projection onto $|\Phi\rangle$ leads to the state for the first register

$$\begin{aligned}
\hat{R}' &= \frac{1}{d} \sum_{ij} \mathscr{F}(|v_i\rangle \langle v_j|) \text{Tr}(|v_i\rangle \langle v_j| \otimes \hat{\rho}) |\Phi\rangle \langle \Phi| \\
&= \frac{1}{d} \sum_{ij} \mathscr{F}(|v_i\rangle \langle v_j|) \frac{1}{d} \sum_{kl} (|v_i\rangle \langle v_j| \otimes \hat{\rho})(|v_k\rangle \langle v_l| \otimes |v_k\rangle \langle v_l|) \\
&= \frac{1}{d^2} \sum_{ij} \mathscr{F}(|v_i\rangle \langle v_j|) \langle v_i| \hat{\rho} |v_j\rangle
\end{aligned} \tag{F.22}$$

By the linearity of the supermap $\mathscr{F}$, one obtains

$$\hat{R}' = \frac{1}{d^2} \mathscr{F} \left(\sum_{ij} |v_i\rangle \langle v_i| \hat{\rho} |v_j\rangle \langle v_j| \right) = \frac{1}{d^2} \mathscr{F}(\hat{\rho}). \tag{F.23}$$

The factor of $1/d^2$ indicates the success probability of the protocol.

Problem 5.10 We first prove (5.245). From the singular-value decomposition of $\hat{A}$

$$\hat{A} = \sum_j |u_j\rangle a_j \langle v_j| \quad (a_j \geq 0), \tag{F.24}$$

we have

$$\|\hat{A}\|_{\text{HS}}^2 = \sum_j a_j^2, \tag{F.25}$$

and

$$\|\hat{A}^\dagger \hat{A}\|_{\text{HS}}^2 = \sum_j a_j^4 \leq \left(\sum_j a_j^2 \right)^2 = \|\hat{A}\|_{\text{HS}}^2, \tag{F.26}$$

which proves (5.245).

Next, we prove (5.244). We note that

$$\|\hat{A}\hat{B}\|_{\text{HS}}^2 = \text{Tr}\hat{B}^\dagger \hat{A}^\dagger \hat{A}\hat{B} = \text{Tr}\hat{A}^\dagger \hat{A}\hat{B}\hat{B}^\dagger = \langle \hat{A}^\dagger \hat{A}, \hat{B}\hat{B}^\dagger \rangle. \tag{F.27}$$

Using the Cauchy-Schwarz inequality and (5.245), we have

$$\|\hat{A}\hat{B}\|_{\text{HS}}^2 = \langle \hat{A}^\dagger \hat{A}, \hat{B}\hat{B}^\dagger \rangle \leq \|\hat{A}^\dagger \hat{A}\|_{\text{HS}} \|\hat{B}\hat{B}^\dagger\|_{\text{HS}} \leq \|\hat{A}\|_{\text{HS}}^2 \|\hat{B}\|_{\text{HS}}^2. \tag{F.28}$$

Problem 5.11 We first note that

$$\mathrm{Tr}\hat{G} = \sum_\mu \mathrm{Tr}\hat{F}_\mu \hat{F}_\mu^\dagger = \sum_\mu \mathrm{Tr}\hat{F}_\mu^\dagger \hat{F}_\mu = \mathrm{Tr}\hat{I} = \dim \mathcal{V}. \tag{F.29}$$

From (5.168), we have

$$|\mathrm{Tr}\hat{G}| = \dim \mathcal{V} \le \|\hat{G}\|_{\mathrm{tr}}. \tag{F.30}$$

On the other hand, from (5.183), we have

$$\|\hat{G} = \mathscr{F}(\hat{I})\|_{\mathrm{tr}} \le \|\hat{I}\|_{\mathrm{tr}} = \dim \mathcal{V}. \tag{F.31}$$

Problem 5.12 Consider the singular-value decomposition (Theorem A.22 and Eq. (A.67)) of $\hat{A}$ in the form

$$\hat{A} = \sum_j |u_j\rangle a_j \langle v_j|, \tag{F.32}$$

where $a_j \ge 0$ are the singular values of $\hat{A}$, $\langle u_i|u_j\rangle = \delta_{ij}$, and $\langle v_i|v_j\rangle = \delta_{ij}$.

(a) Then, it follows that

$$\begin{aligned}
\sum_m \left|\langle m|\hat{A}|m\rangle\right|^2 &= \sum_{mj} |\langle m|u_j\rangle|^2 a_j^2 |\langle v_j|m\rangle|^2 \\
&\le \sum_{mj} |\langle m|u_j\rangle|^2 a_j^2 \\
&= \sum_j a_j^2 \\
&= \|\hat{A}\|^2.
\end{aligned} \tag{F.33}$$

Here we have used the fact that $|u_i\rangle$ and $|v_j\rangle$ are normalized vectors, $|\langle m|u_j\rangle|^2 \le 1$, and $|\langle m|v_j\rangle|^2 \le 1$.

On the other hand, we note that

$$\begin{aligned}
\sum_m \left|\langle m|\hat{A}|m\rangle\right| &= \sum_{mj} |\langle m|u_j\rangle a_j \langle v_j|m\rangle| \\
&= \sum_{mj} a_j |\langle m|u_j\rangle| |\langle v_j|m\rangle| \\
&\le \frac{1}{2} \sum_{mj} a_j \left(|\langle m|u_j\rangle|^2 + |\langle v_j|m\rangle|^2\right),
\end{aligned} \tag{F.34}$$

where we have used the inequality

$$xy \leq \frac{x^2 + y^2}{2} \tag{F.35}$$

for any real numbers x and y. Note that

$$\sum_m \left| \langle m | u_j \rangle \right|^2 = \sum_m \langle u_j | m \rangle \langle m | u_j \rangle = \langle u_j | u_j \rangle = 1, \tag{F.36}$$

and similarly,

$$\sum_m \left| \langle m | v_j \rangle \right|^2 = 1. \tag{F.37}$$

Therefore we have

$$\sum_m \left| \langle m | \hat{A} | m \rangle \right| \leq \sum_j a_j = \| \hat{A} \|_{\mathrm{tr}}. \tag{F.38}$$

(b) In addition to the singular-value decomposition (F.32) of $\hat{A}$, we also consider the spectral decomposition of $\hat{E}_m$

$$\hat{E}_m = \sum_k |\epsilon_{mk}\rangle \, \epsilon_{mk} \, \langle \epsilon_{mk}|. \tag{F.39}$$

Here note that

$$0 \leq \epsilon_{mk} \leq 1 \tag{F.40}$$

because the POVM elements satisfy the closure relation

$$\sum_m \hat{E}_m = \hat{I}. \tag{F.41}$$

Therefore, it follows that

$$\begin{aligned}
\sum_m (\mathrm{Tr}\hat{E}_m \hat{A})^2 &= \sum_{mkj} \epsilon_{mk}^2 \left| \langle \epsilon_{mk} | u_j \rangle \right|^2 a_j^2 \left| \langle v_j | \epsilon_{mk} \rangle \right|^2 \\
&\leq \sum_{mkj} \epsilon_{mk} \left| \langle \epsilon_{mk} | u_j \rangle \right|^2 a_j^2 \\
&= \sum_j \sum_{mk} \langle u_j | \epsilon_{mk} \rangle \epsilon_{mk} \langle \epsilon_{mk} | u_j \rangle a_j^2 \\
&= \sum_j \sum_m \langle u_j | \hat{E}_m | u_j \rangle a_j^2 \\
&= \sum_j a_j^2 \\
&= \| \hat{A} \|^2.
\end{aligned} \tag{F.42}$$

On the other hand, we note that

$$\sum_m \left| \text{Tr} \hat{E}_m \hat{A} \right| = \sum_{mkj} \left| \epsilon_{mk} \langle \epsilon_{mk} | u_j \rangle a_j \langle v_j | \epsilon_{mk} \rangle \right|$$

$$= \sum_{mkj} a_j \epsilon_{mk} \left| \langle \epsilon_{mk} | u_j \rangle \right| \left| \langle v_j | \epsilon_{mk} \rangle \right|$$

$$\leq \frac{1}{2} \sum_{mkj} a_j \epsilon_{mk} \left(\left| \langle \epsilon_{mk} | u_j \rangle \right|^2 + \left| \langle v_j | \epsilon_{mk} \rangle \right|^2 \right). \qquad \text{(F.43)}$$

Again, we have used the inequality (F.35). Note that

$$\sum_{mk} \epsilon_{mk} \left| \langle \epsilon_{mk} | u_j \rangle \right|^2 = \sum_{mk} \langle u_j \epsilon_{mk} \rangle \epsilon_{mk} \langle \epsilon_{mk} | u_j \rangle$$

$$= \sum_m \langle u_j | \hat{E}_m | u_j \rangle = \langle u_j | u_j \rangle = 1, \qquad \text{(F.44)}$$

and similarly,

$$\sum_{mk} \epsilon_{mk} \left| \langle \epsilon_{mk} | v_j \rangle \right|^2 = 1. \qquad \text{(F.45)}$$

Therefore we have

$$\sum_m |\text{Tr} \hat{E}_m \hat{A}| \leq \sum_j a_j = \|\hat{A}\|_{\text{tr}}. \qquad \text{(F.46)}$$

Problem 5.7 Consider a polar decomposition of $\hat{A}$,

$$\hat{A} = V \sqrt{\hat{A}^\dagger \hat{A}}. \qquad \text{(F.47)}$$

Then we have

$$|\text{Tr} \hat{A} \hat{U}| = \left| \text{Tr} \hat{U} \hat{V} \sqrt{\hat{A}^\dagger \hat{A}} \right|. \qquad \text{(F.48)}$$

Define a unitary operator $\hat{W} := \hat{U} \hat{V}$ and consider its spectral decomposition

$$\hat{W} = \sum_j |w_j\rangle e^{i\phi_j} \langle w_j|. \qquad \text{(F.49)}$$

Putting it into (F.48), we have

$$|\text{Tr}\hat{A}\hat{U}| = \left| \sum_j \langle w_j | \sqrt{\hat{A}^\dagger \hat{A}} | w_j \rangle e^{i\phi_j} \right|$$

$$\leq \sum_j \left| \langle w_j | \sqrt{\hat{A}^\dagger \hat{A}} | w_j \rangle e^{i\phi_j} \right| \tag{F.50}$$

$$= \text{Tr}\sqrt{\hat{A}^\dagger \hat{A}}$$

$$= \|\hat{A}\|_{\text{tr}}.$$

F.6   Quantum Error-Correction Codes

Problem 6.3 Let $\left| \hat{G}_1 \right\rangle, \left| \hat{G}_2 \right\rangle, \ldots, \left| \hat{G}_k \right\rangle$ be the Gottesman vectors of the generators $\hat{G}_1, \hat{G}_2, \ldots, \hat{G}_k$. As the generators are independent, the Gottesman vectors are linearly independent of each other. Construct a matrix M from the rows of $\left\langle \hat{G}_1 \right|, \left\langle \hat{G}_2 \right|, \ldots, \left\langle \hat{G}_k \right|$,

$$M = \begin{bmatrix} \left\langle \hat{G}_1 \right| \\ \left\langle \hat{G}_2 \right| \\ \vdots \\ \left\langle \hat{G}_k \right| \end{bmatrix} = \begin{bmatrix} M_{11} & M_{12} & \cdots & M_{1n} \\ & & \vdots & \\ M_{k1} & M_{k2} & \cdots & M_{kn} \end{bmatrix}. \tag{F.51}$$

Suppose that we want to find an operator $\hat{G}$ that anti-commutes with, say, $\hat{G}_1$ but commute with all others. Because the rows of M are linearly independent by construction, the equation

$$\begin{bmatrix} M_{11} & M_{12} & \cdots & M_{1n} \\ & & \vdots & \\ M_{k1} & M_{k2} & \cdots & M_{kn} \end{bmatrix} \begin{bmatrix} y_1 \\ y_2 \\ y_3 \\ \vdots \\ y_n \end{bmatrix} = \begin{bmatrix} 1 \\ 0 \\ 0 \\ \vdots \\ 0 \end{bmatrix}. \tag{F.52}$$

has at least one solution. Define a Gottesman vector $\left\langle \hat{G} \right| := (y_1, y_2, y_3, \ldots, y_n)\hat{J}$ where the operator $\hat{J}$ on the Gottesman vector space is defined in Eq. (6.43). Then we have

$$
\begin{bmatrix} M_{11} & M_{12} & \cdots & M_{1n} \\ & \vdots & & \\ M_{k1} & M_{k2} & \cdots & M_{kn} \end{bmatrix} \hat{J} \left| \hat{G} \right\rangle = \begin{bmatrix} 1 \\ 0 \\ 0 \\ \vdots \\ 0 \end{bmatrix}. \tag{F.53}
$$

That is,

$$
\langle \hat{G}_1 | \hat{G} \rangle = 1 , \quad \langle \hat{G}_2 | \hat{G} \rangle = 0 , \quad \ldots, \quad \langle \hat{G}_k | \hat{G} \rangle = 0 . \tag{F.54}
$$

Therefore, the operator $\hat{G}$ corresponding to the Gottesman vector $\left| \hat{G} \right\rangle$ must anti-commute with $\hat{G}_1$ but commute with $\hat{G}_2, \ldots, \hat{G}_k$.

Problem 6.5 See Problem 2.4.

Problem 6.6 Let

$$
\hat{Z}' := \hat{U}(\hat{Z} \otimes \hat{J})\hat{U}^\dagger , \quad \hat{X}' := \hat{U}(\hat{X} \otimes \hat{J})\hat{U}^\dagger . \tag{F.55}
$$

Note that they anti-commute. Therefore, there exists at least one qubit on which $\hat{Z}'$ and $\hat{X}'$ anti commute. We rearrange the qubits so that the particular qubit comes at the first place. Now $\hat{Z}'$ and $\hat{X}'$ must be of the form

$$
\hat{Z}' = \hat{P} \otimes \hat{A} , \quad \hat{X}' = \hat{Q} \otimes \hat{B} \tag{F.56}
$$

with $\left\{ \hat{P}, \hat{Q} \right\} = 0$. Then, one can apply the conjugation by a single-qubit operation in $\mathcal{C}(1)$ on the first qubit so that

$$
\hat{P} \to \hat{X} , \quad \hat{Q} \to \hat{Z} . \tag{F.57}
$$

Problem 6.7 Let $\hat{P}$ be an element of the n-qubit Pauli group $\mathcal{P}(n)$. Consider

$$
\begin{aligned}
\hat{P}\hat{V} | y \rangle &= \sum_x \hat{P} | x \rangle \left(\langle 0 | \otimes \langle x | \rangle \hat{U} (| 0 \rangle \otimes | y \rangle) \right) \\
&= \sum_x \hat{P} | x \rangle \left(\langle 0 | \otimes \langle x | \hat{P}^\dagger)(\hat{I} \otimes \hat{P})\hat{U} (| 0 \rangle \otimes | y \rangle) \right) \\
&= \sum_x | x \rangle \left(\langle 0 | \otimes \langle x |)(\hat{I} \otimes \hat{P})\hat{U} (| 0 \rangle \otimes | y \rangle) \right),
\end{aligned}
$$

where we have used the fact that $\hat{P}$ is invertible, and accordingly changed the dummy variable $\hat{P} | x \rangle \to | x \rangle$. Now as $\hat{U}$ is an element of the Clifford group and $\hat{I} \otimes \hat{P} \in \mathcal{P}(n + 1)$, there must exist $\hat{P}' \in \mathcal{P}(n + 1)$ such that $(\hat{I} \otimes \hat{P})\hat{U} = \hat{U}\hat{P}'$.

$$\hat{P}\hat{V}\,|y\rangle = \sum_x |x\rangle\,((\langle 0| \otimes \langle x|)\hat{U}\hat{P}'(|0\rangle \otimes |y\rangle)).$$

(i) Suppose that $\hat{P}' = \hat{Z} \otimes \hat{P}''$ or $\hat{P} = \hat{I} \otimes \hat{P}''$ for some $\hat{P}'' \in \mathcal{P}(n)$. Then,

$$\hat{P}\hat{V}\,|y\rangle = \sum_x |x\rangle\,((\langle 0| \otimes \langle x|)\hat{U}(|0\rangle \otimes \hat{P}''\,|y\rangle))$$
$$= \hat{V}\hat{P}''\,|y\rangle$$

for all $|y\rangle$. Therefore, $\hat{V} \in \mathcal{C}(n)$. (ii) Next, suppose that $\hat{P}' = \hat{X} \otimes \hat{P}''$; the case $\hat{P}' = \hat{Y} \otimes \hat{P}''$ can be treated by combining with the above argument. Then, using the property

$$\hat{U}(\hat{X} \otimes \hat{J})\hat{U}^\dagger = \hat{Z} \otimes \hat{P}''', \quad \hat{J} := \hat{I}^{\otimes n} \tag{F.58}$$

for some $\hat{P}''' \in \mathcal{P}(n)$, one can see that

$$\hat{P}\hat{V}\,|y\rangle = \sum_x |x\rangle\,((\langle 0| \otimes \langle x|)(\hat{Z} \otimes \hat{P}''')\hat{U}(|0\rangle \otimes \hat{P}''\,|y\rangle)).$$

This implies that for any $\hat{P} \in \mathcal{P}(n)$, there exist $\hat{P}''$, $\hat{P}''' \in \mathcal{P}(n)$ such that

$$\hat{P}'''\hat{P}\hat{V} = \hat{V}\hat{P}''. \tag{F.59}$$

As $\hat{P}'''$ is invertible, $\hat{P}'''\hat{P}$ covers whole $\mathcal{P}(n)$, and Eq. (F.59) proves the statement.

Problem 6.10 We start with the quantum circuit

$$\tag{F.60}$$

where the states $|x'\rangle$ with $x = 0, 1$ are defined by

$$|0'\rangle \equiv |+\rangle := \frac{|0\rangle + |1\rangle}{\sqrt{2}}, \quad |1'\rangle \equiv |-\rangle := \frac{|0\rangle - |1\rangle}{\sqrt{2}}. \tag{F.61}$$

As $\hat{H}^2 = \hat{I}$, it is equivalent to the following quantum circuit

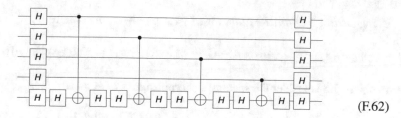

$$(F.62)$$

Now we use the identity in Eq. (2.44) to get

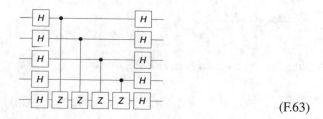

$$(F.63)$$

Since CZ gates are symmetric about control and target qubits, the above quantum circuit is equivalent to the following

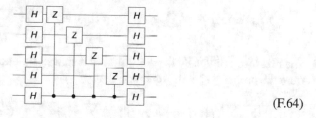

$$(F.64)$$

Finally, we use the identity in Eq. (2.44) again to arrive at the following quantum circuit

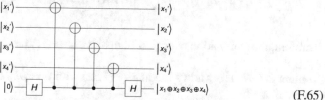

$$(F.65)$$

F.7 Quantum Information Theory

Problem 7.2 Since N_{typical} is a huge number for large n, it is convenient to tame it with the logarithm function,

$$\log N_{\text{typical}} \approx \log n! - \log (np)! - \log (n - np)!. \qquad (F.66)$$

Using *Stirling's approximation*, $\log x! \approx x \log x - x$, we obtain

$$\log N_{\text{typical}} \approx n \left[-p \log p - (1-p) \log(1-p)\right] = nH(\{p, 1-p\}). \tag{F.67}$$

Problem 7.3 Consider the spectral decompositions of $\hat{A}$ and $\hat{B}$,

$$\hat{A} = \sum_i |\alpha_i\rangle a_i \langle \alpha_i|, \quad \hat{B} = \sum_j |\beta_j\rangle b_j \langle \beta_j|. \tag{F.68}$$

For later use, define

$$P_{ij} := \left|\langle \alpha_i | \beta_j\rangle\right|^2. \tag{F.69}$$

Note that P_{ij} are *doubly stochastic*, that is,

$$0 \le P_{ij} \le 1, \quad \sum_i P_{ij} = \sum_j P_{ij} = 1. \tag{F.70}$$

As $f(x)$ is convex, one has the inequalities

$$\sum_i f(a_i) P_{ij} \ge f(\sum_i a_i P_{ij}) \ge f(b_j) + f'(b_j) \left(\sum_i a_i P_{ij} - b_j\right). \tag{F.71}$$

The first inequality is by the definition of a convex function, and the second follows from the property (7.95). Now note that

$$\sum_j \sum_i f(a_i) P_{ij} = \sum_i f(a_i) \sum_j P_{ij} = \sum_i f(a_i) = \text{Tr} f(\hat{A}), \tag{F.72}$$

$$\sum_j f'(b_j) \sum_{ij} a_i P_{ij} = \sum_{ij} f'(b_j)\langle \beta_j|\alpha_i\rangle a_i \langle \alpha_i|\beta_j\rangle = \text{Tr} f'(\hat{B})\hat{A}. \tag{F.73}$$

Therefore, the summation over j of both sides of (F.71) proves the inequality.

Problem 7.5 We use Klein's inequality (7.52) with $\hat{\rho} = \hat{\rho}_{AB}$ and $\hat{\sigma} = \hat{\rho}_A \otimes \hat{\rho}_B$. We observe that

$$\begin{aligned}
S(\hat{\rho}_{AB}) &\le -\text{Tr}\hat{\rho} \log \hat{\sigma} \\
&= -\text{Tr}\hat{\rho}_{AB} \left(\log \hat{\rho}_A + \log \hat{\rho}_B\right) \\
&= -\text{Tr}\hat{\rho}_A \log \hat{\rho}_A - \text{Tr}\hat{\rho}_B \log \hat{\rho}_B \\
&= S(\hat{\rho}_A) + S(\hat{\rho}_B)
\end{aligned} \tag{F.74}$$

Bibliography

Aaronson, S., & Gottesman, D. (2004). Improved simulation of stabilizer circuits. *Physical Review A, 70*(5). https://doi.org/10.1103/physreva.70.052328. arXiv:quant-ph/0406196.

Aharonov, Y., & Anandan, J. (1987). Phase change during a cyclic quantum evolution. *Physical Review Letters, 58*(16), 1593. https://doi.org/10.1103/PhysRevLett.58.1593.

Alicea, J., Oreg, Y., Refael, G., von Oppen, F., & Fisher, M. P. A. (2011). Non-Abelian statistics and topological quantum information processing in 1D wire networks. *Nature Physics, 7*(5), 412. https://doi.org/10.1038/nphys1915.

Anandan, J. (1988). Non-adiabatic non-abelian geometric phase. *Physics Letters A, 133*(4-5), 171. https://doi.org/10.1016/0375-9601(88)91010-9.

Aspect, A., Grangier, P., & Roger, G. (1981). Experimental tests of realistic local theories via Bell's theorem. *Physical Review Letters, 47*(7), 460.

Barenco, A., Bennett, C. H., Cleve, R., et al. (1995). Elementary gates for quantum computation. *Physical Review A, 52*(5), 3457. https://doi.org/10.1103/physreva.52.3457. arXiv:quant-ph/9503016.

Bell, J. S. (1966). On the problem of hidden variables in quantum mechanics. *Reviews of Modern Physics, 38*(3), 447.

Bennett, C. H., DiVincenzo, D. P., Smolin, J. A., & Wootters, W. K. (1996). Mixed-state entanglement and quantum error correction. *Physical Review A, 54*(5), 3824. https://doi.org/10.1103/PhysRevA.54.3824. arXiv:quant-ph/9604024.

Bennett, C. H., & Wiesner, S. J. (1992). Communication via one- and two-particle operators on Einstein-Podolsky-Rosen states. *Physical Review Letters, 69*(20), 2881.

Bergou, J. A., Herzog, U., & Hillery, M. (2004). Discrimination of quantum states. In Paris & Rehacek (Chap. 11, pp. 417–465). https://doi.org/10.1007/978-3-540-44481-7_11.

Bernstein, E., & Vazirani, U. (1993). Quantum complexity theory. In *Proceedings of the 25th Annual ACM Symposium on the Theory of Computing* (pp. 11–20). New York: ACM Press.

Bernstein, E., & Vazirani, U. (1997). Quantum complexity theory. *SIAM Journal on Computing, 26*(5), 1411. https://doi.org/10.1137/s0097539796300921.

Berry, M. V. (1984). Quantal phase factors accompanying adiabatic changes. *Proceedings of the Royal Society London A, 392*, 45.

Blum, K. (2012). *Density matrix theory and applications*, Springer series on atomic, optical, and plasma physics (Vol. 64, 3rd ed.). Berlin, Heidelberg: Springer. ISBN 978-3-642-20560-6.

Bohr, N. (1949). Discussion with Albert Einstein on epistemological problems in atomic physics. In P. A. Schilpp (Ed.), *Albert Einstein, philosopher-scientist*, The library of living philosophers (Vol. VII, pp. 200–241, 1st ed.). Evanston: Harper.

Born, M. (1926). Zur Quantenmechanik der Stoß"vorgänge. *Zeitschrift für Physik, 37*(12), 863.

© The Editor(s) (if applicable) and The Author(s), under exclusive license to Springer
Nature Switzerland AG 2022, corrected publication 2023
M.-S. Choi, *A Quantum Computation Workbook*,
https://doi.org/10.1007/978-3-030-91214-7

Bouwmeester, D., Pan, J.-W., Daniéll, M., Weinfurter, H., & Zeilinger, A. (1999). Observation of three-photon Greenberger-Horne-Zeilinger entanglement. *Physical Review Letters, 82*(7), 1345.

Bravyi, S. B., & Kitaev, A. Y. (1998). Quantum codes on a lattice with boundary. arXiv:quant-ph/9811052.

Breuer, H.-P., & Petruccione, F. (2002). *The theory of open quantum systems*. New York: Oxford University Press.

Browne, D., & Briegel, H. (2016) One-way quantum computation. In D. Bruß, & G. Leuchs (Eds.), *Quantum information: From foundations to quantum technology applications* (pp. 449–473, 2nd ed.). Wiley. https://doi.org/10.1002/9783527805785.ch21. arXiv:quant-ph/0603226.

Calderbank, A. R., & Shor, P. W. (1996). Good quantum error-correcting codes exist. *Physical Review A, 54*(2), 1098.

Caves, C. M. (1981). Quantum-mechanical noise in an interferometer. *Physical Review D, 23*(8), 1693.

Chefles, A. (2004). Quantum states: Discrimination and classical information transmission. A review of experimental progress. In Paris & Rehacek (Chap. 12, pp. 467–511). https://doi.org/10.1007/978-3-540-44481-7_12.

Chiaverini, J. (2005). Implementation of the semiclassical quantum Fourier transform in a scalable system. *Science, 308*(5724), 997. https://doi.org/10.1126/science.1110335.

Choi, M.-S. (2003). Geometric quantum computation in solid-state qubits. *Journal of Physics: Condensed Matter, 15*(46), 7823. arXiv:quant-ph/0111019.

Clauser, J. F., Horne, M. A., Shimony, A., & Holt, R. A. (1969). Proposed experiment to test local hidden-variable theories. *Physical Review Letters, 23*, 880.

Cleve, R., Ekert, A., Macchiavello, C., & Mosca, M. (1998). Quantum algorithms revisited. *Proceedings of the Royal Society A, 454*(1969), 339. https://doi.org/10.1098/rspa.1998.0164. arXiv:quant-ph/9708016.

Cleve, R., & Gottesman, D. (1997). Efficient computations of encodings for quantum error correction. *Physical Review A, 56*(1), 76. https://doi.org/10.1103/physreva.56.76. arXiv:quant-ph/9607030.

Cornwell, J. F. (1984). *Group theory in physics* (Vol. I). Orlando: Academic Press.

Cornwell, J. F. (1997). *Group theory in physics: An introduction*. San Diego: Academic Press.

Crease, R. P. (2002). The most beautiful experiment. *Physics World, 15*(9), 19. https://doi.org/10.1088/2058-7058/15/9/22.

Das, A., Ronen, Y., Most, Y., Oreg, Y., Heiblum, M., & Shtrikman, H. (2012). Zero-bias peaks and splitting in an Al-InAs nanowire topological superconductor as a signature of Majorana fermions. *Nature Physics, 8*(12), 887.

Deng, M. T., Yu, C. L., Huang, G. Y., Larsson, M., Caroff, P., & Xu, H. Q. (2012). Anomalous Zero-Bias Conductance Peak in a Nb–InSb Nanowire–Nb Hybrid Device. *Nano Letters, 12*(12), 6414.

Dennis, E., Kitaev, A., Landahl, A., & Preskill, J. (2002). Topological quantum memory. *Journal of Mathematical Physics, 43*(9), 4452. https://doi.org/10.1063/1.1499754. arXiv:quant-ph/0110143.

Deutsch, D. (1985). Quantum theory, the Church-Turing principle and the universal quantum computer. *Proceedings of the Royal Society of London A, 400*, 97. https://doi.org/10.1098/rspa.1985.0070.

Deutsch, D., & Jozsa, R. (1992). Rapid solution of problems by quantum computation. *Proceedings of the Royal Society A: Mathematical, Physical and Engineering Sciences, 439*(1907), 553. https://doi.org/10.1098/rspa.1992.0167.

Dirac, P. A. M. (1958). *The principles of quantum mechanics* (4th ed.). Oxford: Oxford University Press.

DiVincenzo, D. P. (2000). The physical implementation of quantum computation. *Fortschritte der Physik, 48*, 771. https://doi.org/10.1002/1521-3978(200009)48:9/11<771::AID-PROP771>3.0.CO;2-E. arXiv:quant-ph/0002077.

Dum, R., Parkins, A. S., Zoller, P., & Gardiner, C. W. (1992). Monte Carlo simulation of master equations in quantum optics for vacuum, thermal, and squeezed reservoirs. *Physical Review A, 46*(7), 4382.

Einstein, A., Podolsky, B., & Rosen, N. (1935). Can quantum-mechanical description of physical reality be considered complete? *Physical Review, 47*, 777.

Feynman, R., Leighton, R. B., & Sands, M. L. (1963). *The Feynman lectures on physics* (Vol. III, 1st ed.). Redwood City: Addison-Wesley.

Fowler, A. G., Mariantoni, M., Martinis, J. M., & Cleland, A. N. (2012). Surface codes: Towards practical large-scale quantum computation. *Physical Review A, 86*(3), 032324. https://doi.org/10.1103/physreva.86.032324. arXiv:1208.0928.

Freedman, M. H. (2001). Quantum computation and the localization of modular functors. *Foundations of Computational Mathematics, 1*(2), 183. https://doi.org/10.1007/s102080010006.

Giovannetti, V., Lloyd, S., & Maccone, L. (2006). Quantum metrology. *Physical Review Letters, 96*(1), 010401. https://doi.org/10.1103/PhysRevLett.96.010401. arXiv:quant-ph/0509179.

Gisin, N. (1989). Stochastic quantum dynamics and relativity. *Helvetica Physica Acta, 62*, 363. https://doi.org/10.5169/seals-116034.

Goldstein, S. (1994). Nonlocality without inequalities for almost all entangled states for two particles. *Physical Review Letters, 72*(13), 1951.

Gottesman, D. (1996). Class of quantum error-correcting codes saturating the quantum Hamming bound. *Physical Review A, 54*(3), 1862. https://doi.org/10.1103/physreva.54.1862. arXiv:quant-ph/9604038.

Gottesman, D. (1997). *Stabilizer Codes and Quantum Error Correction*, Ph.D. thesis, California Institute of Technology, Pasadena, California. arXiv:quant-ph/9705052.

Gottesman, D. (1998). Theory of fault-tolerant quantum computation. *Physical Review A, 57*(1), 127. https://doi.org/10.1103/PhysRevA.57.127. arXiv:quant-ph/9702029.

Gottesman, D. (1999). The Heisenberg representation of quantum computers. In S. P. Corney, R. Delbourgo, & P. D. Jarvis (Eds.), *Group22: Proceedings of XXII International Colloquium on Group Theoretical Methods in Physics: Hobart, July 13–17, 1998*. Cambridge, MA: International Press. ISBN 978-1571460547. arXiv:quant-ph/9807006.

Greenberger, D. M., Horne, M. A., Shimony, A., & Zeilinger, A. (1990). Bell's theorem without inequalities. *American Journal of Physics, 58*, 1131. https://doi.org/10.1119/1.16243.

Greenberger, D. M., Horne, M. A., & Zeilinger, A. (1989). Going beyond Bell's theorem. In M. Kafatos (Ed.), *Bell's theorem, quantum theory, and conceptions of the universe*. Dordrecht, The Netherlands: Kluwer Academic. arXiv:0712.0921.

Griffiths, R. B., & Niu, C.-S. (1996). Semiclassical Fourier transform for quantum computation. *Physical Review Letters, 76*(17), 3228. https://doi.org/10.1103/physrevlett.76.3228. arXiv:quant-ph/9511007.

Grover, L. K. (1996). A fast quantum mechanical algorithm for database search. In *Proceedings of the 28th Annual ACM Symposium on the Theory of Computing* (p. 212). New York: ACM Press. arXiv:quant-ph/9605043.

Grover, L. K. (1997). Quantum mechanics helps in searching for a needle in a haystack. *Physical Review Letters, 79*(2), 325.

Hardy, L. (1992). Quantum mechanics, local realistic theories, and Lorentz-invariant realistic theories. *Physical Review Letters, 68*(20), 2981.

Hardy, L. (1993). Nonlocality for two particles without inequalities for almost all entangled states. *Physical Review Letters, 71*, 1665.

Higgins, B. L., Berry, D. W., Bartlett, S. D., Wiseman, H. M., & Pryde, G. J. (2007). Entanglement-free Heisenberg-limited phase estimation. *Nature, 450*(7168), 393. https://doi.org/10.1038/nature06257. arXiv:0709.2996.

Horodecki, M., Horodecki, P., & Horodecki, R. (1996). Separability of mixed states: Necessary and sufficient conditions. *Physics Letters A, 223*(1), 1. https://doi.org/10.1016/0375-9601(95)00930-2.

Horodecki, R., Horodecki, P., Horodecki, M., & Horodecki, K. (2009). Quantum entanglement. *Reviews of Modern Physics, 81*(2), 865. https://doi.org/10.1103/RevModPhys.81.865.

Hughston, L. P., Jozsa, R., & Wootters, W. K. (1993). A complete classification of quantum ensembles having a given density matrix. *Physics Letters A, 183*(1), 14. https://doi.org/10.1016/0375-9601(93)90880-9.

Jiang, M., Luo, S., & Fu, S. (2013). Channel-state duality. *Physical Review A, 87*(2). https://doi.org/10.1103/physreva.87.022310.

Jönsson, C. (1961). Electron diffraction at multiple slits. *Zeitschrift für Physik, 161*, 454. https://doi.org/10.1007/BF01342460.

Kitaev, A. Y. (1996). Quantum measurements and the Abelian stabilizer problem. *Electronic Colloquium on Computational Complexity, 3*, 3. arXiv:quant-ph/9511026.

Kitaev, A. Y. (1997). Quantum computations: Algorithms and error correction. *Russian Mathematical Surveys, 52*(6), 1191.

Kitaev, A. Y. (2001). Unpaired Majorana fermions in quantum wires. *Physics-Uspekhi, 44*(10S), 131. https://doi.org/10.1070/1063-7869/44/10S/S29. arXiv:cond-mat/0010440.

Kitaev, A. Y. (2003). Fault-tolerant quantum computation by anyons. *Annals of Physics, 303*(1), 2. https://doi.org/10.1016/S0003-4916(02)00018-0. arXiv:quant-ph/9707021.

Laflamme, R., Miquel, C., Paz, J. P., & Zurek, W. H. (1996). Perfect quantum error correcting code. *Physical Review Letters, 77*(1), 198. https://doi.org/10.1103/physrevlett.77.198. arXiv:quant-ph/9602019.

Landauer, R. (1991). Information is physical. *Physics Today, 44*(5), 23.

Lang, S. (1986). *Introduction to linear algebra*, Undergraduate texts in mathematics (2nd ed.). New York: Springer. ISBN 9781461210702. https://doi.org/10.1007/978-1-4612-1070-2.

Lang, S. (1987). *Linear algebra* (3rd ed.). Berlin: Springer. ISBN 978-1-4757-1949-9. https://doi.org/10.1007/978-1-4757-1949-9.

Loss, D., & DiVincenzo, D. P. (1998). Quantum computation with quantum dots. *Physical Review A, 57*(1), 120.

Lundeen, J. S., Sutherland, B., Patel, A., Stewart, C., & Bamber, C. (2011). Direct measurement of the quantum wavefunction. *Nature, 474*(7350), 188. https://doi.org/10.1038/nature10120.

Mourik, V., Zuo, K., Frolov, S. M., Plissard, S. R., Bakkers, E. P. A. M., & Kouwenhoven, L. P. (2012). Signatures of Majorana fermions in hybrid superconductor-semiconductor nanowire devices. *Science, 336*(6084), 1003.

Nadj-Perge, S., Drozdov, I. K., Li, J., et al. (2014). Observation of Majorana fermions in ferromagnetic atomic chains on a superconductor. *Science, 346*(6209), 602. https://doi.org/10.1126/science.1259327. arXiv:http://www.sciencemag.org/content/346/6209/602.full.pdf.

Nakazato, H., Hida, Y., Yuasa, K., Militello, B., Napoli, A., & Messina, A. (2006). Solution of the Lindblad equation in the Kraus representation. *Physical Review A, 74*(6), 062113. https://doi.org/10.1103/physreva.74.062113. arXiv:quant-ph/0606193.

Nielsen, M., & Chuang, I. L. (2011). *Quantum computation and quantum information* (10th anniversary ed.). New York: Cambridge University Press. ISBN 978-1107002173.

Ozawa, M. (2000). Entanglement measures and the Hilbert–Schmidt distance. *Physics Letters A, 268*(3), 158. https://doi.org/10.1016/s0375-9601(00)00171-7.

Pan, J.-W., Bouwmeester, D., Daniell, M., Weinfurter, H., & Zeilinger, A. (2000). Experimental test of quantum nonlocality in three-photon Greenberger-Horne-Zeilinger entanglement. *Nature, 403*, 515.

Paris, M., & Rehacek, J. (Eds.). (2004). *Quantum state estimation*, Lecture notes in physics (Vol. 649). Berlin, Heidelberg: Springer. ISBN 9783540444817. https://doi.org/10.1007/b98673.

Peres, A. (1996). Separability criterion for density matrices. *Physical Review Letters, 77*(8), 1413. https://doi.org/10.1103/PhysRevLett.77.1413. arXiv:quant-ph/9604005.

Pérez-García, D., Wolf, M. M., Petz, D., & Ruskai, M. B. (2006). Contractivity of positive and trace-preserving maps under Lp norms. *Journal of Mathematical Physics, 47*(8), 083506. https://doi.org/10.1063/1.2218675. arXiv:math-ph/0601063.

Plenio, M. B., & Knight, P. L. (1998). The quantum-jump approach to dissipative dynamics in quantum optics. *Reviews of Modern Physics, 70*(1), 101.

Plenio, M. B., & Virmani, S. (2007). An introduction to entanglement measures. *Quantum Information & Computation, 7*(1&2), 1. arXiv:quant-ph/0504163.

Preskill, J. (1998). Lecture Notes on Quantum Information and Computation, unpublished.

Raussendorf, R., & Briegel, H. J. (2001). A one-way quantum computer. *Physical Review Letters, 86*(22), 5188.

Raussendorf, R., Browne, D., & Briegel, H. (2002). The one-way quantum computer–a non-network model of quantum computation. *Journal of Modern Optics, 49*(8), 1299. https://doi.org/10.1080/09500340110107487. arXiv:quant-ph/0108118.

Raussendorf, R., Browne, D. E., & Briegel, H. J. (2003). Measurement-based quantum computation on cluster states. *Physical Review A, 68*(2), 022312. https://doi.org/10.1103/PhysRevA.68.022312. arXiv:quant-ph/0301052.

Schwinger, J. (1959). The algebra of microscopic measurement. *Proceedings of the National Academy of Sciences, 45*(10), 1542. https://doi.org/10.1073/pnas.45.10.1542.

Shor, P. W. (1994). Algorithms for quantum computation: Discrete logarithms and factoring. In *Proceedings of the 35th Annual Symposium on Foundations of Computer Science* (pp. 124–134). Washington, DC, USA: IEEE Computer Society. SFCS '94. https://doi.org/10.1109/SFCS.1994.365700.

Shor, P. W. (1997). Polynomial-time algorithms for prime factorization and discrete logarithms on a quantum computer. *SIAM Journal on Computing, 26*(5), 1484. arXiv:quant-ph/9508027.

Simon, D. R. (1997). On the power of quantum computation. *SIAM Journal on Computing, 26*(5), 1474. https://doi.org/10.1137/s0097539796298637.

Sjöqvist, E., Tong, D. M., Mauritz Andersson, L., Hessmo, B., Johansson, M., & Singh, K. (2012). Non-adiabatic holonomic quantum computation. *New Journal of Physics, 14*(10), 103035. https://doi.org/10.1088/1367-2630/14/10/103035. arXiv:1107.5127.

Smolin, J. A., & DiVincenzo, D. P. (1996). Five two-bit quantum gates are sufficient to implement the quantum Fredkin gate. *Physical Review A, 53*(4), 2855. https://doi.org/10.1103/PhysRevA.53.2855.

Steane, A. M. (1996). Error correcting codes in quantum theory. *Physical Review Letters, 77*(5), 793.

Størmer, E. (2013). *Positive linear maps of operator algebras*. Berlin: Springer. ISBN 9783642343698. https://doi.org/10.1007/978-3-642-34369-8.

Tonomura, A., Endo, J., Matsuda, T., Kawasaki, T., & Ezawa, H. (1989). Demonstration of single-electron buildup of an interference pattern. *American Journal of Physics, 57*(2), 117. https://doi.org/10.1119/1.16104.

Vallone, G., & Dequal, D. (2016). Strong measurements give a better direct measurement of the quantum wave function. *Physical Review Letters, 116*(4), 040502. https://doi.org/10.1103/physrevlett.116.040502. arXiv:1504.06551.

Vedral, V., Barenco, A., & Ekert, A. (1996). Quantum networks for elementary arithmetic operations. *Physical Review A, 54*(1), 147. https://doi.org/10.1103/physreva.54.147. arXiv:quant-ph/9511018.

Vedral, V., Plenio, M. B., Rippin, M. A., & Knight, P. L. (1997). Quantifying entanglement. *Physical Review Letters, 78*(12), 2275. https://doi.org/10.1103/physrevlett.78.2275.

Wang, C., Harrington, J., & Preskill, J. (2003). Confinement-Higgs transition in a disordered gauge theory and the accuracy threshold for quantum memory. *Annals of Physics, 303*(1), 31. https://doi.org/10.1016/s0003-4916(02)00019-2.

Wehrl, A. (1978). General properties of entropy. *Reviews of Modern Physics, 50*(2), 221. https://doi.org/10.1103/revmodphys.50.221.

Weyl, H. (1931). *The theory of groups and quantum mechanics*. London: Dover.

Wigner, E. P. (1959). *Group theory and its application to the quantum mechanics of atomic spectra* (English translation ed.). New York: Academic Press.

Wilczek, F., & Zee, A. (1984). Appearance of Gauge structure in simple dynamical systems. *Physical Review Letters, 52*(24), 2111. https://doi.org/10.1103/PhysRevLett.52.2111.

Wilmut, I., Schnieke, A. E., McWhir, J., Kind, A. J., & Campbell, K. H. S. (1997). Viable offspring derived from fetal and adult mammalian cells. *Nature, 385*(6619), 810.

Wooters, W. K., & Zurek, W. H. (1982). A single quantum cannot be cloned. *Nature, 299*, 802.

Zanardi, P., & Rasetti, M. (1999). Holonomic quantum computation. *Physics Letters A, 264*(2-3), 94. https://doi.org/10.1016/S0375-9601(99)00803-8. arXiv:quant-ph/9904011.

Zurek, W. H. (1991). Decoherence and the transition from quantum to classical. *Physics Today, 44*(10), 36.

Zurek, W. H. (2000). Quantum cloning: Schrodinger's sheep. *Nature, 404*, 130. https://doi.org/10.1038/35004684.

Zurek, W. H. (2002). Decoherence and the transition from quantum to classical: Revisited. *Los Alamos Science, 27*, 2.

Index

© The Editor(s) (if applicable) and The Author(s), under exclusive license to Springer
Nature Switzerland AG 2022, corrected publication 2023
M.-S. Choi, *A Quantum Computation Workbook*,
https://doi.org/10.1007/978-3-030-91214-7

Printed in the United States
by Baker & Taylor Publisher Services